AF902687

TARGETING CELL SURVIVAL PATHWAYS TO ENHANCE RESPONSE TO CHEMOTHERAPY

VOLUME 3

TARGETING CELL SURVIVAL PATHWAYS TO ENHANCE RESPONSE TO CHEMOTHERAPY

VOLUME 3

Edited by

DANIEL E. JOHNSON, PHD
Professor, Department of Otolaryngology-Head and Neck Surgery,
University of California at San Francisco, San Francisco, CA, United States

ELSEVIER

ACADEMIC PRESS
An imprint of Elsevier

Academic Press is an imprint of Elsevier
125 London Wall, London EC2Y 5AS, United Kingdom
525 B Street, Suite 1800, San Diego, CA 92101-4495, United States
50 Hampshire Street, 5th Floor, Cambridge, MA 02139, United States
The Boulevard, Langford Lane, Kidlington, Oxford OX5 1GB, United Kingdom

Copyright © 2018 Elsevier Inc. All rights reserved.

No part of this publication may be reproduced or transmitted in any form or by any means, electronic or mechanical, including photocopying, recording, or any information storage and retrieval system, without permission in writing from the publisher. Details on how to seek permission, further information about the Publisher's permissions policies and our arrangements with organizations such as the Copyright Clearance Center and the Copyright Licensing Agency, can be found at our website: www.elsevier.com/permissions.

This book and the individual contributions contained in it are protected under copyright by the Publisher (other than as may be noted herein).

Notices
Knowledge and best practice in this field are constantly changing. As new research and experience broaden our understanding, changes in research methods, professional practices, or medical treatment may become necessary.

Practitioners and researchers must always rely on their own experience and knowledge in evaluating and using any information, methods, compounds, or experiments described herein. In using such information or methods they should be mindful of their own safety and the safety of others, including parties for whom they have a professional responsibility.

To the fullest extent of the law, neither the Publisher nor the authors, contributors, or editors, assume any liability for any injury and/or damage to persons or property as a matter of products liability, negligence or otherwise, or from any use or operation of any methods, products, instructions, or ideas contained in the material herein.

Library of Congress Cataloging-in-Publication Data
A catalog record for this book is available from the Library of Congress

British Library Cataloguing-in-Publication Data
A catalogue record for this book is available from the British Library

ISBN: 978-0-12-816432-7

For information on all Academic Press publications visit our website at https://www.elsevier.com/books-and-journals

ELSEVIER Book Aid International
Working together to grow libraries in developing countries
www.elsevier.com • www.bookaid.org

Publisher: Fedor, John (ELS-HBE)
Acquisition Editor: TEIXEIRA, RAFAEL E. (ELS-SPA)
Editorial Project Manager: Kuhl, Mariana L. (ELS-SPA)
Production Project Manager: Mohanapriyan Rajendran
Designer: Greg Harris

Typeset by Thomson Digital

This book is dedicated to Becky Johnson,
whose intrepid and resolute struggle with her disease
has inspired more than she may know.

Contents

6. PIK3CA Mutations in Colorectal and Breast Cancer: Impact on Oncogenesis and Response to Nonsteroidal Anti-Inflammatory Drugs

RICHARD T. ZHU, J. SILVIO GUTKIND, DANIEL E. JOHNSON, JENNIFER R. GRANDIS

7. STAT3 as a Major Contributor to Chemoresistance

DAVID S. LEE, JENNIFER R. GRANDIS, DANIEL E. JOHNSON

8. Targeting the Hippo Pathway to Improve Response to Chemotherapy

YULEI ZHAO, XIAOLONG YANG

9. Modulation of the Epigenome (Methylome) to Improve Chemotherapeutic Efficacy

WILLIAM B. COLEMAN

10. Targeting the Ataxia Telangiectasia and Rad3 Signaling Pathway to Overcome Chemoresistance in Cancer

MARWAN KWOK, TATJANA STANKOVIC

11. PARP Inhibition to Enhance Response to Chemotherapy

NAOKO TAKEBE, MARY QUINN, GARIMA GUPTA, ALICE P. CHEN

12. Autophagy Inhibition and Chemosensitization in Cancer Therapy

EMMANUEL K. CUDJOE, JR., S. LAUREN KYTE, TAREQ SALEH, JOSEPH W. LANDRY, DAVID A. GEWIRTZ

13. Targeting Necroptosis in Antitumor Therapy

FRANCIS KA-MING CHAN, HIMANI NAILWAL, KENTA MORIWAKI

List of Contributors

Andrea Cavazzoni University of Parma, Parma, Italy

Francis Ka-Ming Chan University of Massachusetts Medical School, Worcester, MA, USA

Alice P. Chen National Cancer Institute, National Institutes of Health, Bethesda, MD, United States

William B. Coleman Department of Pathology and Laboratory Medicine, Curriculum in Toxicology, UNC Program in Translational Medicine, UNC Lineberger Comprehensive Cancer Center, University of North Carolina School of Medicine, Chapel Hill, NC, United States

Emmanuel K. Cudjoe, Jr. Massey Cancer Center, Virginia Commonwealth University, Richmond, VA, United States

Deborah DeRyckere Emory University Department of Pediatrics; Aflac Cancer and Blood Disorders Center, Children's Healthcare of Atlanta; Winship Cancer Institute, Emory University, Atlanta, GA, United States

Ingrid Garajova Sant'Orsola-Malpighi Hospital, University of Bologna; Interdepartmental Centre of Cancer Research "Giorgio Prodi", University of Bologna, Bologna, Italy

David A. Gewirtz Massey Cancer Center, Virginia Commonwealth University, Richmond, VA, United States

Elisa Giovannetti VU University Medical Center, Amsterdam, The Netherlands; University of Pisa, Pisa, Italy

Douglas K. Graham Emory University Department of Pediatrics; Aflac Cancer and Blood Disorders Center, Children's Healthcare of Atlanta; Winship Cancer Institute, Emory University, Atlanta, GA, United States

Jennifer R. Grandis University of California at San Francisco, San Francisco, CA, United States

Garima Gupta Jewish Hospital of Cincinnati, Cincinnati, OH, United States

J. Silvio Gutkind Moores Cancer Center, University of California at San Diego, La Jolla, CA, United States

Cinta Hierro Vall d'Hebron University Hospital; Molecular Therapeutics Research Unit, Vall d'Hebron Institute of Oncology (VHIO), Universitat Autonoma de Barcelona (UAB), Barcelona, Spain

Lanlin Hu Sichuan University, Chengdu, Sichuan, PR China

Daniel E. Johnson University of California at San Francisco, San Francisco, CA, United States

Marwan Kwok Institute of Cancer and Genomic Sciences, University of Birmingham; Centre for Clinical Haematology, Queen Elizabeth Hospital Birmingham, United Kingdom

S. Lauren Kyte Massey Cancer Center, Virginia Commonwealth University, Richmond, VA, United States

Joseph W. Landry Massey Cancer Center, Virginia Commonwealth University, Richmond, VA, United States

David S. Lee University of California at San Francisco, San Francisco, CA, United States

Katherine A. Minson Emory University Department of Pediatrics; Aflac Cancer and Blood Disorders Center, Children's Healthcare of Atlanta, Atlanta, GA, United States

Kenta Moriwaki Osaka University, Japan

Himani Nailwal University of Massachusetts Medical School, Worcester, MA, USA

Rachel A. O'Keefe University of California at San Francisco, San Francisco, CA, United States

Mary Quinn National Cancer Institute, National Institutes of Health, Bethesda, MD, United States

Christian Rolfo Center for Oncological Research, University of Antwerp, Antwerp, Belgium

Tareq Saleh Massey Cancer Center, Virginia Commonwealth University, Richmond, VA, United States

Tatjana Stankovic Institute of Cancer and Genomic Sciences, University of Birmingham, United Kingdom

Ryan J. Summers Emory University Department of Pediatrics; Aflac Cancer and Blood Disorders Center, Children's Healthcare of Atlanta, Atlanta, GA, United States

Josep Tabernero Vall d'Hebron University Hospital; Molecular Therapeutics Research Unit, Vall d'Hebron Institute of Oncology (VHIO), Universitat Autonoma de Barcelona (UAB), Barcelona, Spain

Naoko Takebe National Cancer Institute, National Institutes of Health, Bethesda, MD, United States

Nele Van Der Steen VU University Medical Center, Amsterdam, The Netherlands; Center for Oncological Research, University of Antwerp, Antwerp, Belgium

Xiaolong Yang Queen's University, Kingston, ON, Canada

Yulei Zhao Queen's University, Kingston, ON, Canada

Richard T. Zhu The Warren Alpert Medical School of Brown University, Providence, RI, United States

Fangdong Zou Sichuan University, Chengdu, Sichuan, PR China

Author Biography

Daniel E. Johnson received his undergraduate degrees in chemistry and mathematics from North Park University in Chicago, IL, USA. He received his doctoral degree in molecular biology from Princeton University under the mentorship of Dr. Mark A. Bothwell. His postdoctoral research was done under the mentorship of Dr. Lewis T. Williams at the University of California at San Francisco (UCSF). He joined the faculty at the University of Pittsburgh and the University of Pittsburgh Cancer Institute in 1993, where he served as Professor of Medicine and Scientific Director of the Acute Leukemia Working Group. In 2016, he moved to UCSF where he is currently Professor in the Department of Otolaryngology – Head and Neck Surgery. Dr. Johnson has served as a standing member on National Institutes of Health and American Cancer Society study sections and has been a Section Editor for the journal Leukemia since 2001. His research is focused on anticancer drug development and the elucidation of cell death and cell-survival signaling pathways in head and neck cancer and leukemia. He has additional interests in the development of chemopreventive agents and strategies. Dr. Johnson places particular emphasis on translating laboratory findings to the clinic and has collaborated extensively with physician scientists to initiate and conduct clinical trials in head and neck cancer and acute myeloid leukemia.

Preface

Diagnosis of cancer can be devastating and often creates fear in the heart of the patient. Part of this fear may stem from the known toxic and debilitating effects of current cancer treatments. Chemotherapy drugs have been a mainstay of anticancer treatments for several decades. In certain forms of cancer, chemotherapy drugs are highly effective, whereas in other malignancies, only modest or partially effective chemotherapy regimens have been identified. Unfortunately, treatment with chemotherapy, as well as radiation, is inevitably associated with adverse effects on normal cells, leading to undesirable toxicities and side effects for the patient. Even molecular targeting agents, which hold the promise of more precise targeting of tumor cells, are limited by nonspecific activities and adverse toxicities. The limitations of chemotherapy, radiation, and molecular targeting agents are exacerbated by the inherent resistance of many cancer cells to these agents and by the development of acquired resistance. These resistance mechanisms directly contribute to reduced anticancer efficacy and increased side effects in patients. Thus, a key goal in cancer research is the development of new agents and strategies that can be used to overcome either inherent or acquired resistance to standard of care therapies.

Efforts to rationally devise approaches for overcoming resistance to chemotherapy, radiation, and molecular targeting agents will benefit from a thorough understanding of the mechanisms responsible for resistance. In recent years, a great deal has been learned about the proteins, pathways, and processes that contribute to resistance to anticancer agents. This book will highlight current knowledge of the cell-survival pathways that are integral to drug resistance in cancer. Particular emphasis is placed on preclinical and clinical studies that have examined the impact of targeting these pathways as a means of overcoming either inherent or acquired resistance. The book begins by describing cell-survival pathways mediated by growth factor receptors and how these pathways promote drug resistance. Signaling emanating from members of the epidermal growth factor receptor and fibroblast growth factor receptor families, as well as hepatocyte growth factor receptor and AXL and MERTK will be discussed. The impact of targeting these cell-surface receptors in preclinical models and in clinical trials using recently described inhibitors will be described. The role of intracellular signaling molecules, including JNKs, PI3K, STAT3, and the Hippo pathway, and the benefit of targeting these molecules and pathways to promote sensitization to chemotherapy are then discussed. Lastly, recent breakthroughs in our understanding of the cellular processes of DNA repair, genome methylation, autophagy, and necroptosis are reviewed. Aberrant alteration or activation of these processes can play important roles in the development of drug resistance. Successes in restoring tumor sensitivity to chemotherapy in preclinical models and cancer patients using epigenetic modulators or therapies targeting DNA repair, autophagy, or necroptosis are summarized in these final chapters.

Targeting Cell Survival Pathways to Enhance Response to Chemotherapy will be of particular interest to cancer biologists, biochemists, and molecular biologists interested in the mechanisms and pathways that regulate the sensitivity and resistance of cancer cells and tumors to standard of care therapies. Translational researchers and physician scientists will benefit from the emphasis placed in each chapter on findings from clinical trials and analysis of primary patient specimens. Pharmaceutical industry and academic scientists focused on anticancer drug discovery will gain deeper understanding of current opportunities, as well as challenges, in the development of new molecular targeting agents aimed at combating chemotherapy resistance.

This book summarizes work from a broad array of dedicated scientists. I am thankful for their excellence. I also wish to gratefully acknowledge those whose work was not presented, or whose research area did not fit within the specific focus of this book. Special thanks go to my scientific mentors, Mark A. Bothwell and Lewis T. Williams, for guiding me in my early scientific studies and introducing me to cell-survival signaling. I am particularly grateful for the encouragement and inspiration of fellow scientists and friends Jennifer Grandis, Bob Redner, and Richard Steinman. Finally, I wish to express much appreciation to my wife Sylvia and my children, Rachel, Josiah, and Matthew, for the care and delight they bring to myself and to others.

Daniel E. Johnson, Ph.D.

Department of Otolaryngology – Head and Neck Surgery, Helen Diller Family Comprehensive Cancer Center, University of California at San Francisco, San Francisco, CA, USA

CHAPTER

1

Targeting Members of the Epidermal Growth Factor Receptor Family to Improve Response to Chemotherapy

Rachel A. O'Keefe, Jennifer R. Grandis, Daniel E. Johnson

University of California at San Francisco, San Francisco, CA, United States

Abstract

Decades of research on the four receptor tyrosine kinases (RTKs) that comprise the human epidermal growth factor receptor (HER) family have established these proteins as key oncogenes in many cancer types. The strong evidence supporting pro-tumorigenic roles of HER3, HER4, and especially epidermal growth factor receptor (EGFR) and HER2 in cancer has led to the generation and use of HER family-targeted agents that can inhibit one or more of these RTKs. Although these inhibitors are used as a single agent in some contexts, they are often used in combination with chemotherapeutic agents and have been shown to enhance the response to chemotherapy in many cancer types. In this chapter, we outline the basic biology of the HER family receptors before examining each of the four receptors, focusing on mutations and other alterations that occur in cancer and on preclinical and clinical studies in which HER family-targeted agents were used to delay and/or overcome chemoresistance. Throughout, we address open questions in the field and discuss current and future areas of research.

ABBREVIATIONS

ADCC	Antibody-dependent cell-mediated cytotoxicity
AREG	Amphiregulin
BTC	Betacellulin
CDK	Cyclin-dependent kinase
CDKN1A	Cyclin-dependent kinase inhibitor p21Cip1
EGF	Epidermal growth factor
EGFR	Epidermal growth factor receptor
EPGN	Epigen
Erα	Estrogen receptorα
EREG	Epiregulin
ERK	Extracellular signal-regulated kinase
FDA	Food and Drug Administration
HB-EGF	Heparin-binding epidermal growth factor-like growth factor

Targeting Cell Survival Pathways to Enhance Response to Chemotherapy. http://dx.doi.org/10.1016/B978-0-12-813753-6.00001-9
Copyright © 2018 Elsevier Inc. All rights reserved.

HER	Human epidermal growth factor receptor
HNSCC	Head and neck squamous cell carcinoma
JAK	Janus kinase
LCC	Large cell carcinoma
LUAD	Lung adenocarcinoma
LUSC	Lung squamous cell carcinoma
MEK	Mitogen-activated protein kinase kinase
mTOR	Mechanistic target of rapamycin
NRG	Neuregulin
NSCLC	Non-small cell lung cancer
PI3K	Phosphoinositide 3-kinase
PTEN	Phosphatase and tensin homolog deleted on chromosome ten
RTK	Receptor tyrosine kinase
STAT	Signal transducer and activator of transcription
TACE	Tumor necrosis factor-α converting enzyme
TGF-α	Transforming growth factor-α
TKI	Tyrosine kinase inhibitor

1.1 INTRODUCTION

For many years, the backbone of nonsurgical cancer treatment comprised chemotherapy and radiation. Although effective in some cases, many tumors proved refractory to these interventions, and low survival rates in many cancer types reflected the inadequacy of available cancer treatments. As significant research efforts have led to the identification of oncogenic drivers and unique vulnerabilities in cancer cells, cancer treatment is increasingly moving toward precision medicine, in which alterations specific to each tumor guide the selection of treatment. A key component of precision medicine is the use of targeted agents that inhibit the oncogenic drivers upon which cancer cells depend for their survival and proliferation. Targeted agents can be used alone or in combination with chemotherapy and/or other treatment modalities and have demonstrated efficacy in many cancer types.

Among the oncogenic targets against which inhibitors have been developed are the four members of the human epidermal growth factor receptor (HER) family. These receptor tyrosine kinases (RTKs) are overexpressed and/or mutated in many cancer types and have been shown to promote tumorigenesis, tumor maintenance, and metastasis. Targeting HER family members is an attractive prospect not only because of their demonstrated role as oncogenic drivers, but also because they have been shown to mediate resistance to chemotherapy [1–8]. Although employed as a single agent for certain indications, HER family-targeted agents are often used in combination with chemotherapy, and the addition of these inhibitors to the treatment arsenal has improved outcomes for patients with many different cancer types.

In this chapter, we first provide a background on HER family members and the roles of each of the four receptors in cancer, then discuss examples of how HER family-targeted agents can improve response to chemotherapy and the mechanisms by which the enhanced responses are achieved. We also discuss current and future avenues of research that will inform efforts to more effectively harness the therapeutic potential of combining chemotherapy with HER family-targeted agents in the era of precision medicine.

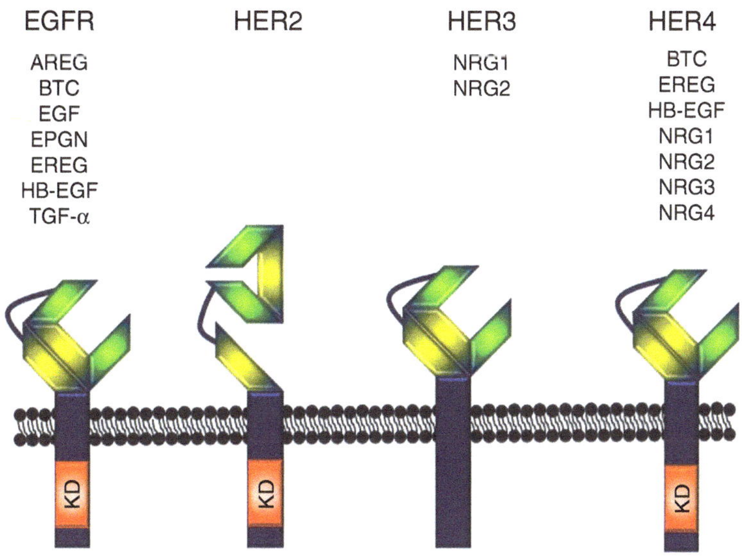

FIG. 1.1 HER family receptors. The four members of the HER family of RTKs are highly homologous in both sequence and structure, but they exhibit functional differences and are regulated by different ligands. The extracellular region of each HER family receptor contains two leucine-rich domains (green) and two cysteine-rich domains (yellow). Ligand binding to the leucine-rich domains of EGFR, HER3, and HER4 induces a conformational change that frees the N-terminal cysteine-rich domain to bind to that of another receptor molecule, facilitating homo- or heterodimerization. HER2 is constitutively available for dimerization, and no ligands have been identified for this receptor. EGFR, HER2, and HER4 each contain an active intracellular kinase domain (KD), while the catalytic activity of HER3 is greatly reduced compared to that of the other HER family receptors.

1.2 HER FAMILY RECEPTORS

The HER family of proteins, also called the epidermal growth factor (EGF) or ErbB family, comprises four RTKs: epidermal growth factor receptor (EGFR, HER1, ErbB1), HER2 (ErbB2, neu in rodents), HER3 (ErbB3), and HER4 (ErbB4) (Fig. 1.1). Residing primarily in the plasma membrane, the HER proteins recognize extracellular cues and transmit these signals to downstream signaling pathways that mediate a number of cellular processes [9–11]. In normal (nontransformed) cells, HER family receptors play a role in organ development and wound healing, among other physiological processes [12–15]. In cancer, HER family members are commonly implicated in tumorigenesis and tumor progression, and aberrant activation of one or more of these RTKs is often observed in cancer cells.

Activation of HER family members is initiated when ligand binds to the extracellular domain of its cognate receptor [9,12,14,16]. Ligand binding triggers a conformational change that allows for homodimerization or heterodimerization with other HER family members or non-HER-family RTKs [9,12,14–24]. (An exception is HER2, which has no known ligands and is thought to be constitutively available for dimerization [1,9,14,15,19,20,23,25–29].) Dimerization facilitates allosteric activation of the kinase domains of the HER family receptors, promoting transphosphorylation of tyrosine residues in the intracellular domains (ICDs) of their dimerization partners [9,14,16,17,19,20,23]. HER3 does not contain a catalytically active kinase, but is still able to allosterically activate its dimerization partners [1,14,16, 18–20,22,23,25,26,28–30].

The impact of phosphorylation on HER family receptor function depends on which sites are phosphorylated. Phosphorylation of certain residues promotes internalization and subsequent proteasomal degradation or endocytic recycling of the receptors, which ensures that activation of these receptors is transient and tightly regulated [14,31,32]. However, phosphorylation of most of the residues creates plasma membrane-proximal binding sites for downstream signaling molecules [14,17,26,29]. Binding of these molecules to HER family receptors leads to activation of a number of signaling pathways, including the phosphoinositide 3-kinase/Akt/mechanistic target of rapamycin (PI3K/Akt/mTOR), RAS/RAF/mitogen-activated protein kinase kinase/extracellular signal-regulated kinase (RAS/RAF/MEK/ERK), and Janus kinase/signal transducer and activator of transcription (JAK/STAT) pathways [9,11,12,21,33,34]. The ability of HER family members to activate these signaling pathways is co-opted by cancer cells, particularly by those in which one or more of the genes encoding HER family members are amplified and/or mutated, to drive tumorigenesis, to promote tumor cell survival and metastasis, to enable immune evasion, and to mediate resistance to chemotherapy.

Although the precise mechanisms of resistance to chemotherapy depend on the chemotherapeutic agent(s) used, the cancer type, and other factors specific to a particular tumor, in general, resistance to chemotherapy arises when cancer cells are able to overcome the pro-death signaling induced by chemotherapeutic agents. HER family receptors are among the factors that have been shown to promote tumor cell survival in the presence of chemotherapy and are thereby implicated in chemoresistance [1–8]. More convincing evidence for a role for HER family RTKs in resistance to chemotherapy comes from reports of increased expression and/or activation of these RTKs in chemorefractory tumors and both preclinical and clinical data demonstrating that inhibiting one or more HER family receptors can overcome resistance to chemotherapy [1,2,7,8,35–37]. Thus, combining HER family-targeted inhibitors with chemotherapy is a rational approach to improving responses.

Other features that make HER family receptors attractive targets for drug development include their subcellular location and the presence of targetable kinase domains. HER family receptors reside primarily on the cell surface and are thus accessible to antibodies in the tumor microenvironment. Monoclonal antibody therapies that bind and inhibit HER family receptors have demonstrated efficacy in a multitude of cancer types. Small-molecule tyrosine kinase inhibitors (TKIs) that can inhibit EGFR, HER2, and/or HER4 have also proved effective in the clinic. (HER3 contains a catalytically impaired kinase domain; thus, kinase inhibition is not thought to be an effective approach to inhibit HER3 [1,15,16,26,30,38].) Some of these agents target more than one HER family member, while others are selective for a single receptor. Inhibitor selectivity and type (i.e., monoclonal antibody vs. TKI) can have important implications for treatment efficacy, as will be discussed in subsequent sections.

Although the four HER family receptors are highly homologous at both the sequence and structural levels and often cooperate to support the proliferation and/or survival of cancer cells, they also possess distinct, nonoverlapping functions. Moreover, different HER family receptors are implicated in different types of cancer, and the oncogenic alterations that occur vary among cancer types. Thus, we will next cover each of the HER family receptors in turn, providing an overview of the specific alterations observed in each receptor and the cancer types in which they are implicated before describing methods to target these RTKs. Finally, we will, when possible, discuss clinical outcomes when HER family-targeted agents are combined with chemotherapy and preclinical studies elucidating the mechanisms by which these combinations elicit their antitumor effects.

1.2.1 EGFR

EGFR was the first member of the HER family of RTKs to be discovered and remains perhaps the most well-studied HER family member in cancer [9,39]. Evidence of a role for EGFR in cancer came from the seminal finding that the kinase domain of EGFR is homologous to the viral oncogene v-erb-B, which had previously been shown to induce transformation [9,40]. These and other discoveries, along with studies demonstrating that targeting EGFR can inhibit cancer cell proliferation and induce apoptosis in certain preclinical cancer models, established EGFR as a proto-oncogene and a potential drug target in cancer.

Activation of EGFR occurs when one of its ligands binds and elicits a conformational change that enables dimerization [12,33,34]. EGFR binds seven ligands: amphiregulin (AREG), betacellulin (BTC), EGF, epigen (EPGN), epiregulin (EREG), heparin-binding EGF-like growth factor (HB-EGF), and transforming growth factor-α (TGF-α) (Fig. 1.1) [20,25,29,34]. EGFR is able to both bind ligands and phosphorylate substrates and can thus form functional homodimers [5,13,24]. In addition, EGFR can dimerize with other HER family members or, possibly, with non-HER family RTKs, and the formation of dimers with different partners can modulate the outcome of EGFR activation [5,13,17]. Transphosphorylation of tyrosine residues in the EGFR ICD by its dimerization partners creates binding sites for downstream effectors, facilitating their activation [9,14,17]. Among the signaling pathways downstream of EGFR that have been shown to mediate EGFR-induced transformation are the RAS/RAF/MEK/ERK, PI3K/Akt/mTOR, and JAK/STAT pathways [11–13,33,34,41].

Aberrant activation of EGFR occurs in many cancer types, although the alterations that lead to this hyperactivation vary among cancer types. Some cancers exhibit overexpression and/or hyperactivation of wild-type EGFR, whereas activating mutations are observed in others [12,13,42]. Activating mutations, sometimes co-occurring with amplification of the *EGFR* gene, have been observed in non-small cell lung cancer (NSCLC) and glioblastoma [13,22,43,44]. In NSCLC, the most common mutations are exon 19 deletions and the L858R point mutation, both of which occur in the kinase domain and confer ligand-independent activation of EGFR [17,18,32,43,44]. In glioblastoma, mutations in the extracellular domain, including point mutations and in-frame deletions that lead to expression of a constitutively active truncated version of EGFR called EGFRvIII, are more common [43,45]. In other cancer types, activating mutations in *EGFR* are rarely observed, but wild-type EGFR serves as an oncogenic driver. For example, in head and neck squamous cell carcinoma (HNSCC), squamous cell carcinoma of the lung (LUSC), and breast cancer, amplification of wild-type *EGFR* occurs and often results in overexpression of the EGFR protein [12,13,15,17,42]. Transcriptional upregulation of *EGFR* in the absence of gene amplification has also been observed [13]. In addition, oncogenic fusion proteins containing the N-terminal domain of EGFR were recently identified in lung cancer [32].

In addition to alterations in *EGFR* itself, mutations in other genes involved in the regulation of EGFR can result in aberrant activation of this RTK. While the activation of EGFR is tightly regulated in normal (nontransformed) cells, these regulatory mechanisms can dissolve in cancer cells, leading to unchecked activation of the receptor and its downstream signaling molecules [46]. One mechanism for aberrant activation of EGFR is the loss of negative feedback pathways that, in normal cells, prevent sustained activation of these receptors and dampen signaling downstream. For example, hyperactivation of EGFR signaling can occur due to a loss of tumor suppressive phosphatases, such as phosphatase and tensin homolog deleted on chromosome 10 (PTEN), which inhibits EGFR signaling and can promote degradation of EGFR [45]. Increased autocrine and/or paracrine secretion of EGFR ligands by tumor cells and/or other cells in the tumor microenvironment is another mechanism by which EGFR can be aberrantly activated in cancer cells [34,47].

The importance of EGFR in cancer is demonstrated not only by the frequency of alterations that lead to increased expression of EGFR and/or activation of EGFR signaling, but also by the efficacy of EGFR-targeted therapies, both alone and in combination with chemotherapy and other treatment modalities, in several cancer types. The primary classes of EGFR inhibitors that have been employed in the clinic thus far are small-molecule TKIs and neutralizing monoclonal antibodies [48,49]. In general, EGFR-targeting TKIs are used to treat cancers harboring activating mutations in EGFR, whereas antibody-based therapies are more commonly used in the treatment of tumors that overexpress wild-type EGFR, although there are exceptions to this. Although EGFR-targeted therapies are used as a single agent for some indications, they are often combined with chemotherapy, surgery, and/or radiation.

A number of small-molecule TKIs have been approved by the United States Food and Drug Administration (FDA) for cancer treatment. Activating mutations in *EGFR* tend to confer sensitivity to EGFR-targeted TKIs; thus, these inhibitors are used primarily for the treatment of *EGFR*-mutant tumors (Table 1.1) [11,16,18,28,47,50]. Erlotinib (Tarceva) and gefitinib (Iressa) are reversible adenosine triphosphate (ATP)-competitive (type I) EGFR-specific inhibitors [18,32,51]. Afatinib (Gilotrif) is a second-generation, irreversible ATP-competitive TKI that targets HER2 and HER4 in addition to EGFR [9,18,32,51,52]. Third-generation TKIs that can overcome resistance to first- and second-generation TKIs are also being used in the clinic. Osimertinib (Tagrisso) is the only third-generation EGFR-targeted TKI that has been approved by the FDA thus far [12,51]. It has demonstrated efficacy against tumors bearing the T790M mutation in EGFR, the expression of which confers resistance to first- and second-generation inhibitors [12,51]. Osimertinib is selective for mutant EGFR (i.e., wild-type EGFR is largely spared) [51].

Three monoclonal antibodies targeting EGFR have been approved by the FDA for cancer treatment: cetuximab (Erbitux), panitumumab (Vectibix), and necitumumab (Portrazza) (Table 1.1) [12,33,53,54]. These monoclonal antibodies are often used in combination with chemotherapeutic agents and/or radiation. Although the efficacy of TKIs rests primarily on their ability to inhibit kinase activity, EGFR-targeted monoclonal antibodies work via a number of mechanisms, including outcompeting endogenous ligands, preventing dimerization, and inducing internalization and subsequent degradation of EGFR [12,13,33]. In addition, cetuximab and necitumumab, as IgG1-isotype antibodies, can induce antibody-dependent cell-mediated cytotoxicity (ADCC), in which immune cells expressing the FcγRIII receptor (primarily natural killer [NK] cells) recognize the Fc fragment of an antibody and release

TABLE 1.1 HER Family-Targeted Agents Approved by the FDA

Inhibitor	Trade name	Type of inhibitor	Target(s)	Cancer type(s)
Ado-trastuzumab emtansine	Kadcyla	Antibody-drug conjugate	HER2	HER2+ breast cancer
Afatinib	Gilotrif	TKI	EGFR, HER2, HER4	EGFR-mutant NSCLC
Cetuximab	Erbitux	mAb (IgG1 κ)	EGFR	HNSCC; KRAS wild-type, EGFR-expressing CRC
Erlotinib	Tarceva	TKI	EGFR	EGFR-mutant NSCLC; pancreatic cancer
Gefitinib	Iressa	TKI		EGFR-mutant NSCLC
Lapatinib	Tykerb	TKI	EGFR, HER2, HER4	HER2+ breast cancer
Necitumumab	Portrazza	mAb (IgG1 κ)	EGFR	LUSC
Osimertinib	Tagrisso	TKI	EGFR T790M	EGFR-mutant (T790M) NSCLC
Panitumumab	Vectibix	mAb (IgG2 κ)	EGFR	EGFR-expressing CRC
Pertuzumab	Perjeta	mAb (IgG1 κ)	HER2	HER2+ breast cancer
Trastuzumab	Herceptin	mAb (IgG1 κ)	HER2	HER2+ breast cancer; HER2+ gastric and GEJ adenocarcinoma

Abbreviations: TKI, tyrosine kinase inhibitor; mAb, monoclonal antibody; IgG, immunoglobulin G; NSCLC, non-small cell lung cancer; HNSCC, head and neck squamous cell carcinoma; LUSC, lung squamous cell carcinoma; CRC, colorectal cancer; GEJ, gastroesophageal junction.

cytolytic proteins that can kill tumor cells [12,13,53–57]. (It is thought that panitumumab, as an IgG2-isotype antibody, is less effective at inducing NK cell-mediated ADCC than cetuximab or necitumumab [33,49,53].) Induction of ADCC has been shown to contribute to the antitumor effects of these antibodies in preclinical studies [53].

EGFR alterations are commonly observed in NSCLC, with the frequency of alterations varying among the three NSCLC subtypes (adenocarcinoma [LUAD], LUSC, and large cell carcinoma [LCC] [12,17]). Approximately 10%–25% of lung adenocarcinomas bear activating mutations in *EGFR* [44,58]. As mentioned above, the most common of these activating mutations are exon 19 deletions and the L858R mutation, which occur in the kinase domain and result in constitutive activation of the kinase [17,18,32,44,51]. The current standard of care for first-line treatment of patients with *EGFR*-mutant NSCLC is an EGFR-targeted TKI used as a single agent [11]. The EGFR-targeted TKIs afatinib, erlotinib, gefitinib, and osimertinib have received approval from the FDA for the treatment of *EGFR*-mutant NSCLC [12,17,18,36,51,59,60].

Although these TKIs are primarily used as a single agent in *EGFR*-mutant NSCLC, they have been used in combination with chemotherapy in settings in which combining EGFR-targeted inhibitors with chemotherapeutic agents has demonstrated enhanced efficacy compared to monotherapy. For example, results of a recent clinical trial demonstrated that the addition of afatinib to the investigator's choice of chemotherapy improved the objective response rate and progression-free survival in NSCLC patients whose tumors had acquired resistance to first-generation EGFR TKIs (erlotinib and/or gefitinib) and afatinib [36]. However, results from another clinical trial in NSCLC in patients whose tumors had progressed on first-line gefinitib showed that continuing gefitinib treatment in combination with platinum-based doublet chemotherapy did not improve progression-free survival compared to chemotherapy alone [60]. These and other studies demonstrate that combining chemotherapy with EGFR-targeted TKIs may improve outcomes, but only for certain indications.

In contrast to LUAD, overexpression of EGFR is more common than mutation in LUSC [12,17,54]. In this subtype, both TKIs and antibody-based therapies have demonstrated efficacy in the clinic [12,54]. However, the use of EGFR-targeted TKIs is largely limited to NSCLC tumors bearing activating mutations in *EGFR*, which are rare in LUSC [12,17]. In 2015, the EGFR-targeted monoclonal antibody necitumumab, in combination with gemcitabine and cisplatin, was approved by the FDA for first-line treatment of metastatic LUSC [12,54].

Activating mutations in EGFR are also rare in HNSCC [13]. However, EGFR overexpression occurs in up to 90% of HNSCC tumors, and EGFR has been established as an oncogenic driver in this disease [13,17,48]. Cetuximab is the only oncogene-targeted agent that has been approved by the FDA for treatment of HNSCC [12,48,61]. Although approved as a single agent for the treatment of patients with recurrent/metastatic HNSCC whose tumors have progressed on platinum-based chemotherapy, cetuximab is more commonly used in combination with chemotherapy and/or radiation [59,61]. FDA approval for cetuximab in combination with chemotherapy was granted based on results from a clinical trial in which 442 patients with recurrent/metastatic HNSCC were assigned to be treated with platinum-based chemotherapy plus fluorouracil alone or in combination with cetuximab [62]. The addition of cetuximab to this chemotherapy regimen improved overall survival compared to chemotherapy alone [62]. It should be noted that cetuximab has been shown to improve response in combination with chemotherapy in HNSCC, whereas other EGFR inhibitors have not fared as well in clinical trials in this disease, despite a number of studies demonstrating that erlotinib and panitumumab are effective in preclinical models of HNSCC [13,48,49,59,62]. It has been suggested, though not verified, that the enhanced clinical efficacy of cetuximab compared to other EGFR inhibitors tested in HNSCC may be explained, at least in part, by its ability to promote immune-mediated tumor clearance via induction of ADCC [49,59].

As in LUSC and HNSCC, *EGFR* is commonly overexpressed, but rarely mutated, in colorectal adenocarcinoma [24,47]. However, in contrast to HNSCC, both cetuximab and panitumumab are FDA approved for the treatment of metastatic colorectal cancer [12,24,33,61]. While initially approved broadly for the treatment of colorectal cancer on the basis of clinical trials in unselected patient populations, it was subsequently demonstrated that the clinical benefit of cetuximab in this cancer type is largely limited to patients with *KRAS* wild-type tumors, as activating mutations in *KRAS* maintain activation of oncogenic signaling pathways even in the presence of cetuximab [33,35,41,47]. As a result, EGFR-targeted antibodies are largely ineffective against *KRAS*-mutant colorectal tumors; thus, cetuximab is now approved for the treatment of only *KRAS* wild-type tumors [33]. More recently, activating mutations in *BRAF* or *NRAS* were also shown to be negative predictive biomarkers for response to cetuximab and panitumumab in colorectal cancer [33,41]. Official recommendations from the FDA do not currently reflect these still-recent findings.

Nonetheless, cetuximab has demonstrated efficacy in combination with chemotherapy in some *KRAS*-mutant colorectal tumors [63]. Recently, Pozzi et al. [63] hypothesized that this effect may be due to cetuximab's ability to induce immune-mediated tumor clearance and set out to uncover the mechanism by which the combination of cetuximab and chemotherapy could produce an antitumor effect even in *KRAS*-mutant colorectal cancer cells. The authors reported in 2016 that the combination of cetuximab and the chemotherapy regimen FOLFIRI (leucovorin calcium, fluorouracil, and irinotecan hydrochloride) could trigger immunogenic cell death in colorectal cancer cells, thereby enhancing phagocytosis of these cells by dendritic cells and triggering an antitumor immune response mediated by $CD8^+$ T cells [63]. The induction of immunogenic cell death by the combination of cetuximab and FOLFIRI occurred in both *KRAS* wild-type and *KRAS*-mutant cells, although cells containing activating mutations in *BRAF* did not undergo immunogenic cell death when treated with this combination [63]. This study highlights the possibility that inhibition of EGFR (and perhaps other HER family receptors) in combination with chemotherapy may enhance the efficacy of chemotherapy by inducing immune-mediated tumor clearance.

These findings emphasize the need for more research to assess the impact of EGFR inhibition both alone and in combination with chemotherapy on tumor-immune cell interactions in order to identify approaches that can both inhibit tumor cell proliferation and survival and promote antitumor immune responses. Identifying such strategies will further the goal of improving outcomes for patients.

1.2.2 HER2

Following the discovery of EGFR and subsequent identification of its role as a proto-oncogene, researchers sought to add to the growing list of oncogenic drivers. HER2 was discovered in 1982, when the protein encoded by DNA that had been shown to induce transformation in mouse embryonic fibroblasts (NIH/3T3 cells) was found to be a 185-kDa phosphoprotein [64,65]. (This DNA had been isolated from carcinogen-induced rat neuroblastomas; thus, HER2 is referred to as *neu* in rodents [14,64,66,67].) HER2 was quickly implicated in human cancers, as it was shown to be overexpressed in breast tumors and correlated with poor prognosis in breast cancer [9,19,38,66,68]. Given that HER2 was initially discovered in transformed cells, it is unsurprising that the bulk of research on HER2 has focused on its roles in cancer and the development and use of HER2-targeted agents for cancer treatment.

HER2 is unique among the HER family receptors in that it does not require a ligand to be available for dimerization, as it is fixed in an open (untethered) confirmation (Fig. 1.1) [9,14–16]. Due to its constitutive availability for dimerization, HER2 is thought to be the preferred dimerization partner for the other HER family members, with HER2/HER3 heterodimers being particularly potent [1,8,9,14–16,19,20]. HER2/HER3 dimers have been implicated in tumorigenesis, tumor progression, and resistance to chemotherapy [1,8,14,25].

The *ERBB2* gene is altered in a number of cancer types. The most common alteration of *ERBB2* in cancer is gene amplification, which leads to overexpression of HER2 [14,15,69].

Particularly high frequencies of *ERBB2* amplification are observed in gastric/esophageal and breast cancers [1,3,9,14,69–75]. Activating mutations in HER2, many of which cluster in the kinase domain, have also been identified in breast cancer and lung adenocarcinoma and have been shown to promote tumor growth in preclinical models [9,16,22,28,50,74,76]. However, such mutations are uncommon [14,74].

Evidence cementing HER2 as an oncogenic driver in several cancer types has supported the development of HER2-targeted inhibitors. Both TKIs and monoclonal antibodies targeting HER2 have been approved by the FDA for the treatment of HER2-expressing cancers (Table 1.1). The first FDA-approved HER2-targeted agent was trastuzumab (Herceptin), a humanized IgG1 monoclonal antibody [19,57,71,74]. The monoclonal antibody pertuzumab (Perjeta) has also been FDA approved for the treatment of HER2-positive breast cancer [9,71,74]. Although trastuzumab and pertuzumab share a target, they bind to different epitopes in the HER2 extracellular domain [9,19,77,78]. It has been shown that although pertuzumab can inhibit ligand-induced HER2/HER3 dimerization, trastuzumab primarily inhibits ligand-independent dimerization of these RTKs, conferring efficacy in HER2-overexpressing breast cancer cells in which HER2/HER3 dimers can form even in the absence of a ligand [15,19,38,74,78]. Thus, the combination of trastuzumab and pertuzumab can be more effective at inhibiting HER2 than either antibody alone [19,77]. The combination of trastuzumab, pertuzumab, and docetaxel has been shown to improve overall and progression-free survival compared to trastuzumab and docetaxel in patients with HER2-positive metastatic breast cancer [9,19,74,77]. HER2 is also overexpressed at a high frequency in gastric cancer, and trastuzumab has been approved by the FDA in combination with cisplatin and either capecitabine or 5-fluorouracil for the treatment of adenocarcinomas of the stomach and gastroesophageal junction [9,73]. In addition, the antibody-drug conjugate ado-trastuzumab emtansine (Kadcyla), which contains trastuzumab linked to the microtubule polymerization inhibitor emtansine (DM1), has been approved for the treatment of HER2-positive metastatic breast cancer that had been previously treated with trastuzumab and/or a taxane [9,74,75]. To date, lapatinib (Tykerb) is the only HER2-targeted TKI that has been approved by the FDA for cancer treatment [71,74]. Lapatinib is used in combination with letrozole or capecitabine for the treatment of HER2-positive breast cancer [8,9,72,74,79]. Although not currently approved for cancer treatment, the EGFR- and HER2-targeted TKI neratinib is currently being tested in clinical trials and has demonstrated efficacy in patients with HER2-positive breast cancer [9,18,71,72].

HER2-targeted agents have transformed the treatment of HER2-positive breast cancer. Before these agents became available, patients whose tumors expressed HER2 faced a poor prognosis compared to those with HER2-negative tumors [16,68,69]. When trastuzumab, the first FDA-approved HER2-targeted agent, was introduced, the trend reversed: patients with HER2-positive tumors treated with trastuzumab now achieved better outcomes than those with HER2-negative tumors, for whom trastuzumab was not a treatment option [69]. The fact that HER2 expression augured a poor prognosis prior to the development and use of HER2-targeted therapy suggests that HER2 may play a role in intrinsic resistance to chemotherapy, as chemotherapeutic agents, often in combination with surgery, were the backbone of breast cancer treatment prior to the introduction of targeted therapy.

Preclinical research has identified a number of mechanisms that may explain why HER2-positive tumors are less likely to respond to chemotherapy and how targeting HER2 may enhance the efficacy of chemotherapy. A series of papers in the late 1990s and early 2000s demonstrated that HER2 can mediate resistance to the chemotherapeutic drug paclitaxel (Taxol) by inhibiting the cyclin-dependent kinase CDK1 (also referred to as Cdc2) [3,70,80]. Activation of CDK1 is required for entry into mitosis, and nonmitotic cells are largely protected from chemotherapeutic agents; thus, inhibition of CDK1 confers resistance to paclitaxel-induced apoptosis [3]. It has been reported that HER2 can phosphorylate (and thereby inactivate) CDK1 [3,80]. HER2 can also induce transcriptional upregulation of *CDKN1A*, which encodes the cyclin-dependent kinase inhibitor $p21^{Cip1}$ [2,3,70,80]. Activation of Src downstream of HER2 was shown to activate signal transducer and activator of transcription 3 (STAT3), a transcription factor that can induce expression of *CDKN1A* [80]. p21, in turn, can promote resistance to paclitaxel by inducing mitotic arrest via inhibition of the CDK1/cyclin B complex [2,70,80]. HER2 is also associated with expression of efflux pumps that can remove drugs from the cell, thereby promoting resistance to chemotherapeutic agents [2,81].

Just as HER2 inhibition has been shown to improve outcomes when added to chemotherapeutic regimens, chemotherapeutic agents can also enhance the efficacy of HER2-targeted agents. Like the EGFR-targeting antibody cetuximab, trastuzumab and pertuzumab are IgG1-isotype antibodies and have been shown to induce ADCC [9,19,55–57,78]. This is believed to contribute to the antitumor efficacy of these antibodies [55,56,73]. It has been shown that the chemotherapy drug paclitaxel can recruit NK cells to breast cancer tumors, thereby promoting trastuzumab-induced ADCC [56]. In addition, docetaxel can enhance trastuzumab-mediated ADCC by inducing both expression of NK cell-activating ligands on breast cancer cells and upregulation of their cognate receptor (NKG2D) on NK cells [57]. It should be noted, however, that others have reported that the ability of trastuzumab to induce ADCC is not affected by chemotherapy [55].

Thus far, HER2-targeted agents have only been approved by the FDA for the treatment of HER2-positive breast cancer and gastric adenocarcinoma [71,82]. However, HER2 has been shown to be overexpressed and/or mutated in a number of other cancer types, including ovarian and lung cancers [9,14,28]. Although HER2-targeted agents have not yet been approved for treatment of these cancer types, one might predict that HER2-targeted agents will soon be used in the treatment of patients with other cancer types in which HER2 is overexpressed and/or mutated, especially those for which preclinical studies suggest that using HER2 inhibitors, either alone or in combination with chemotherapy, is beneficial.

1.2.3 HER3

In contrast to EGFR and HER2, for which targeted agents have been in use for over a decade, HER3 has only recently been recognized as a promising drug target [15,25,29,38]. HER3 is increasingly being studied in the context of cancer, and roles for HER3 are being identified in a number of cancer types by basic and translational scientists. This coincides with the development and clinical testing of HER3-targeted agents that seek to capitalize on this research to improve outcomes for patients in whose tumors HER3 promotes tumor cell proliferation and survival and/or resistance to chemotherapy [15,25].

Amino acid substitutions in the HER3 kinase domain that were sustained following the gene duplication events that led to the generation of the four homologous HER family members greatly reduce the kinase activity of HER3 [23,26,38]. As a result, unlike the other HER family receptors, HER3 lacks appreciable kinase activity and is thus dependent on other RTKs to phosphorylate the tyrosine residues in its C-terminal tail [14,15,18,20,22,23,29,30,38]. Because of this, HER3 homodimers are thought to be enzymatically inactive [14,30]. However, HER3 can form functional heterodimers with other RTKs upon binding of one of its ligands (neuregulin 1 [NRG1] and NRG2) (Fig. 1.1) [6,14,15,18,20,25,29]. As mentioned above, the mechanism of dimerization-induced activation of HER family members involves allosteric rather than kinase-dependent activation; thus, HER3 can promote phosphorylation by its dimerization partners even though it lacks inherent kinase activity [23,38]. HER2 is the preferred binding partner for HER3, but HER3 has also been shown to dimerize with EGFR [1,8,16,18,23,30,38,83].

Once phosphorylated, HER3 is a potent activator of the PI3K/Akt/mTOR signaling pathway, due to enrichment in PI3K-binding sites in its ICD [8,14]. Tyrosine-phosphorylated HER3 contains six PI3K docking sites and promotes PI3K activity [8,25,29,30,38]. The ability of HER3 to serve as a plasma membrane-adjacent binding partner for PI3K contributes to its role as an active participant in tumorigenesis and tumor progression in a number of cancer types, despite its lack of kinase activity [25,29].

HER3 overexpression has been observed in many cancer types, including colorectal, breast, lung, ovarian, and pancreatic cancers; melanoma; and human papillomavirus-positive (HPV^+) HNSCC, and has been associated with poor prognosis [25,38,84–87]. In addition, point mutations in HER3, many of which are thought to promote its activation, have been identified [15,25,29,30]. In contrast to HER2, in which most of the activating mutations identified occur in the kinase domain, activating mutations in HER3 largely cluster in the extracellular domain and are hypothesized to promote adoption of the untethered (open) conformation, thereby fostering dimerization (although this hypothesis has yet to be rigorously tested) [15,29]. However, mutations in the kinase domain of HER3 have also been identified, although the functional significance of these mutations has not yet been definitively determined [15,29]. HER3 can also be aberrantly activated in cancer due to autocrine and/or paracrine secretion of its ligand neuregulin 1 (NRG1) [7,25,83,87]. In addition, HER3 can cooperate with its dimerization partners to promote activation of oncogenic signaling pathways, particularly the PI3K/Akt/mTOR pathway. This is especially evident in HER2-expressing tumors, in which HER3, as a dimerization partner for HER2, has been shown not only to drive tumorigenesis, but also to play a role in both intrinsic and acquired resistance to chemotherapy [1,2,4,8].

To date, no drugs designed to specifically target HER3 have been approved by the FDA for cancer treatment. However, several HER3 inhibitors are in advanced clinical development. Because HER3 lacks apparent kinase activity, it is thought that using TKIs is not an effective method of HER3 inhibition [38]. However, HER3 is susceptible to inhibition by extracellular neutralizing antibodies [25,29,38]. Several such antibodies have been developed in recent years, and some of these are now being tested in clinical trials. The HER3-targeted antibody CDX-3379 (formerly KTN3379) has demonstrated efficacy in preclinical cancer models and has been tested in clinical trials in HNSCC and thyroid cancer (NCT02473731 and NCT02456701, respectively) [86]. Another anti-HER3 monoclonal antibody, patritumab, is also slated to be tested in upcoming clinical trials, including the I-SPY 2 trial in breast cancer (NCT01042379). It should be noted that although specifically inhibiting HER3 is one method to target HER3-mediated signaling, targeting RTKs with which HER3 dimerizes can also result in inhibition of HER3-mediated signaling [18,19,29,71]. For example, treating gastrointestinal carcinoma cell lines or patient-derived xenografts with the TKI afatinib or a combination of the HER2 inhibitors lapatinib and trastuzumab can reduce HER3 phosphorylation [71].

Although targeting HER3 in cancer types in which it has been shown to be an oncogenic driver is one potential use of HER3 inhibitors, preclinical data also support a role for HER3 targeting to overcome resistance to chemotherapy. HER3 was shown to mediate resistance to paclitaxel in HER2-overexpressing breast cancer cells by promoting expression of survivin, a protein that prevents apoptosis [1,2]. PI3K/Akt signaling downstream of HER3 has also been shown to promote tumor cell survival in the presence of chemotherapeutic agents [4,8]. A 2003 study assessing the role of HER family members in resistance to chemotherapy identified HER2/HER3-mediated activation of the PI3K/Akt signaling pathway as a mechanism of resistance to a number of chemotherapeutic agents [4]. In this study, exogenous overexpression of HER2 in the HER3-expressing breast cancer cell line MCF7, which expresses a low level of HER2 at baseline, induced PI3K/Akt signaling [4]. HER2 overexpression promoted resistance to the chemotherapeutic agents camptothecin, doxorubicin, etoposide, 5-fluorouracil, and paclitaxel via activation of the PI3K/Akt pathway [4]. In 2012, Bezler et al. [8] reported that treatment of ovarian cancer cells with either doxorubicin or cisplatin induced expression of HER3 ligands, leading to activation of HER3/PI3K/Akt signaling in a HER2-dependent manner. Upregulation of this signaling axis reduced doxorubicin- and cisplatin-induced apoptosis [8]. Collectively, these findings support a role for the HER2/HER3 heterodimer in chemoresistance and suggest that targeting HER3 may delay and/or overcome resistance to chemotherapy.

These and other preclinical studies and the observation of HER3 overexpression in a number of cancer types have established HER3 as an emerging therapeutic target [15,38,82]. As HER3-targeted agents are being tested in clinical trials and becoming more readily available for preclinical research, additional roles for HER3 in chemoresistance may be identified and may lead to the use of HER3-targeted agents in combination with chemotherapeutic agents in the coming years.

1.2.4 HER4

HER4 was the fourth member of the HER family to be discovered (Fig. 1.1); the gene encoding HER4 (*ERBB4*) was cloned in 1993. HER4 has been well studied in the context of development of the brain, heart, and other organs, and as an essential participant in a number of physiological processes, including lactation (which is particularly relevant to studies on the roles of HER4 in breast cancer) [20,26,88–91]. However, the role of HER4 in cancer is not as well established as for the other three HER family members, in part due to conflicting reports on the prognostic implications of HER4 expression [10,15,21,26,89,92–94]. Most studies on HER4 in cancer report a tumor-suppressive role for this RTK, as it can promote differentiation and apoptosis of cancer cells, and several studies have reported reduced HER4 expression in tumors compared to tumor-adjacent normal tissue [21,82,95]. On the other hand, overexpression of and activating mutations in HER4 have also been identified [15,16,21,38,52,92,93,96,97], and HER4 has been shown to induce transformation and proliferation in some preclinical studies [10,21,52,93]. These discrepancies highlight the need for more research on the role of HER4 in cancer and the importance of considering the context in which the roles of HER4 are assessed (including the cancer type, HER4 isoform(s)

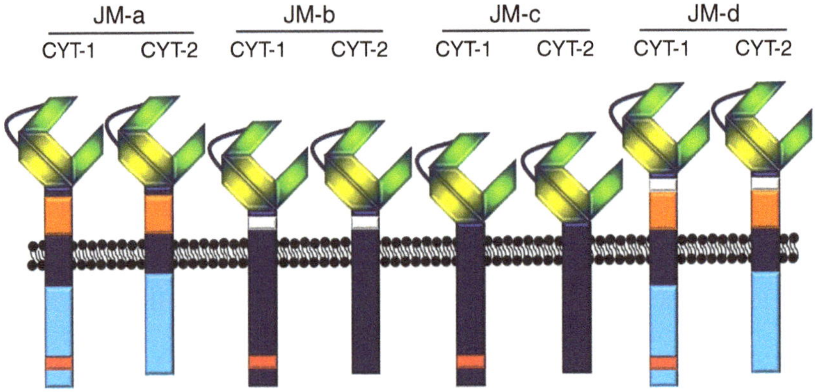

FIG. 1.2 HER4 isoforms. Alternative splicing leads to the generation of HER4 isoforms that vary in the juxtamembrane (JM) and cytosolic (CYT) domains. The four juxtamembrane variants (JM-a, JM-b, JM-c, and JM-d) can pair with either of the two cytosolic variants (CYT-1 and CYT-2). The JM-c isoform lacks both the JM-a and JM-b sequences (orange and white rectangles, respectively), whereas JM-d isoforms contain both these sequences. HER4 receptors containing JM-a (i.e., the JM-a and JM-d isoforms) are susceptible to proteolytic cleavage by TACE and subsequent cleavage of the receptor by γ-secretase. This leads to the release of the HER4 ICD (light blue rectangles). CYT-1 and CYT-2 differ in that CYT-1 is generated by inclusion of an additional exon that encodes a PI3K-binding site (red rectangle).

expressed, and coexpression of other proteins) in order to understand the functions of HER4 and the potential impacts of targeting this RTK in cancer.

HER4, like EGFR, can both bind ligands and phosphorylate substrates. Upon binding of one of its seven ligands, HER4 can form functional homodimers or heterodimerize with other RTKs, especially HER2 [22,26,82,97,98]. HER4 is activated by BTC, ERG, HB-EGF, and NRG1-4 (Fig. 1.1) [6,20,25,29]. In addition to the traditional ligand binding/dimerization mechanism of activation, HER4 is further regulated by alternative splicing and proteolytic cleavage [6,10,21,26,82,89,92,98–100]. Alternative splicing leads to the expression of several HER4 isoforms that vary in the juxtamembrane and cytosolic domains (Fig. 1.2) [6,10,21,26,82,89,92,99,101]. It is important to consider the isoform(s) expressed when assessing the functions of HER4, as these isoforms exhibit functional differences [6,21,26,93].

The four different splice variants of the (extracellular) juxtamembrane domain are referred to as JM-a, JM-b, JM-c, and JM-d [6,26,101]. (Often, only the JM-a and JM-b variants are mentioned.) JM-c-containing isoforms lack the juxtamembrane domain [6,101]. JM-a- and JM-d-containing isoforms possess longer juxtamembrane domains than JM-b-containing isoforms and are susceptible to proteolytic cleavage by tumor necrosis factor α-converting enzyme (TACE, also called ADAM17) upon the binding of certain ligands [10,20,21,26,27,89–91,93,94,99–101]. TACE-mediated cleavage of JM-a- and JM-d-containing isoforms leads to shedding of the HER4 ectodomain and subsequent cleavage of the remaining membrane-associated receptor by γ-secretase to generate a soluble form of the HER4 ICD [10,20,21,26,27,82,89–94,97,100]. The ICD contains a constitutively active kinase domain and a nuclear localization sequence and has been shown to perform a number of functions (Fig. 1.3) [26,27,91,94]. The ICD can localize to mitochondria and act as a BH3-only (pro-apoptotic) protein [26,90,94,102]. The

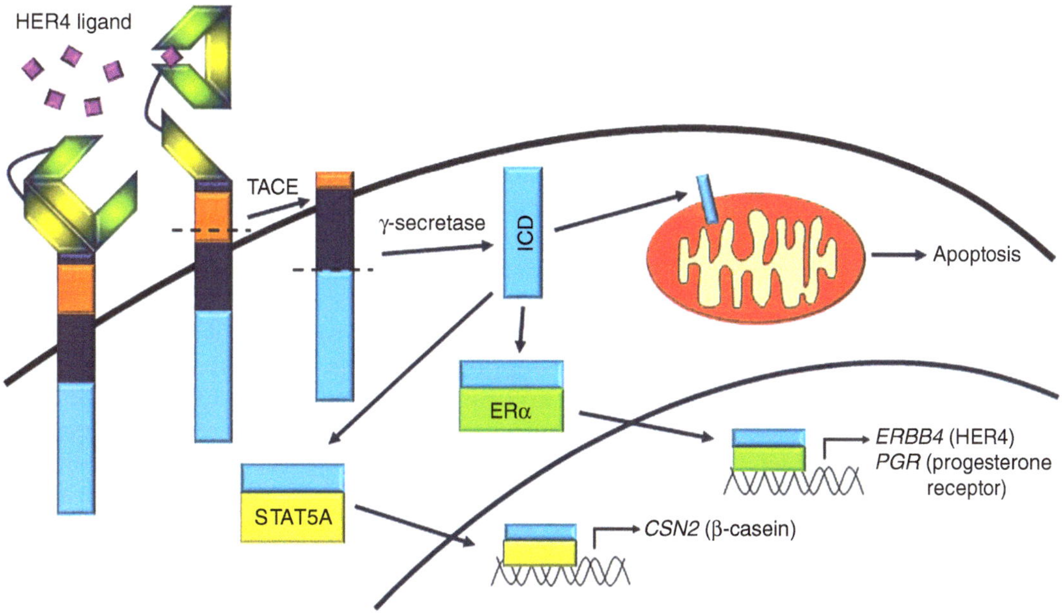

FIG. 1.3 Generation and functions of the HER4 ICD. Ligand binding to HER4 JM-a-containing isoforms induces TACE-mediated cleavage of the HER4 ectodomain. The remaining HER4 molecule can then be cleaved by γ-secretase to release the ICD, an 80-kDa molecule that performs a number of functions. The ICD contains a nuclear localization sequence (NLS) and can facilitate nuclear translocation of signal transducer and activator of transcription 5A (STAT5A). The ICD/STAT5A complex binds to the promoter of the gene *CSN2*, which encodes β-casein, stimulating its transcription. The ICD also serves as a co-activator of estrogen receptor α (ERα)-induced transcription, leading to the expression of genes such as *PGR* (progesterone receptor) and *ERBB4* (HER4). In addition, the ICD can migrate to mitochondria and promote activation of the pro-apoptotic protein BAK, a member of the Bcl-2 protein family (not shown). This results in cytochrome c release and induction of apoptosis.

ICD can also bind to the transcription factor signal transducer and activator of transcription 5A (STAT5A) and can promote STAT5A-induced gene expression [20,26,90,91,93]. In addition, the ICD has been shown to serve as a coactivator of estrogen receptor alpha (ERα)-mediated transcription, promoting expression of genes such as *PGR* (which encodes the progesterone receptor) and *ERBB4* (HER4). Overexpression of the JM-a CYT-2 isoform of HER4 (which can be cleaved to generate the ICD) was shown to promote proliferation of the ER-positive breast cancer cell line MCF7 [10,90,93,94,102]. Thus, the HER4 ICD has been shown to play both tumor-suppressive and pro-tumorigenic roles.

Alternative splicing affecting the cytosolic region of HER4 supplies an additional mechanism of regulation. CYT-1 and CYT-2 are variants in the cytosolic domain of HER4 [10,82,92–94]. Although similar, CYT-1 is generated by inclusion of an additional exon that encodes a binding site for PI3K [6,10,26,82,89,92–94]. In preclinical models of breast cancer, shifting expression in favor of CYT-2-containing isoforms reduced proliferation of breast cancer cells [92]. However, the CYT-1 isoform itself has been shown to inhibit proliferation and promote differentiation in mammary epithelial cells [92–94]. Thus, a blanket statement regarding the implications of CYT-1 versus CYT-2 expression would not be supported by the literature at this point.

HER4 alterations observed in cancer include those that promote expression and/or activation of HER4 and those that lead to reduced HER4 levels in tumors. Activating mutations in HER4 have been reported in melanoma, breast cancer, NSCLC, and HNSCC [16,22,38,52,92,93,96,97]. Some of these activating mutations occur in the kinase domain and promote activation of downstream signaling molecules and/or HER4 cleavage. On the other hand, expression of both HER4 and one of its ligands, neuregulin 4 (NRG4), was found to be decreased in gastric cancer tumors compared to adjacent normal tissue [82]. Moreover, hypermethylation of the *ERBB4* promoter, resulting in reduced expression of HER4, has been observed in breast cancer [95]. Collectively, these studies suggest that while HER4 acts as an oncogene in certain settings, it plays a tumor-suppressive role in other contexts.

In some cancers, such as colorectal cancer and gastrointestinal stromal tumors originating from the stomach, expression of HER4 has been shown to be a negative prognostic marker, whereas in bladder cancer, HER4 expression has been associated with a favorable outcome [103,104]. Even within a particular cancer type, the role of HER4 is somewhat controversial, as expression of HER4 has been reported to be both a negative and a positive prognostic marker in breast cancer [26,105]. For example, expression of HER4 on the cell surface was associated with increased survival in patients with ER-positive breast cancer, but not in those with ER-negative breast tumors [10,90]. In the same study [10], it was reported that nuclear localization of the HER4 ICD, as opposed to HER4 expression on the cell membrane, was associated with decreased survival in both ER-positive and -negative breast cancer [10].

Several factors may explain the apparent discrepancies in studies investigating HER4 function. The unique functions of the different HER4 isoforms likely contribute to the confusion; however, because many studies do not assess which isoform is expressed, it is not possible to definitively determine whether differential isoform expression can explain the discrepancies [10,21,26,89,92–94]. The subcellular localization of HER4 also has important implications for its function and its ability to serve as a predictive prognostic marker [10,89,92,94]. Moreover, it would be prudent to take into account the expression of proteins that are known to interact with HER4, either directly or as components of a common signaling pathway, when assessing the function of HER4 [92]. This is exemplified in the differences in the prognostic implications of HER4 expression in ER-positive and -negative breast cancer, but could also apply to coexpression with other molecules [10]. For example, heterodimerization with other HER family members may influence HER4 function and the effect of HER4 expression on prognosis. In addition, Chuu et al. [26] have suggested that the medical history of breast cancer patients (and of the patients from whom preclinical models were derived) should be considered when studying HER4, as the functions of HER4 can be affected by such factors as pregnancy, lactation, and hormone treatments.

Perhaps because expression of HER4 has been shown to correlate with opposing outcomes (e.g., increased vs. decreased overall survival), the development of HER4-targeting agents lags behind that of other three HER family members [93]. However, targeted agents that can inhibit HER4 have been used in the clinic. Although the primary targets of the TKIs lapatinib, afatinib, and neratinib are EGFR and HER2, HER4 is also inhibited by these agents [9,18,52,72,96]. Naturally, it is not possible to determine if or to what degree HER4 inhibition contributes to the efficacy of these drugs in patients, as the relative influence of EGFR,

HER2, and HER4 inhibition cannot be dissected experimentally in this context. Thus, a role for HER4 inhibition in the efficacy of lapatinib, afatinib, and neratinib cannot be definitively determined.

Some preclinical data suggest that inhibition of HER4 may enhance response to chemotherapy. In 2013, Hegde et al. [7] reported that tumor cells that persisted after chemotherapy in a *Kras*^{G12D}-expressing genetically engineered mouse model of NSCLC and in human NSCLC cell line xenografts expressed increased levels of HER3, HER4, and neuregulin 1 (NRG1, a ligand for HER3 and HER4) compared to treatment-naïve tumors. The authors also found that inhibition of HER4 could delay or prevent tumor relapse following chemotherapy [7]. Notably, the tumors that grew in the genetically engineered mouse model used in this study were engineered to express *Kras*^{G12D}, not HER4 or another HER family member, suggesting that inhibition of HER4 could be a strategy to delay and/or overcome chemoresistance even in tumors in which HER4 is not thought to be the primary oncogenic driver. This warrants future investigation.

Continued research on the functions of HER4 in different contexts and the patterns of expression of HER4 isoforms in tumors will likely improve the ability of physicians and researchers to use HER4 expression as a prognostic indicator and could help to identify cancer types in which a HER4-targeted agent (either one that specifically targets HER4 or one that can inhibit multiple HER family members, such as afatinib and lapatinib) may demonstrate efficacy [52,96]. On the other hand, the identification of antitumorigenic roles for HER4 in several cancer types also raises the question of whether HER4 inhibition by HER family-targeted TKIs currently in use in the clinic might be hampering the efficacy of these drugs. This should be considered when using TKIs that can target HER4, as the use of such an inhibitor for the treatment of a tumor in which HER4 is playing a tumor suppressive role might not be advisable. Again, however, more research is needed on this topic.

1.3 CONCLUSIONS

EGFR, HER2, HER3, and HER4 have, to varying degrees, been implicated in tumorigenesis and tumor maintenance. Monoclonal antibodies and TKIs targeting EGFR, HER2, and, more recently, HER3, have been developed and used in the treatment of cancer types in which these RTKs have been shown to play an oncogenic role. Encouragingly, the use of HER family inhibitors has resulted in improved outcomes for patients, and HER family-targeted agents continue to be developed and tested in clinical settings, expanding the number of indications for which HER family-targeted agents are approved by the FDA. In addition to their roles in tumor cell proliferation and survival, HER family receptors have been shown to mediate resistance to chemotherapy, making inhibition of these RTKs a rational cotargeting strategy in the cancer types in which HER family members are expressed. However, even combining HER family inhibitors with chemotherapy does not necessarily lead to durable or even short-term response to treatment, highlighting the need for more basic and translational research on HER family receptors and HER family-targeted agents, both alone and in combination with chemotherapy, to identify more effective treatment strategies.

Given the many roles of HER family members and their downstream effectors in cancer, there are many potential future avenues of research that may further this goal. Of these, it will be particularly interesting and timely to assess the effects of HER family inhibitors on antitumor immune responses and to identify drug combinations that enhance immune-mediated tumor clearance. Although not discussed extensively in this chapter, resistance to HER family-targeted agents remains a major obstacle in cancer treatment, and devising methods to overcome resistance will improve patient outcomes. In addition, continued research on the functions of HER4 could help to identify cancer types in which HER4 targeting may be effective and avoid the potential consequences of utilizing agents that can target HER4 for the treatment of tumors in which HER4 is acting as a tumor suppressor. Finally, as advances in DNA sequencing and other methods to characterize patients' tumors are making these analyses more economically feasible and more common, there may be opportunities to uncover additional cancer types in which HER family members are playing an oncogenic role. Using this information to identify and treat patients who might otherwise not be considered candidates for treatment with HER family-targeting agents could improve outcomes, thus realizing the goal of precision medicine.

Acknowledgements

This work was supported by the National Institutes of Health grants R01 DE024728 (D.E.J.), P50CA097190 (D.E.J. and J.R.G.), and R01 DE023685 (J.R.G.).

Conflict of Interest: No potential conflicts of interest were disclosed.

References

[1] Wang S, Huang X, Lee C-K, Liu B. Elevated expression of erbB3 confers paclitaxel resistance in erbB2-overexpressing breast cancer cells via upregulation of survivin. Oncogene 2010;29(29):4225–36. Available from: http://dx.doi.org/10.1038/onc.2010.180.

[2] de Hoon JPJ, Veeck J, Vriens BEPJ, Calon TGA, van Engeland M, Tjan-Heijnen VCG. Taxane resistance in breast cancer: a closed HER2 circuit? Biochim Biophys Acta 2012;1825(2):197–206. Available from: http://dx.doi.org/10.1016/j.bbcan.2012.01.001.

[3] Tan M, Jing T, Lan KH, Neal CL, Li P, Lee S, et al. Phosphorylation on tyrosine-15 of p34Cdc2 by ErbB2 inhibits p34Cdc2 activation and is involved in resistance to taxol-induced apoptosis. Mol Cell 2002;9(5):993–1004.

[4] Knuefermann C, Lu Y, Liu B, Jin W, Liang K, Wu L, et al. HER2/PI-3K/Akt activation leads to a multidrug resistance in human breast adenocarcinoma cells. Oncogene 2003;22:3205–12.

[5] Vouri M, Croucher DR, Kennedy SP, An Q, Pilkington GJ, Hafizi S. Axl-EGFR receptor tyrosine kinase hetero-interaction provides EGFR with access to pro-invasive signalling in cancer cells. Oncogenesis 2016;5(10):e266. Available from: http://www.nature.com/doifinder/10 1038/oncsis.2016.66.

[6] Gilmour LM, Macleod KG, McCaig A, Gullick WJ, Smyth JF, Langdon SP. Expression of erbB-4/HER-4 growth factor receptor isoforms in ovarian cancer. Cancer Res 2001;61:2169–76.

[7] Hegde GV, de la Cruz CC, Chiu C, Alag N, Schaefer G, Crocker L, et al. Blocking NRG1 and other ligand-mediated Her4 signaling enhances the magnitude and duration of the chemotherapeutic response of non-small cell lung cancer. Sci Transl Med 2013;5:171ra18. Available from: http://www.ncbi.nlm.nih.gov/pubmed/23390248.

[8] Bezler M, Hengstler JG, Ullrich A. Inhibition of doxorubicin-induced HER3-PI3K-AKT signalling enhances apoptosis of ovarian cancer cells. Mol Oncol 2012;6(5):516–29.

[9] Iqbal N, Iqbal N. Human epidermal growth factor receptor 2 (HER2) in cancers: overexpression and therapeutic implications. Mol Biol Int 2014;2014:852748. doi: 10.1155/2014/852748.

[10] Junttila TT, Sundvall M, Lundin M, Lundin J, Tanner M, Harkonen P, et al. Cleavable ErbB4 isoform in estrogen receptor—regulated growth of breast cancer cells. Cancer Res 2005;65(4):1384–94.

[11] Carcereny E, Morán T, Capdevila L, Cros S, Vilà L, de Los Llanos Gil M, et al. The epidermal growth factor receptor (EGFR) in lung cancer. Transl Respir Med 2015;3:1. Available from: http://www.pubmedcentral.nih.gov/articlerender.fcgi?artid=4366432&tool=pmcentrez&rendertype=abstract.

[12] Genova C, Hirsch FR. Clinical potential of necitumumab in non-small cell lung carcinoma. Onco Targets Ther 2016;9:5427–37.

[13] Sharafinski ME, Ferris RL, Ferrone S, Grandis JR. Epidermal growth factor receptor targeted therapy of squamous cell carcinoma of the head and neck. Head Neck 2010;32(10):1412–21.
[14] Moasser MM. The oncogene HER2: its signaling and transforming functions and its role in human cancer pathogenesis. Oncogene 2007;26(45):6469–87.
[15] Jaiswal BS, Kljavin NM, Stawiski EW, Chan E, Parikh C, Durinck S, et al. Oncogenic ERBB3 mutations in human cancers. Cancer Cell 2013;23(5):603–17. Available from: http://dx.doi.org/10.1016/j.ccr.2013.04.012.
[16] Boulbes DR, Arold ST, Chauhan GB, Blachno KV, Deng N, Chang WC, et al. HER family kinase domain mutations promote tumor progression and can predict response to treatment in human breast cancer. Mol Oncol 2015;9(3):586–600. Available from: http://dx.doi.org/10.1016/j.molonc.2014.10.011.
[17] Prabhakar CN. Epidermal growth factor receptor in non-small cell lung cancer. Transl Lung Cancer Res 2015;4(2):110–8. Available from: http://www.ncbi.nlm.nih.gov/pubmed/25870793%0Ahttp://www.pubmedcentral.nih.gov/articlerender.fcgi?artid=PMC4384217.
[18] Hirsh V. Next-generation covalent irreversible kinase inhibitors in NSCLC: focus on afatinib. BioDrugs 2015;29(3):167–83.
[19] Baselga J, Cortes J, Kim SB, Im SA, Hegg R, Im Y-H, et al. Pertuzumab plus trastuzumab plus docetaxel for metastatic breast cancer. N Engl J Med 2012;366(2):109–19.
[20] Muraoka-Cook RS, Sandahl M, Husted C, Hunter D, Miraglia L, Feng S, et al. The intracellular domain of ErbB4 induces differentiation of mammary epithelial cells. Mol Biol Cell 2006;17 (September):4118–29.
[21] Maatta JA, Sundvall M, Junttila TT, Peri L, Laine VJO, Isola J, et al. Proteolytic cleavage and phosphorylation of a tumor-associated ErbB4 isoform promote ligand-independent survival and cancer cell growth. Mol Biol Cell 2006;17:67–79.
[22] Soung YH, Lee JW, Kim SY, Wang YP, Jo KH, Moon SW, et al. Somatic mutations of the ERBB4 kinase domain in human cancers. Int J Cancer 2006;118(6):1426–9.
[23] Jura N, Shan Y, Cao X, Shaw DE, Kuriyan J. Structural analysis of the catalytically inactive kinase domain of the human EGF receptor 3. Proc Natl Acad Sci USA 2009;106(51):21608–13.
[24] Cunningham D, Humblet Y, Siena S, Khayat D, Bleiberg H, Santoro A, et al. Cetuximab monotherapy and cetuximab plus irinotecan in irinotecan-refractory metastatic colorectal cancer. N Engl J Med 2004;351(4):337–45.
[25] Mujoo K, Choi B-K, Huang Z, Zhang N, An Z. Regulation of ERBB3/HER3 signaling in cancer. Oncotarget 2014;5(21):10222–36. Available from: http://www.pubmedcentral.nih.gov/articlerender.fcgi?artid=4279368&tool=pmcentrez&rendertype=abstract.
[26] Chuu C-P, Chen R-Y, Barkinge JL, Ciaccio MF, Jones RB. Systems-level analysis of ErbB4 signaling in breast cancer: a laboratory to clinical perspective. Mol Cancer Res 2008;6(6):885–91.
[27] Linggi B, Cheng QC, Rao AR, Carpenter G. The ErbB-4 s80 intracellular domain is a constitutively active tyrosine kinase. Oncogene 2006;25(1):160–3. Available from: http://www.ncbi.nlm.nih.gov/entrez/query.fcgi?db=pubmed&cmd=Retrieve&dopt=AbstractPlus&list_uids=16170367.
[28] Shigematsu H, Takahashi T, Nomura M, Majmudar K, Suzuki M, Lee H, et al. Somatic mutations of the HER2 kinase domain in lung adenocarcinomas. Cancer Res 2005;65(5):1642–6.
[29] Gala K, Chandarlapaty S. Molecular pathways: HER3 targeted therapy. Clin Cancer Res 2014;20(6):1410–6.
[30] Umelo I, Noeparast A, Chen G, Renard M, Geers C, Vansteenkiste J, et al. Identification of a novel HER3 activating mutation homologous to EGFR-L858R in lung cancer. Oncotarget 2016;7(3):3068–83. Available from: http://www.ncbi.nlm.nih.gov/pubmed/26689995%5Cnhttp://www.pubmedcentral.nih.gov/articlerender.fcgi?artid=PMC4823091.
[31] Levkowitz G, Waterman H, Ettenberg SA, Katz M, Tsygankov AY, Alroy I, et al. Ubiquitin ligase activity and tyrosine phosphorylation underlie suppression of growth factor signaling by c-Cbl/Sli-1. Mol Cell 1999;4(6):1029–40. Available from: http://linkinghub.elsevier.com/retrieve/pii/S1097276500802312.
[32] Konduri K, Gallant J-N, Chae YK, Giles FJ, Gitlitz BJ, Gowen K, et al. EGFR fusions as novel therapeutic targets in lung cancer. Cancer Discov 2016;6(6):601–11.
[33] Hocking CM, Price TJ. Panitumumab in the management of patients with KRAS wild-type metastatic colorectal cancer. Therap Adv Gastroenterol 2014;7(1):20–37. Available from: http://journals.sagepub.com/doi/10.1177/1756283X13498660.
[34] Gao J, Ulekleiv CH, Halstensen TS. Epidermal growth factor (EGF) receptor-ligand based molecular staging predicts prognosis in head and neck squamous cell carcinoma partly due to deregulated EGF-induced amphiregulin expression. J Exp Clin Cancer Res 2016;35(1):151–2. Available from: http://jeccr.biomedcentral.com/articles/10.1186/s13046-016-0422-z.

[35] Van Cutsem E, Kohne C-H, Hitre E, Zaluski J, Chien C-RC, Makhson A, et al. Cetuximab and chemotherapy as initial treatment for metastatic colorectal cancer. N Engl J Med 2009;360(14):1408–17.
[36] Schuler M, Yang JCH, Park K, Kim JH, Bennouna J, Chen YM, et al. Afatinib beyond progression in patients with non-small-cell lung cancer following chemotherapy, erlotinib/gefitinib and afatinib: phase III randomized LUX-Lung 5 trial. Ann Oncol 2016;27(3):417–23.
[37] Slamon DJ, Leyland-Jones B, Shak S, Fuchs H, Paton V, Bajamonde A, et al. Use of chemotherapy plus a monoclonal antibody against HER2 for metastatic breast cancer that overexpresses HER2. N Engl J Med 2001;344(11):783–92. Available from: http://www.nejm.org/doi/full/10.1056/NEJM200103153441101.
[38] Campbell MR, Amin D, Moasser MM. HER3 comes of age: new insights into its functions and role in signaling, tumor biology, and cancer therapy. Clin Cancer Res 2010;16(5):1373–83.
[39] Carpenter G, King LJ, Cohen S. Epidermal growth factor stimulates phosphorylation in membrane preparations in vitro. Nature 1978;276:409–10.
[40] Downward J, Yarden Y, Mayes E, Scrace G, Totty N, Stockwell P, et al. Close similarity of epidermal growth factor receptor and v-erb-B oncogene protein sequences. Nature 1984;307:521–7.
[41] Hsu H-C, Thiam TK, Lu Y-J, Yeh CY, Tsai W-S, You JF, et al. Mutations of KRAS/NRAS/BRAF predict cetuximab resistance in metastatic colorectal cancer patients. Oncotarget 2016;7(16):22257–70.
[42] Bhargava R, Gerald WL, Li AR, Pan Q, Lal P, Ladanyi M, et al. EGFR gene amplification in breast cancer: correlation with epidermal growth factor receptor mRNA and protein expression and HER-2 status and absence of EGFR-activating mutations. Mod Pathol 2005;18(8):1027–33. Available from: http://www.ncbi.nlm.nih.gov/pubmed/15920544.
[43] Vivanco I, Ian Robins H, Rohle D, Campos C, Grommes C, Nghiemphu PL, et al. Differential sensitivity of glioma- versus lung cancer-specific EGFR mutations to EGFR kinase inhibitors. Cancer Discov 2012;2(5):458–71.
[44] Li AR, Chitale D, Riely GJ, Pao W, Miller VA, Zakowski MF, et al. EGFR mutations in lung adenocarcinomas. J Mol Diagn 2008;10(3):242–8. Available from: http://linkinghub.elsevier.com/retrieve/pii/S1525157810601560.
[45] Vivanco I, Rohle D, Versele M, Iwanami A, Kuga D, Oldrini B, et al. The phosphatase and tensin homolog regulates epidermal growth factor receptor (EGFR) inhibitor response by targeting EGFR for degradation. Proc Natl Acad Sci 2010;107(14):6459–64. Available from: http://www.pnas.org/cgi/doi/10.1073/pnas.0911188107.
[46] Krasinskas AM. EGFR signaling in colorectal carcinoma. Patholog Res Int 2011;2011. 932932. Available from: http://www.pubmedcentral.nih.gov/articlerender.fcgi?artid=3042643&tool=pmcentrez&rendertype=abstract.
[47] Khambata-Ford S, Garrett CR, Meropol NJ, Basik M, Harbison CT, Wu S, et al. Expression of epiregulin and amphiregulin and K-ras mutation status predict disease control in metastatic colorectal cancer patients treated with cetuximab. J Clin Oncol 2007;25(22):3230–7.
[48] Quesnelle KM, Wheeler SE, Ratay MK, Grandis JR. Preclinical modeling of EGFR inhibitor resistance in head and neck cancer. Cancer Biol Ther 2012;13(10):935–45.
[49] Kruser TJ, Armstrong EA, Ghia AJ, Huang S, Wheeler DL, Radinsky R, et al. Augmentation of radiation response by panitumumab in models of upper aerodigestive tract cancer. Int J Radiat Oncol Biol Phys 2008;72(2):534–42.
[50] Stephens P, Hunter C, Bignell G, Edkins S, Davies H, Teague J, et al. Lung cancer: intragenic ERBB2 kinase mutations in tumours. Nature 2004;431(7008):525–6. Available from: http://www.nature.com/nature/journal/v431/n7008/full/431525a.html%5Cnhttp://www.nature.com/nature/journal/v431/n7008/pdf/431525a.pdf%5Cnhttp://www.ncbi.nlm.nih.gov/pubmed/15457249.
[51] Sullivan I, Planchard D. Next-generation EGFR tyrosine kinase inhibitors for treating EGFR-mutant lung cancer beyond first line. Front Med 2017;3 (January):1–13. Available from: http://journal.frontiersin.org/article/10.3389/fmed.2016.00076/full.
[52] Nakamura Y, Togashi Y, Nakahara H, Tomida S, Banno E, Terashima M, et al. Afatinib against esophageal or head-and-neck cell squamous carcinoma: significance of activating oncogenic HER4 mutations in HNSCC. Mol Cancer Ther 2016;15 (August):1988–97. Available from: http://www.ncbi.nlm.nih.gov/pubmed/27207775.
[53] Schneider-Merck T, Lammerts van Bueren JJ, Berger S, Rossen K, van Berkel PHC, Derer S, et al. Human IgG2 antibodies against epidermal growth factor receptor effectively trigger antibody-dependent cellular cytotoxicity but, in contrast to IgG1, only by cells of myeloid lineage. J Immunol 2010;184(1):512–20.
[54] Thatcher N, Hirsch FR, Luft AV, Szczesna A, Ciuleanu TE, Dediu M, et al. Necitumumab plus gemcitabine and cisplatin versus gemcitabine and cisplatin alone as first-line therapy in patients with stage IV squamous non-small-cell lung cancer (SQUIRE): an open-label, randomised, controlled phase 3 trial. Lancet Oncol 2015;16(7):763–74.

[55] Petricevic B, Laengle J, Singer J, Sachet M, Fazekas J, Steger G, et al. Trastuzumab mediates antibody-dependent cell-mediated cytotoxicity and phagocytosis to the same extent in both adjuvant and metastatic HER2/neu breast cancer patients. J Transl Med 2013;11:307. Available from: http://www.ncbi.nlm.nih.gov/pubmed/24330813%5Cnhttp://www.pubmedcentral.nih.gov/articlerender.fcgi?artid=PMC4029549.

[56] Miura D, Yoneyama K. Paclitaxel enhances antibody-dependent cell-mediated cytotoxicity of trastuzumab by rapid recruitment of natural killer cells in HER2-positive breast. J Nippon Med 2014;81(4):211–20. Available from: http://www.ncbi.nlm.nih.gov/pubmed/25186575%5Cnhttp://www.nms.ac.jp/jnms/2014/081040211.pdf.

[57] Di Modica M, Sfondrini L, Regondi V, Varchetta S, Oliviero B, Mariani G, et al. Taxanes enhance trastuzumab-mediated ADCC on tumor cells through NKG2D-mediated NK cell recognition. Oncotarget 2016;7(1):255–65. Available from: http://www.ncbi.nlm.nih.gov/pubmed/26595802%5Cnhttp://www.pubmedcentral.nih.gov/articlerender.fcgi?artid=PMC4807996.

[58] Collisson EA, Campbell JD, Brooks AN, Berger AH, Lee W, Chmielecki J, et al. Comprehensive molecular profiling of lung adenocarcinoma. Nature 2014;511(7511):543–50. Available from: http://www.nature.com/doifinder/10.1038/nature13385.

[59] Martins RG, Parvathaneni U, Bauman JE, Sharma AK, Raez LE, Papagikos MA, et al. Cisplatin and radiotherapy with or without erlotinib in locally advanced squamous cell carcinoma of the head and neck: a randomized phase II trial. J Clin Oncol 2013;31(11):1415–21.

[60] Soria JC, Wu YL, Nakagawa K, Kim SW, Yang JJ, Ahn MJ, et al. Gefitinib plus chemotherapy versus placebo plus chemotherapy in EGFR-mutation-positive non-small-cell lung cancer after progression on first-line gefitinib (IMPRESS): a phase 3 randomised trial. Lancet Oncol 2015;16(8):990–8. Available from: http://dx.doi.org/10.1016/S1470-2045(15)00121-7.

[61] Vermorken JB, Stöhlmacher-Williams J, Davidenko I, Licitra L, Winquist E, Villanueva C, et al. Cisplatin and fluorouracil with or without panitumumab in patients with recurrent or metastatic squamous-cell carcinoma of the head and neck (SPECTRUM): an open-label phase 3 randomised trial. Lancet Oncol 2013;14(8):697–710.

[62] Vermorken JB, Mesia R, Rivera F, Remenar E, Kawecki A, Rottey S, et al. Platinum-based chemotherapy plus cetuximab in head and neck cancer. N Engl J Med 2008;359(11):1116–27.

[63] Pozzi C, Cuomo A, Spadoni I, Magni E, Silvola A, Conte A, et al. The anti-EGFR monoclonal antibody cetuximab combined with chemotherapy triggers immunogenic cell death. Nat Med 2016;22(6):624–31. Available from: http://dx.doi.org/10.1038/nm.4078.

[64] Padhy LC, Shih C, Cowing D, Finkelstein R, Weinberg RA. Identification of a phosphoprotein specifically induced by the transforming DNA of rat neuroblastomas. Cell 1982;28(4):865–71.

[65] Shih C, Padhy LC, Murray M, Weinberg RA. Transforming genes of carcinomas and neuroblastomas introduced into mouse fibroblasts. Nature 1981;290:261–4.

[66] Venter DJ, Kumar S, Tuzi NL, Gullick WJ. Overexpression of the c-erbB-2 oncoprotein in human breast carcinomas: immunohistological assessment correlates with gene amplification. Lancet 1987;330(8550):69–72.

[67] Schubert D, Heinemann S, Carlisle W, Tarikas H, Kimes B, Patrick J, et al. Clonal cell lines from the rat central nervous system. Nature 1974;249(454):224–7. Available from: http://www.ncbi.nlm.nih.gov/pubmed/4151463.

[68] Slamon DJ, Clark GM, Wong SG, Levin WJ, Ullrich A, McGuire WL. Human breast cancer: correlation of relapse and survival with amplification of the HER-2/neu oncogene. Science 1987;235(4785):177–82.

[69] Dawood S, Broglio K, Buzdar AU, Hortobagyi GN, Giordano SH. Prognosis of women with metastatic breast cancer by HER2 status and trastuzumab treatment: an institutional-based review. J Clin Oncol 2010;28(1):92–8.

[70] Yu D, Jing T, Liu B, Yao J, Tan M, McDonnell TJ, et al. Overexpression of ErbB2 blocks Taxol-induced apoptosis by upregulation of p21Cip1, which inhibits p34Cdc2 kinase. Mol Cell 1998;2(5):581–91.

[71] Leto SM, Sassi F, Catalano I, Torri V, Migliardi G, Zanella ER, et al. Sustained inhibition of HER3 and EGFR is necessary to induce regression of HER2-amplified gastrointestinal carcinomas. Clin Cancer Res 2015;21(24):5519–31.

[72] Burstein HJ, Sun Y, Dirix LY, Jiang Z, Paridaens R, Tan AR, et al. Neratinib, an irreversible ErbB receptor tyrosine kinase inhibitor, in patients with advanced ErbB2-positive breast cancer. J Clin Oncol 2010;28(8):1301–7.

[73] Bang Y-J, Van Cutsem E, Feyereislova A, Chung HC, Shen L, Sawaki A, et al. Trastuzumab in combination with chemotherapy versus chemotherapy alone for treatment of HER2-positive advanced gastric or gastro-oesophageal junction cancer (ToGA): a phase 3, open-label, randomised controlled trial. Lancet 2010;376(9742):687–97. Available from: http://dx.doi.org/10.1016/S0140-6736(10)61121-X.

[74] Drakaki A, Hurvitz SA. HER2-positive breast cancer: update on new and emerging agents. Am J Hematol Oncol 2015;11(4):209–29. Available from: http://www.ncbi.nlm.nih.gov/pubmed/27239239%5Cnhttp://www.pubmedcentral.nih.gov/articlerender.fcgi?artid=PMC4872255%5Cnhttp://tam.sagepub.com/cgi/doi/10.1177/1758834016639873.

[75] Verma S, Miles D, Gianni L, Krop IE, Welslau M, Baselga J, et al. Trastuzumab emtansine for HER2-positive advanced breast cancer. N Engl J Med 2012;367(19):1783–91.

[76] Bose R, Kavuri SM, Searleman AC, Shen W, Shen D, Koboldt DC, et al. Activating HER2 mutations in HER2 gene amplification negative breast cancer. Cancer Discov 2013;3(2):224–37.

[77] Swain SM, Baselga J, Kim S-B, Ro J, Semiglazov V, Campone M, et al. Pertuzumab, trastuzumab, and docetaxel in HER2-positive metastatic breast cancer. N Engl J Med 2015;372(8):724–34. Available from: http://www.nejm.org/doi/10.1056/NEJMoa1413513.

[78] Junttila TT, Akita RW, Parsons K, Fields C, Phillips GDL, Friedman LS, et al. Ligand-independent HER2/HER3/PI3K complex is disrupted by trastuzumab and is effectively inhibited by the PI3K inhibitor GDC-0941. Cancer Cell 2009;15(5):429–40. Available from: http://dx.doi.org/10.1016/j.ccr.2009.03.020.

[79] Geyer CE, Forster J, Lindquist D, Chan S, Romieu CG, Pienkowski T, et al. Lapatinib plus capecitabine for HER2-positive advanced breast cancer. N Engl J Med 2006;355(26):2733–43. Available from: http://www.nejm.org/doi/abs/10.1056/NEJMoa064320.

[80] Hawthorne VS, Huang W-C, Neal CL, Tseng L-M, Hung M-C, Yu D. ErbB2-mediated Src and signal transducer and activator of transcription 3 activation leads to transcriptional up-regulation of p21Cip1 and chemoresistance in breast cancer cells. Mol Cancer Res 2009;7(4):592–600. Available from: http://www.pubmedcentral.nih.gov/articlerender.fcgi?artid=2689096&tool=pmcentrez&rendertype=abstract.

[81] Perry J, Ghazaly E, Kitromilidou C, McGrowder EH, Joel S, Powles T. A synergistic interaction between lapatinib and chemotherapy agents in a panel of cell lines is due to the inhibition of the efflux pump BCRP. Mol Cancer Ther 2010;9(12):3322–9. Available from: http://www.ncbi.nlm.nih.gov/pubmed/20889729.

[82] Nielsen TO, Friis-Hansen L, Poulsen SS, Federspiel B, Sorensen BS. Expression of the EGF family in gastric cancer: downregulation of HER4 and its activating ligand NRG4. PLoS ONE 2014;9(4):1–9.

[83] Sheng Q, Liu X, Fleming E, Yuan K, Piao H, Chen J, et al. An activated ErbB3/NRG1 autocrine loop supports in vivo proliferation in ovarian cancer cells. Cancer Cell 2010;17(3):298–310. Available from: http://dx.doi.org/10.1016/j.ccr.2009.12.047.

[84] Reschke M, Mihic-Probst D, Van der Horst EH, Knyazev P, Wild PJ, Hutterer M, et al. HER3 is a determinant for poor prognosis in melanoma. Clin Cancer Res 2008;14(16):5188–97.

[85] Pollock NI, Wang L, Wallweber G, Gooding WE, Huang W, Chenna A, et al. Increased expression of HER2, HER3, and HER2:HER3 heterodimers in HPV-positive HNSCC using a novel proximity-based assay: implications for targeted therapies. Clin Cancer Res 2015;21(20):4597–606.

[86] Brand TM, Hartmann S, Bhola NE, Peyser ND, Li H, Zeng Y, et al. Human papillomavirus regulates HER3 expression in head and neck cancer: implications for targeted HER3 therapy in HPV^+ patients. Clin Cancer Res 2016;16(16). Available from: http://ovidsp.ovid.com/ovidweb.cgi?T=JS&CSC=Y&NEWS=N&PAGE=fulltext&D=medp&AN=27986750%0Ahttp://sfxhosted.exlibrisgroup.com/mayo?sid=OVID:medline&id=pmid:27986750&id=doi:10.1158%2F1078-0432.CCR-16-2203&issn=1078-0432&isbn=&volume=&issue=&spage=&pages=&date.

[87] Gilmour LMR, Macleod KG, McCaig A, Sewell JM, Gullick WJ, Smyth JF, et al. Neuregulin expression, function, and signaling in human ovarian cancer cells. Clin Cancer Res 2002;8 (December):3933–42.

[88] Plowman GD, Culouscou JM, Whitney GS, Green JM, Carlton GW, Foy L, et al. Ligand-specific activation of HER4/p180erbB4, a fourth member of the epidermal growth factor receptor family. Proc Natl Acad Sci USA 1993;90(5):1746–50.

[89] Wali VB, Gilmore-Hebert M, Mamillapalli R, Haskins JW, Kurppa KJ, Elenius K, et al. Overexpression of ERBB4 JM-a CYT-1 and CYT-2 isoforms in transgenic mice reveals isoform-specific roles in mammary gland development and carcinogenesis. Breast Cancer Res 2014;16(6):501–2. Available from: http://breast-cancer-research.com/content/16/6/.

[90] Zhu Y, Sullivan LL, Nair SS, Williams CC, Pandey AK, Marrero L, et al. Coregulation of estrogen receptor by ERBB4/HER4 establishes a growth-promoting autocrine signal in breast tumor cells. Cancer Res 2006;66(16):7991–8.

[91] Williams CC, Allison JG, Vidal GA, Burow ME, Beckman BS, Marrero L, et al. The ERBB4/HER4 receptor tyrosine kinase regulates gene expression by functioning as a STAT5A nuclear chaperone. J Cell Biol 2004;167(3):469–78.

[92] Nielsen TO, Sorensen S, Dagnæs-Hansen F, Kjems J, Sorensen BS. Directing HER4 mRNA expression towards the CYT2 isoform by antisense oligonucleotide decreases growth of breast cancer cells in vitro and in vivo. Br J Cancer 2013;108(11):2291–8. Available from: http://dx.doi.org/10.1038/bjc.2013.247.
[93] Wali VB, Haskins JW, Gilmore-Herbert M, Platt JT, Liu Z, Stern DF. Convergent and divergent cellular responses by ErbB4 isoforms in mammary epithelial cells. Mol Cancer Res 2014;12(8):1140–55.
[94] Fujiwara S, Yamamoto-ibusuk M, Yamamoto Y. The localization of HER4 intracellular domain and expression of ER+HER2-breast cancer. Oncotarget 2014;5(11):3919–30.
[95] Das PM, Thor AD, Edgerton SM, Barry SK, Chen DF, Jones FE. Reactivation of epigenetically silenced HER4/ERBB4 results in apoptosis of breast tumor cells. Oncogene 2010;29(37):5214–9. Available from: http://www.nature.com/doifinder/10.1038/onc.2010.271.
[96] Prickett TD, Agrawal NS, Wei X, Yates KE, Lin JC, Wunderlich J, et al. Analysis of the tyrosine kinome in melanoma reveals recurrent mutations in ERBB4. Nat Genet 2010;41(10):1127–32.
[97] Kurppa KJ, Denessiouk K, Johnson MS, Elenius K. Activating ERBB4 mutations in non-small cell lung cancer. Oncogene 2016;35(10):1283–91. Available from: http://dx.doi.org/10.1038/onc.2015.185%5Cnhttp://www.nature.com/doifinder/10.1038/onc.2015.185.
[98] Cheng QC, Tikhomirov O, Zhou W, Carpenter G. Ectodomain cleavage of ErbB-4: characterization of the cleavage site and m80 fragment. J Biol Chem 2003;278(40):38421–7.
[99] Elenius K, Corfas G, Paul S, Choi CJ, Rio C, Plowman GD, et al. A novel juxtamembrane domain isoform of HER4/ErbB4: isoform-specific tissue distribution and differential processing in response to phorbol ester. J Biol Chem 1997;272(42):26761–8.
[100] Ni C-Y, Murphy MP, Golde TE, Carpenter G. gamma-Secretase cleavage and nuclear localization of ErbB-4 receptor tyrosine kinase. Science 2001;294(5549):2179–81. Available from: http://www.sciencemag.org/cgi/doi/10.1126/science.1065412.
[101] Gilbertson R, Hernan R, Pietsch T, Pinto L, Scotting P, Allibone R, et al. Novel ERBB4 juxtamembrane splice variants are frequently expressed in childhood medulloblastoma. Genes Chromosom Cancer 2001;31(3):288–94.
[102] Naresh A, Long W, Vidal GA, Wimley WC, Marrero L, Sartor CI, et al. The ERBB4/HER4 intracellular domain 4ICD is a BH3-only protein promoting apoptosis of breast cancer cells. Cancer Res 2006;66(12):6412–20.
[103] Zhao W-Y, Zhuang C, Xu J, Wang M, Zhang Z-Z, Tu L, et al. HER4 is a novel prognostic biomarker in gastrointestinal stromal tumor specifically originated from stomach. Am J Cancer Res 2014;4(6):838–49. Available from: http://www.pubmedcentral.nih.gov/articlerender.fcgi?artid=4266716&tool=pmcentrez&rendertype=abstract.
[104] Memon AA, Sorensen BS, Melgard P, Fokdal L, Thykjaer T, Nexo E. Expression of HER3, HER4 and their ligand heregulin-4 is associated with better survival in bladder cancer patients. Br J Cancer 2004;91(12):2034–41.
[105] Lodge AJ, Anderson JJ, Gullick WJ, Haugk B, Leonard RCF, Angus B. Type 1 growth factor receptor expression in node positive breast cancer: adverse prognostic significance of c-erbB-4. J Clin Pathol 2003;56:300–4.

CHAPTER

2

Targeting the Hepatocyte Growth Factor Receptor to Overcome Resistance to Targeted Therapies

Nele Van Der Steen[*,**], *Ingrid Garajova*[†,‡], *Christian Rolfo*[**], *Andrea Cavazzoni*[§], *Elisa Giovannetti*[*,¶]

*VU University Medical Center, Amsterdam, The Netherlands
**Center for Oncological Research, University of Antwerp, Antwerp, Belgium
†Sant'Orsola-Malpighi Hospital, University of Bologna, Bologna, Italy
‡Interdepartmental Centre of Cancer Research "Giorgio Prodi", University of Bologna, Bologna, Italy
§University of Parma, Parma, Italy
¶University of Pisa, Pisa, Italy

Abstract

The hepatocyte growth factor (HGF) and its tyrosine kinase receptor cMET (MET proto-oncogene) entered into the spotlight mainly as a bypassing pathway for other targeted therapies (e.g., therapies targeting epidermal growth factor receptor). However, the HGF–cMET signaling axis can be oncogenic by itself. Aberrations in HGF–cMET occur in many cancer types, such as non-small-cell lung cancer, pancreatic cancer, and renal carcinoma. This resulted in the development of several inhibitors targeting this signaling axis. Biomarkers for these targeted therapies include cMET amplification and cMET exon 14 skipping. Activation of cMET by HGF results in the activation of several downstream pathways, among which are the mitogen-activated protein kinase cascade and phosphatidylinositol-3-kinase-Akt signaling. During embryogenesis, they control the development of tubules and are involved in the migration and invasion of several cell types. These functions are mirrored in cancer growth, whereby cMET is a known activator of cell migration and metastasis. Therefore, in this chapter, we start by describing the road to discovery, the functions in development, and explain the HGF–cMET signaling in detail. Next we focus on HGF–cMET in cancer, describing the possible aberrations and providing an overview of inhibitors (under development). Finally, we pay attention to the role of HGF–cMET as a resistance mechanism against other (targeted) therapies.

Targeting Cell Survival Pathways to Enhance Response to Chemotherapy. http://dx.doi.org/10.1016/B978-0-12-813753-6.00002-0
Copyright © 2018 Elsevier Inc. All rights reserved.

ABBREVIATIONS

ADAM	A disintegrin and metalloproteinase
BMP-2	Bone morphogenetic protein-2
ccRCC	clear cell renal cell cancer
CDA	Cytidine deaminase
CRC	Colorectal cancer
CSC	Cancer stem cell
EGFR	Epidermal growth factor receptor
EMT	Epithelial-mesenchymal transition
ERK	Extracellular signal-regulated kinase
ERM	Ezrin, radixin, moesin protein family
FAK	Focal adhesion kinase
FGF	Fibroblast growth factor
Gab1	Grb2-associated binding protein 1
GGA3	Golgi-localized gamma-ear containing Arf-binding protein 3
Grb2	Growth factor receptor bound protein 2
GSK3ß	Glycogen synthase kinase 3 beta
HGF	Hepatocyte growth factor
iNOS	inducible NOS
JNK	C-Jun N-terminal kinase
MAPK	Mitogen-activated protein kinase
MET	Tyrosine-protein kinase Met
MKK1	Mitogen-activated protein kinase kinase 1
MSP	Macrophage-stimulating protein
mTOR	mammalian target of rapamycin
NF-kB	Nuclear factor-kB
NK	N-terminal and kringle
NSCLC	Non-small cell lung cancer
PDAC	Pancreatic ductal adenocarcinoma
PI3K	Phosphoinositide 3 kinase
PKC	Protein kinase C
pRCC	papillary renal cell cancer
PSC	Pancreatic stellate cells
RCC	Renal cell carcinoma
SF	Scatter factor
SFR	Scatter factor receptor
SHC	Src homology 2 domain containing
Shp2	Protein-tyrosine phosphatase 2C
STAT3	Signal transducer and activator of transcription 3
TKI	Tyrosine kinase inhibitor
VEGF	Vascular endothelial growth factor

2.1 DISCOVERY AND STRUCTURE OF HEPATOCYTE GROWTH FACTOR

The first report of hepatocyte growth factor (HGF) dated back to 1984. Almost simultaneously, two articles described the isolation of a new growth factor in rats, named HGF or hepatotropin [1,2]. This growth factor was isolated from serum and enhanced DNA synthesis and mitogenesis in hepatocytes [1,2]. Platelet-derived serum showed only 50% activity when compared to serum from whole blood. The addition of platelet lysates restored this decreased activity, suggesting an interaction with platelets [2]. The other article showed that HGF displayed high affinity for heparin, suggesting association with platelets, but they were not able to confirm this [1]. Reduction of the disulfide bonds inactivated HGF, as well as heat and increased acidity. In 1988, human HGF was isolated for the first time. Purification from plasma of a patient with fulminant hepatic failure resulted in a protein of 76–91 kDa, which consisted of two chains of 31–34 and 54–65 kDa, respectively [3]. HGF was shown to activate the cMET receptor [4].

In 1973, a scatter factor (SF) that promotes cell migration and overgrowth in SV40-transformed cells was described in Balb/c3T3 cells [5]. This factor was later shown to be produced by embryonic fibroblasts and stimulated migration in epithelial cells only, suggesting a paracrine effect. This SF affected locomotion in cell colonies, whereas in single cells it only leads to membrane ruffling [6]. Further research into these effects showed that cancer cell lines neither produce nor respond to this factor [7]. Purification of the SF succeeded in 1989. The purified factor retained all the previously described features of the fraction, isolated from a conditioned medium. This SF increased motility and modulated morphology. It inhibits junction formation of single epithelial cells and promotes cell scattering in epithelial cell sheets, without influencing cell growth [8]. The SF receptor was identified as the product of the cMET proto-oncogene [9].

Further study of both factors led to the conclusion that HGF and SF are one and the same growth factor [10]. Both factors were mapped onto the same chromosomal region (7q11.2-21), and their protein sequence was indistinguishable. Moreover, SF was able to promote hepatocyte growth, whereas HGF showed SF activity [10,11]. Both SF and HGF were shown to bind the cMET β-chain with high affinity and heparansulfate proteoglycans with low affinity. Cell responses to HGF were also shown to be mediated by cMET activation [12].

The HGF gene was mapped to chromosome 7 at the 7q21.1 position [13]. The gene contains 18 exons and spans over 70 kbp [14]. Exon 1 encodes a signal sequence and the 5′-UTR. The α-chain is encoded by exons 2–11, exon 12 results in a spacer sequence, and the β-chain is encoded by exons 13–18. HGF is translated as a single-chain 83 kDa protein: pre-pro-HGF. In the first step, the signal sequence is removed, resulting in pro-HGF. In the second step, pro-HGF is cleaved between Arg494 and Val495, resulting in a 69 kDa α- and 34 kDa β-chain that are connected by disulfide bonds [15].

HGF is part of the plasminogen-related growth factor family, together with HGF-like/macrophage-stimulating protein (HGF1/MSP). HGF and HGF1 share a common ancestor protein with plasminogen and apo(A) and evolved most likely through gene duplication [16]. This evolution is visible when comparing the very similar exon–intron structure coding for the first three kringle domains of plasminogen and the α-chain of HGF and HGF1. In the case of HGF, the fourth kringle domain most likely resulted from internal duplication, since its organization is very different from the fourth kringle domain in plasminogen. The serine protease domain of the β-chain of HGF is not active. This inactivity is caused by point mutations resulting in replacement of two out of three amino acids of the catalytic triad (His534Gln and Ser673Tyr) [17]. In plasminogen, this catalytic triad is formed by His–Asp–Ser, whereas the triad consists of Gln–Asp–Tyr in HGF and Gln–Gln–Tyr in HGF1. This suggests that mutation of the serine protease domain happened before the duplication of HGF and HGF1, but after the duplication of plasminogen, which retained its activity.

In contrast to plasminogen, the N-terminal domain of HGF is not cleaved [16] but is necessary for receptor binding. The N-terminal and kringle 1 domain of HGF (NK1) is able to associate with the receptor, but with reduced affinity. Adding the second kringle domain (NK2) doubles this binding affinity. The binding affinity of NK2 when compared to complete HGF is still fivefold lower. The addition of kringle domains 3 and 4 did not lead to further improvement of binding affinity. This points toward a 2:2 complex of HGF and cMET [18]. The cleavage of pro-HGF results in an open form of HGF [19]. The 7-bladed propeller sema-domain of cMET contains the binding site for HGF [20]. A first high-affinity binding site is occupied by the N-terminal domain and the first kringle domain of HGF [19]. A second low-affinity binding site of cMET is bound by the serine protease domain on the β-chain of HGF [21]. Hereby, the mutations of the catalytic triad play a role in the association of HGF to cMET [21]. Complete HGF, consisting of both the α- and β-chains, is needed to activate cMET signaling [19].

Two naturally occurring splice variants of HGF have been described: NK1 [22] and NK2 [23,24]. The NK1 variant consists of the N-terminal region and the first kringle domain. This variant is capable of binding to both heparin and cMET. This splice variant is able to induce phosphorylation of cMET, but a 16× higher concentration is needed to induce the same effect as full-length HGF [22]. In a concentration ranging from 5 to 10 nM, NK1 is capable of inducing scattering, however, it is about 50 times less potent in comparison to HGF. When adding NK1 in 40-times molar excess to HGF, it reduces activity of HGF by 70% and is able to inhibit DNA synthesis [22]. The second splice variant consists of the N-terminal domain and the first two kringle domains (NK2) [24]. This variant competes with HGF for cMET binding and inhibits HGF-dependent mitogenesis [24].

2.2 ACTIVATION OF THE HGF-MET SIGNALING PATHWAY

HGF is the only known ligand of cMET. Mature HGF, consisting of an alpha and beta chains, binds cMET and causes receptor dimerization (Fig. 2.1). This is a paracrine process, since HGF is mainly secreted by mesenchymal cells, and cMET is mainly expressed on epithelial cells [6,7,11,25]. Binding of HGF leads to cross-phosphorylation of cMET at positions Tyr1003, Tyr1234, Tyr1235, Tyr1349, and Tyr1356. Once Tyr1349 and Tyr1356 are phosphorylated, they serve as docking sites for the adaptor proteins Grb2 (Growth Factor Receptor Bound protein 2), Gab1 (Grb2-associated binding protein 1), and SHC (Src homology 2 domain containing) [26]. These adaptor proteins link the receptor to its downstream signaling effectors [26–28]. In this process, several coreceptors are involved. The most important coreceptor of cMET is CD44v6, a splice variant of CD44 containing exon 6. A complex is formed between cMET-HGF and CD44v6 [29]. Orian-Rousseau et al. [30] revealed a two-step process. In the first step, this complex is formed and HGF-induced phosphorylation of cMET and its adaptor proteins takes place. In the second step, the cytoplasmic domain of CD44v6 is necessary to induce phosphorylation of MEK and ERK through HGF–cMET signaling [30]. The cytoplasmic domain of CD44v6 is also necessary to induce internalization of the HGF–cMET signaling complex. Phosphorylated Tyr1003 is the binding site for the E3-ligase c-Cbl [31,32]. Internalization of HGF–cMET is dependent on the association of the cytoplasmic domain of CD44v6 with ERM proteins (Ezrin, Radixin, Moesin) and their connection to the cytoskeleton [33]. A second coreceptor/enhancer is α6β4 integrin [34]. This integrin forms a complex with cMET that functions as a signal amplifier for HGF–cMET

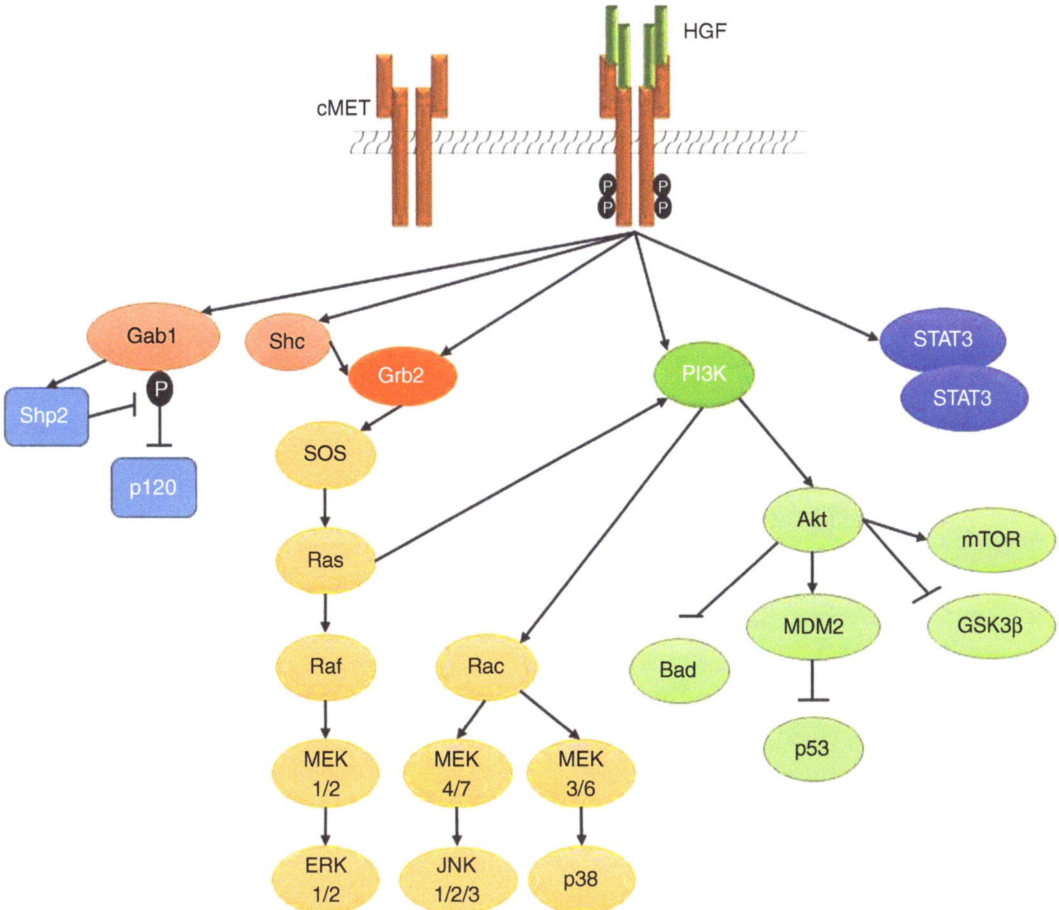

FIG. 2.1 **Schematic of HGF–cMET signaling.** Abbreviations: HGF, hepatocyte growth factor; Gab1, Grb2-associated binding protein 1; Grb2, growth factor receptor bound protein 2; Shp2/PTPN11, tyrosine protein phosphatase nonreceptor type 11; p120, p120 catenin; Shc, Src homology 2 domain containing; SOS, Son of Sevenless; MEK, mitogen-activated protein kinase kinase; ERK, extracellular signal-regulated kinase; JNK, c-Jun N-terminal kinase; p38, p38 mitogen-activated protein kinase; PI3K, phosphoinositide 3 kinase; Akt, protein kinase B; MDM, mouse double minute 2; GSK3β, glycogen synthase kinase 3 beta; mTOR, mammalian target of rapamycin: STAT3, signal transducer and activator of transcription 3.

signaling. Active cMET phosphorylates the β4 subunit, which results in the recruitment of SHC and activation of phosphatidylinositol-3-kinase (PI3K) [34]. A third coreceptor/amplifier is the Semaphorin 4D ligand and its receptor Plexin B1 [35,36]. Association of this complex with cMET is involved in cMET-induced angiogenesis [37]. Semaphorin 4D association with cMET is able to induce phosphorylation without the binding of HGF. However, this stimulation is weaker and delayed compared to HGF activation of cMET. Moreover, Semaphorin 4D–cMET is able to induce cell motility and morphogenesis, but is not able to stimulate cell proliferation. It also has no mitogenic effect on endothelial cells and is not capable of inducing VEGF expression [37]. This association with cMET has been shown for all class B plexins [36], but more studies are warranted to elucidate downstream effects of these interactions. However, this interaction is reported to be antimigratory in the case of breast carcinoma [38] and melanocytes [39].

2.2.1 Mitogen-Activated Protein Kinase Cascades

HGF activation of cMET leads to the induction of several mitogen-activated protein kinase (MAPK) cascades. Signaling from the Grb2 adaptor protein leads to activation of Ras, either through direct association of Grb2 to cMET or indirectly by SHC binding to cMET [40]. In the meantime, activation of Ras is stimulated indirectly. The adaptor protein Gab1 recruits Shp2, which in turn dephosphorylates a phosphorylation site on Gab1 involved in the activation of p120-Ras-GAP. Inhibition of the p120 Ras GTPase activating protein leads to increased levels of active Ras coupled to GTP [41]. Activation of Ras continues through several signaling pathways. First, Ras activates the Raf–MEK1/2–ERK1/2 cascade. This results in cell proliferation and transformation. Second, Ras activates PI3K signaling. This stimulates Rac that phosphorylates MEKK1/2/3/4. On the one hand, MEK4/7 are phosphorylated, leading to activation of JNK1/2/3 [42] and, on the other hand, MEK3/6 activation leads to activation of p38 $\alpha/\beta/\gamma/\delta$ [43,44]. The activation of both JNK and p38 signaling is associated with cell proliferation, transformation, differentiation, and apoptosis [42].

2.2.2 PI3K-Akt

Binding of HGF to cMET also results in the recruitment of the p85 subunit of PI3K to the receptor [45]. This leads to the activation of PI3K-Akt signaling. PI3K has several functions. First, it has an antiapoptotic effect [46] by inhibiting Bad, an effector of apoptosis [47], and at the same time activating MDM2, an inhibitor of p53 [48]. Second, it stimulates cell growth and proliferation through activation of mTOR [49], resulting in protein translation and cell growth, and by inhibiting GSK3β that activates cell-cycle regulators [50], leading to cell proliferation.

2.2.3 STAT3

STAT3 is another downstream effector of cMET. STAT3 is recruited directly to the cytoplasmic tail of cMET and upon phosphorylation forms homodimers that translocate to the nucleus [51]. In the nucleus, STAT3 binds to the promoters of several genes. STAT3-regulated genes are reported to play a role in tubulogenesis, leading to cell proliferation and differentiation [51]. Furthermore, STAT3 also plays a role in pro-cancer immunity. IL6 is upregulated, activating a feed-forward loop by increasing STAT3 activation. Through the sequestering of RelA in the nucleus, NF-κB is activated. Moreover, STAT3 signaling is necessary for regulatory T-cell expansion, T_h17 cells development, and the immunosuppressive effects of myeloid-derived suppressor cells [52].

2.2.4 HGF-Independent Activation and cMET Internalization

Besides activation by HGF, cMET can also be activated through cell adhesion. Hereby, $\alpha5\beta1$ integrin plays a role. Once this integrin associates with type IV collagen, it forms a complex with cMET, thus activating the receptor. Next, cMET phosphorylates Src, which in turn leads to the activation of FAK [53–56].

As already briefly mentioned, Tyr1003 in the juxtamembrane domain of cMET is phosphorylated upon activation. Once phosphorylated, this Tyr is part of a docking site for the E3-ligase c-Cbl [31]. This starts the breakdown process for cMET. Hereby, c-Cbl has two tasks. First, it recruits an E2-ligase that in turn ubiquinates cMET, marking cMET for internalization through clathrin-coated pits in a dynamin-dependent manner [57,58]. Second, c-Cbl attracts the CIN85–endophilin complex. This complex is involved in the invagination and scission of these clathrin-coated pits [59]. From here, intracellular cMET trafficking goes over the microtubules from the early endosomal complex to the late perinuclear complex and finally cMET is broken down by the 26S proteasome [60].

Recycling of cMET from endosomes has also been reported. Hereby, GGA3 (Golgi-localized gamma-ear containing Arf-binding protein 3) plays an important role. After internalization, cMET is contained in Rab4/Rab5-positive endosomes. GGA3 colocalizes with these endosomes and stimulates recycling to the plasma membrane [61]. GGA3 has been shown to bind to Crk, which in turn binds the Gab1 adaptor protein that is associated to the cMET receptor [62]. Loss of GGA3 results in increased breakdown of HGF-bound cMET, a decrease in ERK signaling, and a decrease in cell migration [61]. This decrease in ERK signaling can be explained by the continuous signaling of the cMET receptor once it is internalized [63]. PKCα is involved in the microtubule-dependent trafficking of cMET to the late endosome, whereas PKCε, on the other hand, controls sustained phosphorylation of ERK1/2 by cMET from the early endosome. Once internalized, cMET continues to signal and phosphorylate ERK1/2. PKCε is involved in the translocation of ERK1/2 from the cytoplasm to the focal complexes at the plasma membrane [58]. Here, ERK1/2 phosphorylates paxillin, thus contributing to cell migration [64].

Signaling from HGF–cMET continues from the perinuclear endosome. Rac1 associates with cMET from the moment of internalization, however signaling from the perinuclear endosome is necessary for sustained activation of Rac1. This is caused by the cMET-dependent activation of Vav2, a Rho GEF able to activate Rac1. Next to Vav2, activation of PI3K is also necessary for sustained Rac1 activation [65]. The p85 subunit associates with cMET in the perinuclear endosome, and the PI3K product PIP3 regulates Vav2 activity [66]. Active Rac1 on the edges of the cell leads to membrane ruffling, cell migration, and invasion [65].

Activation of STAT3 also requires HGF–cMET signaling from the perinuclear endosome. Although phosphorylated STAT3 is assumed to diffuse freely to the nucleus, there is a long period of exposure to phosphatases between the cell membrane and the nucleus. By activating STAT3 in proximity to the nucleus, this exposure is considerably more limited, resulting in a strong STAT3 signal and translocation to the nucleus [67].

2.2.5 Receptor Shedding

The first report of cMET shedding dates back to 1995. Herein, Galvani et al. [68] described a reduction of cMET expression of 50% in reaction to the growth factor inhibitor suramin. Shedding of cMET can be caused by stress signals like TNF-α-actinomycin D and camptothecin [69], by activation of another receptor tyrosine kinase, e.g., epidermal growth factor receptor (EGFR) [70] or by antibody treatment, e.g., DN30 [71]. This shedding has an extra advantage for treatment with antibodies, as it also functions as a decoy receptor for HGF, thus preventing cMET activation by two distinct mechanisms. However, not all antibodies cause cMET shedding, as is illustrated by onartuzumab that does not induce shedding or internalization of the receptor and prevents clearance of s-cMET [72]. ADAM10 or ADAM17 recognize the 998-SVD-1000 motif in cMET, which is mainly dependent on D1000 [69]. Next, a second cleavage is performed in the membrane by γ-secretase [73], resulting in two intracellular parts of cMET that are degraded by the proteasome [74]. This process is independent of the kinase activity of cMET. The two remaining intracellular parts are pro-apoptotic in a kinase-dependent manner [69]. Soluble-cMET has been reported to be a biomarker for pre-eclampsia [75] and hepatocellular damage [76] and points toward a poorer outcome in non-small cell lung cancer (NSCLC), where it correlates with cMET expression [77].

2.3 HGF FUNCTIONS

HGF has a critical role in the development and organogenesis of several tissues, as described in the following sections.

2.3.1 Placenta

The importance of HGF in placenta formation is illustrated by $HGF^{-/-}$ or $cMET^{-/-}$ mice. Homozygous mice die *in utero* around E13.5–E15.5 and show deformation of the placenta, among other defects [78]. These homozygous-deficient embryos show reduced numbers of labyrinthine trophoblast cells, along with a decrease in labyrinth area size and poorly developed embryonic vessels and maternal sinuses. These defects can be rescued by stimulation with HGF *in vitro*. During embryogenesis, HGF is produced by the allantoic mesenchyme, whereas the cMET receptor is expressed on the extraembryonic ectoderm. Upon contact between these tissues, mitotic activity and scattering of trophoblasts are induced [78]. HGF is also needed in an earlier stage, the early postimplantation growth. Here, HGF does not stimulate the scattering of trophoblasts, but stimulates cell division of trophoblasts [79]. Pre-eclampsia is associated with impaired trophoblast invasion [80]. A study using human placenta explants showed that in the case of pre-eclampsia, the production of HGF was 25% less over 24 h [80]. This is confirmed by the results of a second study, where the level of soluble cMET (s-cMET) in serum of pregnant women was determined, in both healthy subjects and patients with pre-eclampsia [81]. Here a significant lower concentration of s-cMET was measured in women suffering from pre-eclampsia when compared to healthy subjects, both in the first and second trimester of the pregnancy. Even more, a significantly lower s-cMET concentration was detected in the serum of women with early and serious pre-eclampsia when compared to women with later and milder pre-eclampsia [81].

HGF induces trophoblast motility by activation of the PI3K and MAPK pathways. Activation of p42/44 MAPK is necessary for motility and for induction of inducible NOS (iNOS) [82]. This activation of iNOS can be inhibited by transforming growth factor-β (TGF-β) [83]. Downstream of PI3K, activation of p70S6 [84] and Akt [85] are needed to induce trophoblast motility. PI3K signaling also causes phosphorylation of GSK3β, a member of the Wnt signaling cascade [86]. In the case of GSK3β, phosphorylation causes inhibition and not activation. Inhibition of GSK3β leads to the activation of NF-κB and β-catenin. In turn, β-catenin activates the Lef/Tcf transcription factor [87], leading to cell survival.

There are four different isoforms of HGF identified that play a role during growth of the placenta [88]: the NK1 and NK2 isoforms and the dNK1 and dNK2 isoforms. The last two contain a deletion of five amino acids in the first kringle domain of HGF. All of these isoforms are expressed in the endometrium, but dNK2 is expressed at higher levels than NK2. Presumably, expression of the HGF antagonists NK1 and dNK1 moderates the activity of HGF in the endometrium. In contrast, NK1 and dNK1 are not expressed in the placenta, whereas NK2 and dNK2 are expressed in equal amounts [88].

2.3.2 Muscle

The development of muscles can be divided into several steps. First, the precursor cell pool in the dermomyotome needs to be established. Second, these precursors need to be delaminated and migrate to their final position. Third, the precursor cells proliferate and finally differentiate to form skeletal muscles [89]. HGF is one of the three important players in this process.

Pax3 is the first gene to become active. It is responsible for the establishment of the precursor pool and the delamination and migration of the cells [90]. This is visible in the *Splotch* mutant. This mutant contains a deletion in Pax3 [91] and lacks the limb muscles, whereas development of the axial and body wall musculature is not affected [92]. Pax3 induces cMET expression. This is shown in the *Splotch* embryos wherein cMET expression is lacking [93,94]. Activation of the cMET–HGF axis is important to induce the delamination and migration of the myogenic precursor cells from the somites into the limb buds [95]. In mice embryos that are deficient in either cMET or HGF, these precursor cells develop normally in the dermomyotome, but fail to start the migration to the limb buds, resulting in the lack of hypaxial muscle groups. On the other hand, skeletal muscles outside the limb develop normally [95], showing that cMET signaling is not necessary in further differentiation of muscle cells. In normal embryos, cMET expression is detected in the migrating myogenic precursor cells [96]. HGF, on the other hand, is expressed in the mesenchyme [97]. It has been shown that HGF expression is stimulated by fibroblast growth factor (FGF) in the apical ectodermal ridge. When this zone is removed, HGF expression was detected until the 18/19th stage, when the apical ectodermal ridge is formed. This shows that FGF is necessary for the continuation of HGF production [96]. The zone of polarizing activity, on the other hand, controls HGF expression by expression of BMP2. The removal of this zone leads to ectopic and enhanced expression of HGF [96]. The homeobox gene Lbx1 is induced before delamination, expressed during migration and downregulated during differentiation. This gene is responsible for the correct guiding of the precursor cells during migration [89]. Before the final differentiation of the muscle progenitor cells, a proliferation step takes place. Here, cMET also plays a role, since a mutation of the Grb2 binding site in cMET leads to a decrease in the size of skeletal limb muscles [98]. In these homozygotic $cMET^{Grb2/Grb2}$ mice, the placenta and liver are developed normally. In contrast, muscles are reduced in size, although no effect on migration or differentiation of myoblasts are observed during embryogenesis. Even more, defects in the appendicular and axial muscles are observed. These muscle groups do not derive from migratory precursor cells. A reduction in the secondary muscle fibers is observed in all muscle groups, pointing toward a role for cMET during proliferation of myoblasts forming secondary muscle fibers [98].

2.3.3 Liver

During embryogenesis, the liver develops from the ventral endoderm under the influence of the HNF3β and GATA-4 transcription factors [99]. Interaction of the liver bud with the cardiac mesoderm leads to hepatic induction. The cardiac mesoderm produces FGF-8 before hepatic induction [100] and FGF-1 and FGF-2 at the time of hepatic induction [101]. These FGFs have been shown to be necessary and sufficient for hepatic induction, but not sufficient for further liver development [101]. Next to the influence of the cardiac mesoderm, the septum-derived mesenchyme also plays a role in liver development. Here, bone morphogenetic protein (BMP) is produced that, in parallel to FGF, induces liver development [102]. BMP4 is necessary during hepatic specification and to inhibit the expression of pancreatic genes. This is shown in BMP4 knockout mice that present with a delay in liver bud formation. Also the use of Noggin, a BMP inhibitor, inhibits the outgrowth and migration of hepatic cells [102]. HGF is also expressed by the septum transversum mesenchyme, whereas its receptor cMET is expressed on hepatocytes [103]. A knockout of either HGF or cMET leads to death *in utero*, due to strongly reduced liver growth and disturbed hematopoiesis [103,104]. When culturing hepatocytes from $HGF^{-/-}$ embryos *in vitro* in medium supplemented with growth factors supporting epithelial growth, they showed no difference on the survival or growth of these remaining hepatic cells [103], confirming the mitogenic influence of HGF on hepatocytes. Activation of cMET leads to the expression of the c-Jun transcription factor. Knockout c-Jun embryos display the same phenotype as HGF or cMET knockout embryos, due to reduced hematopoiesis that is not capable of sustaining the growing embryo [104,105]. These matching phenotypes point toward the involvement of the HGF/cMET/c-Jun axis in liver development.

2.3.4 Kidney

During the development of the renal system, contacts between the ureteric bud and the metanephrogenic mesenchyme starts the branching process [106]. During branching and invasion into the mesenchyme, several growth factors work together to coordinate this process [107]. In the first step, GDNF**31** and formin IV [108,109] are important to induce ureteric bud formation. Disruption of the formin IV isoform has been shown to lead to renal agenesis/dysgenesis due to failure in ureteric bud outgrowth and branching [108]. Formin IV, a known actin nucleator, colocalizes with the actin cytoskeleton and in the perinuclear zone in quiescent epithelial cells. After stimulation with HGF, formin IV transiently translocated in these cells to the nucleus in a MAPK-dependent manner [110]. However, its exact role in the nucleus is not known to date.

Embryos from $HGF^{-/-}$ mice are reported to have normal renal development. However, homozygous deletion of HGF or cMET leads to death *in utero*, with 12% of embryos reaching the E14.5 stage and 2% the E16.5 stage [103], whereas the induction of renal branching is situated around stage E13.5 [111]. This urges for caution on drawing conclusions on the necessity of HGF for renal development based on an unconditional HGF knockout model. Several findings point toward an important role for HGF during renal development. First, glycosaminoglycans are necessary for ureteric bud development. When these glycosaminoglycans are removed, the development of the ureteric bud stops, and this effect is rescued by addition of HGF only [112]. Second, contact with the mesenchyme is necessary to start the development of the kidney. When the mesenchyme and epithelium are separated, development stops, whereas restoration *in vitro* rescued this effect [106]. With the mesenchyme being the main producer of HGF and the epithelium mainly expressing cMET [113], this also points toward an important role for the HGF–cMET axis in kidney development. Third, when growing developing murine kidneys (d11.5–12.5) were cultured *in vitro*, addition of anti-HGF serum resulted in significant growth inhibition, a poorly defined ureteric duct and poorly developed branches [111,113].

To control the branching of the nephrotic duct, BMPs serve as negative signals. They prevent the branching into already colonized regions of the mesenchyme and stimulate branching into new regions [114]. BMP4 has been shown to inhibit branching, but stimulates tubule elongation [115]. In high concentrations, BMP7 inhibits branching, whereas in low concentration it stimulates branching [116]. Finally, MMP9 is needed to facilitate the growth of the branches into the mesenchyme [117].

The first critical pathway for HGF-induced branching is the MAPK pathway [118]. HGF induces phosphorylation of MKK1 through activation of cMET, whereas EGF induces phosphorylation of both MKK1 and ERK5. Inhibition of MKK1 by PD98059 abolishes HGF-induced cell migration and branching completely, whereas it only partially inhibited EGF-induced motility and branching. Phosphorylation studies showed that this inhibitor is able to inhibit MKK1 completely, whereas ERK5 is still partially active. Using the U0126 inhibitor, which inhibits both MKK1 and ERK5, the effects of both HGF and EGF were abolished completely. Expression of a dominant-negative ERK5 only influenced cell motility and branching after activation of EGF and did not affect the HGF-induced effects [118]. The manner of signaling, either transiently or endured, is important for the correct induction of branching. The cell-surface heparan sulfate Gpc4 is needed for sustained signaling through MAPK by HGF [119]. Without Gpc4, HGF is still able to induce cell migration and proliferation, but fails to induce branching. Branching can be rescued by addition of EGF [119]. Although the exact role of Gpc4 in HGF signaling has not been revealed yet, possibly this role is similar to that of Gpc1. Gpc1 is able to induce lipid rafts formation, thus stabilizing the HGF–cMET signaling complex, which leads to sustained downstream signaling [120].

Next to the MAPK pathway, activation of the PI3K pathway is also involved in branching [119,121,122]. Addition of PD98059 (MAPK inhibitor) or LY294002 (PI3K inhibitor) leads to inhibition of scattering, but morphologically different outcomes. A constitutively active form of PI3K was able to induce scattering, but active MAPK was needed during this process. Activation of Akt of Rac did not lead to scattering of the cells, whereas induction of scattering by active Ras or PI3K was sensitive to both the LY294002 and PD98059 inhibitors. These findings point toward a role for PI3K act signaling in branching, but this signaling does not include downstream activation of Akt or Rac, suggesting a different downstream signalization path [122].

2.3.5 Nervous System

Nervous tissue arises from the node of Hensen, which gives rise to the nervous streak during early stages of embryogenesis. HGF is expressed in this node of Hensen [123] and plays a role in retaining the neural competence of this tissue [124]. A marker of this competence is $L5^{220}$. When the expression of $L5^{220}$ decreases, the level of competence of the node decreases. Addition of HGF after its normal expression window leads to a longer expression of $L5^{220}$ and thus a prolonged time of complete competence. Ectopic addition of HGF leads to the formation of a neural streak/neural plate-like structure, which does not possess the completely undetermined character of the normal neural plate [124]. However, HGF or cMET knockout embryos do not show large defects in the neural system up to their death *in utero* at E13.5–E15.5, which suggests that HGF is involved but not necessary in these early stages of neuronal development [103].

In later stages, HGF is involved in neuronal survival, proliferation, and outgrowth. First, HGF enhances the uptake of dopamine by the dopaminergic neurons in the brain [125]. In the brain, HGF is mainly produced by microglia [126], where it functions as a neurotrophic factor (together with bFGF) through the induction of Ras signaling [126]. In contrast, cMET is expressed in microglia cells, astrocytes, and neurons [126]. Second, in the case of sympathetic neurons, HGF plays a role in survival and axonal outgrowth [127]. Both HGF and cMET are expressed during the development of the sympathetic ganglia [127]. When anti-HGF antibodies are added, a reduced number of sympathetic neuroblasts is observed, along with reduced survival [127]. Expression of a mutated form of cMET also leads to increased sympathetic neuroblast apoptosis *in vivo* [127]. Third, signaling mutants of cMET lead to a reduction in sensory nerves in the skin of limbs and the thorax [128]. Together with NGF, HGF is responsible for the axonal outgrowth from the dorsal root ganglion (DRG) neurons and sustains a part of these DRG neurons. No synergy with other neurotrophic factors as BDNF and NT3 is detected. Disruption of the Grb2 docking site of cMET does not lead to defects in the development of these sensory neurons, pointing toward downstream signaling that is independent of Grb2 [128]. Finally, HGF acts as a mitogen for Schwann cells [129]. However, anti-HGF antibodies do not lead to impaired development of Schwann cells. This suggests that HGF is not necessary for Schwann cell development [129].

2.3.6 Testis

During the development of the testis, one of the first steps is the migration of cells from the mesonephric mesenchyme toward the celomic epithelium [130]. During this stage, HGF is expressed by the celomic epithelium, whereas cMET expression is detected in the mesonephric mesenchyme [131]. HGF functions as a chemoattractant for the migrating cells from the mesonephric mesenchyme [132]. These migrating cells are necessary for the testicular cord morphogenesis [130]. At stage E12.5, cMET is expressed in the developing testicular cords, whereas HGF production is situated in the stroma and myoid cells [131,132]. Organ culture *ex vivo* demonstrated that HGF is sufficient for testis differentiation and testicular cord formation from undifferentiated male urogenital ridges [132]. Around E17.5, cMET expression in the testicular cord is downregulated, cMET expression increased in interstitial fetal Leydig cells, and HGF is now being produced by the interstitial compartment [133]. During this stage, HGF promotes testosterone production [133] and Leydig cell survival and differentiation [134].

2.4 DYSREGULATION IN CANCER

2.4.1 cMET Mutations

The cMET receptor has several domains that are crucial for its function. A mutation in each one of these domains can deregulate its activity and have oncogenic potential. Mutations in the ligand-binding domain can affect HGF affinity, whereas mutations in the kinase domain might lead to a constitutively active receptor. Mutations in the juxtamembrane domain, containing the E3-ligase c-Cbl docking site, can lead to impaired cMET breakdown. Many mutations are known in cMET, but only a few of them are carcinogenic. Therefore, we will limit our description to mutations that influence treatment and resistance; for the complete list of mutations, we refer the reader to the COSMIC and ClinVar databases.

Several splice site mutations are known that cause exon 14 skipping, which occur in around 3% of adenocarcinomas [135]. Exon 14 forms part of the juxtamembrane domain and contains the c-Cbl docking area. Therefore, a deletion of this exon leads to impaired breakdown of cMET and prolongs signaling. These mutations can be present in the 5′ or 3′ splice sites, the branching A or the polypyrimidine tract that are all necessary for the splicing process. Known types of mutations that cause exon 14 skipping are missense mutations and deletions [136,137]. Patients with detected exon 14 skipping have been shown to respond to several cMET small-molecule inhibitors like crizotinib, capmatinib, and cabozantinib [137].

Hereditary papillary renal cell carcinoma is the only cancer type with known germline mutations in cMET to date. These mutations are mostly situated in the kinase domain of cMET and lead to an autosomal-dominant syndrome. Known mutated codons include V1110L/I, H1112R/L/Y, M1149T, V1238I, D1246N/H, V1238I, and M1268T [138–140]. Some of these mutations show homology to previously known mutations in oncogenes. The D1246N/H missense mutations are compatible with the D816V mutation in cKit [141,142], whereas the Y1248C/H/D mutations are related to the M918T mutation in RET [143]. All of these above-mentioned mutations have transforming capabilities and lead to constitutive, activated cMET.

Mutations causing resistance to cMET-targeted therapies have been reported: Y1230C [144] and D1228N/V [145,146]. The activation loop of cMET controls the activity of the receptor. This loop closes the ATP binding pocket of the receptor by interacting through salt bridges with amino acids at positions K1110 and D1228. Y1230 is located on the activation loop and is turned inwards through interaction with A1226 [147]. Type I small-molecule cMET inhibitors (crizotinib, capmatinib, tepotinib) interact with the receptor through the formation of π-stacks with the Y1230 residue. These π-stacks are abolished by the substitution of Tyr to Cys at this position [144]. Type II cMET inhibitors (cabozantinib, glesatinib) have an extra interaction with cMET through the association with a hydrophobic pocket on the receptor, thus being less dependent on these π-stacks [148]. A missense mutation at D1228, on the other hand, leads to a destabilization of the activation loop. These mutations have been reported in both patients presenting with cMET exon 14 skipping and cMET amplification after EGFR-TKI treatment [144–146] and lead to resistance against type I cMET small-molecule inhibitors, but can be overcome by type II inhibitors.

2.4.2 Amplification and Fusion Protein of cMET

Amplification of cMET has been described to occur in approximately 1.5% of NSCLC adenocarcinomas [135] and in around 5% of gastric carcinoma [149,150] and glioblastomas [151]. A consensus of scoring criteria is lacking to determine the exact cut-off for cMET amplification, leading to the reporting of varying percentages. However, this cut-off might prove to be important for treatment outcomes as has been reported in the first results of the crizotinib trial in NSCLC [152]. Herein, patients with a cMET/CEN7 ratio >5 showed a partial response to crizotinib in 50% of cases, when compared to 20% in the intermediate amplified population (>2.2 to <5) and 0% in the low amplified population (>1.8 to ≤2.2) [152]. Moreover, Noonan et al. [153] showed that patients with a low and intermediate ratio showed concomitant driver mutations in other genes in 52% and 50% of cases, respectively, whereas for the high amplification this was 0%. It is important to make a distinction between cMET amplification and polysomy, given that besides cMET, the genes for EGFR and HGF are also located on chromosome 7. In NSCLC patients that originally presented with a sensitizing mutation in EGFR and that received an EGFR-TKI, the percentage of cMET amplification rises to approximately 20% [154,155]. We will discuss this in more detail in the next section. In addition to amplification, a fusion protein of cMET is also known: TPR-MET [156,157]. This fusion protein was originally identified in the MNNG-HOS cell line and is the result of a translocation between the TPR gene on chromosome 1 and the cMET gene [158]. The fusion protein is located intracellularly, and through the bindings of the coiled coils of TPR the kinase domain of cMET autophosphorylates and is constitutively active [159]. Moreover, the juxtamembrane domain of cMET is missing in the fusion protein, resulting in impaired breakdown of the fusion protein, thus contributing to its oncogenicity [160]. Because the kinase domain of cMET is retained, TPR-MET is inhibited by small-molecule inhibitors, like SU11274 [161]. In addition to NSCLC, TPR-MET has also been reported in gastric carcinoma [157].

2.4.3 cMET Overexpression

Overexpression can result from several dysregulations. First, epigenetic regulation through the loss of DNA methylation might result in higher promotor activity, causing cMET overexpression [162]. Second, miRNA regulation of the promotor can also result in increased expression [163]. Finally, the transcription of cMET can be upregulated, leading to higher translational activity and expression [164]. In the case of NSCLC, cMET overexpression has not proved to be a good biomarker for response to cMET small-molecule inhibitors, as was shown in the onartuzumab phase III trials [165]. The percent of cMET-overexpressing tumors varied substantially between the different studies, ranging from 14% to over 50% of cases in NSCLC in various patient populations [166–169]. The same variation is seen in gastric carcinoma [150,170], glioma [151,171], and pancreatic cancer [172,173].

2.4.4 HGF Overexpression

Like cMET overexpression, HGF overexpression can be caused by transcriptional [174] or epigenetic changes [162]. However, autocrine or paracrine loops can also cause overexpression of HGF and thus increase cMET activity [175]. Therefore, a distinction between HGF expression between the stroma and the tumor cells themselves is important. The reported percentages of cells with HGF overexpression range from 25% to 82% in tumor cells [176,177]. These loops have been described in multiple cancer types: hepatoma [178], lung carcinoma [179], and multiple myeloma [180], among others.

2.5 HGF INHIBITORS FOR CANCER THERAPY

The HGF–cMET signaling pathway represents a promising drug target for novel anticancer strategies. Targeting the HGF ligand or the cMET receptor are two possible ways to influence the activation of this pathway. In general, agents against the HGF ligand can be further divided into HGF activation inhibitors or HGF inhibitors [181]. HGF activation inhibitors prevent the cleavage of pro-HGF into the active form and HGF inhibitors block the direct binding of HGF to the cMET receptor. These inhibitors exist as monoclonal antibodies or small molecules. Moreover, blocking the cMET receptor by cMET antagonists or cMET tyrosine kinase inhibitors (TKIs) represents alternative therapeutic strategies for modifying HGF–cMET signaling pathway [181,182]. HGF inhibitors can be used as single agents or in combination with other anticancer drugs such as cytostatic drugs or targeted compounds [182,183]. Most of the HGF–cMET inhibitors have shown encouraging results when used in drug combination studies.

2.5.1 HGF Activation Inhibitors

The inactive precursor of HGF (pro-HGF) transforms to its active form, thanks to the balance between the activators (HGFAs) and the inhibitors (HAIs) [184,185]. Several authors investigated the role of HAIs in solid tumors. HAI-1 was identified as a prognostic marker for prostate cancer and might be a novel target for the treatment of this malignancy. The authors described that the level of HAI-1 changes in prostate cancer compared to benign prostate hyperplasia. Further, a low level of HAI-1 correlated with a high Gleason score, a more advanced pathological stage as well as worse clinical outcome [186]. The same trend was observed by Tsai et al. [187] for HAI-2 in human prostate cancer progression. The expression of HAI-2 decreased throughout progression in cell invasion capability, contributing to tumor genesis and metastasis. Therefore, HAI-1 and HAI-2 might represent novel prognostic markers as well as therapeutic targets. To date, however, no clinical tests for HAIs are available.

2.5.2 HGF Inhibitors

HGF inhibitors bind to HGF, blocking binding to the cMET receptor and resulting in subsequent effects on downstream activation of the pathway. Several monoclonal antibodies against HGF have been tested in preclinical studies, such as rilotumumab, ficlatuzumab, and TAK701. Rilotumumab is a fully humanized IgG2 monoclonal antibody [188]. Early preclinical data showed synergistic cytotoxic effects with temozolomide and docetaxel [189,190]. In a phase II study evaluating the efficacy of rilotumumab in combination with capecitabine vs placebo plus capecitabine in patients affected by gastric or esophagogastric junction tumors, the patients with cMET overexpression experienced better progression-free survival (PFS) and overall survival (OS) with rilotumumab and capecitabine combination [191]. Ficlatuzumab is a humanized anti-HGF IgG1 monoclonal antibody. Mok et al. [192] evaluated its efficacy in combination with gefitinib. In the overall Asian population, no significant benefit from the addition of ficlatuzumab to gefitinib in patients with advanced stage lung adenocarcinoma was reported, though patients classified as "VeriStrat poor" may benefit from ficlatuzumab combination therapy [192]. TAK701 is a humanized monoclonal antibody directed against HGF [193]. Addition of TAK-701 to gefitinib is a promising strategy to overcome EGFR-TKI resistance induced by HGF in NSCLC with activating EGFR mutation [194].

2.5.3 MET Antagonists

cMET antagonists compete with HGF for binding to cMET which causes cMET degradation and inactivation of the pathway [195]. Several cMET antagonists have been investigated in preclinical studies. However, disappointing results from recent clinical studies have been obtained with onartuzumab, in particular, for colorectal cancer (CRC), gastric cancer, glioblastoma, and NSCLC, such as in the phase III study NCT01456325 METlung.

This study analyzed the efficacy and safety of combination therapy with onartuzumab and erlotinib in patients with locally advanced or metastatic NSCLC, after progression on first-line with a platinum-based chemotherapy regimen. Moreover, patient selection according to MET expression by immunohistochemistry was provided. The authors concluded that the combination of onartuzumab and erlotinib did not improve clinical outcomes and, on the contrary, shorter OS was observed in the onartuzumab arm, compared with erlotinib in patients with cMET-positive NSCLC [196]. Further, Wakelee and coworkers [197] observed that onartuzumab does not provide any clinical benefit when given in combination with current first-line standard-of-care chemotherapy regimen for nonsquamous NSCLC. Recently, the results of a randomized phase II trial investigating the role of onartuzumab in metastatic CRC were published. Onartuzumab combined with FOLFOX (5-FU and oxaliplatin regimen) and bevacizumab did not significantly improve efficacy outcomes in this setting and, moreover, MET expression was not a predictive biomarker [198]. The same disappointing results were obtained in the METGastric Randomized Clinical Trial where the addition of onartuzumab to FOLFOX did not significantly improve OS, PFS, or ORR vs FOLFOX and placebo. Moreover, as in previous trials, cMET expression did not predict the clinical response to onartuzumab treatment in patients affected by metastatic gastric cancer [199]. Finally, in another phase II study, there was no evidence of further clinical benefit with the addition of onartuzumab to bevacizumab compared with bevacizumab plus placebo in unselected patients with recurrent glioblastoma [200].

2.5.4 MET Kinase Inhibitors

Targeting intracellular cMET with cMET kinase inhibitors represents another approach for HGF–cMET signaling pathway deactivation. These kinase inhibitors can be divided into two groups: selective cMET kinase inhibitors (tivantinib, savolitinib) and nonselective cMET kinase inhibitors (crizotinib, cabozantinib, foretinib). A phase II study evaluated the efficacy of the tivantinib with erlotinib in patients with locally advanced or metastatic EGFR mutation-positive NSCLC after progression on EGFR-TKI (gefitinib or erlotinib) [201]; this study enrolled 45 NSCLC patients with acquired resistance to EGFR-TKIs and who continued treatment with the combination of tivantinib and erlotinib. The median PFS and median OS were 2.7 months and 18.0 months, respectively. Interestingly, better clinical outcomes were obtained in patients whose tumors overexpressed cMET [201]. In another study, the combination of tivantinib and erlotinib was compared to erlotinib monotherapy in nonsquamous NSCLC patients with one or two lines of prior chemotherapy. No statistically significant difference in OS between the arms was observed, although a significant improvement in PFS was observed in the group of patients with KRAS-mutant tumors ($P = 0.006$) [202]. Eng et al. [203] evaluated the role of tivantinib in combination with CETIRI (irinotecan and cetuximab) in patients with metastatic KRAS wild-type CRC, as a second-line treatment. The combination of tivantinib and CETIRI was well tolerated but did not significantly improve PFS. Subgroup analyses trended in favor of tivantinib in patients with overexpression of cMET, though further studies are needed to confirm these results [203]. In HCC patients, a significant improvement in OS was observed in patients with cMET-overexpressed tumors in a phase II study; the results of a phase III study in the subgroup of cMET-overexpressed HCC tumors after sorafenib failure will be soon to come [204].

Crizotinib is a small-molecule inhibitor of the anaplastic lymphoma kinase (ALK) with additional activity against the cMET, ROS, and RON receptors. Crizotinib was approved for use in ALK-rearranged advanced NSCLC [205–207]. In a case report recently published by Pietrantonio et al. [208], the authors described a patient affected by metastatic CRC, BRAF-mutated. This patient initially responded to combination treatment with anti-EGFR (panitumumab) and anti-BRAF (vemurafenib). At progression, rebiopsy was provided and cMET amplification was determined (not found in the pretreatment specimen). Probably, ectopic cMET overexpression led to resistance to panitumumab and vemurafenib. Based on these data, the patient was treated with crizotinib and vemurafenib, with rapid and marked effectiveness [208].

Cabozantinib is a small-molecule inhibitor of cMET, VEGFR2, KIT, and RET followed by AXL and FLT3 [209] and is approved for the treatment of metastatic medullary thyroid cancer and for the second-line therapy for metastatic renal cell carcinoma (RCC) [210,211]. A phase II trial compared cabozantinib with sunitinib as the first-line therapy in patients with metastatic RCC. Cabozantinib demonstrated a significant clinical benefit in PFS and ORR over standard-of-care sunitinib as the first-line therapy in patients with intermediate- or poor-risk metastatic RCC [212]. Daud et al. [213] evaluated the role of cabozantinib in patients affected by malignant melanoma. Median PFS was 4.1 months with cabozantinib and 2.8 months with placebo (hazard ratio of 0.59; $P = 0.284$). The authors concluded that cabozantinib has clinical activity in patients with metastatic melanoma, including uveal melanoma, although further clinical investigation is warranted [213]. Interestingly, a small, randomized phase II trial evaluated the efficacy of erlotinib or cabozantinib monotherapy, or combination treatment with erlotinib and cabozantinib as second-line or third-line treatment of patients with wild-type EGFR and advanced NSCLC. This trial showed that cabozantinib alone or combined with erlotinib had superior efficacy to that of erlotinib alone [214]. The efficacy of cabozantinib was also evaluated in a cohort of metastatic breast cancer patients in a phase II study and demonstrated clinical activity including objective response and disease control [215].

2.6 TARGETING HGF–CMET PATHWAY IN SOLID TUMORS TO OVERCOME DRUG RESISTANCE

HGF–cMET signaling pathway is implicated in mechanisms of acquired resistance to inhibitors against EGFR (gefitinib), VEGF (sunitinib), HER-2 (trastuzumab and lapatinib), and BRAF (vemurafenib) [216–220]. In the following paragraphs, we describe how targeting of the HGF–cMET signaling pathway can overcome the drug resistance in NSCLC, RCC, pancreatic cancer (PDAC), and CRC.

2.6.1 Lung Cancer

Lung cancer is the leading cancer worldwide, both in incidence and in mortality [221]. Eighty percent of lung cancers are NSCLC. In turn, NSCLC can be divided into two main histological subtypes: adenocarcinoma and squamous cell carcinoma. Adenocarcinomas exhibit glandular differentiation and mostly arise in the lung periphery, whereas squamous cell carcinomas often present with keratinization and intercellular bridges. Squamous cell carcinomas occur mostly in the main and lobular bronchi [222]. The 5-year survival rate for NSCLC is very low: around 15%. The current treatment regime [223] is mainly dependent on the stage. For early stages of NSCLC (stage I/II/IIIa), local surgery is the preferred treatment option, with adjuvant chemotherapy for stage II patients. This chemotherapy mostly consists of a two-drug combination with cisplatin. If surgery is not an option, radiotherapy is a suited treatment option. In locally advanced NSCLC, chemotherapy is given to patients that can tolerate it. This regimen consists mostly of a cisplatin–vinorelbine or cisplatin–etoposide doublet; combinations with carboplatin are used to a lesser extent. Resectable tumor surgery is an option after induction therapy [223]. Recently, immunotherapy has also become available for treatment of NSCLC [224].

Several oncogenic drivers have been identified in NSCLC and the list is continuously expanding: EGFR, ALK, cMET, ROS1, RET, FGFR, and NTRK [225]. These drivers are mainly expressed in the adenocarcinoma subtype. In the case of EGFR and ALK, multiple generations of inhibitors have been approved for therapy, whereas for the other drivers, clinical trials are mostly ongoing [226].

Focusing on the HGF–cMET axis, cMET amplification and cMET exon 14 skipping are known as oncogenic drivers in NSCLC. Both occur in around 2% of NSCLC and in both the adenocarcinoma and squamous cell carcinoma subtypes. They are not mutually exclusive and both are responsive to cMET small-molecule inhibitors [136,137]. Several TKIs are currently in clinical trials for the treatment of cMET aberrations, with the dual ALK-cMET inhibitor crizotinib being the most advanced [227].

In the context of resistance, cMET in NSCLC is mostly known to cause resistance against EGFR-TKIs (e.g., erlotinib, gefitinib, afatinib, osimertinib). Primarily cMET amplification is reported in patients with resistance [216,228]. The percentage of cMET amplification increases in this population of EGFR-TKI-treated patients to approximately 20% [154,155]. The resistance mechanism of cMET activation can be explained by the extensive cross-talk between EGFR and cMET signaling [229], with largely overlapping downstream signaling effectors like MAPK and PI3K-Akt. In total, a core network of around 50 proteins can be identified [229]. Moreover, activated EGFR is able to stimulate cMET activation and vice versa [230]. Active EGFR influences cMET in several ways. It increases cMET phosphorylation and expression and decreases ubiquitination of cMET. This effect is enhanced by Her3. Active cMET, in turn, activates EGFR indirectly through MAPK activation [230]. This extensive cross-talk also explains the many clinical trials that immediately focus on combinations of cMET- and EGFR-targeted therapies [227]. However, not all of these combinations have proved to be successful, as was illustrated in the onartuzumab + erlotinib trial that was stopped due to futility [165,231] and the tivantinib + erlotinib trial that did not reach its primary endpoint [232].

Activation of cMET also plays a role in combination with conventional therapies (chemotherapy, radiation therapy). First, radiation therapy is known to upregulate cMET expression [233]. This effect can be explained by the involvement of cMET in tissue repair [234] and homologous recombination [235] that are both activated in a reaction to restore damage by radiation therapy. However, the combination of cMET inhibition and radiotherapy still needs to be tested in patients. Second, cMET inhibition also has an effect in combination with chemotherapy. This interplay is more complex and seems to be dependent on the type of chemotherapy and the form of cancer. cMET is reported to play a role in chemoresistance in both pancreatic cancer (gemcitabine) [236] and ovarian cancer (cisplatin and paclitaxel) [237]. Inhibition of cMET is able to overcome this resistance in ovarian cancer (carboplatin, paclitaxel) [238] and gastric carcinoma (irinotecan) [239]. *In vitro* results show that combining cMET and cisplatin did not lead to synergism [240], whereas functional studies show that addition of HGF increases chemoresistance toward cisplatin in NSCLC cells through downregulation of AIF [241]. Reports on cMET and chemotherapy are often contradicting, and clinical trials are necessary to elucidate the profit of combining cMET inhibitors and chemotherapy in patients.

2.6.2 Renal Cancer

Renal cancer (RCC) is the third most frequent cancer originating from the genitourinary organs [242]. Based on the place of origin of cancer (proximal tubule of the kidney or collecting duct), four main histological types can be recognized: clear cell (ccRCC), papillary (pRCC, type 1 which consists of predominantly basophilic cells and type 2 which consists of mostly eosinophilic cells), chromophobe, and collecting duct tumor. The most frequent type is ccRCC. Division into histological types influences the patient outcome: in particular, metastatic pRCC has a worse prognosis than ccRCC [243]. Moreover, type 1 and type 2 pRCC have different clinical features. Type 1 pRCC is characterized by an indolent clinical course, and type 2 pRCC by more aggressive clinical behavior [244]. The HGF–cMET signaling pathway plays an important role in RCC pathogenesis. In particular, mutation or functional inactivation of the von Hippel–Lindau (VHL) gene is present in the majority of ccRCC cases. Absence or decreased levels of the VHL protein leads to transcriptional activation of HIF-targeted genes, including VEGF, PDGF, TGF-α, HGF, and cMET [245–247]. The expression of cMET and HGF in RCC is 70% and 60%, respectively [248]. Moreover, some authors reported that cMET expression is higher in pRCC than in ccRCC, and higher in type 1 pRCC compared to type 2 pRCC [249]. pRCC often shows trisomy of chromosome 7, where the cMET gene is located [250]. Further, both sporadic and hereditary forms show cMET gene mutations [251]. Interestingly, some RCC cancer stem cells (CSCs) overexpress the cMET receptor which might play a role in the pathogenesis of bone metastases induced by RCC CSCs in mice and humans [252].

Nowadays, first-line treatment for patients affected by metastatic RCC includes several anti-VEGF inhibitors (sunitinib, pazopanib, and bevacizumab) and the mTOR inhibitor (temsirolimus) [211]. VEGF-targeted therapies have demonstrated a survival benefit in RCC patients with PFS of about 9–11 months [211]. Cross-talk between the VEGFR and HGF–cMET pathways is implicated in the resistance to anti-VEGFR therapies and therefore the clinical testing of this combination is very promising [253,254]. In fact, Ciamporcero et al. [247] evaluated the effects of either monotherapy or a combination treatment strategy targeting the VEGF (axitinib) and cMET (crizotinib) pathways in animal models affected by ccRCC. As expected, the combination treatment was more effective than the monotherapy with crizotinib. In another study, increased expression of HGF was observed in mouse models resistant to sunitinib. Combination with a cMET inhibitor was able to overcome sunitinib resistance [217]. Interestingly, foretinib—a dual inhibitor of cMET and VEGFR showed efficacy in pRCC with germline cMET mutations [255]. Another cMET inhibitor, savolitinib, is undergoing clinical development for various cancer types, including pRCC [256]. This agent can suppress the cMET signaling pathway and the duration of target inhibition is dose related. The authors have suggested that savolitinib could have therapeutic potential in sunitinib-resistant pRCC patients [257]. Temsirolimus is an mTOR inhibitor and is used in RCC patients with poor prognosis [211]. mTOR inhibitors such as temsirolimus and everolimus might overcome HGF-dependent resistance to EGFR-TKIs in preclinical studies [258]. Microphthalmia transcription factor (MITF)-associated (MiT) tumors are a family of rare malignancies, including translocation-associated renal cell carcinoma (tRCC) that show dysregulation in MITF family proteins. The cMET gene is transcriptionally activated by MITF family proteins which provides a rationale for the use of cMET inhibitors in this setting. A small study investigated the activity of tivantinib in patients with MiT-associated RCC patients ($n = 6$) with modest antitumor activity [259].

Simultaneous blocking of cMET and VEGFR pathways represents a promising treatment approach for RCC treatment since this will target multiple pathways involved in angiogenesis, tumor survival, and metastasis.

2.6.3 Pancreatic Ductal Adenocarcinoma

Pancreatic ductal adenocarcinoma (PDAC) is a highly aggressive malignancy and the fourth leading cause of cancer-related death in developed countries [260]. In the metastatic setting, cytostatic drugs in different combinations are the gold standard, while no targeted therapy is being used in clinical practice. The chemotherapy choices include gemcitabine as monotherapy, gemcitabine in combinations with nab-paclitaxel or the combination of 5-FU, leucovorin, irinotecan, and oxaliplatin (FOLFIRINOX) or recently nanoliposomal irinotecan [261,262]. The expression of cMET and HGF is increased in >70% and >35% of PDAC patients, respectively [248]. cMET overexpression correlates with worse clinical outcome of PDAC patients (worse TNM stage, presence of lymph node invasion, poor tumor differentiation) [172,263,264]. Zhang et al. [265] developed a novel c-MET CTC assay for detecting cMET

CTCs in patients with cMET amplifications. HGF–cMET overexpression causes chemoresistance by different mechanisms of action. PDAC stroma is the predominant source of HGF with a subsequent increase in VEGF production by stromal cells [266,267]. Interestingly, some preclinical studies have demonstrated that overexpression of cMET is correlated with EMT-like changes in PDAC cells with acquired resistance to gemcitabine [236]. Pancreatic stellate cells (PSCs) are cells that produce the PDAC stroma. They secrete HGF and promote cancer cell growth but do not express the cMET receptor which is present in cancer cells. Pothula et al. [268] evaluated the effect of HGF inhibition, using a neutralizing antibody AMG102, in monotherapy or in combination with gemcitabine *in vitro* (PDAC cells) and *in vivo* (orthotopic model). The authors concluded that HGF inhibition was as effective as standard chemotherapy with gemcitabine at inhibiting local tumor growth. Moreover, HGF inhibition reduced metastasis, although this antimetastatic effect was lost when combined with gemcitabine. Therefore, carefully combining HGF inhibitors with existing treatment modalities is warranted [268]. Additionally, cMET as a marker of pancreatic CSCs was associated with PDAC aggressiveness, metastatic behavior, and intrinsic resistance to chemotherapy [269]. Avan et al. [270] evaluated the synergism of gemcitabine and crizotinib in an orthotopic mouse model of primary PDAC. Crizotinib prolonged survival and promoted gemcitabine uptake, accompanied by an increased activity of the human equilibrative and concentrative nucleoside transporters (hENT1 and hCNT1), and a decreased cytidine deaminase (CDA). Interestingly, crizotinib targets CSC-like subpopulations, interferes with cell proliferation and cell survival, decreases cell migration, and synergistically interacts with gemcitabine [271]. Similar results were obtained with cabozantinib which targets pancreatic CSCs and increases the cytotoxic effect of gemcitabine in a human pancreatic cancer model grown orthotopically in NOD SCID mice [264]. Hage et al. [272] demonstrated that cabozantinib interferes with PDAC cell survival and increased the efficacy of gemcitabine in high-gemcitabine-resistant PDAC, suggesting a capacity to overcome gemcitabine resistance. Brandes et al. [273] evaluated the efficacy of INC280 (cMET inhibitor) *in vitro* and *in vivo*. The combination of gemcitabine with INC280 significantly prolonged survival in an orthotopic syngeneic tumor model and, therefore, this combination warrants further clinical evaluation.

Previous studies have shown that HGF must dimerize to activate cMET. Small-molecule antagonists with homology to a "hinge" region within the putative dimerization domain of HGF have been developed that bind to HGF and block dimerization with subsequent inhibition of cMET signaling. Apart from cMET signaling, which leads to cancer progression, the MSP/Ron (MSP receptor) systems characterized by structural and sequence homology have the same role in PDAC cells. Church et al. [274] hypothesized that the inhibition of HGF by the hinge analogs may be extended to MSP, resulting in more efficient blocking of tumor progression. As expected, hinge analog compounds inhibited HGF and MSP activity and resulted in decreased cMET and Ron activation which may represent a new therapeutic approach for the treatment of metastatic PDAC [274].

2.6.4 Colorectal Cancer

Only about 2% of metastatic CRC tumors show cMET amplification [275,276]. cMET amplification in CRC is correlated with increased aggressiveness and worse clinical prognosis. Moreover, Bardelli et al. [277] showed that cMET amplification is associated with acquired resistance in CRC patients that do not develop KRAS mutations during anti-EGFR therapy. His functional studies demonstrated that cMET activation confers resistance to anti-EGFR therapy both *in vitro* and *in vivo*. Interestingly, cMET amplifications were present in circulating tumor DNA before clinically evident radiologic relapse was observed [277]. The efficacies of several agents targeting the HGF–cMET signaling pathway have been investigated in preclinical studies or clinical trials in metastatic CRC patients, as, e.g., NK4. NK4 is an HGF antagonist that binds to cMET and inhibits HGF-induced tyrosine phosphorylation of the receptor [278]. Wen et al. [278] demonstrated that NK4 has inhibitory effects on angiogenesis in CRC cells. In a randomized phase II clinical trial, the combination of rilotumumab with panitumumab in previously pretreated patients with wild-type KRAS metastatic CRC was evaluated. This combination led to a higher response rate and showed a trend for a better outcome in the population with cMET overexpression [279]. Recently, Bendell et al. [198] published the results of a phase II randomized trial exploring the efficacy of cMET inhibitor onartuzumab in combination with FOLFOX and bevacizumab. The addition of onartuzumab to the standard first-line treatment did not significantly improve efficacy outcomes. Further, cMET expression by immunohistochemistry was not predictive in this setting [198]. Lastly, a phase II randomized study of biweekly CETIRI (cetuximab and irinotecan) plus tivantinib or placebo was restricted to patients who had received only one prior line of chemotherapy and were characterized by wild-type KRAS. Subgroup analyses showed a trend in favor of tivantinib in patients with cMET overexpressing tumors, with low expression of PTEN, or with those pretreated with oxaliplatin, though the subgroups were too small [203].

2.7 BIOMARKERS FOR HGF/MET INHIBITORS

Nowadays, personalized cancer treatment for every patient is an important aim of clinical oncologists. As the efficacy of anticancer treatment is always reserved to a certain group of patients, it is important to identify biomarkers that could predict response or resistance to a specific class of agents, including HGF/cMET inhibitors. Several predictive biomarkers of HGF/cMET inhibitors have been investigated.

Circulating HGF and cMET were evaluated as pharmacodynamic biomarkers of cMET inhibition in different clinical trials, although the results have been inconclusive and it seems that their potential as predictive biomarkers of response depends on the type of cMET inhibition [280–286]. cMET protein expression might be prognostic of clinical outcome in selected cancer types with specific molecular aberrations. Some authors have demonstrated that in certain cancer types, overexpression of cMET protein, as determined by immunohistochemistry, may be associated with poor prognosis. Iveson et al. evaluated the subgroup of cMET-overexpressing gastric tumors treated with a chemotherapy regimen containing epirubicin, cisplatin, and capecitabine with or without rilotumumab. In this phase II study, cMET overexpression was correlated with better clinical outcome in the patients treated with rilotumumab [287]. These results were not confirmed in the phase III RILOMET-1 study [288]. In another study exploring the role of tivantinib in NSCLC, it was shown that overexpression of cMET might have predictive potential only in cMET-positive patients with nonsquamous histology [289]. Amplification of the cMET gene locus with overexpression of the receptor on the cell surface or cMET gene mutations might activate the HGF–cMET signaling pathway. cMET gene amplification, copy number, and mutations appear to be relatively conservative biomarkers, though they are rare. High levels of cMET amplification correlated with responsiveness to crizotinib in esophagogastric adenocarcinoma patients [290]. Germline cMET mutations were predictive of clinical response in pRCC [255]. Interestingly, some authors analyzed changes in cMET phosphorylation before and after treatment with several anti-cMET drugs, although it remains unclear whether changes in these markers are predictive of clinical responses [291–293].

2.8 CONCLUSIONS AND FUTURE DIRECTIONS

With over 25 years since its first discovery, the cMET receptor is emerging as an important target for personalized cancer therapy. cMET is a unique RTK, expressed in the epithelial cells of many organs during embryogenesis and in adulthood, with a versatile role in control and regulation of several biological functions in response to HGF. In addition to its importance in normal physiology, a wide variety of human malignancies have sustained cMET stimulation, overexpression, or genetic aberrations, including lung cancer, as well as upper gastrointestinal cancers such as pancreatic, gastric, and hepatocellular cancers. This aberrant activation of the cMET/HGF signaling pathway is associated with tumor development, progression and aggressive phenotype, and poor clinical outcome.

As summarized in this review, a variety of different strategies to inhibit this signaling pathway have been developed, and several cMET inhibitors are now under clinical investigation in different tumors, with encouraging results. Crizotinib has been approved for ALK-rearranged NSCLC, while cabozantinib is approved for the treatment of metastatic medullary thyroid cancer and for second-line therapy for metastatic RCC. However, the clinical efficacy of several other cMET inhibitors needs to be ultimately validated in ongoing phase II and III randomized trials. Moreover, recent studies suggested the potential role of crizotinib and other cMET TKIs for the treatment of selected upper gastrointestinal cancers, including gastric cancers with cMET amplification and pancreatic tumors, where the HGF–cMET axis plays a pivotal role in progression and invasive growth.

Several important questions remain to be answered. Most clinical studies testing anti-cMET agents have been conducted in combination with other anticancer drugs, such as erlotinib in the context of mutated or wild-type EGFR NSCLC patients, as a means of overcoming acquired EGFR-TKI resistance. Further preclinical and clinical studies to evaluate a possible role for cMET monotherapy and for combination with other agents are warranted. Knowledge gained from these studies should be complemented with molecular and biochemical studies on the function of cMET and related pathways in response to cMET inhibition, as well as studies on the acquired resistance to this treatment. Finally, identification and selection of the optimal patient populations that will benefit from treatment would provide valuable direction and innovative strategies for the clinical development of cMET targeted therapies.

Acknowledgments

The authors are thankful for support received from grants of Netherlands Organization for Scientific Research, NWO-Veni grant; Italian Association for Cancer Research; Cancer Center Amsterdam Foundation, and KWF Dutch Cancer Society (the Netherlands).

Conflict of Interest: No potential conflicts of interest were disclosed.

References

[1] Nakamura T, Nawa K, Ichihara A. Partial purification and characterization of hepatocyte growth factor from serum of hepatectomized rats. Biochem Biophys Res Commun 1984;122:1450–9.

[2] Russell E, McGowan J, Bucher N. Partial characterization of a hepatocyte growth factor from rat platelets. J Cell Physiol 1984;119:183–92.

[3] Gohda E, Tsubouchi H, Nakayama H, Hirono S, Sakiyama O, Takahashi K, et al. Purification and partial characterization of hepatocyte growth factor from plasma of a patient with fulminant hepatic failure. J Clin Invest 1988;81:414–9.

[4] Naldini L, Vigna E, Narsimhan R, Gaudino G, Zarnegar R, Michalopoulos G, et al. Hepatocyte growth factor (HGF) stimulates the tyrosine kinase activity of the receptor encoded by the proto-oncogene cMET. Oncogene 1991;6:501–4.

[5] Bürk RR. A factor from a transformed cell line that affects cell migration. Proc Natl Acad Sci USA, 70. p. 369-72.

[6] Stoker M, Perryman M. An epithelial scatter factor released by embryo fibroblasts. J Cell Sci 1985;77:209–23.

[7] Stoker M, Gherardi E, Perryman M, Gray J. Scatter factor is a fibroblast-derived modulator of epithelial cell mobility. Nature 1987;327:239–42.

[8] Gherardi E, Gray J, Stoker M, Perryman M, Furlong R. Purification of scatter factor, a fibroblast-derived basic protein that modulates epithelial interactions and movement. PNAS 1989;86:5844–8.

[9] Bottaro D, Rubin J, Faletto D, Chan A, Kmiecik T, Vande Woude G, et al. Identification of the hepatocyte growth factor receptor as the cMET proto-oncogene product. Science 1991;251:802–4.

[10] Weidner KM, Arakakit N, Hartmann G, Vandekerckhove J, Weingart S, Rieder H, et al. Evidence for the identity of human scatter factor and human hepatocyte growth factor (motility factor/tumor cell invasion/chromosomal localization/c-met tyrosine kinase receptor/lung fibroblast-derived mitogen). Cell Biol 1991;88:7001–5.

[11] Gherardi E, Stoker M. Hepatocytes and scatter factor. Nature 1990;346:228.

[12] Naldini L, Weidner KM, Vigna E, Gaudino G, Bardelli A, Ponzetto C, et al. Scatter factor and hepatocyte growth factor are indistinguishable ligands for the MET receptor. EMBO 1991;10:2867–78.

[13] Fukuyama R, Ichijoh Y, Minoshima S, Kitamura N, Shimizu N. Regional localization of the hepatocyte growth factor (HGF) gene to human chromosome 7 band q21.1. Genomics 1991;11:410–5.

[14] Fajardo-Puerta AB, Mato Prado M, Frampton AE, Jiao LR. Gene of the month: *HGF*. J Clin Pathol 2016;69:575–9.

[15] Schultz JM, Khan SN, Ahmed ZM, Riazuddin S, Waryah AM, Chhatre D, et al. Noncoding mutations of HGF are associated with nonsyndromic hearing loss, DFNB39. Am J Hum Genet 2009;85:25–39.

[16] Donate LE, Gherardi E, Srinivasan N, Sowdhamini R, Aparicio S, Blundell TL. Molecular evolution and domain structure of plasminogen-related growth factors (HGF/SF and HGF1/MSP). Protein Sci 1994;3:2378–94.

[17] Seki T, Hagiya M, Shimonishi M, Nakamura T, Shimizu S. Organization of the human hepatocyte growth factor-encoding gene. Gene 1991;102:213–9.

[18] Niemann HH. Structural insights into Met receptor activation. Eur J Cell Biol 2011;90:972–81.

[19] Gherardi E, Sandin S, Petoukhov MV, Finch J, Youles ME, Ofverstedt L-G, et al. Structural basis of hepatocyte growth factor/scatter factor and MET signalling. PNAS 2006;103:4046–51.

[20] Stamos J, Lazarus RA, Yao X, Kirchhofer D, Wiesmann C. Crystal structure of the HGF beta-chain in complex with the Sema domain of the Met receptor. EMBO J 2004;23:2325–35.

[21] Kirchhofer D, Yao X, Peek M, Eigenbrot C, Lipari MT, Billeci KL, et al. Structural and functional basis of the serine protease-like hepatocyte growth factor beta-chain in met binding and signaling. J Biol Chem 2004;279:39915–24.

[22] Cioce V, Csaky KG, Chan AM, Bottaro DP, Taylor WG, Jensen R, et al. Hepatocyte growth factor (HGF)/NK1 is a naturally occurring HGF/scatter factor variant with partial agonist/antagonist activity. J Biol Chem 1996;271:13110–5.

[23] Chan A, Rubin J, Bottaro D, Hirschfield D, Chedid M, Aaronson S. Identification of a competitive HGF antagonist encoded by an alternative transcript. Science 1991;254:1382–5.

[24] Miyazawa K, Kitamura A, Naka D, Kitamura N. An alternatively processed mRNA generated from human hepatocyte growth factor gene. Eur J Biochem 1991;197:15–22.

[25] Sonnenberg E, Meyer D, Weidner KM, Birchmeier C. Scatter factor/hepatocyte growth factor and its receptor, the c-met tyrosine kinase, can mediate a signal exchange between mesenchyme and epithelia during mouse development. J Cell Biol 1993;123:223–35.

[26] Ponzetto C, Bardelli A, Zhen Z, Maina F, Zonca PD, Giordano S, et al. A multifunctional docking site mediates signaling and transformation by the hepatocyte growth factor/scatter factor receptor family. Cell 1994;77:261–71.
[27] Fixman ED, Fournier TM, Kamikura DM, Naujokas MA, Park M. Pathways downstream of Shc and Grb2 are required for cell transformation by the Tpr-Met oncoprotein. J Biol Chem 1996;271:13116–22.
[28] Weidner KM, Di Cesare S, Sachs M, Brinkmann V, Behrens J, Birchmeier W. Interaction between Gab1 and the c-Met receptor tyosine kinase is responsible for epithelial morphogenesis. Nature 1996;384:173–6.
[29] Volz Y, Koschut D, Matzke-Ogi A, Dietz MS, Karathanasis C, Richert L, et al. Direct binding of hepatocyte growth factor and vascular endothelial growth factor to CD44v6. Biosci Rep 2015;35:e00236.
[30] Orian-Rousseau V, Chen L, Sleeman JP, Herrlich P, Ponta H. CD44 is required for two consecutive steps in HGF/c-Met signaling. Genes Dev 2002;16:3074–86.
[31] Peschard P, Fournier TM, Lamorte L, Naujokas MA, Band H, Langdon WY, et al. Mutation of the c-Cbl TKB domain binding site on the Met receptor tyrosine kinase converts it into a transforming protein. Mol Cell 2001;8:995–1004.
[32] Cho K-W, Park JH, Park C-W, Lee D, Lee E, Kim DJ, et al. Identification of a pivotal endocytosis motif in c-Met and selective modulation of HGF-dependent aggressiveness of cancer using the 16-mer endocytic peptide. Oncogene 2012;32(8):1018–29.
[33] Hasenauer S, Malinger D, Koschut D, Pace G, Matzke A, von Au A, et al. Internalization of Met requires the co-receptor CD44v6 and its link to ERM proteins. PLoS ONE 2013;8(4):e62357.
[34] Trusolino L, Bertotti A, Comoglio PM. A signaling adapter function for alpha6beta4 integrin in the control of HGF-dependent invasive growth. Cell 2001;107:643–54.
[35] Giordano S, Corso S, Conrotto P, Artigiani S, Gilestro G, Barberis D, et al. The semaphorin 4D receptor controls invasive growth by coupling with Met. Nat Cell Biol 2002;4:720–4.
[36] Conrotto P, Corso S, Gamberini S, Comoglio PM, Giordano S. Interplay between scatter factor receptors and B plexins controls invasive growth. Oncogene 2004;23:5131–7.
[37] Conrotto P, Valdembri D, Corso S, Serini G, Tamagnone L, Comoglio PM, et al. Sema4D induces angiogenesis through Met recruitment by Plexin B1. Blood 2005;105:4321–9.
[38] Swiercz JM, Worzfeld T, Offermanns S. ErbB-2 and met reciprocally regulate cellular signaling via plexin-B1. J Biol Chem 2008;283:1893–901.
[39] Soong J, Chen Y, Shustef EM, Scott GA. Sema4D, the ligand for Plexin B1, suppresses c-Met activation and migration and promotes melanocyte survival and growth. J Invest Dermatol 2012;132:1230–8.
[40] Pelicci G, Giordano S, Zhen Z, Salcini A, Lanfrancone L, Bardelli A, et al. The motogenic and mitogenic responses to HGF are amplified by the Shc adaptor protein. Oncogene 1995;10:1631–8.
[41] Montagner A, Yart A, Dance M, Perret B, Salles JP, Raynal P. A novel role for Gab1 and SHP2 in epidermal growth factor-induced Ras activation. J Biol Chem 2005;280:5350–60.
[42] Rodrigues GA, Park M, Schlessinger J. Activation of the JNK pathway is essential for transformation by the Met oncogene. EMBO J 1997;16:2634–45.
[43] Recio JA, Merlino G. Hepatocyte growth factor/scatter factor activates proliferation in melanoma cells through p38 MAPK, ATF-2 and cyclin D1. Oncogene 2002;21:1000–8.
[44] Coltella N, Rasola A, Nano E, Bardella C, Fassetta M, Filigheddu N, et al. p38 MAPK turns hepatocyte growth factor to a death signal that commits ovarian cancer cells to chemotherapy-induced apoptosis. Int J Cancer 2006;118:2981–90.
[45] Graziani A, Gramaglia D, Cantley LC, Comoglio PM. The tyrosine-phosphorylated hepatocyte growth factor/scatter factor receptor associates with phosphatidylinositol 3-kinase. J Biol Chem 1991;266(33): 22087–90.
[46] Xiao GH, Jeffers M, Bellacosa A, Mitsuuchi Y, Vande Woude GF, Testa JR. Anti-apoptotic signaling by hepatocyte growth factor/Met via the phosphatidylinositol 3-kinase/Akt and mitogen-activated protein kinase pathways. Proc Natl Acad Sci USA, 98. p. 247-52.
[47] del Peso L, Gonzalez-Garcia M, Page C, Herrera R, Nuiiez G. Interleukin-3 induced phosphorylation of BAD through the protein kinase Akt. Science 1997;278:687–9.
[48] Moumen A, Patané S, Porras A, Dono R, Maina F. Met acts on Mdm2 via mTOR to signal cell survival during development. Development (Cambridge, Engl) 2007;134:1443–51.
[49] Zoncu R, Sabatini DM, Efeyan A. mTOR: from growth signal integration to cancer, diabetes and ageing. Nat Rev Mol Cell Biol 2011;12:21–35.

[50] Hoeflich KP, Luo J, Rubie EA, Tsao M, Jin O, Woodgett JR. Requirement for glycogen synthase kinase-3beta in cell survival and NF-kB activation. Nature 2000;406:2–6.
[51] Boccaccio C, Andò M, Tamagnone L, Bardelli A, Michieli P, Battistini C, et al. Induction of epithelial tubules by growth factor HGF depends on the STAT pathway. Nature 1998;391:285–8.
[52] Yu H, Pardoll D, Jove R. STATs in cancer inflammation and immunity: a leading role for STAT3. Nat Rev Cancer 2009;9:798–809.
[53] Wang R, Kobayashi R, Bishop JM. Cellular adherence elicits ligand-independent activation of the Met cell-surface receptor. Proc Natl Acad Sci USA, 93. p. 8425-30.
[54] Nakamura Y, Matsubara D, Goto A, Ota S, Sachiko O, Ishikawa S, et al. Constitutive activation of c-Met is correlated with c-Met overexpression and dependent on cell-matrix adhesion in lung adenocarcinoma cell lines. Cancer Sci 2008;99:14–22.
[55] Hui AY, Meens JA, Schick C, Organ SL, Qiao H, Tremblay EA, et al. Src and FAK mediate cell-matrix adhesion-dependent activation of met during transformation of breast epithelial cells. J Cell Biochem 2009;107:1168–81.
[56] Mitra A, Sawada K, Tiwari P, Mui K, Gwin K, Lengyel E. Ligand independent activation of c-Met by fibronectin and α5 β1-integrin regulates ovarian cancer invasion and metastasis. Oncogene 2011;30:1566–76.
[57] Hammond DE, Urbé S, Vande Woude GF, Clague MJ. Down-regulation of MET, the receptor for hepatocyte growth factor. Oncogene 2001;20:2761–70.
[58] Kermorgant S, Zicha D, Parker PJ. PKC controls HGF-dependent c-Met traffic, signalling and cell migration. EMBO J 2004;23:3721–34.
[59] Petrelli A, Gilestro GF, Lanzardo S, Comoglio PM, Migone N, Giordano S. The endophilin–CIN85–Cbl complex mediates ligand-dependent downregulation of c-Met. Nature 2002;416:187–90.
[60] Jeffers M, Taylor GA, Weidner KM, Omura S, Vandewoude G. Degradation of the Met tyrosine kinase receptor by the ubiquitin proteasome pathway. Mol Cell Biol 1997;17:799–808.
[61] Parachoniak CA, Luo Y, Abella JV, Keen JH, Park M. GGA3 functions as a switch to promote Met receptor recycling, essential for sustained ERK and cell migration. Dev Cell 2011;20:751–63.
[62] Lamorte L, Royal I, Naujokas M, Park M. Crk adapter proteins promote an epithelial–mesenchymal-like transition and are required for HGF-mediated cell spreading and breakdown of epithelial adherens junctions. Mol Biol Cell 2002;13:1449–61.
[63] Kermorgant S, Parker PJ. c-Met signalling: spatio-temporal decisions. Cell Cycle 2005;4:352–5.
[64] Ishibe S, Joly D, Zhu X, Cantley LG, Haven N. Association mediates hepatocyte growth factor-stimulated epithelial morphogenesis. Mol Cell 2003;12:1275–85.
[65] Ménard L, Parker PJ, Kermorgant S. Receptor tyrosine kinase c-Met controls the cytoskeleton from different endosomes via different pathways. Nat Commun 2014;5:3907.
[66] Aoki K, Nakamura T, Fujikawa K, Matsuda M. Local phosphatidylinositol 3,4,5-trisphosphate accumulation recruits Vav2 and Vav3 to activate Rac1/ Cdc42 and initiate neurite outgrowth in nerve growth factor-stimulated PC12 cells. Mol Biol Cell 2005;16:2207–17.
[67] Kermorgant S, Parker PJ. Receptor trafficking controls weak signal delivery: a strategy used by c-Met for STAT3 nuclear accumulation. J Cell Biol 2008;182:855–63.
[68] Galvani AP, Cristiani C, Carpinelli P, Landonio A, Bertolero F. Suramin modulates cellular levels of hepatocyte growth factor receptor by inducing shedding of a soluble form. Biochem Pharmacol 1995;50:959–66.
[69] Tulasne D, Deheuninck J, Calheiros F, Lamballe F, Ji Z, Leroy C, et al. Proapoptotic function of the MET tyrosine kinase receptor through caspase cleavage proapoptotic function of the MET tyrosine kinase receptor through caspase cleavage. Mol Cell Biol 2004;24:10328–39.
[70] Nath D, Williamson NJ, Jarvis R, Murphy G. Shedding of c-Met is regulated by crosstalk between a G-protein coupled receptor and the EGF receptor and is mediated by a TIMP-3 sensitive metalloproteinase. J Cell Sci 2001;114:1213–20.
[71] Petrelli A, Circosta P, Granziero L, Mazzone M, Pisacane A, Fenoglio S, et al. Ab-induced ectodomain shedding mediates hepatocyte growth factor receptor down-regulation and hampers biological activity. Proc Natl Acad Sci USA 2006;103(13):5090–5.
[72] Mai E, Zheng Z, Chen Y, Peng J, Severin C, Filvaroff E, et al. Nonclinical evaluation of the serum pharmacodynamic biomarkers HGF and shed MET following dosing with the anti-MET monovalent monoclonal antibody onartuzumab. Mol Cancer Ther 2014;13:540–52.
[73] Foveau B, Ancot F, Leroy C, Petrelli A, Reiss K, Vingtdeux V, et al. Down-regulation of the Met receptor tyrosine kinase by presenilin-dependent regulated intramembrane proteolysis. Mol Biol Cell 2009;20:2495–507.

[74] Ancot F, Leroy C, Muharram G, Lefebvre J, Vicogne J, Lemiere A, et al. Shedding-generated Met receptor fragments can be routed to either the proteasomal or the lysosomal degradation pathway. Traffic 2012;13:1261–72.
[75] Yang Y, Wang Y, Zeng X, Ma XJ, Zhao Y, Qiao J, et al. Self-control of HGF regulation on human trophoblast cell invasion via enhancing c-met receptor shedding by ADAM10 and ADAM17. J Clin Endocrinol Metab 2012;97:1390–401.
[76] Chalupsky K, Kanchev I, Zbodakova O, Buryova H, Jirouskova M, Korinek V, et al. ADAM10/17-dependent release of soluble c-Met correlates with hepatocellular damage. Folia Biol 2013;59:76–86.
[77] Fu L, Guo W, Liu B, Sun L, Bi Z, Zhu L, et al. Shedding of c-Met ectodomain correlates with c-Met expression in non-small cell lung cancer. Biomarkers 2013;18:126–35.
[78] Uehara Y, Minowa O, Mori C, Shiota K, Kuno J, Noda T, et al. Placental defect and embryonic lethality in mice lacking hepatocyte growth factor/scatter factor. Nature 1995;373:702–5.
[79] Patel Y, Kim H, Rappolee DA. A role for hepatocyte growth factor during early postimplantation growth of the placental lineage in mice. Biol Reprod 2000;62:904–12.
[80] Kauma SW, Bae-Jump V, Walsh SW. Hepatocyte growth factor stimulates trophoblast invasion: a potential mechanism for abnormal placentation in preeclampsia. J Clin Endocrinol Metab 2014;84(11):4092–6.
[81] Naghshvar F, Torabizadeh Z, Moslemi Zadeh N, Mirbaha H, Gheshlaghi P. Investigating the relationship between serum level of s-Met (soluble hepatic growth factor receptor) and preeclampsia in the first and second trimesters of pregnancy. ISRN Obstet Gynecol 2013;2013:925062.
[82] Cartwright JE, Holden DP, Whitley GS. Hepatocyte growth factor regulates human trophoblast motility and invasion: a role for nitric oxide. Br J Pharmacol 1999;128:181–9.
[83] Tse WK, Whitley GS, Cartwright JE. Transforming growth factor-beta1 regulates hepatocyte growth factor-induced trophoblast motility and invasion. Placenta 2002;23:699–705.
[84] Cartwright J. Hepatocyte growth factor induced human trophoblast motility involves phosphatidylinositol-3-kinase, mitogen-activated protein kinase, and inducible nitric oxide synthase. Exp Cell Res 2002;279:219–26.
[85] Dash PR, Whitley GSJ, Ayling LJ, Johnstone AP, Cartwright JE. Trophoblast apoptosis is inhibited by hepatocyte growth factor through the Akt and beta-catenin mediated up-regulation of inducible nitric oxide synthase. Cell Signal 2005;17:571–80.
[86] Frame S, Cohen P. GSK3 takes centre stage more than 20 years after its discovery. Biochem J 2001;359:1–16.
[87] Behrens J, von Kries JP, Kühl M, Bruhn L, Wedlich D, Grosschedl R, et al. Functional interaction of beta-catening with the transcription factor LEF-1. Nature 1996;382:638–42.
[88] Lindsey JS, Brenner RM. Novel hepatocyte growth factor/scatter factor isoform transcripts in the macaque endometrium and placenta. Mol Hum Reprod 2002;8:81–7.
[89] Birchmeier C, Brohmann H. Genes that control the development of migrating muscle precursor cells. Curr Opin Cell Biol 2000;12:725–30.
[90] Goulding M, Lumsden A, Paquette AJ. Regulation of Pax-3 expression in the dermomyotome and its role in muscle development. Development 1994;120:957–71.
[91] Epstein DJ, Vekemans M, Grost P. Splotch (Sp''), a mutation affecting development of the mouse neural tube shows a deletion within the paired homeodomain of Pax-3. Cell 1991;67:767–74.
[92] Franz T, Kothary R, Surani M, Halata Z, Grim M. The Splotch mutation interferes with muscle development in the limbs. Anat Embryol 1993;187:153–60.
[93] Epstein JA, Shapiro DN, Cheng J, Lam PY, Maas RL. Pax3 modulates expression of the c-Met receptor during limb muscle development. Proc Natl Acad Sci USA, 93. p. 4213-8.
[94] Yang XM, Vogan K, Gros P, Park M. Expression of the met receptor tyrosine kinase in muscle progenitor cells in somites and limbs is absent in Splotch mice. Development 1996;122:2163–71.
[95] Dietrich S, Abou-Rebyeh F, Brohmann H, Bladt F, Sonnenberg-Riethmacher E, Yamaai T, et al. The role of SF/HGF and c-Met in the development of skeletal muscle. Development 1999;126:1621–9.
[96] Scaal M, Bonafede A, Dathe V, Sachs M, Cann G, Christ B, et al. SF/HGF is a mediator between limb patterning and muscle development. Development 1999;126:4885–93.
[97] Bladt F, Riethmacher D, Isenmann S, Aguzzi A, Birchmeier C. Essential role for the c-met receptor in the migration of myogenic precursor cells into the limb bud. Nature 1995;376:768–71.
[98] Maina F, Casagranda F, Audero E, Simeone A, Comoglio PM, Klein R, et al. Uncoupling of Grb2 from the Met receptor *in vivo* reveals complex roles in muscle development. Cell 1996;87:531–42.
[99] Duncan SA. Transcriptional regulation of liver development. Dev Dyn 2000;219:131–42.

[100] Crossley PH, Martin GR. The mouse Fgf8 gene encodes a family of polypeptides and is expressed in regions that direct outgrowth and patterning in the developing embryo. Development 1995;121:439–51.
[101] Jung J, Zheng M, Goldfarb M, Zaret K. Initiation of mammalian liver development from endoderm by fibroblast growth factors. Science 1999;284:1998–2003.
[102] Rossi JM, Dunn NR, Hogan BLM, Zaret KS. Distinct mesodermal signals, including BMPs from the septum, transversum mesenchyme, are required in combination for hepatogenesis from the endoderm. Genes Dev 2001;15:1998–2009.
[103] Schmidt C, Bladt F, Goedecke S, Brinkmann V, Zschiesche W, Sharpe M, et al. Scatter factor/hepatocyte growth factor is essential for liver development. Nature 1995;373:699–702.
[104] Zaret KS. Liver specification and early morphogenesis. Mech Dev 2000;92:83–8.
[105] Hilberg F, Aguzzi A, Howells N, Wagner EF. c-Jun is essential for normal mouse development and hepatogenesis. Nature 1993;363:179–81.
[106] Grobstein C. Mechanimsms of organogenetic tissue interaction. Natl Cancer Inst Monogr 1967;26:279–99.
[107] Davies JA. Morphogenesis of the metanephric kidney. Sci World J 2002;2:1937–50.
[108] Wynshaw-Boris A, Ryan G, Deng CX, Chan DC, Jackson-Grusby L, Larson D, et al. The role of a single formin isoform in the limb and renal phenotypes of limb deformity. Mol Med 1997;3:372–84.
[109] Dettenhofer M, Zhou F, Leder P. Formin 1-isoform IV deficient cells exhibit defects in cell spreading and focal adhesion formation. PLoS ONE 2008;3:1–8.
[110] O'Rourke DA, Liu ZX, Sellin L, Spokes K, Zeller R, Cantley LG. Hepatocyte growth factor induces MAPK-dependent formin IV translocation in renal epithelial cells. J Am Soc Nephrol 2000;11:2212–21.
[111] Santos O, Barros E, Yang X-M, Matsumoto K, Nakamura T, Park M, et al. Involvement of hepatocyte growth factor in kidney development. Dev Biol 1994;163(2):525–9.
[112] Davies J, Lyon M, Gallagher J, Garrod D. Sulphated proteoglycan is required for collecting duct growth and branching but not nephron formation during kidney development. Development 1995;121:1507–17.
[113] Woolf AS, Kolatsi-Joannou AS, Hardman P, Andermarcher E, Moorby C, Fine LG, et al. Roles of hepatocyte growth factor/scatter factor and the met receptor in the early development of the metanephros. J Cell Biol 1995;128:171–84.
[114] Davies JA, Fisher CE. Genes and proteins in renal development. Exp Nephrol 2002;10:102–13.
[115] Miyazaki Y, Oshima K, Fogo A, Hogan BLM, Ichikawa I. Bone morphogenetic protein 4 regulates the budding site and elongation of the mouse ureter. J Clin Invest 2000;105:863–73.
[116] Gupta IR, Piscione TD, Grisaru S, Phant T, Macias-Silva M, Zhou X, et al. Protein kinase A is a negative regulator of renal branching morphogenesis and modulates inhibitory and stimulatory bone morphogenetic proteins. J Biol Chem 1999;274:26305–14.
[117] Lelongt B, Trugnan G, Murphy G, Ronco PM. Matrix metalloproteinases MMP2 and MMP9 are produced in early stages of kidney morphogenesis but only MMP9 is required for renal organogenesis *in vitro*. J Cell Biol 1997;136:1363–73.
[118] Karihaloo A, O'Rourke DA, Nickel C, Spokes K, Cantley LG. Differential MAPK pathways utilized for HGF- and EGF-dependent renal epithelial morphogenesis. J Biol Chem 2001;276:9166–73.
[119] Karihaloo A, Kale S, Rosenblum ND, Cantley LG. Hepatocyte growth factor-mediated renal epithelial branching morphogenesis is regulated by glypican-4 expression hepatocyte growth factor-mediated renal epithelial branching morphogenesis is regulated by glypican-4 expression. Mol Cell Biol 2004;24:8745–52.
[120] Gutiérrez J, Cabrera D, Brandan E. Glypican-1 regulates myoblast response to HGF via Met in a lipid raft-dependent mechanism: effect on migration of skeletal muscle precursor cells. Skelet Muscle 2014;4:5.
[121] Derman MP, Cunha MJ, Barros EJ, Nigam SK, Cantley LG. HGF-mediated chemotaxis and tubulogenesis require activation of the phosphatidylinositol 3-kinase. Am J Physiol 1995;268:F1211–7.
[122] Khwaja A, Lehmann K, Barbara M, Chem JB, Marte BM, Downward J. Phosphoinositide 3-kinase induces scattering and tubulogenesis in epithelial cells through a novel pathway phosphoinositide 3-kinase induces scattering and tubulogenesis in epithelial cells through a novel pathway. J Biol Chem 1998;273:18793–801.
[123] Streit A, Stern CD, Théry C, Ireland GW, Aparicio S, Sharpe MJ, et al. A role for HGF/SF in neural induction and its expression in Hensen's node during gastrulation. Development 1995;121:813–24.
[124] Stern CD, Ireland GW, Herrick SE, Gherardi E, Gray J, Perryman M, et al. Epithelial scatter factor and development of the chick embryonic axis. Development 1990;110:1271–84.
[125] Hamanoue M, Takemoto N, Matsumoto K, Nakamura T, Nakajima K, Kohsaka S. Neurotrophic effect of hepatocyte growth factor on central nervous system neurons *in vitro*. J Neurosci Res 1996;43:554–64.

[126] Yamagata T, Muroya K, Mukasa T, Igarashi H, Momoi M, Tsukahara T, et al. Hepatocyte growth factor specifically expressed in microglia activated Ras in the neurons, similar to the action of neurotrophic factors. Biochem Biophys Res Commun 1995;210:231–7.
[127] Maina F, Hilton M, Andres R, Wyatt S, Klein R, Davies A. Multiple roles for hepatocyte growth factor in sympathetic neuron development. Neuron 1998;20:835–46.
[128] Maina F, Hilton M, Ponzzetto C, Davies A, Klein R. Met receptor signaling is required for sensory nerve developments and HGF promote axonal growth and survival of sensory neurons. Ganes Dev 1997;11:334–3350.
[129] Krasnoselsky A, Massay MJ, DeFrances MC, Michalopoulos G, Zarnegar R, Ratner N. Hepatocyte growth factor is a mitogen for Schwann cells and is present in neurofibromas. J Neurosci 1994;14:7284–90.
[130] Merchant-Larios H, Moreno-Mendoza N, Buehr M. The role of the mesonephros in cell differentiation and morphogenesis of the mouse fetal testis. Int J Dev Biol 1993;37:407–15.
[131] Ricci G, Catizone A, Innocenzi A, Galdieri M. Hepatocyte growth factor (HGF) receptor expression and role of HGF during embryonic mouse testis development. Dev Biol 1999;216:340–7.
[132] Ricci G, Catizone A, Galdieri M. Embryonic mouse testis development: role of platelet derived growth factor (PDGF-BB). J Cell Physiol 2004;200:458–67.
[133] Ricci G, Catizone A, Galdieri M. Expression and functional role of hepatocyte growth factor and its receptor (c-met) during fetal mouse testis development. J Endocrinol 2006;191:559–70.
[134] Ricci G, Guglielmo MC, Caruso M, Ferranti F, Canipari R, Galdieri M, et al. Hepatocyte growth factor is a mouse fetal Leydig cell terminal differentiation factor. Biol Reprod 2012;87:146.
[135] Onozato R, Kosaka T, Kuwano H, Sekido Y, Yatabe Y, Mitsudomi T. Activation of MET by gene amplification or by splice mutations deleting the juxtamembrane domain in primary resected lung cancers. J Thorac Oncol 2009;4:5–11.
[136] Awad MM, Oxnard GR, Jackman DM, Savukoski DO, Hall D, Shivdasani P, et al. MET exon 14 mutations in non-small-cell lung cancer are associated with advanced age and stage-dependent MET genomic amplification and c-Met overexpression. J Clin Oncol 2016;34:721–30.
[137] Van Der Steen N, Giovannetti E, Pauwels P, Peters GJ, Hong DS, Cappuzzo F, et al. cMET exon 14 skipping: from the structure to the clinic. J Thorac Oncol 2016;11:1423–32.
[138] Schmidt L, Duh F-M, Chen F, Kishida T, Glenn G, Choyke P, et al. Germline and somatic mutations in the tyrosine kinase domain of the MET proto-oncogene in papillary renal carcinomas. Nat Genet 1997;16:68–73.
[139] Schmidt L, Junker K, Weirich G, Glenn G, Choyke P, Lubensky I, et al. Two North American families with hereditary papillary renal carcinoma and identical novel mutations in the MET proto-oncogene. Cancer Res 1998;58:1719–22.
[140] Schmidt L, Junker K, Nakaigawa N, Kinjerski T, Weirich G, Miller M, et al. Novel mutations of the MET proto-oncogene in papillary renal carcinomas. Oncogene 1999;18:2343–50.
[141] Nagata H, Worobec AS, Oh CK, Chowdhury BA, Tannenbaum S, Suzuki Y, et al. Identification of a point mutation in the catalytic domain of the protooncogene c-kit in peripheral blood mononuclear cells of patients who have mastocytosis with an associated hematologic disorder. Proc Natl Acad Sci USA, 92. p. 10560-4.
[142] Piao X, Bernstein A. A point mutation in the catalytic domain of c-kit induces growth factor independence, tumorigenicity, and differentiation of mast cells. Blood 1996;87:3117–23.
[143] Hofstra RMW, Landsvater RM, Ceccherini I, Stulp RP, Stelwagen T, Luo Y, et al. A mutation in the RET proto-oncogene associated with multiple endocrine neoplasia type 2B and sporadic medullary thyroid carcinoma. Nature 1994;367:375–6.
[144] Ignatius Ou S-H, Young L, Schrock AB, Johnson A, Klempner SJ, Zhu VW, et al. Emergence of pre-existing MET Y1230C mutation as a resistance mechanism to crizotinib in NSCLC with MET exon 14 skipping. J Thorac Oncol 2016;12:137–40.
[145] Bahcall M, Sim T, Paweletz CP, Patel JD, Alden RS, Kuang Y, et al. Acquired MET D1228V mutation and resistance to MET inhibition in lung cancer. Cancer Discov 2016;6:1334–41.
[146] Heist RS, Sequist LV, Borger D, Gainor JF, Arellano RS, Le LP, et al. Acquired resistance to crizotinib in non-small cell lung cancer with MET exon 14 skipping. J Thorac Oncol 2016;11:1242–5.
[147] Cui JJ. Targeting receptor tyrosine kinase MET in cancer: Small molecule inhibitors and clinical progress. J Med Chem 2014;57:4427–53.
[148] Yan SB, Peek VL, Ajamie R, Buchanan SG, Graff JR, Heidler SA, et al. LY2801653 is an orally bioavailable multi-kinase inhibitor with potent activity against MET, MST1R, and other oncoproteins, and displays anti-tumor activities in mouse xenograft models. Invest New Drugs 2013;31:833–44.

[149] Yang Y, Wu N, Shen J, Teixido C, Sun X, Lin Z, et al. MET overexpression and amplification define a distinct molecular subgroup for targeted therapies in gastric cancer. Gastric Cancer 2016;19:778–88.
[150] Zhang J, Guo L, Liu X, Li W, Ying J, Zhang J, et al. MET overexpression, gene amplification and relevant clinicopathological features in gastric adenocarcinoma. Oncotarget 2016;5:10264–73.
[151] Kwak Y, Kim S, Park C-K, Paek SH, Lee S, Park S. C-MET overexpression and amplification in gliomas. Int J Clin Exp Pathol 2015;8:14932–8.
[152] Camidge DR, Ou SI, Shapiro GI, Otterson GA, Villaruz LC, Villalona-Calero MA, et al. Efficacy and safety of crizotinib in patients with advanced c-MET-amplificed non-small cell lung cancer (NSCLC). J Clin Oncol 2014;32. abstr8001.
[153] Noonan SA, Berry L, Lu X, Gao D, Baron AE, Chesnut P, et al. Identifying the appropriate FISH criteria for defining MET copy number-driven lung adenocarcinoma through oncogene overlap analysis. J Thorac Oncol 2016;11:1293–304.
[154] Bean J, Brennan C, Shih J-Y, Riely G, Viale A, Wang L, et al. MET amplification occurs with or without T790M mutations in EGFR mutant lung tumors with acquired resistance to gefitinib or erlotinib. Proc Natl Acad Sci USA, 104. p. 20932-7.
[155] Chen HJ, Mok TS, Chen ZH, Guo AL, Zhang XC, Su J, et al. Clinicopathologic and molecular features of epidermal growth factor receptor T790M mutation and c-MET amplification in tyrosine kinase inhibitor-resistant chinese non-small cell lung cancer. Pathol Oncol Res 2009;15:651–8.
[156] Cooper CS, Blair DG, Oskarsson MK, Tainsky MA, Eader LA, Vande Woude GF. Characterization of human transforming genes from chemically transformed, teratocarcinoma, and pancreatic carcinoma cell lines. Cancer Res 1984;44:1–10.
[157] Soman NR, Correa P, Ruiz BA, Wogan GN. The TPR-MET oncogenic rearrangement is present and expressed in human gastric carcinoma and precursor lesions. Proc Natl Acad Sci USA, 88. p. 4892-6.
[158] Dean M, Park M, Vande Woude GF. Characterization of the rearranged tpr-met oncogene breakpoint. Mol Cell Biol 1987;7:921–4.
[159] Gonzatti-Haces M, Seth A, Park M, Copeland T, Oroszlan S, Vande Woude GF. Characterization of the TPR-MET oncogene p65 and the MET protooncogene p140 protein-tyrosine kinases. Proc Natl Acad Sci USA, 85. p. 21-5.
[160] Vigna E, Gramaglia D, Longati P, Bardelli A, Comoglio PM. Loss of the exon encoding the juxtamembrane domain is essential for the oncogenic activation of TPR-MET. Oncogene 1999;18:4275–81.
[161] Sattler M, Pride YB, Ma P, Kinase TT, Gramlich JL, Chu SC, et al. A novel small molecule met inhibitor induces apoptosis in cells transformed by the oncogenic TPR-MET tyrosine kinase. Cancer Res 2003;63:5462–9.
[162] Ogunwobi OO, Puszyk W, Dong HJ, Liu C. Epigenetic upregulation of HGF and c-Met drives metastasis in hepatocellular carcinoma. PLoS ONE 2013;8:1–12.
[163] Garofalo M, Romano G, Di Leva G, Nuovo G, Jeon YJ, Ngankeu A, et al. EGFR and MET receptor tyrosine kinase-altered microRNA expression induces tumorigenesis and gefitinib resistance in lung cancers. Nat Med 2012;18:74–82.
[164] Elia G, Ren Y, Lorenzoni P, Zarnegar R, Burger MM, Rusciano D. Mechanisms regulating c-met overexpression in liver-metastatic B16-LS9 melanoma cells. J Cell Biochem 2001;81:477–87.
[165] Rolfo C, Van Der Steen N, Pauwels P, Cappuzzo F. Onartuzumab in lung cancer: the fall of Icarus? Expert Rev Anticancer Ther 2015;15:487–9.
[166] Park S, Choi YL, Sung CO, An J, Seo J, Ahn MJ, et al. High MET copy number and MET overexpression: poor outcome in non-small cell lung cancer patients. Histol Histopathol 2012;27:197–207.
[167] Tsuta K, Kozu Y, Mimae T, Yoshida A, Kohno T, Sekine I, et al. c-MET/phospho-MET protein expression and MET gene copy number in non-small cell lung carcinomas. J Thorac Oncol 2012;7:331–9.
[168] Sun W, Song L, Ai T, Zhang Y, Gao Y, Cui J. Prognostic value of MET, cyclin D1 and MET gene copy number in non-small cell lung cancer. J Biomed Res 2013;27:220–30.
[169] Preusser M, Streubel B, Berghoff AS, Hainfellner JA, von Deimling A, Widhalm G, et al. Amplification and overexpression of CMET is a common event in brain metastases of non-small cell lung cancer. Histopathology 2014;65:684–92.
[170] Yildiz Y, Sokmensuer C, Yalcin S. Evaluation of c-Met, HGF, and HER-2 expressions in gastric carcinoma and their association with other clinicopathological factors. OncoTargets Ther 2016;9:5809–17.
[171] Kong DS, Song SY, Kim DH, Kyeung MJ, Yoo JS, Jong SK, et al. Prognostic significance of c-Met expression in glioblastomas. Cancer 2009;115:140–8.

[172] Zhu GH, Huang C, Qiu ZJ, Liu J, Zhang ZH, Zhao N, et al. Expression and prognostic significance of CD151, c-Met, and integrin alpha3/alpha6 in pancreatic ductal adenocarcinoma. Dig Dis Sci 2011;56:1090–8.
[173] Neuzillet C, Couvelard A, Tijeras-Raballand A, de Mestier L, de Gramont A, Bédossa P, et al. High c-Met expression in stage I-II pancreatic adenocarcinoma: proposal for an immunostaining scoring method and correlation with poor prognosis. Histopathology 2015;67:664–76.
[174] Ma J, Defrances MC, Zou C, Johnson C, Ferrell R, Zarnegar R. Somatic mutation and functional polymorphism of a novel regulatory element in the HGF gene promoter causes its aberrant expression in human breast cancer. J Clin Invest 2009;119:478–91.
[175] Rahimi N, Tremblay E, McAdam L, Park M, Schwall R, Elliott B. Identification of a hepatocyte growth factor autocrine loop in a murine mammary carcinoma. Cell Growth Differ 1996;7:263–70.
[176] Masuya D, Huang C, Liu D, Nakashima T, Kameyama K, Haba R, et al. The tumour–stromal interaction between intratumoral c-Met and stromal hepatocyte growth factor associated with tumour growth and prognosis in non-small-cell lung cancer patients. Br J Cancer 2004;90:1555–62.
[177] Trovato M, Vitarelli E, Grosso M, Alesci S, Benvenga S, Trimarchi F, et al. Immunohistochemical expression of HGF, c-MET and transcription factor STAT3 in colorectal tumors. Eur J Histochem 2004;48:291–7.
[178] Miura Y, Kozuki Y, Yagasaki K. Potentiation of invasive activity of hepatoma cells by reactive oxygen species is mediated by autocrine/paracrine loop of hepatocyte growth factor. Biochem Biophys Res Commun 2003;305:160–5.
[179] Tokunou M, Niki T, Eguchi K, Iba S, Tsuda H, Yamada T, et al. c-MET Expression in myofibroblasts: role in autocrine activation and prognostic significance in lung adenocarcinoma. Am J Pathol 2001;158: 1451–63.
[180] Ferrucci A, Moschetta M, Frassanito MA, Berardi S, Catacchio I, Ria R, et al. A HGF/cMET autocrine loop is operative in multiple myeloma bone marrow endothelial cells and may represent a novel therapeutic target. Clin Cancer Res 2014;20:5796–807.
[181] Gherardi E, Birchmeier W, Birchmeier C, Vande Woude G. Targeting MET in cancer: rationale and progress. Nat Rev Cancer 2012;12:89–103.
[182] Garajová I, Giovannetti E, Biasco G, Peters GJ. c-Met as a target for personalized therapy. Transl Oncogenomics 2015;7:13–31.
[183] Zhang Y, Jain R, Zhu M. Recent progress and advances in HGF/MET-targeted therapeutic agents for cancer treatment. Biomedicines 2015;3:149–81.
[184] Naldini L, Vigna E, Bardelli A, Follenzi A, Galili F, Comoglio PM. Biological activation of pro-HGF (hepatocyte growth factor) by urokinase is controlled by a stoichiometric reaction. J Biol Chem 1995;270:603–11.
[185] Parr C, Sanders A, Jiang W. Hepatocyte growth factor activation inhibitors—therapeutic potential in cancer. Anticancer Agents Med Chem 2010;10:47–57.
[186] Hu C, Jiang N, Wang G, Zheng J, Yang W, Yang J. Expression of hepatocyte growth factor activator inhibitor-1 (HAI-1) gene in prostate cancer: clinical and biological significance. J BUON 2014;19:215–20.
[187] Tsai C-H, Teng C-H, Tu Y-T, Cheng T-S, Wu S-R, Ko C-J, et al. HAI-2 suppresses the invasive growth and metastasis of prostate cancer through regulation of matriptase. Oncogene 2014;33:4643–52.
[188] Giordano S. Rilotumumab, a mAb against human hepatocyte growth factor for the treatment of cancer. Curr Opin Mol Ther 2009;11:448–55.
[189] Jun HT, Sun J, Rex K, Radinsky R, Kendall R, Coxon A, et al. AMG 102, a fully human anti-hepatocyte growth factor/scatter factor neutralizing antibody, enhances the efficacy of temozolomide or docetaxel in U-87 MG cells and xenografts. Clin Cancer Res 2007;13:6735–42.
[190] Buchanan IM, Scott T, Tandle AT, Burgan WE, Burgess TL, Tofilon PJ, et al. Radiosensitization of glioma cells by modulation of Met signalling with the hepatocyte growth factor neutralizing antibody, AMG102. J Cell Mol Med 2011;15:1999–2006.
[191] Zhu M, Tang R, Doshi S, Oliner KS, Dubey S, Jiang Y, et al. Exposure-response analysis of rilotumumab in gastric cancer: the role of tumour MET expression. Br J Cancer 2015;112:429–37.
[192] Mok TS, Geater S, Su W, Tan E, Chang G, Han M, et al. A randomized phase 2 study comparing the combination of ficlatuzumab and gefitinib with gefitinib alone in Asian patients with advanced stage pulmonary adenocarcinoma. J Thorac Oncol 2016;11:1736–44.
[193] Hori A, Kitahara O, Ito Y, Nishizawa S, Misumi Y, Inaoka T. Monotherapeutic and combination antitumor activities of TAK-701, a humanized anti-hepatocyte growth factor neutralizing antibody, against multiple types of cancer. Proc Am Assoc Cancer Res 2009;. Apr 18–22, Abstract nr 1244.

[194] Okamoto W, Okamoto I, Tanaka K, Hatashita E, Yamada Y, Kuwata K, et al. TAK-701, a humanized monoclonal antibody to hepatocyte growth factor reverses gefitinib resistance induced by tumor-derived HGF in non-small cell lung cancer with an EGFR mutation. Mol Cancer Ther 2010;9:2785–92.
[195] Peters S, Adjei AA. MET: a promising anticancer therapeutic target. Nat Rev Clin Oncol 2012;9:314–26.
[196] Spigel D, Edelman M, O'Byrne K, Paz-Ares L, Mocci S, Phan S, et al. Results from the phase III randomized trial of onartuzumab plus erlotinib versus erlotinib in previously treated stage IIIB or IV non-small-cell lung cancer: METLung. J Clin Oncol 2017;35:412–20.
[197] Clément-Duchêne C, Natale RB, Jahan T, Krupitskaya Y, Osarogiagbon R, Sanborn RE, et al. A phase II study of enzastaurin in combination with erlotinib in patients with previously treated advanced non-small cell lung cancer. Lung Cancer 2012;78:57–62.
[198] Bendell J, Hochster H, Hart L, Firdaus I, Mace J, McFarlane J, et al. A phase II randomized trial (GO27827) of first-line FOLFOX plus bevacizumab with or without the MET inhibitor onartuzumab in patients with metastatic colorectal cancer. Oncologist 2017;22:264–71.
[199] Shah M, Bang Y, Lordick F, Alsina M, Chen M, Hack S, et al. Effect of fluorouracil, leucovorin, and oxaliplatin with or without onartuzumab in HER2-negative, MET-positive gastroesophageal adenocarcinoma: the MET-Gastric Randomized Clinical Trial. JAMA Oncol 2017;3:620–7.
[200] Cloughesy T, Finocchiaro G, Belda-Iniesta C, Recht L, Brandes A, Pineda E, et al. Randomized, double-blind, placebo-controlled, multicenter phase II study of onartuzumab plus bevacizumab versus placebo plus bevacizumab in patients with recurrent glioblastoma: efficacy, safety, and hepatocyte growth factor and O6-methylguanine-DNA methyltransferase biomarker analyses. J Clin Oncol 2017;35:343–51.
[201] Azuma K, Hirashima R, Yamamoto N, Okamoto I, Takahashi T, Nishio M, et al. Phase II study of erlotinib plus tivantinib (ARQ 197) in patients with locally advanced or metastatic EGFR mutation-positive non-small-cell lung cancer just after progression on EGFR-TKI, gefitinib or erlotinib. ESMO Open 2016;1:e000063.
[202] Azuma K, Yoshioka H, Yamamoto N, Takahashi T, Nishio M, Katakoma N, et al. Tivantinib plus erlotinib versus placebo plus erlotinib in Asian patients with previously treated nonsquamous NSCLC with wild-type EGFR: First report of a phase III ATTENTION trial. J Clin Oncol 2014;32. a8044.
[203] Eng C, Bessudo A, Hart L, Severtsev A, Gladkov O, Müller L, et al. A randomized, placebo-controlled, phase 1/2 study of tivantinib (ARQ 197) in combination with irinotecan and cetuximab in patients with metastatic colorectal cancer with wild-type KRAS who have received first-line systemic therapy. Int J Cancer 2016;139:177–86.
[204] Trojan K, Zeuzem S. Tivantinib in hepatocellular carcinoma. Expert Opin Invest Drugs 2013;22:141–7.
[205] Ettinger D, Akerley W, Borghaei H, Chang A, Cheney R, Chirieac L, et al. Non-small cell lung cancer—clinical practice guidelines in oncology. J Natl Compr Cancer Netw 2012;10:1236–71.
[206] Reck M, Popat S, Reinmuth N, De Ruysscher D, Kerr K, Peters S, et al. Metastatic non-small-cell lung cancer (NSCLC): ESMO Clinical Practice Guidelines for diagnosis, treatment and follow-up. Ann Oncol 2014;25:iii27–39.
[207] Cappuzzo F, Moro-Sibilot D, Gautschi O, Boleti E, Felip E, Groen HJM, et al. Management of crizotinib therapy for ALK-rearranged non-small cell lung carcinoma: an expert consensus. Lung Cancer 2015;87:89–95.
[208] Pietrantonio F, Oddo D, Gloghini A, Valtorta E, Berenato R, Barault L, et al. MET-driven resistance to dual EGFR and BRAF blockade may be overcome by switching from EGFR to MET inhibition in BRAF-mutated colorectal cancer. Cancer Discov 2016;6:963–71.
[209] Viola D, Cappagli V, Elisei R. Cabozantinib (XL184) for the treatment of locally advanced or metastatic progressive medullary thyroid cancer. Future Oncol 2013;9:1083–92.
[210] Schoffski P, Elisei R, Müller S, Brose M, Shah M, Licitra L, et al. An international, double-blind, randomized, placebo-controlled phase III trial (EXAM) of cabozantinib (XL184) in medullary thyroid carcinoma (MTC) patients (pts) with documented RECIST progression at baseline. J Clin Oncol 2012;30:5508.
[211] Escudier B, Porta B, Schmidinger M, Rioux-Leclercq N, Bex A, Khoo V, et al. Renal cell carcinoma: ESMO Clinical Practice Guidelines for diagnosis, treatment and follow-up. Ann Oncol 2016;27:v58–68.
[212] Choueiri T, Halabi S, Sanford B, Hahn O, Michaelson M, Walsh M, et al. Cabozantinib versus sunitinib as initial targeted therapy for patients with metastatic renal cell carcinoma of poor or intermediate risk: the alliance A031203 CABOSUN Trial. J Clin Oncol 2017;35:591–7.
[213] Daud A, Kluger H, Kurzrock R, Schimmoller F, Weitzman A, Samuel T, et al. Phase II randomised discontinuation trial of the MET/VEGF receptor inhibitor cabozantinib in metastatic melanoma. Br J Cancer 2017;116:432–40.

[214] Neal J, Dahlberg S, Wakelee H, Aisner S, Bowden M, Huang Y, et al. Erlotinib, cabozantinib, or erlotinib plus cabozantinib as second-line or third-line treatment of patients with EGFR wild-type advanced non-small-cell lung cancer (ECOG-ACRIN 1512): a randomised, controlled, open-label, multicentre, phase 2 trial. Lancet Oncol 2016;17:1661–71.
[215] Tolaney S, Nechushtan H, Ron I, Schöffski P, Awada A, Yasenchak C, et al. Cabozantinib for metastatic breast carcinoma: results of a phase II placebo-controlled randomized discontinuation study. Breast Cancer Res Treat 2016;160:305–12.
[216] Engelman JA, Zejnullahu K, Mitsudomi T, Song Y, Hyland C, Park JO, et al. MET amplification leads to gefitinib resistance in lung cancer by activating ERBB3 signaling. Science 2007;316:1039–43.
[217] Shojaei F, Lee JH, Simmons BH, Wong A, Esparza CO, Plumlee PA, et al. HGF/c-Met acts as an alternative angiogenic pathway in sunitinib-resistant tumors. Cancer Res 2010;70:10090–100.
[218] Chapman PB, Hauschild A, Robert C, Haanen JB, Ascierto P, Larkin J, et al. Improved survival with vemurafenib in melanoma with BRAF V600E mutation. New Engl J Med 2011;364:2507–16.
[219] Minuti G, Cappuzzo F, Duchnowska R, Jassem J, Fabi A, O'Brien T, et al. Increased MET and HGF gene copy numbers are associated with trastuzumab failure in HER2-positive metastatic breast cancer. Br J Cancer 2012;107:793–9.
[220] Saito S, Morishima K, Ui T, Hoshino H, Matsubara D, Ishikawa S, et al. The role of HGF/MET and FGF/FGFR in fibroblast-derived growth stimulation and lapatinib-resistance of esophageal squamous cell carcinoma. BMC Cancer 2015;15:82.
[221] Riess J. Shifting paradigms in non-small cell lung cancer: an evolving therapeutic landscape. Am J Manag Care 2013;19(19 Suppl.):s390–7.
[222] Travis WD, Brambilla E, Burke AP, Marx A, Nicholson AG. WHO classification of tumours of the lung, pleura, thymus and heart. 4 WHO; 2015.
[223] Vansteenkiste J, De Ruysscher D, Eberhardt WEE, Lim E, Senan S, Felip E, et al. Early and locally advanced non-small-cell lung cancer (NSCLC): ESMO clinical practice guidelines for diagnosis, treatment and follow-up. Ann Oncol 2013;24(Suppl. 6):vi89–98.
[224] National Cancer Institute. https://www.cancer.gov/news-events/cancer-currents-blog/2016/fda-atezolizumab-pembrolizumab-lung; 2016 [accessed November 2017].
[225] Hirsch FR, Suda K, Wiens J, Bunn PA. New and emerging targeted treatments in advanced non-small-cell lung cancer. Lancet (London, Engl) 2016;388:1012–24.
[226] Mascaux C, Tsao M-S, Hirsch FR. Chapter 18: molecular testing in lung cancer. In: Pass H, Ball D, Scagliotti G, editors. IASLC Thoracic Oncology E-Book. Amsterdam: Elsevier Health Science; 2017. p. 164–77.
[227] Drilon A, Cappuzzo F, Ignatius Ou S-H, Camidge DR. Targeting MET in lung cancer: will expectations finally be MET? J Thorac Oncol 2016;12:15–26.
[228] Ortiz-Cuaran S, Schef M, Plenker D, Scheel AH, Fernandez-cuesta L, Meder L, et al. Heterogeneous mechanisms of primary and acquired resistance to third-generation EGFR inhibitors. Clin Cancer Res 2016;22:4837–47.
[229] Guo A, Villén J, Kornhauser J, Lee KA, Stokes MP, Rikova K, et al. Signaling networks assembled by oncogenic EGFR and c-Met. Proc Natl Acad Sci USA, 105. p. 692-7.
[230] Breindel JL, Haskins JW, Cowell EP, Zhao M, Nguyen DX, Stern DF. EGF receptor activates MET through MAPK to enhance non-small cell lung carcinoma invasion and brain metastasis. Cancer Res 2013;73:5053–65.
[231] Spigel DR, Edelman MJ, O'Byrne K, Paz-Ares L, Shames DS, Yu W, et al. Onartuzumab plus erlotinib versus erlotinib in previously treated stage IIIb or IV NSCLC: results from the pivotal phase III randomized, multicenter, placebo-controlled METLung (OAM4971g) global trial. J Clin Oncol 2014;32. abstr 8000.
[232] Scagliotti GV, Di Maio M. Tivantinib added to Erlotinib in non-small cell lung cancer: the primary endpoint was not MET…. Ann Oncol 2015;75:1–7.
[233] De Bacco F, Luraghi P, Medico E, Reato G, Girolami F, Perera T, et al. Induction of MET by ionizing radiation and its role in radioresistance and invasive growth of cancer. J Natl Cancer Inst 2011;103:645–61.
[234] Thiery JP, Sleeman JP. Complex networks orchestrate epithelial-mesenchymal transitions. Nat Rev 2006;7:131–42.
[235] Medová M, Aebersold DM, Zimmer Y. MET inhibition in tumor cells by PHA665752 impairs homologous recombination repair of DNA double strand breaks. Int J Cancer 2012;130:728–34.
[236] Shah AN, Summy JM, Zhang J, Park SI, Parikh NU, Gallick GE. Development and characterization of gemcitabine-resistant pancreatic tumor cells. Ann Surg Oncol 2007;14:3629–37.

[237] Tang MKS, Zhou HY, Yam JWP, Wong AST. c-Met overexpression contributes to the acquired apoptotic resistance of nonadherent ovarian cancer cells through a cross talk mediated by phosphatidylinositol 3-kinase and extracellular signal-regulated kinase 1/2. Neoplasia (New York, NY) 2010;12:128–38.
[238] Marchion DC, Bicaku E, Xiong Y, Zgheib NB, Al Sawah E, Stickles XB, et al. A novel c-Met inhibitor, MK8033, synergizes with carboplatin plus paclitaxel to inhibit ovarian cancer cell growth. Oncol Rep 2013;29:2011–8.
[239] Yashiro M, Nishii T, Hasegawa T, Matsuzaki T, Morisaki T, Fukuoka T, et al. A c-Met inhibitor increases the chemosensitivity of cancer stem cells to the irinotecan in gastric carcinoma. Br J Cancer 2013;1:1–10.
[240] Van Der Steen N, Deben C, Deschoolmeester V, Wouters A, Lardon F, Rolfo C, et al. Better to be alone than in bad company: the antagonistic effect of cisplatin and crizotinib combination therapy in non-small cell lung cancer. World J Clin Oncol 2016;7:114–22.
[241] Chen JT, Huang CY, Chiang YY, Chen WH, Chiou SH, Chen CY, et al. HGF increases cisplatin resistance via down-regulation of AIF in lung cancer cells. Am J Respir Cell Mol Biol 2008;38:559–65.
[242] Miyata Y, Asai A, Mitsunari K, Matsuo T, Ohba K, Mochizuki Y, et al. Met in urological cancers. Cancers 2014;6:2387–403.
[243] Katayama R, Kobayashi Y, Friboulet L, Lockerman EL, Koike S, Shaw AT, et al. Cabozantinib overcomes crizotinib resistance in ROS1 fusion-positive cancer. Clin Cancer Res 2015;21:166–74.
[244] Delahunt B, Eble J. Papillary renal cell carcinoma: a clinicopathologic and immunohistochemical study of 105 tumors. Mod Pathol 1997;10:537–44.
[245] Gnarra J, Tory K, Weng Y, Schmidt L, Wei M, Li H, et al. Mutations of the VHL tumour suppressor gene in renal carcinoma. Nat Genet 1994;7:85–90.
[246] Pennacchietti S, Michieli P, Galluzzo M, Mazzone M, Giordano S, Comoglio PM. Hypoxia promotes invasive growth by transcriptional activation of the met protooncogene. Cancer Cell 2003;3:347–61.
[247] Ciamporcero E, Miles KM, Adelaiye R, Ramakrishnan S, Shen L, Ku S, et al. Combination strategy targeting VEGF and HGF/c-met in human renal cell carcinoma models. Mol Cancer Ther 2015;14:101–10.
[248] Sierra JR, Tsao M-S. c-MET as a potential therapeutic target and biomarker in cancer. Ther Adv Med Oncol 2011;3:S21–35.
[249] Albiges L, Guegan J, Le Formal A, Verkarre V, Rioux-Leclercq N, Sibony M, et al. MET is a potential target across all papillary renal cell carcinomas: result from a large molecular study of pRCC with CGH array and matching gene expression array. Clin Cancer Res 2014;20:3411–21.
[250] Balint I, Szponar A, Jauch A, Kovacs G. Trisomy 7 and 17 mark papillary renal cell tumours irrespectively of variation of the phenotype. J Clin Pathol 2009;62:892–5.
[251] Giubellino A, Linehan W, Bottaro D. Targeting the Met signaling pathway in renal cancer. Expert Rev Anticancer Ther 2009;9:785–93.
[252] D'Amico L, Belisario D, Migliardi G, Grange C, Bussolati B, D'Amelio P, et al. C-met inhibition blocks bone metastasis development induced by renal cancer stem cells. Oncotarget 2014;7(29):45525–37.
[253] Ebos JML, Lee CR, Cruz-Munoz W, Bjarnason GA, Christensen JG, Kerbel RS. Accelerated metastasis after short-term treatment with a potent inhibitor of tumor angiogenesis. Cancer Cell 2009;15:232–9.
[254] Pàez-Ribes M, Allen E, Hudock J, Takeda T, Okuyama H, Viñals F, et al. Antiangiogenic therapy elicits malignant progression of tumors to increased local invasion and distant metastasis. Cancer Cell 2009;15:220–31.
[255] Choueiri TK, Vaishampayan U, Rosenberg JE, Logan TF, Harzstark AL, Bukowski RM, et al. Phase II and biomarker study of the dual MET/VEGFR2 inhibitor foretinib in patients with papillary renal cell carcinoma. J Clin Oncol 2013;31:181–6.
[256] Gavine PR, Ren Y, Han L, Lv J, Fan S, Zhang W, et al. Volitinib, a potent and highly selective c-Met inhibitor, effectively blocks c-Met signaling and growth in c-MET amplified gastric cancer patient-derived tumor xenograft models. Mol Oncol 2015;9:323–33.
[257] Schuller AG, Barry ER, Jones RDO, Henry RE, Frigault MM, Beran G, et al. The MET inhibitor AZD6094 (Savolitinib HMPL-504) induces regression in papillary renal cell carcinoma patient-derived xenograft models. Clin Cancer Res 2015;21:2811–9.
[258] Ishikawa D, Takeuchi S, Nakagawa T, Sano T, Nakade J, Nanjo S, et al. mTOR inhibitors control the growth of EGFR mutant lung cancer even after acquiring resistance by HGF. PLoS ONE 2013;8(5):e62104.
[259] Wagner AJ, Goldberg JM, Dubois SG, Choy E, Lee R, Pappo A, et al. Tivantinib (ARQ 197), a selective inhibitor of MET, in patients with microphthalmia transcription factor-associated tumors: results of a multicenter phase 2 trial. Cancer 2012;118:5894–902.

[260] Mohammed S, Van Buren G, Fisher WE. Pancreatic cancer: advances in treatment. World J Gastroenterol 2014;20:9354–60.

[261] Garajová I, Le Large TY, Frampton AE, Rolfo C, Voortman J, Giovannetti E. Molecular mechanisms underlying the role of microRNAs in the chemoresistance of pancreatic cancer. BioMed Res Int 2014;2014. 678401.

[262] Caparello C, Meijer LL, Garajova I, Falcone A, Le Large TY, Funel N, et al. FOLFIRINOX and translational studies: towards personalized therapy in pancreatic cancer. World J Gastroenterol 2016;22:6987–7005.

[263] Patel M, Pothula S, Xu Z, Lee A, Goldstein D, Pirola R, et al. The role of the hepatocyte growth factor/c-MET pathway in pancreatic stellate cell–endothelial cell interactions: antiangiogenic implications in pancreatic cancer. Carcinogenesis 2014;35:1891–900.

[264] Neuzillet C, Couvelard A, Tijeras-Raballand A, de Mestier L, de Gramont A, Bédossa P, et al. High c-Met expression in stage I–II pancreatic adenocarcinoma: proposal for an immunostaining scoring method and correlation with poor prognosis. Histopathology 2015;67:664–76.

[265] Zhang T, Boominathan R, Foulk B, Rao C, Kemeny G, Strickler JH, et al. Development of a novel c-MET-based CTC detection platform. Mol Cancer Res 2016;14:539–47.

[266] Xin X, Yang S, Ingle G, Zlot C, Rangell L, Kowalski J, et al. Hepatocyte growth factor enhances vascular endothelial growth factor-induced angiogenesis *in vitro* and *in vivo*. Am J Pathol 2001;158:1111–20.

[267] Neesse A, Michl P, Frese K, Feig C, Cook N, Jacobetz M, et al. Stromal biology and therapy in pancreatic cancer. Gut 2011;60:861–8.

[268] Pothula SP, Xu Z, Goldstein D, Biankin AV, Pirola RC, Wilson JS, et al. Hepatocyte growth factor inhibition: a novel therapeutic approach in pancreatic cancer. Br J Cancer 2016;114:269–80.

[269] Li C, Wu J, Hynes M, Dosch J, Sarkar B, Welling TH, et al. C-Met is a marker of pancreatic cancer stem cells and therapeutic target. Gastroenterology 2011;141:2218–27.

[270] Avan A, Quint K, Nicolini F, Funel N, Frampton A, Maftouh M, et al. Enhancement of the antiproliferative activity of gemcitabine by modulation of c-Met pathway in pancreatic cancer. Curr Pharm Des 2013;19:940–50.

[271] Avan A, Caretti V, Funel N, Galvani E, Maftouh M, Honeywell RJ, et al. Crizotinib inhibits metabolic inactivation of gemcitabine in c-Met-driven pancreatic carcinoma. Cancer Res 2013;73:6745–56.

[272] Hage C, Rausch V, Giese N, Giese T, Schönsiegel F, Labsch S, et al. The novel c-Met inhibitor cabozantinib overcomes gemcitabine resistance and stem cell signaling in pancreatic cancer. Cell Death Dis 2013;4:e627.

[273] Brandes F, Schmidt K, Wagner C, Redekopf J, Schlitt HJ, Geissler EK, et al. Targeting cMET with INC280 impairs tumour growth and improves efficacy of gemcitabine in a pancreatic cancer model. BMC Cancer 2015;15:1–14.

[274] Church K, Vanderwerff B, Riggers R, McMicheal M, Mateo-Victoriano B, Sukumar S, et al. Analogs of the hepatocyte growth factor and macrophage-stimulating protein hinge regions act as Met and Ron dual inhibitors in pancreatic cancer cells. Anticancer Drugs 2016;27:766–79.

[275] Renzo MFDi, Olivero M, Giacomini A, Renzo D. Overexpression and amplification of the met/HGF receptor gene during the progression of colorectal cancer. Clin Cancer Res 1995;1(2):147–54.

[276] Liu Y, Yu X-F, Zou J, Luo Z-H. Prognostic value of c-Met in colorectal cancer: a meta-analysis. World J Gastroenterol 2015;21:3706–10.

[277] Bardelli A, Corso S, Bertotti A, Hobor S, Valtorta E, Siravegna G, et al. Amplification of the MET receptor drives resistance to anti-EGFR therapies in colorectal cancer. Cancer Discov 2013;3:658–73.

[278] Wen J, Matsumoto K, Taniura N, Tomioka D, Nakamura T. Hepatic gene expression of NK4, an HGF-antagonist/angiogenesis inhibitor, suppresses liver metastasis and invasive growth of colon cancer in mice. Cancer Gene Ther 2004;11:419–30.

[279] Van Cutsem E, Eng C, Nowara E, Swieboda-Sadlej A, Tebbutt NC, Mitchell E, et al. Randomized phase Ib/II trial of rilotumumab or ganitumab with panitumumab versus panitumumab alone in patients with wild-type KRAS metastatic colorectal cancer. Clin Cancer Res 2014;20:4240–50.

[280] Amgen A phase 3, multicenter, randomized, double-blind, placebo controlled study of rilotumumab (AMG102) with epirubicin, cisplatin, and capecitabine (ECX) as first-line therapy in advanced MET-positive gastric or gastroesophageal junction adenocarcinoma, 2014. http://clinicaltrials.gov/show/NCT01697072.

[281] Rosen LS, Senzer N, Mekhail T, Ganapathi R, Chai F, Savage RE, et al. A phase I dose–escalation study of tivantinib (ARQ 197) in adult patients with metastatic solid tumors. Clin Cancer Res 2011;17:7754–64.

[282] Schöffski P, Garcia JA, Stadler WM, Gil T, Jonasch E, Tagawa ST, et al. A phase II study of the efficacy and safety of AMG 102 in patients with metastatic renal cell carcinoma. BJU Int 2011;108:679–86.

[283] Tan E, Park K, Lim W, Ahn M, Ng Q, Ahn J, et al. Phase Ib study of ficlatuzumab (formerly AV-299), an anti-hepatocyte growth factor (HGF) monoclonal antibody (MAb) in combination with gefitinib (G) in Asian patients (pts) with NSCLC. J Clin Oncol 2011;29(15):7571.

[284] Wen PY, Schiff D, Cloughesy TF, Raizer JJ, Laterra J, Smitt M, et al. A phase II study evaluating the efficacy and safety of AMG 102 (rilotumumab) in patients with recurrent glioblastoma. Neuro Oncol 2011;13:437–46.

[285] Penuel E, Li C, Parab V, Burton L, Cowan KJ, Merchant M, et al. HGF as a circulating biomarker of onartuzumab treatment in patients with advanced solid tumors. Mol Cancer Ther 2013;12:1122–30.

[286] Ryan CJ, Rosenthal M, Ng S, Alumkal J, Picus J, Gravis G, et al. Targeted MET inhibition in castration-resistant prostate cancer: a randomized phase II study and biomarker analysis with rilotumumab plus mitoxantrone and prednisone. Clin Cancer Res 2013;19:215–24.

[287] Iveson T, Donehower R, Davidenko I, Tjulandin S, Deptala A, Harrison M, et al. Rilotumumab in combination with epirubicin, cisplatin, and capecitabine as first-line treatment for gastric or oesophagogastric junction adenocarcinoma: an open-label, dose de-escalation phase 1b study and a double-blind, randomised phase 2 study. Lancet Oncol 2014;2(15):1007–18.

[288] Cunningham D, Tebbutt NC, Davidenko I, Murad AM, Al-Batran S-E, Ilson DH, et al. Phase III, randomized, double-blind, multicenter, placebo (P)-controlled trial of rilotumumab (R) plus epirubicin, cisplatin and capecitabine (ECX) as first-line therapy in patients (pts) with advanced MET-positive (pos) gastric or gastroesophageal junctiion (G/GEJ) cancer: RILOMET-1 study. J Clin Oncol 2015;33(15). 4000.

[289] Rodig S, Sequist LV, Schiller JH, Chen Y, Halim A, Waghorne C, et al. An exploratory biomarker analysis evaluating the effect of the c-MET inhibitor tivantinib (ARQ 197) and erlotinib in NSCLC patients in a randomized, double-blinded phase 2 study. Proc Am Assoc Cancer Res, 72. p. 1729.

[290] Lennerz JK, Kwak EL, Ackerman A, Michael M, Fox SB, Bergethon K, et al. MET amplification identifies a small and aggressive subgroup of esophagogastric adenocarcinoma with evidence of responsiveness to crizotinib. J Clin Oncol 2011;29:4803–10.

[291] Eder JP, Shapiro GI, Appleman LJ, Zhu AX, Miles D, Keer H, et al. A phase I study of foretinib, a multi-targeted inhibitor of c-Met and vascular endothelial growth factor receptor 2. Clin Cancer Res 2010;16:3507–16.

[292] Klotz M, Schmid E, Steiner-Hahn K, Rose T, Laube J, Roese L, et al. Preclinical evaluation of biomarkers for response monitoring to the MET inhibitor BAY-853474. Biomarkers 2012;17:325–35.

[293] Feldman D, Einhorn L, Quinn D, Loriot Y, Joffe J, Vaughn D, et al. A phase 2 multicenter study of tivantinib (ARQ 197) monotherapy in patients with relapsed or refractory germ cell tumors. Invest New Drugs 2013;31:1016–22.

CHAPTER

3

Roles for AXL and MERTK in Resistance to Cytotoxic and Targeted Therapies

*Ryan J. Summers**,**, *Katherine A. Minson**,**, *Deborah DeRyckere**,**,†, *Douglas K. Graham**,**,†

*Emory University Department of Pediatrics, Atlanta, GA, United States

**Aflac Cancer and Blood Disorders Center, Children's Healthcare of Atlanta, Atlanta, GA, United States

†Winship Cancer Institute, Emory University, Atlanta, GA, United States

Abstract

AXL and MERTK are members of the TAM family (TYRO3, AXL, and MERTK) of receptor tyrosine kinases (RTKs). Ligand binding and subsequent activation of AXL and MERTK in human cancers lead to downstream signaling via pathways that mediate tumor cell survival and proliferation. TAM RTK signaling mediates resistance to conventional cytotoxic chemotherapies across a wide variety of both hematopoietic and nonhematopoietic malignancies. Similarly, TAM RTKs are upregulated and mediate bypass signaling in tumor cells with acquired resistance to a variety of different targeted kinase inhibitors. In addition, TAM RTKs can mediate therapeutic resistance through their physiological anti-inflammatory role in the immune system. These roles for AXL and MERTK in therapeutic resistance implicate them as attractive therapeutic targets, both in combination with other agents and in the context of acquired resistance. Accordingly, a number of TAM RTK-targeted agents are currently in development and several have progressed to clinical application, although challenges such as identification of appropriate biomarkers of TAM RTK inhibition remain.

ABBREVIATIONS

ADC	Antibody-drug conjugate
ALL	Acute lymphoblastic leukemia
AML	Acute myeloid leukemia
BAD	BCL-2-associated agonist of cell death
BAK	BCL-2 homologous killer
BAX	BCL-2 associated X

Targeting Cell Survival Pathways to Enhance Response to Chemotherapy. http://dx.doi.org/10.1016/B978-0-12-813753-6.00003-2
Copyright © 2018 Elsevier Inc. All rights reserved.

BCL-2	B-cell lymphoma 2
BCL-XL	B-cell lymphoma-extra large
CLL	Chronic myeloid leukemia
CML	ezrin, radixin, moesin protein family
EGFR	Epidermal growth factor receptor
EMT	Epithelial-to-mesenchymal transition
ERK	Extracellular signal-related kinase
FLT3-ITD	fms-like tyrosine kinase 3-internal tandem duplication
GAS6	Growth arrest-specific protein 6
GBM	Glioblastoma multiforme
IFN	Interferon; IGF-1R, Insulin-like growth factor 1 receptor
JAK	Janus kinase
LXR	Liver X receptor
MCL1	Myeloid cell leukemia 1
mTOR	mechanistic target of rapamycin
NF-kB	Nuclear factor kappa-light-chain-enhancer of activated B cells
NSCLC	Non-small cell lung cancer; PD1, Programmed cell death protein 1
PKC	Protein kinase C
PI3K	Phosphoinositide-3-kinase
PUMA	p53 upregulated modulator of apoptosis
RTK	Receptor tyrosine kinase
shRNA	Short hairpin RNA
siRNA	Small interfering RNA
SOCS	Suppressor of cytokine signaling
STAT	Signal transducer and activator of transcription
VEGFR	Vascular endothelial growth factor receptor

3.1 INTRODUCTION

Receptor tyrosine kinases (RTKs) encompass a wide range of proteins involved in cellular signaling. Many RTKs have been implicated in human malignancies and some have been identified as oncogenic drivers. The TAM family of RTKs is composed of three transmembrane receptors—TYRO3, AXL, and MERTK [1–3]. These receptors have a similar overall structure consisting of an extracellular domain comprised of two immunoglobulin-like domains and two fibronectin type III domains, a transmembrane domain, and an intracellular kinase domain with a conserved KWIAIES sequence, which is unique to the TAM family [4]. The TAM receptors are ectopically or aberrantly expressed and contribute to oncogenesis in a wide range of human malignancies [3,5–7]. Among the TAM receptors, TYRO3 has been less studied compared to AXL and MERTK and therefore this chapter will focus on the latter two members of the TAM family. Although AXL and MERTK activate pathways that can mediate cell proliferation, such as the protein kinase B (AKT), extracellular signal-related kinase (ERK), and members of the signal transducer and activator of transcription (STAT) family pathways, their activation preferentially promotes tumor cell survival, rather than proliferation [8–11]. As such, the TAM RTKs have been the focus of intense study as potential therapeutic targets in a variety of malignancies and the subject of numerous recent reviews [4,5,12–18].

Activation of the TAM receptors in both physiologic and pathologic states is triggered by ligand binding followed by dimerization and auto-phosphorylation. Maximal TAM stimulation requires binding to phosphatidylserine (PtdSer) in a complex with a protein ligand [4]. The TAM ligands are γ-carboxylated proteins that bind to the receptors via their carboxy-terminal domain and to PtdSer via their amino terminus [19–22]. The first identified TAM ligand was the vitamin K-dependent growth arrest-specific protein 6 (GAS6), which binds to all three TAM RTKs [22], although AXL has the highest affinity for GAS6. There is evidence to suggest that AXL binds GAS6 constitutively *in vivo* but is not activated until PtdSer-mediated dimerization occurs [23]. In contrast, vitamin K-dependent protein S (PROS1) only binds MERTK and TYRO3 [23,24]. In addition, recent data demonstrate downregulation of AXL mRNA and protein in oral squamous cell carcinoma in response to PROS1 knockdown [25]. More recently, Tubby, tubby-like protein 1, and galectin-3 have been identified as MERTK ligands that regulate phagocytosis [26–28]. Although PtdSer is ubiquitously expressed, it is only available to activate the TAM receptors when externalized on apoptotic cell membranes, activated platelets, exosomes, and invading virus envelopes [29–34]. Ligand-independent activation of TAM kinases may also occur in pathologic states as there is evidence to suggest GAS6-independent activation of AXL in the setting of AXL overexpression in tumor cells. In this context, aggregation of AXL extracellular domains on adjacent cells [35] or ligand-independent homodimerization [36] can activate downstream signaling. Furthermore, AXL heterodimerizes with the non-TAM RTKs epidermal growth factor receptor (EGFR) and HER3 to initiate AXL-dependent signaling in breast cancer and glioblastoma cells, even in the absence of the dimerization partner ligand [37–39].

The best studied physiologic TAM RTK function is the role of MERTK in efferocytosis—the clearance of apoptotic material by monocyte- [29] and epithelial-derived cells [40]. TAM RTKs are also involved in regulation of the immune system with a net effect of protecting against a pro-inflammatory state. Their role in this context will be discussed later in the chapter. TAM RTKs also play a role in platelet aggregation by stimulating outside-in signaling via αIIbβ3 integrin [41]. These physiologic roles do not necessarily implicate TAM RTKs as oncogenic drivers. Nonetheless, TAM RTK signaling does promote survival, therapeutic resistance, motility, and invasion in neoplastic cells.

3.2 ROLES FOR TAM RTKs IN TUMORIGENESIS

3.2.1 TAM Receptor Expression and Signaling in Cancer

TAM receptors and their ligands are overexpressed in both hematologic and nonhematologic malignancies and their expression has been correlated with poor prognosis in a number of tumor types including glioblastoma multiforme (GBM), non-small cell lung cancer (NSCLC), oral squamous cell carcinoma, renal cell carcinoma, colon cancer, mesothelioma, and acute myeloid leukemia (AML) [4,42–49]. Interestingly, expression of AXL or GAS6 protein are both markers of poor prognosis in GBM and NSCLC [42,43], but decreased GAS6 mRNA expression has been correlated with poorer prognosis in NSCLC [43]. In one study, GAS6 mRNA levels had no prognostic significance in AML [45], but a larger study demonstrated GAS6 expression as an independent predictor of remission failure and shorter disease-free and overall survival [46]. Given that these two studies defined patients with GAS6 expressing AML differently (i.e., 90% [45] vs 26% [46]), further study is needed to draw firm conclusions on the prognostic role of GAS6 mRNA expression in AML.

Ectopic expression of MERTK was first noted in acute lymphoblastic leukemia (ALL) [1] and is present in 30–50% of pediatric B-cell [50] and T-cell [50,51] ALL patient samples. MERTK is also aberrantly expressed in 70%–90% of pediatric and adult AMLs [52]. Additionally, patients with multiple myeloma have an increased fraction of MERTK-expressing bone marrow plasma cells compared to healthy controls [53]. AXL expression has not been described in ALL, but overexpression is common in AML [45,54–56] and has also been demonstrated in chronic myeloid leukemia (CML) [54] and chronic lymphocytic leukemia (CLL) [57]. A complete list of solid tumors with TAM receptor overexpression can be found in the review by Graham et al. [4]. Most notably, MERTK overexpression has been demonstrated in lung cancer [58], glioma [59,60], and melanoma [61–64], and AXL overexpression has been demonstrated in lung cancer [58,65,66], glioblastoma [42,59,67], breast carcinoma [68–70], ovarian carcinoma [71–73], and melanoma [61,62], among others. It is unknown that what mechanisms underlie aberrant expression of TAM receptors across such a diverse range of tumor types, but it is possible that they reflect the physiologic mechanisms that regulate TAM expression in normal tissues. For example, *MERTK* transcription in macrophages is stimulated by liver X receptor (LXR) and LXR ligands such as 27-hydroxycholesterol, which is produced by both tumor cells and cells in the tumor microenvironment [74,75].

The prevalence of TAM receptor overexpression across such a wide range of malignancies implicates a critical oncogenic role for TAM receptor signaling. While expression of activated oncogenes and loss of tumor suppressor genes are critical to promote unregulated growth of some tumor cells, neoplastic cells can also be dependent on the function of endogenous, nonmutated genes for survival under stress (i.e., nononcogene addiction). For example, MERTK inhibition using short hairpin RNA (shRNA) promotes apoptosis in tumor cell cultures following serum starvation [52,58]. In an *in vivo* setting, limited oxygen and nutrients in the tumor microenvironment may also promote signaling through TAM-dependent survival pathways. As an example, GAS6 and PtdSer on exosomes from osteoblasts and bone marrow stromal cells provide ligands for TAM kinase activation in cancer cells in the bone marrow niche [76,77]. Together these data implicate MERTK signaling as a critical mediator of tumor cell survival.

Activation of the TAM receptors stimulates signaling through numerous downstream survival pathways including mitogen-activated protein kinase kinase (MEK)/ERK, phosphoinositide-3-kinase (PI3K)/AKT, p38, RHO family proteins, nuclear factor kappa-light-chain-enhancer of activated B cells (NF-κB), Janus kinase (JAK)-STAT, and SRC family kinases (Fig. 3.1 and reviewed in [5]). Stimulation with GAS6 activates PI3K–AKT [50–52,58,64,78–80] and RAS–RAF–MEK [7,9,50,51,58,59,78–80] signaling in multiple cancers, and TAM receptor inhibition through both pharmacologic and genetic means can abrogate these signals. Signaling through MERTK also regulates downstream phosphorylation and transcriptional activity of STAT3, STAT5, and STAT6, and AXL similarly activates STAT3 in multiple tumor types [48,51,52,64,81]. In the case of TAM-mediated NF-κB signaling, there is a clear distinction between the effects in tumor cells vs normal tissues. In tumor cells, MERTK activates the NF-κB pathway [82–87] and potentially its apoptotic mechanisms. In other contexts, as part of their role in suppressing the innate immune inflammatory response, AXL and MERTK can inhibit NF-κB signaling, thereby decreasing inflammatory cytokine production [88–90]. This may represent a mechanism by which TAM signaling is co-opted in a context-dependent manner to enhance tumor cell survival. Inhibition of

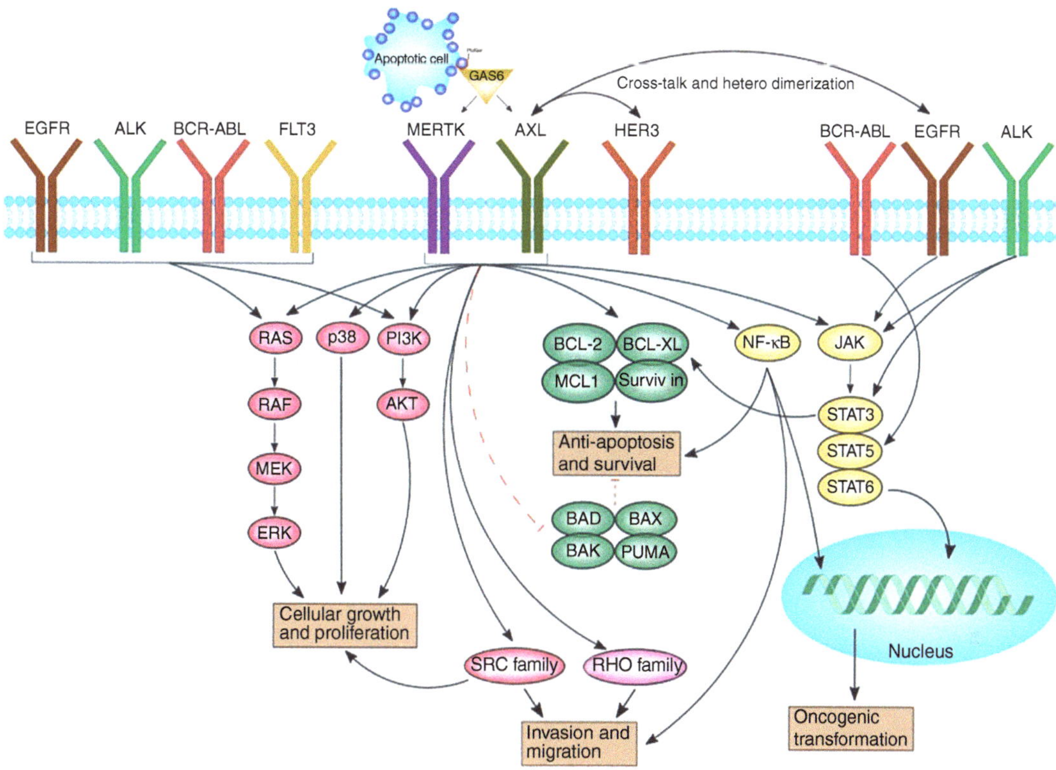

FIG. 3.1 AXL and MERTK activate oncogenic signaling to mediate therapeutic resistance. Activation of TAM RTKs by GAS6 complexed with PtdSer mediates downstream signal transduction through oncogenic pathways to promote tumor cell survival, proliferation, migration, and invasion. These effects are context dependent and vary based on the TAM RTK, ligand, and cell type involved. Oncogenic driver RTKs such as EGFR, ALK, FLT3, and BCR-ABL utilize many of the same downstream signaling pathways and inhibition of these RTKs by targeted therapies can be overcome by compensatory TAM signaling, leading to therapeutic resistance.

AXL or MERTK in tumor cells also regulates expression of numerous apoptotic proteins, including the antiapoptotic proteins B-cell lymphoma 2 (BCL-2), B-cell lymphoma-extra large (BCL-XL), myeloid cell leukemia 1 (MCL1) and survivin [45,57,58] and the pro-apoptotic proteins BCL-2-associated agonist of cell death (BAD), BCL-2 associated X (BAX), BCL-2 homologous killer (BAK), and p53 upregulated modulator of apoptosis (PUMA) [45,50,91]. TAM RTKs modulate these signaling pathways and apoptotic regulators to promote tumor cell survival in response to apoptotic stimuli.

3.2.2 Role of TAM RTKs in Mediating the Epithelial-to-Mesenchymal Transition

The epithelial-to-mesenchymal transition (EMT) is a reversible conversion of epithelial cells to a mesenchymal phenotype and plays an important physiologic role in embryonic development and wound healing. In the context of cancer, EMT is associated with increased cell migration and invasion and ultimately with development of metastatic disease [92]. In cells undergoing EMT, typical epithelial markers such as E-cadherin are suppressed, whereas mesenchymal markers such as N-cadherin and fibronectin are induced. The transcription factors SNAIL1/2, TWIST1/2, and ZEB1/2 are well described as inducers of EMT [16]. Cells that have undergone EMT often exhibit properties characteristic of stem cells, such as decreased proliferation, and, as a result, are more resistant to antiproliferative therapy [93]. As such, intrinsically mesenchymal cells or cells that have undergone EMT are more resistant to both chemotherapy and targeted therapies [94].

AXL, in particular, has been implicated in EMT in a number of studies [94–96] and is one of the primary genes overexpressed in EMT-specific gene expression signatures [97–100]. It remains unclear as to whether AXL expression induces EMT or EMT induces AXL expression. The role of AXL in EMT likely has some relationship to therapeutic resistance, as AXL is upregulated in mesenchymal NSCLC cell lines, which are highly resistant to erlotinib compared to epithelial NSCLC cell lines [98]. Additionally, inhibition of AXL can reverse both the EMT phenotype in cancer stem cells [86] and EMT-associated resistance to EGFR inhibitors [98]. It is likely that the relationship between AXL and EMT is context-dependent, but the available data support a direct role for AXL in maintaining drug resistance and not simply as a biomarker of EMT.

3.2.3 Role of TAM RTKs in Angiogenesis

The TAM receptors are normally expressed in epithelial cells and in this context they function to mediate vascular integrity and promote angiogenesis. In the setting of vascular injury, TAM receptor signaling promotes platelet aggregation and restoration of the endothelial barrier [15,101–103]. In addition, AXL knockdown via shRNA impaired blood vessel formation and function in a mouse model of human angiogenesis [104]. Given this role for TAM RTK signaling, activation of the TAM pathways has been proposed as a mechanism for resistance to therapies targeting the vascular endothelial growth factor receptor (VEGFR) [102,103,105]. Indeed, a recent phase III clinical trial of cabozantinib, an MET, VEGFR, and AXL inhibitor, demonstrated superior outcomes for patients with renal cell carcinoma previously treated with VEGFR inhibitors, supporting a role for AXL in mediating resistance to anti-angiogenic therapies [106].

3.3 RESISTANCE TO THERAPY MEDIATED BY TAM RTKs

Signaling via TAM receptors has been linked to chemotherapy resistance via numerous mechanisms which are discussed in more detail below. TAM RTKs modulate downstream signaling through pathways important in the regulation of cell survival, proliferation, migration, and invasion. It is through these mechanisms that they are thought to promote resistance to traditional cytotoxic chemotherapy. Additionally, TAM RTKs mediate resistance to targeted therapies by allowing tumor cells to bypass the signaling blockade introduced by the targeted therapy (Fig. 3.1).

3.3.1 Role of TAM RTKs in Mediating Resistance to Traditional Chemotherapy

As has been noted, aberrant expression of TAM RTKs occurs in a wide range of both hematologic and nonhematologic malignancies. In many tumor types, higher TAM expression is associated with resistance to cytotoxic chemotherapies. For example, AXL mRNA is increased in cisplatin-resistant ovarian cancer cell lines [107] and AXL protein levels are increased in doxorubicin-resistant breast cancer and CML cell lines compared to nonresistant cell lines [108]. Similarly, AXL expression was upregulated in doxorubicin- and etoposide-resistant NSCLC cell-line derivatives, and ectopic expression of AXL in the parental cell line increased resistance to doxorubicin [83]. Similarly, introduction of a plasmid conferring AXL overexpression doubled the IC_{50} in an esophageal adenocarcinoma cell line treated with cisplatin relative to parental cell lines [109]. In this context, resistance to cisplatin was associated with decreased phosphorylation of c-ABL and p73β, a member of the p53 family. AXL and c-ABL also coprecipitated, suggesting a physical interaction that prevents shuttling of c-ABL to the nucleus where it usually binds and stabilizes p73 protein. Across a panel of cell lines derived from multiple tumor types (pancreatic, breast, prostate, colon, nasopharyngeal, renal cell carcinoma, melanoma, neural cancer, and leukemia), increased AXL protein correlated with increased resistance to doxorubicin, although interpretation of these data is complicated by inclusion of both liquid and solid tumor cell lines in the analysis [83]. AXL was upregulated in AML patient samples collected from patients who failed treatment with doxorubicin and cytarabine relative to diagnostic samples from the same patients, and AXL expression was induced in AML cell lines in the presence of chemotherapy agents in a manner dependent on methylation status of the AXL promoter [79]. Similarly, lymphocytes with ectopic MERTK expression isolated from MERTK transgenic mice were more resistant to dexamethasone than wild-type lymphocytes [7] and stimulation of a B-ALL cell line with GAS6 increased resistance to cytarabine-induced apoptosis by inducing cell-cycle arrest [77]. In an AML cell line, GAS6 stimulation conferred resistance to doxorubicin, etoposide, and cisplatin, and this effect was reversed by addition of a ligand trap to deplete GAS6. In this case, resistance correlated with increased AXL signaling as indicated by increased levels of phosphorylated AXL (pAXL) and concomitant increases in pAKT and pERK. MERTK activation was not examined. Additionally, the pro-survival proteins BCL-2, BCL-XL, and Twist were upregulated in this model [79].

Conversely, inhibition of TAM RTK signaling increases the sensitivity of tumor cells to a variety of chemotherapy agents. A large number of studies have demonstrated increased sensitivity to DNA damaging therapeutic agents in tumor cell lines in response to TAM RTK inhibition. Knockdown of AXL using small interfering RNA (siRNA) increased sensitivity to doxorubicin, vincristine, and paclitaxel in doxorubicin-resistant breast cancer and CML cell lines and abrogated cell migration and invasion induced by treatment with 5-fluorouracil in colon cancer cell lines [48,108]. MERTK knockdown with shRNA also sensitized B-ALL cell lines to dexamethasone, methotrexate, vincristine, and L-asparaginase, resulting in synergistic induction of cell death and increased apoptosis [50]. In T-ALL cell lines, shRNA-mediated MERTK knockdown sensitized cells to cytarabine, etoposide, 6-mercaptopurine, and methotrexate and increased apoptosis in response to treatment with cytarabine, 6-mercaptopurine, and methotrexate [51]. Inhibition of MERTK and AXL increased the sensitivity of astrocytoma cell lines to temozolomide, carboplatin, and vincristine [59] and

MERTK knockdown sensitized GBM cells to etoposide-induced apoptosis [60]. In addition, expression of MERTK was induced in GBM cell lines upon exposure to lomustine, cisplatin, temozolomide, etoposide, or irradiation [60,110]. In NSCLC, shRNA-mediated knockdown of AXL or MERTK increased sensitivity to carboplatin and cisplatin (AXL and MERTK) and etoposide and doxorubicin (AXL only) [58]. In hepatocellular carcinoma and breast cancer models, microRNA-mediated AXL inhibition increased sensitivity to cisplatin and paclitaxel, respectively [111,112]. Finally, AXL inhibition in pancreatic ductal adenocarcinoma enhanced radiation-induced apoptosis [113]. An additional study demonstrated increased sensitivity to methotrexate in an orthotopic B-cell ALL model in response to treatment with a MERTK tyrosine kinase inhibitor (TKI), leading to decreased tumor burden and increased survival in mice treated with combination therapy compared to mice treated with either single agent [114]. Decreased tumor growth was also observed in response to siRNA-mediated AXL inhibition in subcutaneous breast cancer and CML xenograft models generated from doxorubicin-resistant derivative cell lines, although in this case the decrease was independent of treatment with doxorubicin [108].

One major mechanism by which TAM RTK signaling mediates resistance to traditional cytotoxic chemotherapy is by downstream activation of pro-survival pathways. As described above, inhibition of MERTK or AXL signaling sensitizes cancer cells to treatment with chemotherapy in a variety of tumor types, both *in vitro* and *in vivo* (Table 3.1). These data implicate TAM RTKs as attractive targets for therapeutic inhibition in conjunction with cytotoxic agents that are currently the standard of care in many cancers.

3.3.2 Role of TAM RTKs in Mediating Resistance to Targeted Therapies

In addition to their roles in resistance to cytotoxic chemotherapy, TAM RTKs are important in mediating resistance to targeted kinase inhibitor therapies in a number of different tumor types (Table 3.2). Information regarding this aspect of TAM RTK biology is primarily related to AXL and was first demonstrated in response to treatment with BCR-ABL inhibitors. Several studies demonstrated AXL upregulation in CML and gastrointestinal stromal tumor (GIST) cell-line derivatives selected for resistance to imatinib [115], nilotinib [116], or PD-166326 [115] and in tumor samples collected from patients with imatinib-resistant GISTs [117] or CML [116]. In GISTs, both AXL and GAS6 overexpression have been associated with imatinib resistance. Knockdown of AXL with siRNA restored imatinib sensitivity, and transfection of AXL into an imatinib-sensitive CML cell line conferred protection against the antileukemic effects of imatinib, such as alteration of cell metabolism and induction of apoptosis [115]. Molecular modeling indicated that imatinib does not bind to AXL, suggesting that resistance is not mediated by off-target competition for inhibitor binding [117]. Further supporting this hypothesis, several studies have directly demonstrated weak or no binding of AXL and imatinib [118–120]. The mechanism of AXL upregulation in response to BCR-ABL inhibition in CML may be mediated through CBL, an E3 ubiquitin ligase that stabilizes AXL protein and mRNA [116]. Depletion of CBL in nilotinib-sensitive cells increased AXL mRNA and protein; conversely, forced expression of CBL in nilotinib-resistant cells decreased AXL expression and resensitized the cells to nilotinib. Other studies examining AXL in CML have implicated PKCα and β as regulators of AXL expression based on the observation that inhibition of PKC in imatinib-resistant cells via a small-molecule inhibitor or siRNA silencing leads

TABLE 3.1 TAM Receptor-Mediated Resistance to Conventional Therapies

TAM receptor	Cancer type	Therapy	References
AXL	AML	Cisplatin	[79]
		Cytarabine	[79]
		Doxorubicin	[79]
		Etoposide	[79]
	Astrocytoma	Carboplatin	[59]
		Temozolomide	[59]
		Vincristine	[59]
	Breast	Doxorubicin	[83,108]
		Paclitaxel	[108,112]
		Vincristine	[108]
	CML	Doxorubicin	[108]
		Paclitaxel	[108]
		Vincristine	[108]
	Colon	5-FU	[48]
		Doxorubicin	[83]
	Esophageal	Cisplatin	[109]
	Hepatocellular carcinoma	Cisplatin	[111]
	Melanoma	Doxorubicin	[83]
	Neural	Doxorubicin	[83]
	Nasopharyngeal	Doxorubicin	[83]
	NSCLC	Carboplatin	[152]
		Cisplatin	[58]
		Doxorubicin	[58,83]
		Etoposide	[58,83]
		Paclitaxel	[152]
	Ovarian	Cisplatin	[107]
	Pancreatic	Doxorubicin	[83]
		Irradiation	[113]
	Prostate	Doxorubicin	[83]
	Renal cell carcinoma	Doxorubicin	[83]

(*Continued*)

TABLE 3.1 TAM Receptor-Mediated Resistance to Conventional Therapies (*cont.*)

TAM receptor	Cancer type	Therapy	References
MERTK	ALL	6-MP	[51]
		Cytarabine	[51,77]
		Dexamethasone	[7,50]
		Etoposide	[51]
		L-asparaginase	[50]
		Methotrexate	[50,51,114]
		Vincristine	[50]
	Astrocytoma	Carboplatin	[59]
		Temozolomide	[59]
		Vincristine	[59]
	GBM	Cisplatin	[110]
		Etoposide	[60]
		Irradiation	[60,110]
		Lomustine	[110]
		Temozolomide	[110]
	NSCLC	Carboplatin	[58]
		Cisplatin	[58]

Abbreviations: ALL, acute lymphoblastic leukemia; AML, acute myeloid leukemia; CML, chronic myeloid leukemia; NSCLC, non-small cell lung cancer; GBM, glioblastoma multiforme.

to decreased AXL protein levels [115]. Together, these data suggest that PKCα/β mediates increased transcription of AXL, and CBL stabilizes AXL mRNA and protein, leading to overall increased AXL expression and consequent increased downstream survival signaling as a mechanism of resistance to BCR-ABL targeted therapies in CML [12].

AXL also plays a role in resistance to EGFR-family inhibitors. AXL expression was upregulated in EGFR-mutated NSCLC cell-line derivatives and xenografts with acquired resistance to erlotinib, an EGFR inhibitor, and AXL overexpression was both necessary and sufficient to confer resistance [121]. AXL inhibition with shRNA restored erlotinib sensitivity *in vivo*, and treatment with AXL inhibitors restored sensitivity to erlotinib and mediated synergistic antitumor activity in combination with erlotinib treatment in cell-line models. Moreover, upregulation of AXL was also observed in 20% of tumor samples collected from patients with NSCLC after development of resistance to erlotinib compared to matched pretreatment samples, demonstrating the clinical relevance of AXL upregulation as a resistance mechanism. In erlotinib-resistant NSCLC cell lines, activation of downstream signaling through ERK, AKT, and RELA (a subunit of NF-κB) was EGFR independent, suggesting AXL-mediated bypass signaling as a mechanism of resistance [121]. In addition, AXL and EGFR co-immunoprecipitate, suggesting the potential for heterodimerization [37,122]. Overexpression of AXL was also demonstrated in NSCLC cell lines and patient-derived xenografts resistant to cetuximab, and in this context EGFR directly regulates *AXL* mRNA expression via mitogen-activated

TABLE 3.2 TAM Receptor-Mediated Resistance to Targeted Therapies

TAM receptor	Cancer type	Therapy	References
AXL	AML	Midostaurin (FLT3)	[126]
		Quizartinib (FLT3)	[126]
	Breast	Erlotinib (EGFR)	[37]
		Lapatanib (HER2)	[39]
		PD1 inhibition	[143,144]
	CML	Imatinib (BCR-ABL)	[115,116]
		Nilotinib (BCR-ABL)	[116]
		PD-166326 (BCR-ABL)	[115]
	Colon	PD1 inhibition	[145]
	GIST	Imatinib (BCR-ABL)	[115]
		Nilotinib (BCR-ABL)	[116]
		PD-166326 (BCR-ABL)	[115]
	HNSCC	PI3K/AKT inhibition	[132]
	Melanoma	BRAF/MEK inhibition	[97]
		MEK inhibition	[127,128]
		NRAS inhibition	[99]
		PD1 inhibition	[140]
	Neuroblastoma	ALK inhibition	[133,134]
	NSCLC	ALK inhibition	[131]
		Cetuximab (EGFR)	[123]
		Erlotinib (EGFR)	[121,152]
		FGFR inhibition	[130]
		VEGF inhibition	[152]
	Renal cell carcinoma	Sunitinib	[47]
	Rhabdomyosarcoma	IGF-1R inhibition	[129]
MERTK	Breast	PD1 inhibition	[143,144]
	Colon	PD1 inhibition	[145]
	NSCLC	Erlotinib (EGFR)	[124]

Abbreviations: AML, acute myeloid leukemia; CML, chronic myeloid leukemia; GIST, gastrointestinal stromal tumor; HNSCC, head and neck squamous cell carcinoma; NSCLC, non-small cell lung cancer.

protein kinase (MAPK) signaling and the transcription factor c-JUN [123]. Furthermore, cetuximab-sensitive parental cells were rendered resistant by overexpression of AXL. AXL expression has also been associated with relative resistance to EGFR inhibitors in triple-negative breast cancer cell lines [37,122] and is overexpressed in estrogen receptor-positive breast cancer cell lines that express HER-2, an EGFR-family member, but are resistant to lapatinib, an HER-2 inhibitor [39]. In this context, AXL and HER3 heterodimerized to bypass HER-2 signaling inhibition mediated by lapatinib and treatment with foretinib (a multikinase inhibitor targeting AXL, MET, and VEGFR) or AXL siRNA restored lapatinib sensitivity. Although roles for MERTK in resistance to EGFR-targeted therapy have not been thoroughly explored, overexpression of MERTK in an NSCLC cell line was sufficient to confer erlotinib resistance, and treatment with an MERTK kinase inhibitor reversed this phenotype [124]. Further, it was recently demonstrated that MERTK and EGFR co-immunoprecipitate in NSCLC cells, suggesting heterodimerization [125]. Additionally, the combination of small-molecule inhibitors targeting MERTK and EGFR synergized to inhibit cell expansion in NSCLC cell lines.

In AML with an fms-like tyrosine kinase 3-internal tandem duplication (FLT3-ITD) mutation, AXL expression was necessary for constitutive activation of FLT3 [56], and both total AXL and pAXL were increased in FLT3-ITD AML cell lines and patient samples resistant to the FLT3-targeted TKI midostaurin [126]. In addition, treatment with the FLT3-targeted TKIs quizartinib and midostaurin increased pAXL, consistent with a role for AXL in TKI resistance in this model. Further validating this hypothesis, inhibition of AXL with a small-molecule inhibitor, shRNA, or GAS6 ligand-trap resensitized resistant AML cell lines to FLT3-targeted TKIs.

Overexpression of AXL mRNA has also been described in BRAF- and NRAS-mutated melanoma cell lines and patient samples that are resistant to BRAF/MEK inhibitors [97,99]. In this context, two distinct populations of melanoma cells expressing high levels of either AXL or MITF were observed at baseline, but upon relapse after BRAF ± MEK inhibition tumor cells predominantly expressed high levels of AXL [97]. This observation suggests that AXL-expressing cells present from the outset may harbor intrinsic resistance leading to inevitable relapse. Additional studies revealed decreased extracellular shedding of numerous kinase receptors, including AXL, in response to treatment with an MEK inhibitor, leading to increased receptor expression on the surface of tumor cells and consequent increased downstream signaling [127]. This alternative mechanism to increase TAM RTK signaling may be particularly important in patients with melanoma, where progression-free survival following treatment with BRAF/MEK inhibitors is correlated with high serum levels of soluble TAM RTKs before treatment and reduced shedding of TYRO-3, AXL, and MERTK post-treatment [127,128].

Roles for AXL in resistance to other kinase inhibitors have also been described, but have been less well studied. In a derivative of the Rh41 rhabdomyosarcoma cell line selected for resistance to treatment with an anti-insulin-like growth factor 1 receptor (IGF-1R) monoclonal antibody, AXL mRNA and protein were overexpressed compared to parental cells [129]. Interestingly, AXL upregulation was not observed in Rh41 rhabdomyosarcoma cells selected for resistance to a TKI targeting IGF-1R, demonstrating differential mechanisms of resistance to therapeutic agents with different modes of action in the same cell line. Upregulation of AXL has also been observed in NSCLC cell lines with acquired resistance to fibroblast growth factor receptor (FGFR) [130] and anaplastic lymphoma kinase (ALK) [131] inhibitors and in head and neck and esophageal squamous cell carcinomas, where AXL forms a heterodimer with EGFR and induces PKC/mTOR pathway signaling to mediate resistance to PI3K/AKT pathway inhibition [132]. Ligand-dependent activation of AXL decreased sensitivity of ALK-mutated neuroblastoma cells to crizotinib, an ALK TKI, and AXL inhibition via siRNA restored the activity of crizotinib [133]. Additionally, AXL inhibition with a small-molecule inhibitor increased apoptosis in response to treatment with crizotinib in ALK-mutated neuroblastoma cell lines and enhanced the efficacy of ALK inhibition *in vivo*. In ALK-mutated neuroblastoma cells resistant to crizotinib, ligand-dependent AXL activation conferred resistance to the ALK inhibitor ceritinib through upregulation of ERK signaling [134]. Chronic treatment of renal cell carcinoma cell lines with the anti-angiogenic agent sunitinib activated AXL and increased cell invasion, and AXL inhibition with cabozantinib impaired metastatic behavior *in vitro* and rescued acquired sunitinib resistance *in vivo* [47].

3.3.3 Role of TAM RTKs in Immunosuppression

Another mechanism by which TAM RTKs promote therapeutic resistance in cancer is through modulation of the immune system. Tumors arise from "self" tissue and in order to mount an effective antitumor immune response tolerance in the adaptive immune system must be overcome. Immune checkpoint therapeutics represent one approach that has been used successfully, with a goal of reactivating $CD8^+$ cytotoxic T-cells to mediate tumor cell killing. The innate immune system, however, has a built-in "braking" system mediated by TAM RTKs that functions to reduce inflammation and resulting tissue damage in the context of physiologic tissue remodeling and wound healing [15]. AXL and MERTK bind interferon (IFN) α receptors to promote downstream signaling via STAT1 and activate transcription of suppressor of cytokine signaling (SOCS) proteins, which inhibit JAK [15], thereby mediating a negative feedback loop to dampen inflammation [16]. Furthermore, activation of TAM RTKs in macrophages leads to a switch from a pro-inflammatory M1 phenotype to an anti-inflammatory M2 phenotype [135]. M2 macrophages are unable to activate $CD8^+$ cytotoxic T-cells, and increased incidence of M2 macrophages in the tumor microenvironment correlates with poor prognosis in some tumor types [136]. In addition, MERTK signaling in macrophages promotes biosynthesis of immunosuppressive long-chain fatty acid-derived lipids called specialized pro-resolving mediators [137]. Thus, physiologic TAM RTK signaling can lead to blunt innate and adaptive immune responses to cancer. On the contrary, inhibition of TAM RTKs promotes an M1 phenotype in macrophages, which then secrete pro-inflammatory cytokines leading to increased recruitment and/or activation of T-cells in the tumor microenvironment and reduced tumor incidence and growth in syngeneic mouse models [135]. These data validate TAM RTKs as potential therapeutic targets in the tumor microenvironment.

In an effort to overcome adaptive immune tolerance in the tumor microenvironment, immune checkpoint therapeutics targeting programmed cell death protein 1 (PD1) or its ligand (PD-L1) have recently been developed and have shown promise in clinical trials (reviewed in [138]). The PD1 immune checkpoint pathway mediates suppression of cytotoxic T-cells, and PD-L1 and PD-L2 ligands can be expressed on tumor cells to inhibit antitumor immunity [139]. Interestingly, intrinsic resistance to PD1-targeted therapy is associated with increased AXL expression in metastatic melanoma [140]. In addition, MERTK expression in an epithelial cell line was sufficient to induce expression of the checkpoint pathway ligands PD-L1 and PD-L2, and shRNA-mediated inhibition of MERTK suppressed PD-L1 expression in a breast cancer cell line [141]. Similarly, in an NSCLC cell line, PtdSer binding to TAM RTKs mediated upregulation of PD-L1 expression and AKT-dependent therapeutic resistance [142]. Together, these data suggest that therapeutic strategies combining TAM RTK inhibition with immune checkpoint inhibitors may be particularly effective for induction of antitumor immunity. Indeed, several recently reported studies demonstrated synergistic inhibition of tumor growth in mouse models of colon and breast cancer upon dual inhibition of AXL and the PD1 or CTLA4 checkpoint pathways [143–145], lending support to this hypothesis.

3.4 TARGETING AXL AND MERTK

3.4.1 Agents and Approaches to Inhibit TAM RTKs

As the TAM RTKs represent attractive therapeutic targets across a wide range of malignancies, there has been significant effort to identify targeting methods with the potential for clinical translation. Several strategies have been used to target the TAM RTKs. These include disruption of ligand-receptor binding using antibodies or ligand traps, inhibition of kinase activity with ATP-competitive small-molecule inhibitors, and antibody-mediated downregulation of cell-surface TAM RTK expression. One group has used an aptamer approach to target AXL [146] and others have developed ligand traps [21,56,147–149] consisting of the TAM RTK extracellular domain fused to the Fc domain of human IgG, including an AXL decoy receptor that binds GAS6 with higher affinity than AXL itself and abrogates AXL signaling [150,151]. Monoclonal antibodies targeting the extracellular domains of TYRO3, AXL, and MERTK have all been developed. Antibody binding inhibits TAM RTK activity either through disruption of ligand binding [105,152] or through increased receptor internalization and subsequent degradation [153,154]. Several groups have used anti-AXL antibodies in preclinical studies in Kaposi sarcoma, NSCLC, and pancreatic cancer models [80,152,154]. RU-301 and RU-302 bind the extracellular Ig-1 ectodomain of AXL and inhibit binding of GAS6, and modeling suggests they will bind MERTK and TYRO3 as well [155]. Two anti-AXL antibodies [151,156] have been used to generate antibody–drug conjugates (ADCs) and have shown promising results in preclinical studies [156,157]. An antibody targeting MERTK cell-surface expression has also been described and mediates decreased migration in glioblastoma cells [158] and decreased colony-forming potential and therapeutic resistance in NSCLC cells [153]. The most common approach to TAM RTK inhibition, and the tool most likely to translate to clinical application, is the use of ATP-competitive small-molecule TKIs. The first TKIs with TAM RTK-inhibitory activity were multikinase inhibitors. Due to similar ATP binding sites between TAM RTKs and MET, many TKIs target both. Both TAM RTKs and MET play roles in therapeutic resistance, especially in NSCLC, so there may be some advantage to targeting both receptors [16]. The multikinase inhibitor sitravatinib has a low IC_{50} for both AXL and MERTK and was superior to imatinib and crizotinib in preclinical sarcoma models [159]. Clinical trials are ongoing to test the multikinase inhibitors cabozantinib [160], sitravatinib [161], and glesatinib [161] in NSCLC patients, including those with high expression of or genetic changes in AXL. Sitravatinib and glesatinib are being used in combination with the PD-1 inhibitor nivolumab with the goal of bypassing adaptive immune tolerance, as described above [161]. Sitravatinib is also being tested as a single agent in patients with advanced solid tumors demonstrating genetic changes in a number of RTKs, including AXL [162].

There are also several AXL or MERTK selective inhibitors in preclinical and clinical development (Table 3.3 and reviewed in [4,12,16]). The AXL inhibitor TP-0903 has been used in preclinical studies in CLL [163,164], AML [126], and SLC-0211, another AXL

TABLE 3.3 TAM-Selective Targeted Therapies Currently in Development

			Inhibitory activity (IC_{50})						
			AXL		MERTK				
Compound	Mechanism	Tumor types	Enzymatic	Cell-based	Enzymatic	Cell-based	Phase	Other kinase targets	References
CAB-AXL-ADC	ADC	Solid tumors	*	*	*	*	Preclinical	N/A	[157]
HuMax-AXL-ADC	ADC	Solid tumors	*	*	*	*	I	N/A	[156]
Aravive-S6 (MYD1–72)	Ligand trap	AML	*	*	*	*	Preclinical	N/A	[151]
BGB324	TKI	AML, NSCLC, melanoma	14 nM	14 nM	220 nM	700 nM	I/II	RET, VEGFR, FLT3, ABL, TIE2	[78,175]
Gilteritinib (ASP2215)	TKI	NSCLC, AML	0.7 nM	*	2.9 nM	*	II	FLT3, ALK, LKT	[176,177]
MRX-2843	TKI	AML	15 nM	*	1.3 nM	*	Preclinical	FLT3	[169]
TP-0903	TKI	Solid tumors	27 nM	222 nM	<200 nM	*	I	AURKA, AURKB, JAK2, ALK, ABL1	[178]
UNC2025	TKI	ALL, AML	14 nM	122 nM	0.74 nM	2.7 nM	Preclinical	FLT3, TYRO3	[114,179]

Abbreviations: ADC, antibody–drug conjugate; ALK, anaplastic lymphoma kinase; ALL, acute lymphoblastic leukemia; AML, acute myeloid leukemia; AURK, aurora kinase; FLT3, fms-like tyrosine kinase 3; IC_{50}, half maximum inhibitory concentration; JAK, Janus kinase; NSCLC, non-small cell lung cancer; RET, rearranged during transfection; TKI, tyrosine kinase inhibitor; TAM, TYRO3, AXL, and MERTK; VEGFR, vascular endothelial growth factor receptor.

* *Information not published.*

inhibitor, has been effective in NSCLC and CML animal models [16,165]. The AXL inhibitor BGB324 is currently being tested in phase I/II trials in adults with NSCLC and metastatic melanoma and a phase I trial in adults with AML. Preliminary data indicate that the compound was well tolerated and demonstrated an antileukemic effect. Importantly, biologic correlates demonstrated decreased activation of AXL and downstream signaling through ERK and AKT in leukemic blasts after 21 days of treatment [166]. Additional preclinical studies demonstrated therapeutic activity in CML, irrespective of sensitivity to BCR-ABL inhibition [167]. Gilteritinib, a dual FLT3 and AXL inhibitor, is currently being studied in AML clinical trials [160]. At the time of this writing, there are no clinical trials of MERTK-selective TKIs ongoing; however, there are robust preclinical data to support further development of several compounds. UNC2025 and MRX-2843 are dual MERTK and FLT3 inhibitors that were effective as monotherapy in preclinical models of leukemia [168,169], NSCLC [153], and GBM [110] and mediated synergistic antileukemia activity in combination with methotrexate [168].

3.4.2 TAM RTKs as Dual Targets

The roles for TAMs in promoting tumor cell survival and resistance to both targeted and cytotoxic chemotherapies coupled with their roles in immunosuppression in the tumor microenvironment suggest that they could be effective therapeutic targets in many human cancers. Given the context-dependent expression of TAM RTKs, optimum targeting could be modified based on the tumor type. For example, dual targeting of AXL and MERTK could be considered in NSCLC where both are expressed and have independent roles in tumor progression [58], whereas in pediatric ALL, an MERTK-selective approach might be better as AXL is not expressed in this setting [50,51]. Similarly, subtypes of colon cancer and triple-negative breast cancer with a mesenchymal phenotype overexpress AXL, providing a rationale for targeting AXL in combination with chemotherapy in patients with these tumors as a tool to overcome innate resistance associated with the mesenchymal phenotype [170–172].

3.4.3 TAM RTK Biomarkers

One of the challenges in the use of TAM-targeted therapeutics is the lack of identified robust biomarkers. The signaling pathways downstream of TAM kinases can be regulated by a number of other RTKs, making them less valuable as biomarkers. In addition, the signaling pathways downstream of TAM kinases are tissue dependent, further complicating the identification of biomarkers of TAM inhibition that can be applied across multiple tumor types. Because TAM RTK activity is regulated by ligands, their activity also varies in different tumor contexts, e.g., GAS6 can be secreted by bone marrow stromal cells to activate TAM RTK signaling in leukemia [77] or can be affected by the tumor organizational structure in solid tumors [173]. In current clinical trials, AXL expression and activation are measured by assessing AXL protein levels and AXL phosphorylation at Tyr702 [16]. Interestingly, some inhibitors decrease phosphorylation of AXL at the Tyr 702 site [47] and others such as BGB324 [78] and S49076 [174] do not, a finding which remains of undetermined significance. Development of robust biomarkers of TAM RTK inhibition will be critical for effective clinical application of TAM RTK inhibitors and is an active area of research in the field.

3.5 SUMMARY

Resistance to both conventional and targeted therapy represents a significant challenge in the treatment of cancer. Further studies to explore mechanisms of resistance and identification of drivers of resistance will allow for the development of additional agents targeting drug-resistant tumor cell populations. The TAM receptors have been identified as key mediators of resistance to both cytotoxic chemotherapy and targeted agents across a wide range of malignancies. TAM RTKs provide bypass signaling to facilitate resistance to targeted therapies and function to attenuate the immune response to cancer. As such, they represent attractive therapeutic targets. A number of TAM-selective TKIs have been developed. Clinical trials of AXL-selective agents are ongoing and MERTK-selective TKIs have demonstrated promising activity in preclinical studies. Development of adequate biomarkers of TAM RTK inhibition will be critical to facilitate optimal clinical application of these inhibitors. The current data suggest that treatment with TAM RTK inhibitors in combination with conventional chemotherapy, targeted agents, or immune checkpoint inhibitors will prove useful to overcome or prevent therapeutic resistance in a variety of tumor types.

Conflict of Interest: No potential conflicts of interest were disclosed.

References

[1] Graham DK, Dawson TL, Mullaney DL, Snodgrass HR, Earp HS. Cloning and mRNA expression analysis of a novel human protooncogene, c-mer. Cell Growth Differ 1994;5(6):647–57.

[2] Lai C, Gore M, Lemke G. Structure, expression, and activity of Tyro 3, a neural adhesion-related receptor tyrosine kinase. Oncogene 1994;9(9):2567–78.

[3] O'Bryan JP, Frye RA, Cogswell PC, Neubauer A, Kitch B, Prokop C, et al. axl, a transforming gene isolated from primary human myeloid leukemia cells, encodes a novel receptor tyrosine kinase. Mol Cell Biol 1991;11(10):5016–31.

[4] Graham DK, DeRyckere D, Davies KD, Earp HS. The TAM family: phosphatidylserine sensing receptor tyrosine kinases gone awry in cancer. Nat Rev Cancer 2014;14(12):769–85.

[5] Linger RM, Keating AK, Earp HS, Graham DK. TAM receptor tyrosine kinases: biologic functions, signaling, and potential therapeutic targeting in human cancer. Adv Cancer Res 2008;100:35–83.

[6] Jia R, Mayer BJ, Hanafusa T, Hanafusa H. A novel oncogene, v-ryk, encoding a truncated receptor tyrosine kinase is transduced into the RPL30 virus without loss of viral sequences. J Virol 1992;66(10):5975–87.

[7] Keating AK, Salzberg DB, Sather S, Liang X, Nickoloff S, Anwar A, et al. Lymphoblastic leukemia/lymphoma in mice overexpressing the Mer (MerTK) receptor tyrosine kinase. Oncogene 2006;25(45):6092–100.

[8] Guttridge KL, Luft JC, Dawson TL, Kozlowska E, Mahajan NP, Varnum B, et al. Mer receptor tyrosine kinase signaling: prevention of apoptosis and alteration of cytoskeletal architecture without stimulation or proliferation. J Biol Chemistry 2002;277(27):24057–66.

[9] Goruppi S, Ruaro E, Varnum B, Schneider C. Gas6-mediated survival in NIH3T3 cells activates stress signalling cascade and is independent of Ras. Oncogene 1999;18(29):4224–36.

[10] Melaragno MG, Cavet ME, Yan C, Tai LK, Jin ZG, Haendeler J, et al. Gas6 inhibits apoptosis in vascular smooth muscle: role of Axl kinase and Akt. J Mol Cell Cardiol 2004;37(4):881–7.

[11] Bellosta P, Zhang Q, Goff SP, Basilico C. Signaling through the ARK tyrosine kinase receptor protects from apoptosis in the absence of growth stimulation. Oncogene 1997;15(20):2387–97.

[12] Huey MG, Minson KA, Earp HS, DeRyckere D, Graham DK. Targeting the TAM receptors in leukemia. Cancers 2016;8(11):E101. doi:10.3390/cancers8110101.

[13] Davra V, Kimani SG, Calianese D, Birge RB. Ligand activation of TAM family receptors-implications for tumor biology and therapeutic response. Cancers 2016;8(12):E107. doi:10.3390/cancers8120107.

[14] Gay CM, Balaji K, Byers LA. Giving AXL the axe: targeting AXL in human malignancy. Br J Cancer 2017;116(4):415–23.

[15] Rothlin CV, Carrera-Silva EA, Bosurgi L, Ghosh S. TAM receptor signaling in immune homeostasis. Annu Rev Immunol 2015;33:355–91.
[16] Schoumacher M, Burbridge M. Key roles of AXL and MER receptor tyrosine kinases in resistance to multiple anticancer therapies. Curr Oncol Rep 2017;19(3):19.
[17] Akalu YT, Rothlin CV, Ghosh S. TAM receptor tyrosine kinases as emerging targets of innate immune checkpoint blockade for cancer therapy. Immunol Rev 2017;276(1):165–77.
[18] Aguilera TA, Giaccia AJ. Molecular pathways: oncologic pathways and their role in T-cell exclusion and immune evasion—a new role for the AXL receptor tyrosine kinase. Clin Cancer Res 2017;23(12):2928–33.
[19] Varnum BC, Young C, Elliott G, Garcia A, Bartley TD, Fridell YW, et al. Axl receptor tyrosine kinase stimulated by the vitamin K-dependent protein encoded by growth-arrest-specific gene 6. Nature 1995;373(6515):623–6.
[20] Stitt TN, Conn G, Gore M, Lai C, Bruno J, Radziejewski C, et al. The anticoagulation factor protein S and its relative, Gas6, are ligands for the Tyro 3/Axl family of receptor tyrosine kinases. Cell 1995;80(4):661–70.
[21] Nagata K, Ohashi K, Nakano T, Arita H, Zong C, Hanafusa H, et al. Identification of the product of growth arrest-specific gene 6 as a common ligand for Axl, Sky, and Mer receptor tyrosine kinases. J Biol Chem 1996;271(47):30022–7.
[22] Sasaki T, Knyazev PG, Clout NJ, Cheburkin Y, Gohring W, Ullrich A, et al. Structural basis for Gas6-Axl signalling. EMBO J 2006;25(1):80–7.
[23] Lew ED, Oh J, Burrola PG, Lax I, Zagorska A, Traves PG, et al. Differential TAM receptor-ligand-phospholipid interactions delimit differential TAM bioactivities. eLife 2014;3. doi: 10.7554/eLife.03385.
[24] Tsou WI, Nguyen KQ, Calarese DA, Garforth SJ, Antes AL, Smirnov SV, et al. Receptor tyrosine kinases, TYRO3, AXL, and MER, demonstrate distinct patterns and complex regulation of ligand-induced activation. J Biol Chem 2014;289(37):25750–63.
[25] Abboud-Jarrous G, Priya S, Maimon A, Fischman S, Cohen-Elisha M, Czerninski R, et al. Protein S drives oral squamous cell carcinoma tumorigenicity through regulation of AXL. Oncotarget 2017;8(8):13986–4002.
[26] Caberoy NB, Zhou Y, Li W. Tubby and tubby-like protein 1 are new MerTK ligands for phagocytosis. EMBO J 2010;29(23):3898–910.
[27] Caberoy NB, Alvarado G, Li W. Tubby regulates microglial phagocytosis through MerTK. J Neuroimmunol 2012;252(1–2):40–8.
[28] Caberoy NB, Alvarado G, Bigcas JL, Li W. Galectin-3 is a new MerTK-specific eat-me signal. J Cell Physiol 2012;227(2):401–7.
[29] Scott RS, McMahon EJ, Pop SM, Reap EA, Caricchio R, Cohen PL, et al. Phagocytosis and clearance of apoptotic cells is mediated by MER. Nature 2001;411(6834):207–11.
[30] Chen C, Li Q, Darrow AL, Wang Y, Derian CK, Yang J, et al. Mer receptor tyrosine kinase signaling participates in platelet function. Arterioscler Thromb Vasc Biol 2004;24(6):1118–23.
[31] Angelillo-Scherrer A, Burnier L, Lambrechts D, Fish RJ, Tjwa M, Plaisance S, et al. Role of Gas6 in erythropoiesis and anemia in mice. J Clin Investig 2008;118(2):583–96.
[32] Eken C, Martin PJ, Sadallah S, Treves S, Schaller M, Schifferli JA. Ectosomes released by polymorphonuclear neutrophils induce a MerTK-dependent anti-inflammatory pathway in macrophages. J Biol Chem 2010;285(51):39914–21.
[33] Mercer J, Helenius A. Vaccinia virus uses macropinocytosis and apoptotic mimicry to enter host cells. Science (New York, NY) 2008;320(5875):531–5.
[34] Bhattacharyya S, Zagorska A, Lew ED, Shrestha B, Rothlin CV, Naughton J, et al. Enveloped viruses disable innate immune responses in dendritic cells by direct activation of TAM receptors. Cell Host Microbe 2013;14(2):136–47.
[35] Bellosta P, Costa M, Lin DA, Basilico C. The receptor tyrosine kinase ARK mediates cell aggregation by homophilic binding. Mol Cell Biol 1995;15(2):614–25.
[36] Burchert A, Attar EC, McCloskey P, Fridell YW, Liu ET. Determinants for transformation induced by the Axl receptor tyrosine kinase. Oncogene 1998;16(24):3177–87.
[37] Meyer AS, Miller MA, Gertler FB, Lauffenburger DA. The receptor AXL diversifies EGFR signaling and limits the response to EGFR-targeted inhibitors in triple-negative breast cancer cells. Sci Signal 2013;6(287):ra66.
[38] Vouri M, Croucher DR, Kennedy SP, An Q, Pilkington GJ, Hafizi S. Axl-EGFR receptor tyrosine kinase hetero-interaction provides EGFR with access to pro-invasive signalling in cancer cells. Oncogenesis 2016;5(10):e266.

[39] Liu L, Greger J, Shi H, Liu Y, Greshock J, Annan R, et al. Novel mechanism of lapatinib resistance in HER2-positive breast tumor cells: activation of AXL. Cancer Res 2009;69(17):6871–8.
[40] D'Cruz PM, Yasumura D, Weir J, Matthes MT, Abderrahim H, LaVail MM, et al. Mutation of the receptor tyrosine kinase gene Mertk in the retinal dystrophic RCS rat. Hum Mol Genet 2000;9(4):645–51.
[41] Graham DK, DeRyckere D, Davies KD, Earp HS. The TAM family: phosphatidylserine-sensing receptor tyrosine kinases gone awry in cancer. Nat Rev 2014;14:769–85.
[42] Hutterer M, Knyazev P, Abate A, Reschke M, Maier H, Stefanova N, et al. Axl and growth arrest-specific gene 6 are frequently overexpressed in human gliomas and predict poor prognosis in patients with glioblastoma multiforme. Clin Cancer Res 2008;14(1):130–8.
[43] Ishikawa M, Sonobe M, Nakayama E, Kobayashi M, Kikuchi R, Kitamura J, et al. Higher expression of receptor tyrosine kinase Axl, and differential expression of its ligand, Gas6, predict poor survival in lung adenocarcinoma patients. Ann Surg Oncol 2013;20(Suppl 3):S467–76.
[44] Lee CH, Yen CY, Liu SY, Chen CK, Chiang CF, Shiah SG, et al. Axl is a prognostic marker in oral squamous cell carcinoma. Ann Surg Oncol 2012;19(Suppl 3):S500–8.
[45] Ben-Batalla I, Schultze A, Wroblewski M, Erdmann R, Heuser M, Waizenegger JS, et al. Axl, a prognostic and therapeutic target in acute myeloid leukemia mediates paracrine crosstalk of leukemia cells with bone marrow stroma. Blood 2013;122(14):2443–52.
[46] Whitman SP, Kohlschmidt J, Maharry K, Volinia S, Mrozek K, Nicolet D, et al. GAS6 expression identifies high-risk adult AML patients: potential implications for therapy. Leukemia 2014;28(6):1252–8.
[47] Zhou L, Liu XD, Sun M, Zhang X, German P, Bai S, et al. Targeting MET and AXL overcomes resistance to sunitinib therapy in renal cell carcinoma. Oncogene 2016;35(21):2687–97.
[48] Dunne PD, McArt DG, Blayney JK, Kalimutho M, Greer S, Wang T, et al. AXL is a key regulator of inherent and chemotherapy-induced invasion and predicts a poor clinical outcome in early-stage colon cancer. Clin Cancer Res 2014;20(1):164–75.
[49] Pinato DJ, Mauri FA, Lloyd T, Vaira V, Casadio C, Boldorini RL, et al. The expression of Axl receptor tyrosine kinase influences the tumour phenotype and clinical outcome of patients with malignant pleural mesothelioma. Br J Cancer 2013;108(3):621–8.
[50] Linger RM, Lee-Sherick AB, DeRyckere D, Cohen RA, Jacobsen KM, McGranahan A, et al. Mer receptor tyrosine kinase is a therapeutic target in pre-B-cell acute lymphoblastic leukemia. Blood 2013;122(9):1599–609.
[51] Brandao LN, Winges A, Christoph S, Sather S, Migdall-Wilson J, Schlegel J, et al. Inhibition of MerTK increases chemosensitivity and decreases oncogenic potential in T-cell acute lymphoblastic leukemia. Blood Cancer J 2013;3:e101.
[52] Lee-Sherick AB, Eisenman KM, Sather S, McGranahan A, Armistead PM, McGary CS, et al. Aberrant Mer receptor tyrosine kinase expression contributes to leukemogenesis in acute myeloid leukemia. Oncogene 2013;32(46):5359–68.
[53] Waizenegger JS, Ben-Batalla I, Weinhold N, Meissner T, Wroblewski M, Janning M, et al. Role of growth arrest-specific gene 6-Mer axis in multiple myeloma. Leukemia 2015;29(3):696–704.
[54] Neubauer A, Fiebeler A, Graham DK, O'Bryan JP, Schmidt CA, Barckow P, et al. Expression of axl, a transforming receptor tyrosine kinase, in normal and malignant hematopoiesis. Blood 1994;84(6):1931–41.
[55] Rochlitz C, Lohri A, Bacchi M, Schmidt M, Nagel S, Fopp M, et al. Axl expression is associated with adverse prognosis and with expression of Bcl-2 and CD34 in de novo acute myeloid leukemia (AML): results from a multicenter trial of the Swiss Group for Clinical Cancer Research (SAKK). Leukemia 1999;13(9):1352–8.
[56] Park IK, Mishra A, Chandler J, Whitman SP, Marcucci G, Caligiuri MA. Inhibition of the receptor tyrosine kinase Axl impedes activation of the FLT3 internal tandem duplication in human acute myeloid leukemia: implications for Axl as a potential therapeutic target. Blood 2013;121(11):2064–73.
[57] Ghosh AK, Secreto C, Boysen J, Sassoon T, Shanafelt TD, Mukhopadhyay D, et al. The novel receptor tyrosine kinase Axl is constitutively active in B-cell chronic lymphocytic leukemia and acts as a docking site of nonreceptor kinases: implications for therapy. Blood 2011;117(6):1928–37.
[58] Linger RM, Cohen RA, Cummings CT, Sather S, Migdall-Wilson J, Middleton DH, et al. Mer or Axl receptor tyrosine kinase inhibition promotes apoptosis, blocks growth and enhances chemosensitivity of human non-small cell lung cancer. Oncogene 2013;32(29):3420–31.
[59] Keating AK, Kim GK, Jones AE, Donson AM, Ware K, Mulcahy JM, et al. Inhibition of Mer and Axl receptor tyrosine kinases in astrocytoma cells leads to increased apoptosis and improved chemosensitivity. Mol Cancer Ther 2010;9(5):1298–307.

[60] Wang Y, Moncayo G, Morin P Jr, Xue G, Grzmil M, Lino MM, et al. Mer receptor tyrosine kinase promotes invasion and survival in glioblastoma multiforme. Oncogene 2013;32(7):872–82.
[61] Tworkoski K, Singhal G, Szpakowski S, Zito CI, Bacchiocchi A, Muthusamy V, et al. Phosphoproteomic screen identifies potential therapeutic targets in melanoma. Mol Cancer Res 2011;9(6):801–12.
[62] Tworkoski KA, Platt JT, Bacchiocchi A, Bosenberg M, Boggon TJ, Stern DF. MERTK controls melanoma cell migration and survival and differentially regulates cell behavior relative to AXL. Pigment Cell Melanoma Res 2013;26(4):527–41.
[63] Gyorffy B, Lage H. A web-based data warehouse on gene expression in human malignant melanoma. J Invest Dermatol 2007;127(2):394–9.
[64] Schlegel J, Sambade MJ, Sather S, Moschos SJ, Tan AC, Winges A, et al. MERTK receptor tyrosine kinase is a therapeutic target in melanoma. J Clin Invest 2013;123(5):2257–67.
[65] Wimmel A, Glitz D, Kraus A, Roeder J, Schuermann M. Axl receptor tyrosine kinase expression in human lung cancer cell lines correlates with cellular adhesion. Eur J Cancer 2001;37(17):2264–74.
[66] Shieh YS, Lai CY, Kao YR, Shiah SG, Chu YW, Lee HS, et al. Expression of axl in lung adenocarcinoma and correlation with tumor progression. Neoplasia (New York, NY) 2005;7(12):1058–64.
[67] Vajkoczy P, Knyazev P, Kunkel A, Capelle HH, Behrndt S, von Tengg-Kobligk H, et al. Dominant-negative inhibition of the Axl receptor tyrosine kinase suppresses brain tumor cell growth and invasion and prolongs survival. Proc Natl Acad Sci USA 2006;103(15):5799–804.
[68] Berclaz G, Altermatt HJ, Rohrbach V, Kieffer I, Dreher E, Andres AC. Estrogen dependent expression of the receptor tyrosine kinase axl in normal and malignant human breast. Ann Oncol 2001;12(6):819–24.
[69] Meric F, Lee WP, Sahin A, Zhang H, Kung HJ, Hung MC. Expression profile of tyrosine kinases in breast cancer. Clin Cancer Res 2002;8(2):361–7.
[70] Zhang YX, Knyazev PG, Cheburkin YV, Sharma K, Knyazev YP, Orfi L, et al. AXL is a potential target for therapeutic intervention in breast cancer progression. Cancer Res 2008;68(6):1905–15.
[71] Sun W, Fujimoto J, Tamaya T. Coexpression of Gas6/Axl in human ovarian cancers. Oncology 2004;66(6):450–7.
[72] Rankin EB, Fuh KC, Taylor TE, Krieg AJ, Musser M, Yuan J, et al. AXL is an essential factor and therapeutic target for metastatic ovarian cancer. Cancer Res 2010;70(19):7570–9.
[73] Chen PX, Li QY, Yang Z. Axl and prostasin are biomarkers for prognosis of ovarian adenocarcinoma. Ann Diagn Pathol 2013;17(5):425–9.
[74] AG N, Bensinger SJ, Hong C, Beceiro S, Bradley MN, Zelcer N, et al. Apoptotic cells promote their own clearance and immune tolerance through activation of the nuclear receptor LXR. Immunity 2009;31(2):245–58.
[75] Nelson ER, Wardell SE, Jasper JS, Park S, Suchindran S, Howe MK, et al. 27-Hydroxycholesterol links hypercholesterolemia and breast cancer pathophysiology. Science (New York, NY) 2013;342(6162):1094–8.
[76] Shiozawa Y, Pedersen EA, Patel LR, Ziegler AM, Havens AM, Jung Y, et al. GAS6/AXL axis regulates prostate cancer invasion, proliferation, and survival in the bone marrow niche. Neoplasia (New York, NY) 2010;12(2):116–27.
[77] Shiozawa Y, Pedersen EA, Taichman RS. GAS6/Mer axis regulates the homing and survival of the E2A/PBX1-positive B-cell precursor acute lymphoblastic leukemia in the bone marrow niche. Exp Hematol 2010;38(2):132–40.
[78] Holland SJ, Pan A, Franci C, Hu Y, Chang B, Li W, et al. R428, a selective small molecule inhibitor of Axl kinase, blocks tumor spread and prolongs survival in models of metastatic breast cancer. Cancer Res 2010;70(4):1544–54.
[79] Hong CC, Lay JD, Huang JS, Cheng AL, Tang JL, Lin MT, et al. Receptor tyrosine kinase AXL is induced by chemotherapy drugs and overexpression of AXL confers drug resistance in acute myeloid leukemia. Cancer Lett 2008;268(2):314–24.
[80] Leconet W, Larbouret C, Chardes T, Thomas G, Neiveyans M, Busson M, et al. Preclinical validation of AXL receptor as a target for antibody-based pancreatic cancer immunotherapy. Oncogene 2014;33(47):5405–14.
[81] Giles KM, Kalinowski FC, Candy PA, Epis MR, Zhang PM, Redfern AD, et al. Axl mediates acquired resistance of head and neck cancer cells to the epidermal growth factor receptor inhibitor erlotinib. Mol Cancer Ther 2013;12(11):2541–58.
[82] Georgescu MM, Kirsch KH, Shishido T, Zong C, Hanafusa H. Biological effects of c-Mer receptor tyrosine kinase in hematopoietic cells depend on the Grb2 binding site in the receptor and activation of NF-kappaB. Mol Cell Biol 1999;19(2):1171–81.
[83] Lay JD, Hong CC, Huang JS, Yang YY, Pao CY, Liu CH, et al. Sulfasalazine suppresses drug resistance and invasiveness of lung adenocarcinoma cells expressing AXL. Cancer Res 2007;67(8):3878–87.

[84] Tai KY, Shieh YS, Lee CS, Shiah SG, Wu CW. Axl promotes cell invasion by inducing MMP-9 activity through activation of NF-kappaB and Brg-1. Oncogene 2008;27(29):4044–55.

[85] Paccez JD, Vasques GJ, Correa RG, Vasconcellos JF, Duncan K, Gu X, et al. The receptor tyrosine kinase Axl is an essential regulator of prostate cancer proliferation and tumor growth and represents a new therapeutic target. Oncogene 2013;32(6):689–98.

[86] Asiedu MK, Beauchamp-Perez FD, Ingle JN, Behrens MD, Radisky DC, Knutson KL. AXL induces epithelial-to-mesenchymal transition and regulates the function of breast cancer stem cells. Oncogene 2014;33(10): 1316–24.

[87] Lee CH, Liu SY, Chou KC, Yeh CT, Shiah SG, Huang RY, et al. Tumor-associated macrophages promote oral cancer progression through activation of the Axl signaling pathway. Ann Surg Oncol 2014;21(3):1031–7.

[88] Camenisch TD, Koller BH, Earp HS, Matsushima GK. A novel receptor tyrosine kinase, Mer, inhibits TNF-alpha production and lipopolysaccharide-induced endotoxic shock. J Immunol 1999;162(6):3498–503.

[89] Tibrewal N, Wu Y, D'Mello V, Akakura R, George TC, Varnum B, et al. Autophosphorylation docking site Tyr-867 in Mer receptor tyrosine kinase allows for dissociation of multiple signaling pathways for phagocytosis of apoptotic cells and down-modulation of lipopolysaccharide-inducible NF-kappaB transcriptional activation. J Biol Chem 2008;283(6):3618–27.

[90] Wallet MA, Sen P, Flores RR, Wang Y, Yi Z, Huang Y, et al. MerTK is required for apoptotic cell-induced T cell tolerance. J Exp Med 2008;205(1):219–32.

[91] Papadakis ES, Cichon MA, Vyas JJ, Patel N, Ghali L, Cerio R, et al. Axl promotes cutaneous squamous cell carcinoma survival through negative regulation of pro-apoptotic Bcl-2 family members. J Invest Dermatol 2011;131(2):509–17.

[92] Steeg PS. Tumor metastasis: mechanistic insights and clinical challenges. Nat Med 2006;12(8):895–904.

[93] Singh A, Settleman J. EMT, cancer stem cells and drug resistance: an emerging axis of evil in the war on cancer. Oncogene 2010;29(34):4741–51.

[94] Marcucci F, Stassi G, De Maria R. Epithelial–mesenchymal transition: a new target in anticancer drug discovery. Nat Rev Drug Discov 2016;15(5):311–25.

[95] Thomson S, Petti F, Sujka-Kwok I, Mercado P, Bean J, Monaghan M, et al. A systems view of epithelial–mesenchymal transition signaling states. Clin Exp Metastasis 2011;28(2):137–55.

[96] Kohn KW, Zeeberg BR, Reinhold WC, Sunshine M, Luna A, Pommier Y. Gene expression profiles of the NCI-60 human tumor cell lines define molecular interaction networks governing cell migration processes. PLoS ONE 2012;7(5):e35716.

[97] Tirosh I, Izar B, Prakadan SM, Wadsworth MH II, Treacy D, Trombetta JJ, et al. Dissecting the multicellular ecosystem of metastatic melanoma by single-cell RNA-seq. Science (New York, NY) 2016;352(6282): 189–96.

[98] Byers LA, Diao L, Wang J, Saintigny P, Girard L, Peyton M, et al. An epithelial–mesenchymal transition gene signature predicts resistance to EGFR and PI3K inhibitors and identifies Axl as a therapeutic target for overcoming EGFR inhibitor resistance. Clin Cancer Res 2013;19(1):279–90.

[99] Muller J, Krijgsman O, Tsoi J, Robert L, Hugo W, Song C, et al. Low MITF/AXL ratio predicts early resistance to multiple targeted drugs in melanoma. Nat Commun 2014;5:5712.

[100] Mak MP, Tong P, Diao L, Cardnell RJ, Gibbons DL, William WN, et al. A patient-derived, pan-cancer EMT signature identifies global molecular alterations and immune target enrichment following epithelial-to-mesenchymal transition. Clin Cancer Res 2016;22(3):609–20.

[101] Cosemans JM, Van Kruchten R, Olieslagers S, Schurgers LJ, Verheyen FK, Munnix IC, et al. Potentiating role of Gas6 and Tyro3, Axl and Mer (TAM) receptors in human and murine platelet activation and thrombus stabilization. J Thromb Haemost 2010;8(8):1797–808.

[102] Ruan GX, Kazlauskas A. Axl is essential for VEGF-A-dependent activation of PI3K/Akt. EMBO J 2012;31(7):1692–703.

[103] Ruan GX, Kazlauskas A. Lactate engages receptor tyrosine kinases Axl, Tie2, and vascular endothelial growth factor receptor 2 to activate phosphoinositide 3-kinase/Akt and promote angiogenesis. J Biol Chem 2013;288(29):21161–72.

[104] Holland SJ, Powell MJ, Franci C, Chan EW, Friera AM, Atchison RE, et al. Multiple roles for the receptor tyrosine kinase axl in tumor formation. Cancer Res 2005;65(20):9294–303.

[105] Li Y, Ye X, Tan C, Hongo JA, Zha J, Liu J, et al. Axl as a potential therapeutic target in cancer: role of Axl in tumor growth, metastasis and angiogenesis. Oncogene 2009;28(39):3442–55.

[106] Choueiri TK, Escudier B, Powles T, Tannir NM, Mainwaring PN, Rini BI, et al. Cabozantinib versus everolimus in advanced renal cell carcinoma (METEOR): final results from a randomised, open-label, phase 3 trial. Lancet Oncol 2016;17(7):917–27.
[107] Macleod K, Mullen P, Sewell J, Rabiasz G, Lawrie S, Miller E, et al. Altered ErbB receptor signaling and gene expression in cisplatin-resistant ovarian cancer. Cancer Res 2005;65(15):6789–800.
[108] Zhao Y, Sun X, Jiang L, Yang F, Zhang Z, Jia L. Differential expression of Axl and correlation with invasion and multidrug resistance in cancer cells. Cancer Invest 2012;30(4):287–94.
[109] Hong J, Peng D, Chen Z, Sehdev V, Belkhiri A. ABL regulation by AXL promotes cisplatin resistance in esophageal cancer. Cancer Res 2013;73(1):331–40.
[110] Sufit A, Lee-Sherick AB, DeRyckere D, Rupji M, Dwivedi B, Varella-Garcia M, et al. MERTK inhibition induces polyploidy and promotes cell death and cellular senescence in glioblastoma multiforme. PLoS ONE 2016;11(10):e0165107.
[111] Li XY, Wen JY, Jia CC, Wang TT, Li X, Dong M, et al. MicroRNA-34a-5p enhances sensitivity to chemotherapy by targeting AXL in hepatocellular carcinoma MHCC-97L cells. Oncol Lett 2015;10(5):2691–8.
[112] Chang TY, Chen HA, Chiu CF, Chang YW, Kuo TC, Tseng PC, et al. Dicer elicits paclitaxel chemosensitization and suppresses cancer stemness in breast cancer by repressing AXL. Cancer Res 2016;76(13):3916–28.
[113] Song X, Wang H, Logsdon CD, Rashid A, Fleming JB, Abbruzzese JL, et al. Overexpression of receptor tyrosine kinase Axl promotes tumor cell invasion and survival in pancreatic ductal adenocarcinoma. Cancer 2011;117(4):734–43.
[114] DeRyckere D, Lee Sherick AB, Huey MG, Hill AA, Tyner JW, Song X, Jacobsen KM, et al. UNC2025, a MerTK small molecule inhibitor, is therapeutically effective alone and in combination with methotrexate in leukemia models. Clin Cancer Res 2016;23(14):3971–2.
[115] Dufies M, Jacquel A, Belhacene N, Robert G, Cluzeau T, Luciano F, et al. Mechanisms of AXL overexpression and function in Imatinib-resistant chronic myeloid leukemia cells. Oncotarget 2011;2(11):874–85.
[116] Gioia R, Tregoat C, Dumas PY, Lagarde V, Prouzet-Mauleon V, Desplat V, et al. CBL controls a tyrosine kinase network involving AXL, SYK and LYN in nilotinib-resistant chronic myeloid leukaemia. J Pathol 2015;237(1):14–24.
[117] Mahadevan D, Cooke L, Riley C, Swart R, Simons B, Della Croce K, et al. A novel tyrosine kinase switch is a mechanism of imatinib resistance in gastrointestinal stromal tumors. Oncogene 2007;26(27):3909–19.
[118] Adrian FJ, Ding Q, Sim T, Velentza A, Sloan C, Liu Y, et al. Allosteric inhibitors of Bcr-abl-dependent cell proliferation. Nat Chem Biol 2006;2(2):95–102.
[119] Karaman MW, Herrgard S, Treiber DK, Gallant P, Atteridge CE, Campbell BT, et al. A quantitative analysis of kinase inhibitor selectivity. Nat Biotechnol 2008;26(1):127–32.
[120] Metz JT, Johnson EF, Soni NB, Merta PJ, Kifle L, Hajduk PJ. Navigating the kinome. Nat Chem Biol 2011;7(4):200–2.
[121] Zhang Z, Lee JC, Lin L, Olivas V, Au V, LaFramboise T, et al. Activation of the AXL kinase causes resistance to EGFR-targeted therapy in lung cancer. Nat Genet 2012;44(8):852–60.
[122] Gusenbauer S, Vlaicu P, Ullrich A. HGF induces novel EGFR functions involved in resistance formation to tyrosine kinase inhibitors. Oncogene 2013;32(33):3846–56.
[123] Brand TM, Iida M, Stein AP, Corrigan KL, Braverman CM, Luthar N, et al. AXL mediates resistance to cetuximab therapy. Cancer Res 2014;74(18):5152–64.
[124] Xie S, Li Y, Li X, Wang L, Yang N, Wang Y, et al. Mer receptor tyrosine kinase is frequently overexpressed in human non-small cell lung cancer, confirming resistance to erlotinib. Oncotarget 2015;6(11):9206–19.
[125] Yan DWX, Frye SV, Earp SH, Deryckere D, Graham DK. MerTK promotes resistance to irreversible EGFR TKIs by activation of the PI3K–AKT pathway in NSCLCs expressing wild-type EGFR [abstract]. American Association for Cancer Research Annual Meeting, April 1–5, 2017, Washington, DC. Philadelphia, PA: AACR; 2017.
[126] Park IK, Mundy-Bosse B, Whitman SP, Zhang X, Warner SL, Bearss DJ, et al. Receptor tyrosine kinase Axl is required for resistance of leukemic cells to FLT3-targeted therapy in acute myeloid leukemia. Leukemia 2015;29(12):2382–9.
[127] Miller MA, Oudin MJ, Sullivan RJ, Wang SJ, Meyer AS, Im H, et al. Reduced proteolytic shedding of receptor tyrosine kinases is a post-translational mechanism of kinase inhibitor resistance. Cancer Discov 2016;6(4):382–99.
[128] Miller MA, Sullivan RJ, Lauffenburger DA. Molecular pathways: receptor ectodomain shedding in treatment, resistance, and monitoring of cancer. Clin Cancer Res 2017;23(3):623–9.

[129] Huang F, Hurlburt W, Greer A, Reeves KA, Hillerman S, Chang H, et al. Differential mechanisms of acquired resistance to insulin-like growth factor-i receptor antibody therapy or to a small-molecule inhibitor, BMS-754807, in a human rhabdomyosarcoma model. Cancer Res 2010;70(18):7221–31.
[130] Ware KE, Hinz TK, Kleczko E, Singleton KR, Marek LA, Helfrich BA, et al. A mechanism of resistance to gefitinib mediated by cellular reprogramming and the acquisition of an FGF2–FGFR1 autocrine growth loop. Oncogenesis 2013;2:e39.
[131] Kim HR, Kim WS, Choi YJ, Choi CM, Rho JK, Lee JC. Epithelial–mesenchymal transition leads to crizotinib resistance in H2228 lung cancer cells with EML4-ALK translocation. Mol Oncol 2013;7(6):1093–102.
[132] Elkabets M, Pazarentzos E, Juric D, Sheng Q, Pelossof RA, Brook S, et al. AXL mediates resistance to PI3Kalpha inhibition by activating the EGFR/PKC/mTOR axis in head and neck and esophageal squamous cell carcinomas. Cancer Cell 2015;27(4):533–46.
[133] Xu F, Li H, Sun Y. Inhibition of Axl improves the targeted therapy against ALK-mutated neuroblastoma. Biochem Biophys Res Commun 2014;454(4):566–71.
[134] Debruyne DN, Bhatnagar N, Sharma B, Luther W, Moore NF, Cheung NK, et al. ALK inhibitor resistance in ALK(F1174L)-driven neuroblastoma is associated with AXL activation and induction of EMT. Oncogene 2016;35(28):3681–91.
[135] Cook RS, Jacobsen KM, Wofford AM, DeRyckere D, Stanford J, Prieto AL, et al. MerTK inhibition in tumor leukocytes decreases tumor growth and metastasis. J Clin Invest 2013;123(8):3231–42.
[136] DeNardo DG, Brennan DJ, Rexhepaj E, Ruffell B, Shiao SL, Madden SF, et al. Leukocyte complexity predicts breast cancer survival and functionally regulates response to chemotherapy. Cancer Discov 2011;1(1): 54–67.
[137] Cai B, Thorp EB, Doran AC, Subramanian M, Sansbury BE, Lin CS, et al. MerTK cleavage limits proresolving mediator biosynthesis and exacerbates tissue inflammation. Proc Natl Acad Sci USA 2016;113(23):6526–31.
[138] Postow MA, Callahan MK, Wolchok JD. Immune checkpoint blockade in cancer therapy. J Clin Oncol 2015;33(17):1974–82.
[139] Dong H, Strome SE, Salomao DR, Tamura H, Hirano F, Flies DB, et al. Tumor-associated B7-H1 promotes T-cell apoptosis: a potential mechanism of immune evasion. Nat Med 2002;8(8):793–800.
[140] Hugo W, Zaretsky JM, Sun L, Song C, Moreno BH, Hu-Lieskovan S, et al. Genomic and transcriptomic features of response to Anti-PD-1 therapy in metastatic melanoma. Cell 2016;165(1):35–44.
[141] Nguyen KQ, Tsou WI, Calarese DA, Kimani SG, Singh S, Hsieh S, et al. Overexpression of MERTK receptor tyrosine kinase in epithelial cancer cells drives efferocytosis in a gain-of-function capacity. J Biol Chem 2014;289(37):25737–49.
[142] Kasikara C, Kumar S, Kimani S, Tsou WI, Geng K, Davra V, et al. Phosphatidylserine sensing by TAM receptors regulates AKT-dependent chemoresistance and PD-L1 expression. Mol Cancer Res 2017;15(6):753–64.
[143] Soh KK, Kim W, Lee YS, Peterson P, Siddiqui-Jain A, Warner SL, et al. Abstract 235: AXL inhibition leads to a reversal of a mesenchymal phenotype sensitizing cancer cells to targeted agents and immuno-oncology therapies. Cancer Res 2016;76(14 Suppl):235.
[144] Gausdal G, Davidsen K, Wnuk-Lipinska K, Wiertel K, Kang J, Engelsen A, et al. Abstract 566: BGB324, a selective small molecule inhibitor of the receptor tyrosine kinase AXL, enhances immune checkpoint inhibitor efficacy. Cancer Res 2016;76(14 Suppl):566.
[145] Yoshizawa T, Tanaka K, Yasuhiro T, Fujikawa R, Ri S, Kawabata K. Abstract LB-218: development of Axl/Mer inhibitor, ONO-9330547: preclinical evidence supporting the combination with immunotherapeutics. Cancer Res 2016;76(14 Suppl). LB-218.
[146] Cerchia L, Esposito CL, Camorani S, Rienzo A, Stasio L, Insabato L, et al. Targeting Axl with an high-affinity inhibitory aptamer. Mol Ther 2012;20(12):2291–303.
[147] Rankin EB, Fuh KC, Castellini L, Viswanathan K, Finger EC, Diep AN, et al. Direct regulation of GAS6/AXL signaling by HIF promotes renal metastasis through SRC and MET. Proc Natl Acad Sci USA 2014;111(37): 13373–8.
[148] Sather S, Kenyon KD, Lefkowitz JB, Liang X, Varnum BC, Henson PM, et al. A soluble form of the Mer receptor tyrosine kinase inhibits macrophage clearance of apoptotic cells and platelet aggregation. Blood 2007;109(3):1026–33.
[149] Angelillo-Scherrer A, Burnier L, Flores N, Savi P, DeMol M, Schaeffer P, et al. Role of Gas6 receptors in platelet signaling during thrombus stabilization and implications for antithrombotic therapy. J Clin Invest 2005;115(2):237–46.

[150] Kariolis MS, Miao YR, Jones DS II, Kapur S, Mathews II, Giaccia AJ, et al. An engineered Axl 'decoy receptor' effectively silences the Gas6-Axl signaling axis. Nat Chem Biol 2014;10(11):977–83.

[151] Kariolis MS, Miao YR, Diep A, Nash SE, Olcina MM, Jiang D, et al. Inhibition of the GAS6/AXL pathway augments the efficacy of chemotherapies. J Clin Invest 2017;127(1):183–98.

[152] Ye X, Li Y, Stawicki S, Couto S, Eastham-Anderson J, Kallop D, et al. An anti-Axl monoclonal antibody attenuates xenograft tumor growth and enhances the effect of multiple anticancer therapies. Oncogene 2010;29(38):5254–64.

[153] Cummings CT, Linger RM, Cohen RA, Sather S, Kirkpatrick GD, Davies KD, et al. Mer590, a novel monoclonal antibody targeting MER receptor tyrosine kinase, decreases colony formation and increases chemosensitivity in non-small cell lung cancer. Oncotarget 2014;5(21):10434–45.

[154] Liu R, Gong M, Li X, Zhou Y, Gao W, Tulpule A, et al. Induction, regulation, and biologic function of Axl receptor tyrosine kinase in Kaposi sarcoma. Blood 2010;116(2):297–305.

[155] Kimani SG, Kumar S, Bansal N, Singh K, Kholodovych V, Comollo T, et al. Small molecule inhibitors block Gas6-inducible TAM activation and tumorigenicity. Sci Rep 2017;7:43908.

[156] Breij ECWVS, Lingnau A, van den Brink EN, Janmaat M, Houtkamp M, Bleeker W, et al. Preclinical efficacy studies using HuMax-Axl-ADC, a novel antibody–drug conjugate targeting Axl-expressing solid cancers. J Clin Oncol 2015;33(Suppl). abstr 3066.

[157] Chang C, Frey G, Boyle WJ, Sharp LL, Short JM. Abstract 3836: novel conditionally active biologic anti-Axl antibody–drug conjugate demonstrates anti-tumor efficacy and improved safety profile. Cancer Res 2016;76(14 Suppl):3836.

[158] Rogers AE, Le JP, Sather S, Pernu BM, Graham DK, Pierce AM, et al. Mer receptor tyrosine kinase inhibition impedes glioblastoma multiforme migration and alters cellular morphology. Oncogene 2012;31(38):4171–81.

[159] Patwardhan PP, Ivy KS, Musi E, de Stanchina E, Schwartz GK. Significant blockade of multiple receptor tyrosine kinases by MGCD516 (Sitravatinib), a novel small molecule inhibitor, shows potent anti-tumor activity in preclinical models of sarcoma. Oncotarget 2016;7(4):4093–109.

[160] ClinicalTrials.gov. Cabozantinib in patients with RET fusion-positive advanced non-small cell lung cancer and those with other genotypes: ROS1 or NTRK fusions or increased MET or AXL activity 2017. Available from: https://clinicaltrials.gov/ct2/show/NCT01639508?term=NCT01639508&rank=1.

[161] ClinicalTrials.gov. Phase 2 study of glesatinib, sitravatinib or mocetinostat in combination with nivolumab in non-small cell lung cancer 2017. Available from: https://clinicaltrials.gov/ct2/show/NCT02954991?term=NCT02954991&rank=1.

[162] Clinical Trials.gov. Phase 1/1b study of MGCD516 in patients with advanced cancer 2017. Available from: https://clinicaltrials.gov/ct2/show/NCT02219711.

[163] Patel V, Keating MJ, Wierda WG, Gandhi V. Preclinical combination of TP-0903, an AXL inhibitor and B-PAC-1, a procaspase-activating compound with ibrutinib in chronic lymphocytic leukemia. Leuk Lymphoma 2016;57(6):1494–7.

[164] Sinha S, Boysen J, Nelson M, Secreto C, Warner SL, Bearss DJ, et al. Targeted Axl inhibition primes chronic lymphocytic leukemia B cells to apoptosis and shows synergistic/additive effects in combination with BTK inhibitors. Clin Cancer Res 2015;21(9):2115–26.

[165] SignalChemLifeSciences.com. AXL inhibitor program. Available from: http://signalchemlifesciences.com/therapeutic-programs/axl-inhibitors.

[166] Loges S, Gjertsen BT, Heuser M, Ben-Batalla I, Micklem D, Jorg C, et al. A first-in-patient phase I study of BGB324, a selective Axl kinase inhibitor in patients with refractory/relapsed AML and high-risk MDS. J Clin Oncol 2016;34(Suppl.). Abstract 2561.

[167] Ben-Batalla I, Erdmann R, Jorgensen H, Mitchell R, Ernst T, von Amsberg G, et al. Axl blockade by BGB324 inhibits BCR-ABL tyrosine kinase inhibitor-sensitive and -resistant chronic myeloid leukemia. Clin Cancer Res 2017;23(9):2289–300.

[168] DeRyckere D, Lee-Sherick AB, Huey MG, Hill AA, Tyner JW, Jacobsen KM, et al. UNC2025, a MERTK small-molecule inhibitor, is therapeutically effective alone and in combination with methotrexate in leukemia models. Clin Cancer Res 2017;23(6):1481–92.

[169] Minson KA, Smith CC, DeRyckere D, Libbrecht C, Lee-Sherick AB, Huey MG, et al. The MERTK/FLT3 inhibitor MRX-2843 overcomes resistance-conferring FLT3 mutations in acute myeloid leukemia. JCI Insight 2016;1(3):e85630.

[170] Martinelli E, Martini G, Cardone C, Troiani T, Liguori G, Vitagliano D, et al. AXL is an oncotarget in human colorectal cancer. Oncotarget 2015;6(27):23281–96.

[171] Guinney J, Dienstmann R, Wang X, de Reynies A, Schlicker A, Soneson C, et al. The consensus molecular subtypes of colorectal cancer. Nat Med 2015;21(11):1350–6.

[172] Lehmann BD, Bauer JA, Chen X, Sanders ME, Chakravarthy AB, Shyr Y, et al. Identification of human triple-negative breast cancer subtypes and preclinical models for selection of targeted therapies. J Clin Invest 2011;121(7):2750–67.

[173] Loges S, Schmidt T, Tjwa M, van Geyte K, Lievens D, Lutgens E, et al. Malignant cells fuel tumor growth by educating infiltrating leukocytes to produce the mitogen Gas6. Blood 2010;115(11):2264–73.

[174] Burbridge MF, Bossard CJ, Saunier C, Fejes I, Bruno A, Leonce S, et al. S49076 is a novel kinase inhibitor of MET, AXL, and FGFR with strong preclinical activity alone and in association with bevacizumab. Mol Cancer Ther 2013;12(9):1749–62.

[175] Davis MI, Hunt JP, Herrgard S, Ciceri P, Wodicka LM, Pallares G, et al. Comprehensive analysis of kinase inhibitor selectivity. Nat Biotech 2011;29(11):1046–51.

[176] Baladi T, Abet V, Piguel S. State-of-the-art of small molecule inhibitors of the TAM family: the point of view of the chemist. Eur J Med Chem 2015;105:220–37.

[177] Mori M, Kaneko N, Ueno Y, Yamada M, Tanaka R, Saito R, et al. Gilteritinib, a FLT3/AXL inhibitor, shows antileukemic activity in mouse models of FLT3 mutated acute myeloid leukemia. Invest New Drugs 2017;. May 17. doi:10.1007/s10637-017-0470-z. [Epub ahead of print].

[178] Mollard A, Warner SL, Call LT, Wade ML, Bearss JJ, Verma A, et al. Design, synthesis and biological evaluation of a series of novel Axl kinase inhibitors. ACS Med Chem Lett 2011;2(12):907–12.

[179] Zhang W, DeRyckere D, Hunter D, Liu J, Stashko MA, Minson KA, et al. UNC2025, a potent and orally bioavailable MER/FLT3 dual inhibitor. J Med Chem 2014;57(16):7031–41.

CHAPTER

4

The JNK Pathway in Drug Resistance

Lanlin Hu, Fangdong Zou*, Jennifer R. Grandis**, Daniel E. Johnson***

*Sichuan University, Chengdu, Sichuan, PR China

**Department of Otolaryngology – Head and Neck Surgery, University of California at San Francisco, San Francisco, CA, United States

Abstract

The c-JUN NH2-terminal kinase (JNK) pathway plays key roles in cellular proliferation and survival. Aberrant activation of the pathway occurs in multiple malignancies, where it contributes primarily to tumor growth. In addition, JNK pathway activation has been reported to promote resistance to anticancer agents. In this chapter, we will review current understanding of the involvement of the JNK pathway in tumor development and drug resistance. The identification and characterization of JNK pathway inhibitors will be described. Lastly, we will discuss the potential application of JNK inhibitors in human malignancies as a means of enhancing the effects of conventional chemotherapy drugs or overcoming resistance to these agents.

ABBREVIATIONS

ASK1	Apoptosis signal-regulating kinase 1
DMBA	7,12-Dimethylbenz[α]anthracene
DUSP	Dual-specificity phosphatase
EGFR	Epidermal growth factor receptor
ERK	Extracellular signal-regulated kinase
GPCR	G-protein coupled receptor
GRASP-1	GRIP1-associated protein 1
HCC	Hepatocellular carcinoma
HNSCC	Head and neck squamous cell carcinoma
HPK1	Hematopoietic progenitor kinase 1
JAK	Janus kinase
JIP	c-JUN NH2-terminal kinase-interacting protein 1
JNK	c-JUN NH2-terminal kinase
MAPK	Mitogen-activated protein kinase
MAPKK	Mitogen-activated protein kinase kinase
MAP2K	Mitogen-activated protein kinase kinase
MAPKKK	Mitogen-activated protein kinase kinase kinase

Targeting Cell Survival Pathways to Enhance Response to Chemotherapy. http://dx.doi.org/10.1016/B978-0-12-813753-6.00004-4
Copyright © 2018 Elsevier Inc. All rights reserved.

MAP3K	Mitogen-activated protein kinase kinase kinase
MKP	Mitogen-activated protein kinase phosphatase
MLK	Mixed-lineage kinase
POSH	Plenty of SH3s
RACK1	Receptor for activated C kinase 1
SAPK1	Stress-activated mitogen-activated protein kinase
STAT3	Signal transducer and activator of transcription 3
TAK1	Transforming growth factor-β-activated kinase 1
T-ALL	T-cell acute lymphoblastic leukemia
TGF-β	Transforming growth factor-β
TNF	Tumor necrosis factor
TPA	12-*O*-tetradecanoylphorbol-13-acetate
VEGF	Vascular endothelial growth factor
WDR62	WD repeat domain 62

4.1 INTRODUCTION

Cells respond to changes in the environment, including stimulation with cytokines, growth factors, and stresses, by activating intracellular signal transduction pathways. A key signaling pathway is the mitogen-activated protein kinase (MAPK) pathway, which relays, amplifies, and integrates signals from various stimuli and regulates several important physiological processes, including inflammation, proliferation, differentiation, and cell death [1]. As a ubiquitous and highly conserved signaling mechanism, the MAPK pathway consists of three classic tiers of kinase families, MAPKs, MAPK kinases (MAPKK or MAP2K), and MAPKK kinases (MAPKKK or MAP3K) (Fig. 4.1). Within the distal MAPK family, four subfamilies have been identified in eukaryotic cells, including the c-JUN NH2-terminal kinase (JNK) subfamily, the extracellular signal-related kinases (ERK) 1/2 subfamily, p38, and ERK5 [1,2]. Each of these MAPKs is activated by tyrosine and threonine phosphorylation catalyzed by upstream MAPKKs (Fig. 4.1). The ERK proteins contain classical dual phosphorylation motifs, Thr-Glu-Tyr, and are typically activated following stimulation with mitogens [3]. JNK and p38 contain the dual phosphorylation motifs Thr-Pro-Tyr and Thr-Gly-Tyr, respectively, which commonly are phosphorylated in response to cellular stresses or stimulation with proinflammatory cytokines. The dual phosphorylation of MAPKs is mediated by MAP2Ks, which are themselves activated by serine/threonine phosphorylation by the upstream MAP3Ks.

Early studies demonstrated that JNK, also known as stress-activated MAP kinase (SAPK), binds to the N-terminal activation domain of the transcription factor c-JUN [4]. JNK phosphorylates c-JUN on Ser-63 and Ser-73, promoting the transcriptional activity of the protein [5]. The JNK pathway has been intensively investigated for more than three decades, but its perplexing complexity is still not fully understood. The diverse contributions of the JNK pathway to processes such as proliferation and differentiation give rise to complex roles in cancer, where both oncogenic and tumor suppressor roles have been described. Additionally, a considerable body of evidence has emerged, implicating JNK pathway activation in the development of resistance to anticancer drugs [6]. Thus, considerable effort has been invested by academic and pharmaceutical teams to screen chemical libraries or pursue rational drug design as a means of identifying inhibitors of the

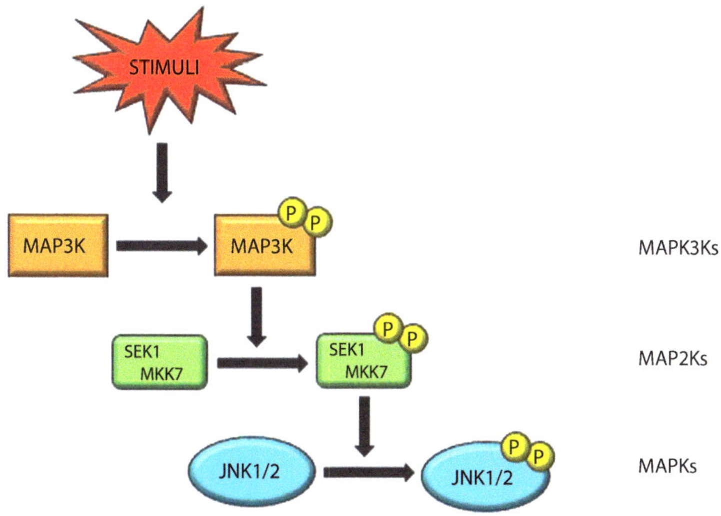

FIG. 4.1 **The MAPK signaling cascade.** In response to cellular stimuli, MAP3Ks become activated via phosphorylation. The phosphorylated/activated MAP3Ks phosphorylate and activate MAP2Ks, which subsequently phosphorylate and activate MAPKs (including JNK1/2).

JNK pathway. In this chapter, we will describe the JNK pathway and discuss its role in the development of cancer and drug resistance. The potential for therapeutic targeting of the JNK pathway will also be discussed.

4.2 THE JNK PATHWAY

4.2.1 Structure and Isoforms

The JNK family comprises three kinases, JNK1, JNK2, and JNK3, encoded by the *jnk1*, *jnk2*, and *jnk3* genes, respectively. The general structure of the JNK proteins consists of N- and C-terminal lobes (Fig. 4.2). The docking site domain, which is comprised of the negatively charged common docking (CD) region and the hydrophobic docking groove, plays a crucial role in physical interactions with binding partners and substrate recruitment. The activation loop (A-loop) contains the Thr-Pro-Tyr motif [7–11]. *Jnk1* and *Jnk2* are ubiquitously expressed, whereas *Jnk3* expression is restricted to brain, heart, and testis [7]. The three *Jnk* genes are also subject to alternative splicing, leading to the production of different isoforms. One alternative splice site lies within the sequence coding for the C-terminal lobe. The use of alternate internal coding exons leads to the generation of two similar isoforms, the α- and β-isoforms. In certain tissues, the expression of tissue-specific splicing factors, such as the Nova family of neuronal splicing factors, results in expression of tissue-specific JNK isoforms. Additionally, alternative splicing in the final coding exon results in a 5-nucleotide shift and the use of different reading frames, producing isoforms with distinct lengths and C-termini [9]. JNK isoforms with different C-termini resulting from frameshift can exhibit distinct cellular

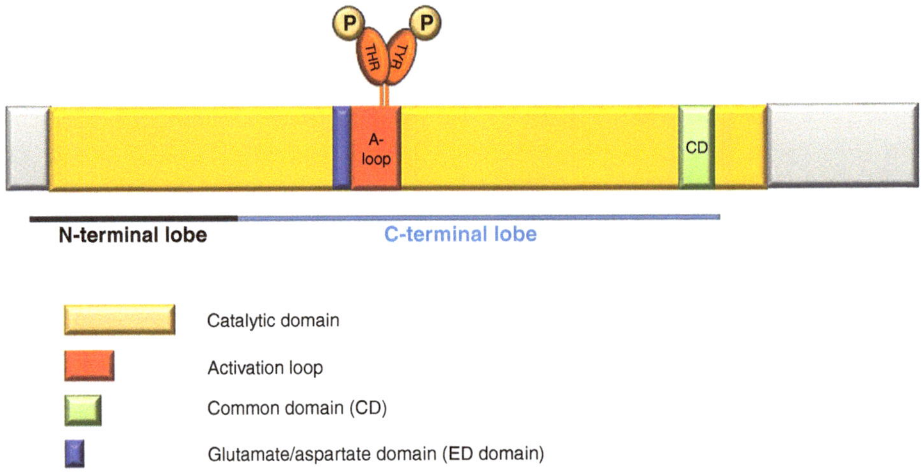

FIG. 4.2 **JNK schematic representation.**

activities. For example, the short isoform JNK1-α1 has an anti-apoptotic function in Colo205 colon cancer cells, whereas the long isoforms JNK1-α2 and -β2 promote apoptosis in these cells [12].

4.2.2 Upstream Activation of the JNK Signaling Pathway

JNK pathways are initiated by an assortment of extracellular and intracellular stimuli, including hormones, pathogens, cytokines, UV radiation, oxidative stress, and DNA damage. Exposure to these stimuli can lead to activation of a broad variety of upstream signaling components, including G-protein coupled receptors (GPCRs), Wnt receptors, transforming growth factor-β (TGF-β) receptors, tumor necrosis factor (TNF) receptors, and the Toll-like receptor complex [9]. Consequently, cascades of kinase signaling are activated that converge on JNKs.

JNKs are activated by phosphorylation of A-loop Thr and Tyr residues by two MAP2ks, MKK7 and SEK1 (also known as MKK4). SEK1 and MKK7 preferentially phosphorylate JNK on Tyr and Thr residues, respectively [13]. In stress-stimulated mouse embryonic stem (ES) cells, JNK1 is first phosphorylated by SEK1 (on Tyr), followed by phosphorylation by MKK7 (on Thr) [14]. Thus, SEK1 and MKK7 appear to cooperate in the activation of JNKs. SEK1 and MMK7 become activated following dual phosphorylation by MAP3Ks. Several MAP3Ks capable of activating the JNK pathway have been identified, including MEKK1 [15], mixed-lineage kinases (MLKs) [16], apoptosis signal-regulating kinase 1 (ASK1) [17], and transforming growth factor β-activated kinase 1 (TAK1) [18].

4.2.3 The Role of Scaffold Proteins

Scaffold proteins play key roles in providing a platform for signaling molecules to assemble, promoting the localization of signaling molecules at specific sites and coordinating positive and negative feedback signals for pathway regulation. Several scaffold proteins have been identified that bind to JNKs and upstream activators. In melanoma cells, filamin was found to interact with SEK1 as a scaffold protein, in response to TNF-α stimulation [19]. The scaffold proteins POSH (plenty of SH3s) promote JNK pathway activation by directly interacting with GTP-bound RAC1 [20]. POSH complexes with and sequentially stimulates RAC1, MLKs, SEK1, MKK7, and JNKs [21]. Scaffolds JIP1 (JNK-interacting protein 1) and JIP3 recruit several different activators and JNKs. JIP1 recruits DLK, MKK7, and JNK as a complex [22], whereas JIP3 binds SEK1 and MEKK1 to stimulate SEK1, leading to JNK3 activation [23]. β-Arrestin-2 contains a MAP kinase docking site and mainly functions as a scaffold protein in activation of JNK3, by tethering ASK1, SEK1, and JNK3 [24]. Crk II activates HPK1 (hematopoietic progenitor kinase 1) and SEK1-dependent activation of the JNK pathway, by recruiting JNK1 to a p130Cas multiprotein complex [25].

DUSP19 (dual-specificity phosphatase 19) is the first MAPK phosphatase (MKP) identified as a scaffold protein. DUSP19 physically interacts with MKK7 and indirectly regulates JNK [26]. In 2016, another MKP, DUSP22, was reported to interact with ASK1, MKK7, and JNK1 to promote JNK activation independent of its phosphatase activity [27]. In neurons, GRASP-1 (GRIP1-associated protein 1) binds both JNK1 and MEKK1. Following cleavage by caspase-3, the C-terminal domain of GRASP1 is released to promote activation of JNK [28]. In response to stress resulting from X-rays and genotoxic drugs, RACK1 (Receptor for Activated C Kinase 1) has been shown to bind and activate JNK1 and MTK1 [29]. Additionally, WDR62 (WD repeat domain 62) has been implicated in noncanonical activation of JNK via association with JNK and MKK7 [30]. Additional scaffold proteins have been identified that participate in the activation of p38 and ERKs. However, the mechanistic details whereby many of the scaffold proteins promote pathway activation remain to be elucidated.

4.2.4 Substrates of JNK

The JNKs were initially described as kinases responsible for the phosphorylation of c-JUN transcription factor. Subsequently, over 100 proteins have been identified as JNK substrates, including several transcription factors and nuclear hormone receptors, as well as proteins that are not transcription factors. JNK1 binds to the activation domain of c-JUN, specifically phosphorylating Ser63 and Ser73 residues near the N-terminus. JNK1-mediated phosphorylation of c-JUN inhibits the ubiquitination and proteasomal degradation of c-JUN, leading to enhanced transcriptional activity of the AP-1 complex containing c-JUN and c-FOS, and subsequently leading to the induction of AP-1 target genes [1]. The impact of JNK on c-JUN function has been explored extensively and is well documented [30]. Below, we will describe the influence of JNKs on three less characterized, albeit key, substrates, the transcription factor STAT3, the anti-apoptotic protein Bcl-2, and the pro-apoptotic protein Bax.

STAT3, or signal transducer and activator of transcription 3, exists in the cytoplasm as an inactive monomer. When cells are stimulated by cytokines or growth factors, STAT3 becomes phosphorylated on Tyr705 by activated tyrosine kinases called JAK enzymes. Phosphorylation of Tyr705 leads to dimerization of STAT3 and translocation to the nucleus, where the active STAT3 dimer induces the expression of a number of target genes that promote cellular proliferation and survival. It has also been reported that JNK1/2 directly phosphorylates Ser727 on STAT3 following UVA irradiation, enhancing the DNA binding activity of STAT3 in epidermal

JB6 cells [31,32]. In BEAS-2B cells, a human bronchial epithelial cell line, carcinogenic arsenic also activates JNK to promote STAT3 Ser727 phosphorylation, followed by STAT3-mediated induction of vascular endothelial growth factor (VEGF) and cell migration [33]. By contrast, in head and neck squamous cell carcinoma (HNSCC) cell lines, JNK phosphorylation of Ser727 has been reported to reduce phosphorylation of Tyr705, leading to inhibition of STAT3 transcriptional activity, decreased cyclin D1 expression, and reduced cellular proliferation and viability [34]. These somewhat discordant findings have raised controversy surrounding the role of JNK-dependent Ser727 phosphorylation. It is possible that the effects of Ser727 phosphorylation on STAT3 functional activity may be dictated by the repertoire of accessory transcriptional modulators expressed by the cell and, hence, are cell-type specific.

Bcl-2 is an anti-apoptotic protein that resides primarily in the mitochondria. Under healthy growing conditions, Bcl-2 prevents apoptosis by binding to the pro-apoptotic Bcl-2 family members Bax and Bak. Additionally, Bcl-2 prevents induction of autophagy by binding to Beclin-1. When cells are subjected to stress conditions, such as nutrient deprivation, JNK1 becomes activated and phosphorylates Ser70, Ser87, and Thr69 in the unstructured loop region of Bcl-2 [35–38]. Phosphorylation of Bcl-2 on these residues causes dissociation from Beclin-1, as well as Bax (Fig. 4.3). Initially, sufficient levels of Beclin-1 are released to trigger induction of autophagy and the cells undergo characteristic biochemical and morphologic features of the autophagy process. If the stress is not removed, and JNK1 remains active, more Bcl-2 is phosphorylated and the amount of released Bax rises to a level that leads to induction of apoptosis and the ultimate death of the cell. Additional mechanisms of Bax regulation by the JNK pathway have also been reported (Fig. 4.3). In one mechanism, activated JNK directly

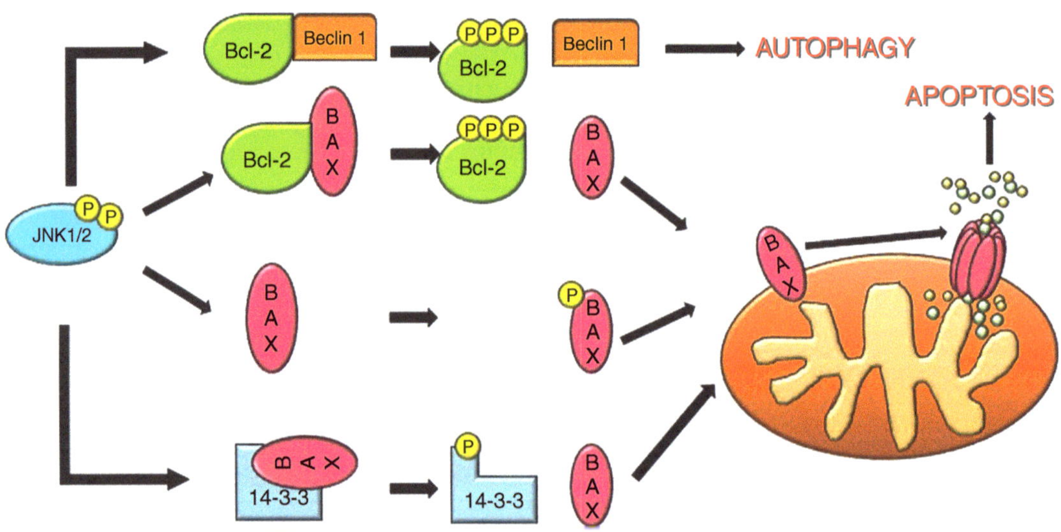

FIG. 4.3 **Induction of apoptosis and autophagy by JNKs.** In response to cellular stimuli, activated JNK1/2: (1) phosphorylate 14-3-3 on serine, causing release of bound Bax, (2) directly phosphorylate Bax on Thr167, and (3) phosphorylate Bcl-2 on Thr69, Ser70, and Ser87, causing the release of Bax (promotes apoptosis) and Beclin-1 (promotes autophagy) from the Bcl-2 protein.

phosphorylates cytosolic Bax on Thr167. This stimulates translocation of Bax to the outer mitochondrial membrane where it becomes available for homo- or hetero-oligomerization with another Bcl-2 family member [38]. In another described mechanism, JNK acts to release cytosolic Bax that is complexed with 14-3-3 protein. Here, Bax is bound by 14-3-3 in a phosphorylation-independent manner, and activated JNK directly phosphorylates the 14-3-3 protein. This allows release of Bax and translocation of Bax to the mitochondria, promoting cytochrome c release and induction of apoptosis [39,40].

4.3 JNK PATHWAY IN CANCER

The roles of JNK enzymes in cancer have been extensively investigated and reported in a large body of scientific literature. While a majority of studies implicate JNKs in promoting cancer, tumor suppressor roles have also been described. It appears that the precise roles of JNKs are context dependent. Thus, it is particularly important to understand the specifics of the cancer model being studied when investigating the functional consequences of JNK activation.

4.3.1 Tumor Suppressive Role

Tumor suppressive roles for JNKs have become readily apparent through the study of knockout mice. Treatment of *JNK1*-deficient mice with 7,12-dimethylbenz[α]anthracene (DMBA) and 12-*O*-tetradecanoylphorbol-13-acetate (TPA) results in enhanced papilloma development relative to treatment of wild-type mice [41]. JNK1 knockout models are also prone to spontaneous development of intestinal tumors [42]. Further studies have shown that the JNK pathway can act to suppress the oncogenic properties of RAS and inhibit the development of tumors of the breast, prostate, and colon [43–47]. The tumor suppressive function of JNKs is likely related to their role in regulating cell death/cell survival processes, including apoptosis and autophagy. For example, when multiple myeloma cells are subjected to stress, JNK activation stimulates the release of pro-apoptotic SMAC from the mitochondria, leading to induction of apoptotic cell death [48]. In addition, treatment with cytotoxic drugs leads to JNK activation in numerous types of cancer cells. In this scenario, the activated JNKs promote apoptosis by phosphorylating transcription factors such as c-JUN, ATF2, or p53, or by phosphorylating anti-apoptotic Bcl-2, causing the release and activation of pro-apoptotic Bax [7,49,50].

4.3.2 Role in Oncogenesis

Emerging evidence indicates that the JNK signaling pathway plays an important role in the transformation of cells, as well as the processes of migration and invasion. Signaling via JNKs has been shown to be important for the transforming activities of several different oncogenes, including the genes encoding RAS, c-FOS, c-MET, and BCR-ABL [51,52]. Interestingly, studies in liver biology have shown that $JNK1^{-/-}$ and $JNK2^{-/-}$ mice exhibit resistance to hepatitis, with JNK1 playing a primary role in the development of hepatitis [53,54]. Additional studies have revealed that JNK1 promotes the development of hepatocellular carcinoma (HCC) [55,56]. The ability of JNK1 signaling to stimulate inflammation may underlie its important role in both hepatitis and HCC [57,58]. Like JNK1, JNK2 has been strongly implicated in promoting the growth of a variety of tumor types, including cancers of the lung, brain, prostate, and skin, as well as multiple myeloma [59–63]. Chemical carcinogenesis in skin caused by the application of DMBA/TPA is significantly reduced in mice lacking normal expression of JNK2 [64]. In the lung, JNK1 and JNK2 have been shown to mediate important oncogenic properties of K-RAS [65–67]. These findings are particularly important in view of the fact that roughly 36% of lung adenocarcinomas exhibit genetic alterations in *K-Ras*. In sum, JNK activation plays a key role in multiple cancers, underscoring the potential value of identifying strategies and agents that can be used to modulate the activity or expression of components of the JNK signaling pathway.

4.4 THE JNK PATHWAY IN DRUG RESISTANCE

As discussed in the last section, the JNK pathway plays an important role in tumor development. Multiple studies have also documented a role for the JNK pathway in mediating resistance to chemotherapy drugs. However, similar to what is seen in tumor development or suppression, roles for the JNK pathway in cellular responses to chemotherapy differ depending on the cancer cell type. Clear roles for the JNK pathway in drug resistance have been demonstrated in several cancer models, including lung adenocarcinoma, squamous cell carcinoma, melanoma, T-cell acute lymphoblastic leukemia (T-ALL), HCC, and colon cancer [68–74]. In breast cancer cell lines, the JNK pathway has been shown to mediate acquired resistance to EGFR/HER2-targeted therapies [73]. In tetrandrine-resistant Jurkat T leukemic cells, AP-1 activity has been shown to be elevated [75]. In lung cancer cells, the JNK pathway mediates cisplatin induction of drug resistance genes [76]. By contrast, inhibition of the JNK pathway in HCC cell lines leads to enhanced resistance to TRAIL (tumor necrosis factor-related apoptosis-inducing ligand) [77]. Despite these divergent roles, on the whole, activation of the JNK is more commonly associated with drug resistance than drug sensitization. Thus, it is not surprising that considerable effort has been invested in developing inhibitors of the JNK pathway. In the next section, we describe several of these inhibitors, including ATP-competitive inhibitors and noncompetitive peptide inhibitors.

4.5 INHIBITORS OF THE JNK PATHWAY

4.5.1 ATP-Competitive Inhibitors in Clinical Trials

First described in 2001, SP600125, an anthrapyrazolone inhibitor of JNK1/2/3 has been used extensively to investigate *in vitro* and *in vivo* roles of the JNK pathway (Table 4.1) [78]. SP600125 is a reversible ATP-competitive inhibitor with >300-fold selectivity for JNKs over the related MAPKs, ERK1, and p38-2 [78]. SP600125 inhibits phosphorylation of c-JUN in a dose-dependent fashion and reduces the expression of pro-inflammatory genes [78]. In some disease models, treatment with SP600125 has been shown to control disease progression [79], whereas in other disease models the inhibitor failed to impact tumor growth and was found to induce significant toxicities [80]. CC-401 is an ATP-competitive inhibitor based

TABLE 4.1 JNK Inhibitors

Name	Structure or sequence	Target	Clinical trials	References
SP600125	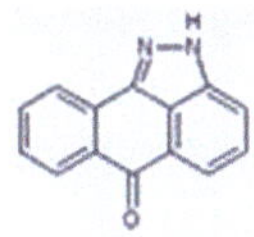	JNK1/2/3 MKK3/4/6 COX-2 IL-2 IFN-y TNF-a	None	[78–80]
CC-401	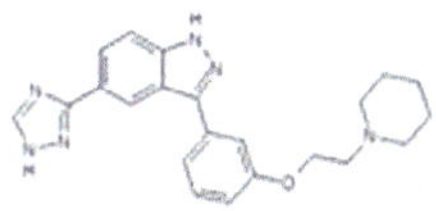	JNK1/2/3	NCT00126893	[69,81,82]
CC-930 (Tanzisertib)	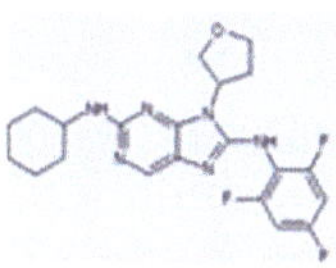	JNK1/2/3	NCT01466725 NCT01203943	[83,84]
AS601245	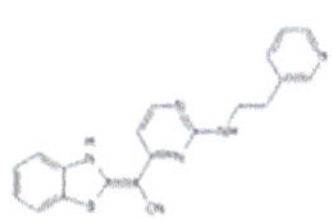	JNK1/2/3		[85,86]
AS602801 (PGL-5001 or Bentamapimod)	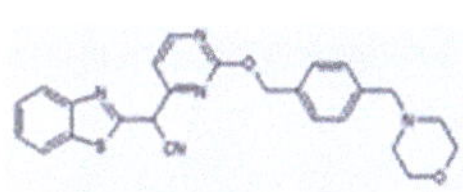	JNK1/2/3	NCT01630252	[87]
D-JNKI (XG-102 or AM-111)	DQSRPVQPFLNLTTPRKP RPPRRRQRRKKRG	JNK1/2/3	NCT01570205 NCT00802425 NCT02235272 NCT02508337 NCT02809118 NCT02561091	[88–95]

on SP600125 that potently inhibits all three JNKs (JNK1/2/3) and exhibits greater than 40-fold selectivity for JNKs compared to other related kinases [69]. CC-401 has shown efficacy in sensitizing colon cancer cells to DNA-damaging agents and also exhibits activity in experimental models of unilateral ureteral obstruction and antiglomerular basement membrane disease [69,81,82]. CC-401 has been investigated in a Phase I clinical trial of high-risk myeloid leukemia patients (NCT00126893; clinicaltrials.gov). Another SP600125-derived JNK inhibitor is CC-930 (Tanzisertib), an orally active aminopurine compound [83]. Phase I evaluation of CC-930 has demonstrated a decrease in UV-induced phosphorylation of c-JUN in the skin of treated patients [84]. However, two Phase II clinical trials of CC-930 in discoid lupus erythematosus (NCT01466725) and idiopathic pulmonary fibrosis (NCT01203943) were prematurely terminated due to a poor benefit–risk profile. AS601245 is a structurally unique ATP-competitive JNK inhibitor with anti-inflammatory and neuroprotective properties [85].

AS601245 inhibits the proliferation of colon cancer cells and has been reported to protect the brain from ischemic injury [85,86]. Phase II testing of AS602801 (also known as PGL-5001 or Bentamapimod), a related orally active inhibitor, has recently concluded in patients with inflammatory endometriosis (NCT01630252). AS602801 has been reported to inhibit pancreatic cancer stem cells *in vitro* and *in vivo* [87]. The structures, activities, and mechanisms of additional ATP-competitive inhibitors of JNKs are described elsewhere [88,89].

4.5.2 Peptide Inhibitors of the JNK Signaling Pathway

The vast majority of inhibitors of protein kinases are directed against the kinase ATP binding site. However, since all kinases contain ATP binding sites, these inhibitors often lack specificity and inhibit multiple different kinases. Thus, it is beneficial to design or screen for noncompetitive inhibitors, including small molecules and peptides that act to disrupt key protein–protein interactions. A cell-permeable peptide encompassing the JNK-binding site of c-JUN has been reported to specifically and efficiently disrupt the c-JUN/JNK complex *in vitro* and *in vivo* [90]. An additional peptide inhibitor, D-JNKI (also known as XG-102, AM-111), interferes with JIP1/JNK interactions (Table 4.1) [91]. This D-amino acid-containing, retro-inverso peptide demonstrates impressive stability in cells and significant biological activity in a variety of *in vivo* models, including preclinical models of cerebral ischemia, hearing loss, liver injury, and cancer [92–95]. Phase I testing has been completed examining the tolerability and pharmacokinetics of a single intravenous infusion of D-JNKI in healthy volunteers (NCT01570205). Clinical trials are currently ongoing, examining the impact of D-JNKI in patients with acute sensorineural hearing loss (NCT00802425, NCT02809118, NCT02561091). In addition, a Phase III clinical trial is investigating the efficacy of D-JNKI in reducing inflammation and pain following cataract surgery (NCT02235272, NCT02508337). In view of the antiproliferative activity of D-JNKI against melanoma cells and its relatively positive safety profile in humans, testing of this novel compound in cancer patients appears warranted [91].

4.6 CONCLUSIONS

To date, the evaluation of JNK inhibitors in cancer clinical trials has been limited, despite ongoing testing in other diseases. The ability to deliver tolerable doses over extended periods of time remains a significant challenge. Moreover, it is doubtful that JNK inhibitors will prove to be highly effective as single agents. The identification of additional, well-tolerated inhibitors and the identification of combinatorial targeting strategies, preferably those that generate synergistic antitumor responses or overcome resistance to conventional anticancer agents, should be key goals of future investigations

Acknowledgments

This work was supported by National Institutes of Health grants R01 DE24728 (DEJ), P50CA097190 (DEJ and JRG), and R01 DE023685 (JRG).

Conflict of Interest: No potential conflicts of interest were disclosed.

References

[1] Weston CR, Davis RJ. The JNK signal transduction pathway. Curr Opin Genet Dev 2002;12(1):14–21.

[2] Lee JD, Ulevitch RJ, Han JH. Primary structure of Bmk1—a new mammalian map kinase. Biochem Biophys Res Commun 1995;213(2):715–24.

[3] Kamakura S, Moriguchi T, Nishida E. Activation of the protein kinase ERK5/BMK1 by receptor tyrosine kinases—identification and characterization of a signaling pathway to the nucleus. J Biol Chem 1999;274(37):26563–71.

[4] Adler V, Polotskaya A, Wagner F, Kraft AS. Affinity-purified C-Jun amino-terminal protein-kinase requires serine threonine phosphorylation for activity. J Biol Chem 1992;267(24):17001–5.

[5] Pulverer BJ, Kyriakis JM, Avruch J, Nikolakaki E, Woodgett JR. Phosphorylation of C-Jun mediated by Map kinases. Nature 1991;353(6345):670–4.

[6] Vasilevskaya I, O'Dwyer PJ. Role of Jun and Jun kinase in resistance of cancer cells to therapy. Drug Resist Update 2003;6(3):147–56.

[7] Davis RJ. Signal transduction by the JNK group of MAP kinases. Cell 2000;103(2):239–52.

[8] Xie XL, Gu Y, Fox T, Coll JT, Fleming MA, Markland W, et al. Crystal structure of JNK3: a kinase implicated in neuronal apoptosis. Struct Fold Des 1998;6(8):983–91.

[9] Zeke A, Misheva M, Remenyi A, Bogoyevitch MA. JNK signaling: regulation and functions based on complex protein–protein partnerships. Microbiol Mol Biol Rev 2016;80(3):793–835.

[10] Shaw D, Wang SM, Villasenor AG, Tsing S, Walter D, Browner MF, et al. The crystal structure of JNK2 reveals conformational flexibility in the MAP kinase insert and indicates its involvement in the regulation of catalytic activity. J Mol Biol 2008;383(4):885–93.

[11] Heo JS, Kim SK, Seo CI, Kim YK, Sung BJ, Lee HS, et al. Structural basis for the selective inhibition of JNK1 by the scaffolding protein JIP1 and SP600125. EMBO J 2004;23(11):2185–95.

[12] Mahalingam D, Keane M, Pirianov G, Mehmet H, Samali A, Szegezdi E. Differential activation of JNK1 isoforms by TRAIL receptors modulate apoptosis of colon cancer cell lines. Br J Cancer 2009;100(9):1415–24.

[13] Lawler S, Fleming Y, Goedert M, Cohen P. Synergistic activation of SAPK1/JNK1 by two MAP kinase kinases *in vitro*. Curr Biol 1998;8(25):1387–90.

[14] Kishimoto H, Nakagawa K, Watanabe T, Kitagawa D, Momose H, Seo J, et al. Different properties of SEK1 and MKK7 in dual phosphorylation of stress-induced activated protein kinase SAPK/JNK in embryonic stem cells. J Biol Chem 2003;278(19):16595–601.

[15] Charlaftis N, Suddason T, Wu XF, Anwar S, Karin M, Gallagher E. The MEKK1 PHD ubiquitinates TAB1 to activate MAPKs in response to cytokines. EMBO J 2014;33(21):2581–96.

[16] Gallo KA, Johnson GL. Mixed-lineage kinase control of JNK and p38 MAPK pathways. Nat Rev Mol Cell Biol 2002;3(9):663–72.

[17] Takeda K, Matsuzawa A, Nishitoh H, Ichijo H. Roles of MAPKKK ASK1 in stress-induced cell death. Cell Struct Funct 2003;28(1):23–9.

[18] Brown K, Vial SCM, Dedi N, Long JM, Dunster NJ, Cheetham GMT. Structural basis for the interaction of TAK1 kinase with its activating protein TAB1. J Mol Biol 2005;354(5):1013–20.

[19] Marti A, Luo Z, Cunningham C, Ohta Y, Hartwig J, Stossel TP, et al. Actin-binding protein-280 binds the stress-activated protein kinase (SAPK) activator SEK-1 and is required for tumor necrosis factor-α activation of SAPK in melanoma cells. J Biol Chem 1997;272(5):2620–8.

[20] Tapon N, Nagata K, Lamarche N, Hall A. A new Rac target POSH is an SH3-containing scaffold protein involved in the JNK and NF-κB signalling pathways. EMBO J 1998;17(5):1395–404.

[21] Xu Z, Kukekov NV, Greene LA. POSH acts as a scaffold for a multiprotein complex that mediates JNK activation in apoptosis. EMBO J 2003;22(2):252–61.

[22] Nihalani D, Wong HN, Holzman LB. Recruitment of JNK to JIP1 and JNK-dependent JIP1 phosphorylation regulates JNK module dynamics and activation. J Biol Chem 2003;278(31):28694–702.

[23] Ito M, Yoshioka K, Akechi M, Yamashita S, Takamatsu N, Sugiyama K, et al. JSAP1, a novel jun N-terminal protein kinase (JNK)-binding protein that functions as a Scaffold factor in the JNK signaling pathway. Mol Cell Biol 1999;19(11):7539–48.

[24] Miller WE, Lefkowitz RJ. Expanding roles for β-arrestins as scaffolds and adapters in GPCR signaling and trafficking. Curr Opin Cell Biol 2001;13(2):139–45.

[25] Girardin SE, Yaniv M. A direct interaction between JNK1 and CrkII is critical for Rac1-induced JNK activation. EMBO J 2001;20(13):3437–46.

[26] Zama T, Aoki R, Kamimoto T, Inoue K, Ikeda Y, Hagiwara M. A novel dual specificity phosphatase SKRP1 interacts with the MAPK kinase MKK7 and inactivates the JNK MAPK pathway implication for the precise regulation of the particular MAPK pathway. J Biol Chem 2002;277(26):23909–18.

[27] Ju A, Cho Y-C, Kim BR, Park SG, Kim J-H, Kim K, et al. Scaffold role of DUSP22 in ASK1-MKK7-JNK signaling pathway. PLoS ONE 2016;11(10):e0164259.

[28] Ye B, Yu W-P, Thomas GM, Huganir RL. GRASP-1 is a neuronal scaffold protein for the JNK signaling pathway. FEBS Lett 2007;581(23):4403–10.

[29] Arimoto K, Fukuda H, Imajoh-Ohmi S, Saito H, Takekawa M. Formation of stress granules inhibits apoptosis by suppressing stress-responsive MAPK pathways. Nat Cell Biol 2008;10(11):1324.

[30] Bogoyevitch MA, Kobe B. Uses for JNK: the many and varied substrates of the c-Jun N-terminal kinases. Microbiol Mol Biol Rev 2006;70(4):1061–95.

[31] Zhang Y, Liu G, Dong Z. MSK1 and JNKs mediate phosphorylation of STAT3 in UVA-irradiated mouse epidermal JB6 cells. J Biol Chem 2001;276(45):42534–42.

[32] Liu J, Chen B, Lu Y, Guan Y, Chen F. JNK-dependent Stat3 phosphorylation contributes to Akt activation in response to arsenic exposure. Toxicol Sci 2012;129(2):363–71.

[33] Sun J, Yu M, Lu Y, Thakur C, Chen B, Qiu P, et al. Carcinogenic metalloid arsenic induces expression of mdig oncogene through JNK and STAT3 activation. Cancer Lett 2014;346(2):257–63.

[34] Gkouveris I, Nikitakis N, Karanikou M, Rassidakis G, Sklavounou A. JNK1/2 expression and modulation of STAT3 signaling in oral cancer. Oncol Lett 2016;12(1):699–706.

[35] Maundrell K, Antonsson B, Magnenat E, Camps M, Muda M, Chabert C, et al. Bcl-2 undergoes phosphorylation by c-Jun N-terminal kinase/stress-activated protein kinases in the presence of the constitutively active GTP-binding protein Rac1. J Biol Chem 1997;272(40):25238–42.

[36] Yamamoto K, Ichijo H, Korsmeyer SJ. BCL-2 is phosphorylated and inactivated by an ASK1/Jun N-terminal protein kinase pathway normally activated at G2/M. Mol Cell Biol 1999;19(12):8469–78.

[37] Wei Y, Pattingre S, Sinha S, Bassik M, Levine B. JNK1-mediated phosphorylation of Bcl-2 regulates starvation-induced autophagy. Mol Cell 2008;30(6):678–88.

[38] Kim B-J, Ryu S-W, Song B-J. JNK-and p38 kinase-mediated phosphorylation of Bax leads to its activation and mitochondrial translocation and to apoptosis of human hepatoma HepG2 cells. J Biol Chem 2006;281(30):21256–65.

[39] Tsuruta F, Sunayama J, Mori Y, Hattori S, Shimizu S, Tsujimoto Y, et al. JNK promotes Bax translocation to mitochondria through phosphorylation of 14-3-3 proteins. EMBO J 2004;23(8):1889–99.

[40] Muscarella DE, Bloom SE. The contribution of c-Jun N-terminal kinase activation and subsequent Bcl-2 phosphorylation to apoptosis induction in human B-cells is dependent on the mode of action of specific stresses. Toxicol Appl Pharmacol 2008;228(1):93–104.

[41] She Q-B', Chen N, Bode AM, Flavell RA, Dong Z. Deficiency of c-Jun-NH2-terminal kinase-1 in mice enhances skin tumor development by 12-*O*-tetradecanoylphorbol-13-acetate. Cancer Res 2002;62(5):1343–8.

[42] Tong C, Yin Z, Song Z, Dockendorff A, Huang C, Mariadason J, et al. c-J un NH 2-terminal kinase 1 plays a critical role in intestinal homeostasis and tumor suppression. Am J Pathol 2007;171(1):297–303.

[43] Kennedy NJ, Sluss HK, Jones SN, Bar-Sagi D, Flavell RA, Davis RJ. Suppression of Ras-stimulated transformation by the JNK signal transduction pathway. Genes Dev 2003;17(5):629–37.

[44] Cellurale C, Weston CR, Reilly J, Garlick DS, Jerry DJ, Sluss HK, et al. Role of JNK in a Trp53-dependent mouse model of breast cancer. PLoS ONE 2010;5(8):e12469.

[45] Cellurale C, Girnius N, Jiang F, Cavanagh-Kyros J, Lu S, Garlick DS, et al. Role of JNK in mammary gland development and breast cancer. Cancer Res 2012;72(2):472–81.

[46] Hübner A, Mulholland DJ, Standen CL, Karasarides M, Cavanagh-Kyros J, Barrett T, et al. JNK and PTEN cooperatively control the development of invasive adenocarcinoma of the prostate. Proc Natl Acad Sci 2012;109(30):12046–51.

[47] Marusiak AA, Stephenson NL, Baik H, Trotter EW, Li Y, Blyth K, et al. Recurrent MLK4 loss-of-function mutations suppress JNK signaling to promote colon tumorigenesis. Cancer Res 2016;76(3):724–35.

[48] Chauhan D, Li G, Hideshima T, Podar K, Mitsiades C, Mitsiades N, et al. JNK-dependent release of mitochondrial protein, Smac, during apoptosis in multiple myeloma (MM) cells. J Biol Chem 2003;278(20):17593–6.

[49] Hayakawa J, Depatie C, Ohmichi M, Mercola D. The activation of c-Jun NH2-terminal kinase (JNK) by DNA-damaging agents serves to promote drug resistance via activating transcription factor 2 (ATF2)-dependent enhanced DNA repair. J Biol Chem 2003;278(23):20582–92.

[50] Oleinik N, Krupenko N, Krupenko S. Cooperation between JNK1 and JNK2 in activation of p53 apoptotic pathway. Oncogene 2007;26(51):7222.
[51] Behrens A, Jochum W, Sibilia M, Wagner EF. Oncogenic transformation by ras and fos is mediated by c-Jun N-terminal phosphorylation. Oncogene 2000;19(22):2657.
[52] Manning AM, Davis RJ. Targeting JNK for therapeutic benefit: from junk to gold? Nat Rev Drug Discov 2003;2(7):554.
[53] Maeda S, Chang L, Li Z-W, Luo J-L, Leffert H, Karin M. IKKβ is required for prevention of apoptosis mediated by cell-bound but not by circulating TNFα. Immunity 2003;19(5):725–37.
[54] Kamata H, Honda S-I, Maeda S, Chang L, Hirata H, Karin M. Reactive oxygen species promote TNFα-induced death and sustained JNK activation by inhibiting MAP kinase phosphatases. Cell 2005;120(5):649–61.
[55] Sakurai T, Maeda S, Chang L, Karin M. Loss of hepatic NF-κB activity enhances chemical hepatocarcinogenesis through sustained c-Jun N-terminal kinase 1 activation. Proc Natl Acad Sci 2006;103(28):10544–51.
[56] Hui L, Zatloukal K, Scheuch H, Stepniak E, Wagner EF. Proliferation of human HCC cells and chemically induced mouse liver cancers requires JNK1-dependent p21 downregulation. J Clin Invest 2008;118(12):3943.
[57] Das M, Garlick DS, Greiner DL, Davis RJ. The role of JNK in the development of hepatocellular carcinoma. Genes Dev 2011;25(6):634–45.
[58] Han MS, Barrett T, Brehm MA, Davis RJ. Inflammation mediated by JNK in myeloid cells promotes the development of hepatitis and hepatocellular carcinoma. Cell Rep 2016;15(1):19–26.
[59] Nitta R, Del Vecchio C, Chu A, Mitra S, Godwin A, Wong A. The role of the c-Jun N-terminal kinase 2-[alpha]-isoform in non-small cell lung carcinoma tumorigenesis. Oncogene 2011;30(2):234.
[60] Cui J, Han S-Y, Wang C, Su W, Harshyne L, Holgado-Madruga M, et al. c-Jun NH2-terminal kinase 2α2 promotes the tumorigenicity of human glioblastoma cells. Cancer Res 2006;66(20):10024–31.
[61] Yang Y-M, Bost F, Charbono W, Dean N, McKay R, Rhim JS, et al. C-Jun NH2-terminal kinase mediates proliferation and tumor growth of human prostate carcinoma. Clin Cancer Res 2003;9(1):391–401.
[62] Ke H, Harris R, Coloff JL, Jin JY, Leshin B, de Marval PM, et al. The c-Jun NH2-terminal kinase 2 plays a dominant role in human epidermal neoplasia. Cancer Res 2010;70(8):3080–8.
[63] Barbarulo A, Iansante V, Chaidos A, Naresh K, Rahemtulla A, Franzoso G, et al. Poly(ADP-ribose) polymerase family member 14 (PARP14) is a novel effector of the JNK2-dependent pro-survival signal in multiple myeloma. Oncogene 2013;32(36):4231.
[64] Chen N, Nomura M, She Q-B, Ma W-Y, Bode AM, Wang L, et al. Suppression of skin tumorigenesis in c-Jun NH2-terminal kinase-2-deficient mice. Cancer Res 2001;61(10):3908–12.
[65] Cellurale C, Sabio G, Kennedy NJ, Das M, Barlow M, Sandy P, et al. Requirement of c-Jun NH2-terminal kinase for Ras-initiated tumor formation. Mol Cell Biol 2011;31(7):1565–76.
[66] Taylor E, Alqadri N, Dodgson L, Mason D, Lyulcheva E, Messina G, et al. MRL proteins cooperate with activated Ras in glia to drive distinct oncogenic outcomes. Oncogene 2017;36(30):4311.
[67] Kim H, Hwang H, Lee H, Hong HJ. L1 cell adhesion molecule promotes migration and invasion via JNK activation in extrahepatic cholangiocarcinoma cells with activating KRAS mutation. Mol Cells 2017;40(5):363.
[68] Chen R, Khatri P, Mazur PK, Polin M, Zheng Y, Vaka D, et al. A meta-analysis of lung cancer gene expression identifies PTK7 as a survival gene in lung adenocarcinoma. Cancer Res 2014;74(10):2892–902.
[69] Vasilevskaya IA, Selvakumaran M, Hierro LC, Goldstein SR, Winkler JD, O'Dwyer PJ. Inhibition of JNK sensitizes hypoxic colon cancer cells to DNA-damaging agents. Clin Cancer Res 2015;21(18):4143–52.
[70] Fallahi-Sichani M, Moerke NJ, Niepel M, Zhang T, Gray NS, Sorger PK. Systematic analysis of BRAF V 600E melanomas reveals a role for JNK/c-Jun pathway in adaptive resistance to drug-induced apoptosis. Mol Syst Biol 2015;11(3):797.
[71] Lim S-C, Jeon HJ, Kee KH, Lee MJ, Hong R, Han SI. Involvement of DR4/JNK pathway-mediated autophagy in acquired TRAIL resistance in HepG2 cells. Int J Oncol 2016;49(5):1983–90.
[72] Lin Y-T, Liu Y-C, Chao CC-K. Inhibition of JNK and prothymosin-alpha sensitizes hepatocellular carcinoma cells to cisplatin. Biochem Pharmacol 2016;122:80–9.
[73] Manole S, Richards EJ, Meyer AS. JNK pathway activation modulates acquired resistance to EGFR/HER2 targeted therapies. Cancer Res 2016;76(18):5219–28.
[74] Yang S, Qiang L, Sample A, Shah P, He Y-Y. NF-κB signaling activation induced by chloroquine requires autophagosome, p62 protein, and c-Jun N-terminal kinase (JNK) signaling and promotes tumor cell resistance. J Biol Chem 2017;292(8):3379–88.

[75] Liou J-T, Lin C-S, Liao Y-C, Ho L-J, Yang S-P, Lai J-H. JNK/AP-1 activation contributes to tetrandrine resistance in T-cell acute lymphoblastic leukaemia. Acta Pharmacol Sin 2017;38(8):1171.
[76] Xu L, Fu Y, Li Y, Han X. Cisplatin induces expression of drug resistance-related genes through c-jun N-terminal kinase pathway in human lung cancer cells. Cancer Chemother Pharmacol 2017;80(2):235–42.
[77] Song IS, Jun SY, Na HJ, Kim HT, Jung SY, Ha GH, et al. Inhibition of MKK7–JNK by the TOR signaling pathway regulator-like protein contributes to resistance of HCC cells to TRAIL-induced apoptosis. Gastroenterology 2012;143(5):1341–51.
[78] Bennett BL, Sasaki DT, Murray BW, O'Leary EC, Sakata ST, Xu W, et al. SP600125, an anthrapyrazolone inhibitor of Jun N-terminal kinase. Proc Natl Acad Sci 2001;98(24):13681–6.
[79] Gross ND, Boyle JO, Du B, Kekatpure VD, Lantowski A, Thaler HT, et al. Inhibition of Jun NH2-terminal kinases suppresses the growth of experimental head and neck squamous cell carcinoma. Clin Cancer Res 2007;13(19):5910–7.
[80] Tanemura S, Momose H, Shimizu N, Kitagawa D, Seo J, Yamasaki T, et al. Blockage by SP600125 of Fcε receptor-induced degranulation and cytokine gene expression in mast cells is mediated through inhibition of phosphatidylinositol 3-kinase signalling pathway. J Biochem 2008;145(3):345–54.
[81] Flanc R, Ma F, Tesch G, Han Y, Atkins R, Bennett B, et al. A pathogenic role for JNK signaling in experimental anti-GBM glomerulonephritis. Kidney Int 2007;72(6):698–708.
[82] Ma FY, Flanc RS, Tesch GH, Han Y, Atkins RC, Bennett BL, et al. A pathogenic role for c-Jun amino-terminal kinase signaling in renal fibrosis and tubular cell apoptosis. J Am Soc Nephrol 2007;18(2):472–84.
[83] Krenitsky VP, Nadolny L, Delgado M, Ayala L, Clareen SS, Hilgraf R, et al. Discovery of CC-930, an orally active anti-fibrotic JNK inhibitor. Bioorg Med Chem Lett 2012;22(3):1433–8.
[84] van der Velden JL, Ye Y, Nolin JD, Hoffman SM, Chapman DG, Lahue KG, et al. JNK inhibition reduces lung remodeling and pulmonary fibrotic systemic markers. Clin Transl Med 2016;5(1):36.
[85] Cerbone A, Toaldo C, Pizzimenti S, Pettazzoni P, Dianzani C, Minelli R, et al. AS601245, an anti-inflammatory JNK inhibitor, and clofibrate have a synergistic effect in inducing cell responses and in affecting the gene expression profile in CaCo-2 colon cancer cells. PPAR Res 2012;2012. 269751.
[86] Gaillard P, Jeanclaude-Etter I, Ardissone V, Arkinstall S, Cambet Y, Camps M, et al. Design and synthesis of the first generation of novel potent, selective, and *in vivo* active (benzothiazol-2-yl) acetonitrile inhibitors of the c-Jun N-terminal kinase. J Med Chem 2005;48(14):4596–607.
[87] Okada M, Kuramoto K, Takeda H, Watarai H, Sakaki H, Seino S, et al. The novel JNK inhibitor AS602801 inhibits cancer stem cells *in vitro* and *in vivo*. Oncotarget 2016;7(19):27021.
[88] Bogoyevitch MA, Arthur PG. Inhibitors of c-Jun N-terminal kinases—JuNK no more? Biochim Biophys Acta 2008;1784(1):76–93.
[89] Messoussi A, Feneyrolles C, Bros A, Deroide A, Daydé-Cazals B, Chevé G, et al. Recent progress in the design, study, and development of c-Jun N-terminal kinase inhibitors as anticancer agents. Chem Biol 2014;21(11):1433–43.
[90] Holzberg D, Knight CG, Dittrich-Breiholz O, Schneider H, Dörrie A, Hoffmann E, et al. Disruption of the c-JUN-JNK complex by a cell-permeable peptide containing the c-JUN δ domain induces apoptosis and affects a distinct set of interleukin-1-induced inflammatory genes. J Biol Chem 2003;278(41):40213–23.
[91] Barr RK, Kendrick TS, Bogoyevitch MA. Identification of the critical features of a small peptide inhibitor of JNK activity. J Biol Chem 2002;277(13):10987–97.
[92] Borsello T, Clarke PG, Hirt L, Vercelli A, Repici M, Schorderet DF, et al. A peptide inhibitor of c-Jun N-terminal kinase protects against excitotoxicity and cerebral ischemia. Nat Med 2003;9(9):1180–6.
[93] Eshraghi AA, Wang J, Adil E, He J, Zine A, Bublik M, et al. Blocking c-Jun-N-terminal kinase signaling can prevent hearing loss induced by both electrode insertion trauma and neomycin ototoxicity. Hearing Res 2007;226(1):168–77.
[94] Gao Y-J, Cheng J-K, Zeng Q, Xu Z-Z, Decosterd I, Xu X, et al. Selective inhibition of JNK with a peptide inhibitor attenuates pain hypersensitivity and tumor growth in a mouse skin cancer pain model. Exp Neurol 2009;219(1):146–55.
[95] Lehnert M, Relja B, Lee VS-Y, Schwestka B, Henrich D, Czerny C, et al. A peptide inhibitor of C-jun N-terminal kinase modulates hepatic damage and the inflammatory response after hemorrhagic shock and resuscitation. Shock 2008;30(2):159–65.

CHAPTER

5

Fibroblast Growth Factor Receptor Inhibitors: Enhancing Therapeutic Strategies for Solid Tumors

Cinta Hierro,**, Josep Tabernero*,***

*Medical Oncology Department, Vall d'Hebron University Hospital, Autonomous University of Barcelona (UAB), Barcelona, Spain

**Molecular Therapeutics Research Group, Vall d'Hebron Institute of Oncology (VHIO), Barcelona, Spain

Abstract

Despite advances in the last decades, cancer continues to be one of the main causes of morbi-mortality worldwide. In fact, cancer remains the second leading cause of death globally. Although chemotherapy had been the mainstay therapeutic option for treating solid tumors, a new era began in the late 1990s with the appearance of targeted agents. Since then, an in-depth knowledge into the molecular biology of certain tumors has allowed the discovery of new predictive gene alterations. In this context, increasing data have shown that the fibroblast growth factor receptor (FGFR) pathway plays a key role in carcinogenesis, by directly interacting with some of the hallmarks of cancer. Thus, the FGFR pathway arises as a promising target for cancer treatment. In this chapter, we will depict the complexity of the FGFR pathway, reviewing the different molecular aberrations described and their clinical implications. We will position the FGFR inhibitors in the early drug development field, for understanding their mechanisms of action, their current development status, and which rational combinations they are undergoing. Only when we have a full comprehensive view of this signaling pathway will we be able to design future personalized options for enhancing the therapeutic strategies for solid tumors.

ABBREVIATIONS

ATP	Adenosine triphosphate
BC	Breast cancer
CRC	Colorectal cancer
EGFR	Epidermal growth factor receptor
ER	Estrogen receptor
ERK	Extracellular signal-regulated kinase
FGF	Fibroblast growth factor

Targeting Cell Survival Pathways to Enhance Response to Chemotherapy. http://dx.doi.org/10.1016/B978-0-12-813753-6.00005-6
Copyright © 2018 Elsevier Inc. All rights reserved.

FGFR	Fibroblast growth factor receptor
GC	Gastric cancer
GI	Gastrointestinal
GIST	Gastrointestinal stromal tumors
GOJ	Gastroesophageal junction cancer
GRB2	Growth factor receptor-bound protein 2
HER2	Human epidermal growth factor receptor 2
HR	Hormone receptor
HPSG	Heparin sulfate proteoglycan
HPV	Human papillomavirus
Ig	Immunoglobulin
LVI	Lymphovascular invasion
MET	Mesenchymal–epithelial transcription factor
mTOR	Mammalian target of rapamycin
MTKI	Multityrosine kinase inhibitor
NSCLC	Non-small cell lung cancer
PDGFR	Platelet-derived growth factor receptor
PI3K	Phosphoinositide 3-kinase
PLC	Phospholipase C
SCLC	Small cell lung cancer
SOS	Son of Sevenless
STAT	Signal transducer and activator of transcription
TKI	Tyrosine kinase inhibitor
VEGF	Vascular endothelial growth factor
WT	Wild-type

5.1 INTRODUCTION

Until the late 1990s, chemotherapy remained the mainstay for cancer therapy, relying on suppression of the proliferative capabilities of tumor cells, without specific blockade of angiogenesis, immune mechanisms, or components of the intracellular signal transduction pathways. Imatinib mesylate, a small-molecule inhibitor of the tyrosine kinase activity of BCR-ABL and *c-kit*, radically changed the landscape, becoming the first targeted agent in demonstrating efficacy for treating molecularly aberrant patients [1]. In 2001 [2], the first reports of *c-kit*-positive metastatic gastrointestinal stromal tumors controlled with imatinib supported the hypothesis that inhibition of oncogene-addicted cells was therapeutically useful and served as a proof-of-principle that specific tyrosine kinase receptors could represent a critical target for certain solid tumors.

Over the last two decades, a deepening into the molecular biology of several tumor types has focused attention toward the multiple genomic abnormalities that can contribute to cancer. The parallel implementation of powerful high-throughput technologies that enable the detailed characterization of these aberrations has led to the discovery of potentially druggable alterations [3]. The identification of these cancer drivers has contributed to an understanding of the specific cellular and genetic contexts in which they act, emerging as potential predictive biomarkers of response for new targeted agents. Since then, significant progress has been made for incorporating different targeted agents as part of the armamentarium for treating cancer patients. For example, BRAF V600 mutant melanomas, which are currently treated with proto-oncogene B-Raf and v-Raf murine sarcoma viral oncogene homolog B (BRAF) inhibitors [4,5], or patients with non-small-cell lung carcinomas (NSCLCs) with mutations in the gene encoding the epidermal growth factor receptor (EGFR), who benefit from receiving anti-EGFR drugs [6,7].

The fibroblast growth factor receptor (FGFR) signaling pathway has increasingly become a major key point in the oncology research field, emerging as a promising target for optimizing the therapeutic options for cancer patients. In this chapter, we will depict the complexity of the FGFR pathway, reviewing the different molecular aberrations described and their clinical implications. We will position the FGFR inhibitors in the current early drug development field, for understanding which hallmarks of cancer they can target, their current development status, and which rational combination strategies are currently under investigation.

5.2 THE FGFR/FGF PATHWAY

The FGFR/FGF axis is an intracellular signaling pathway that mediates several cellular processes, such as proliferation, differentiation, homeostasis, and metabolism. The FGFR/FGF is a complex family that in humans comprises four receptor tyrosine kinases, named FGFR1/2/3/4, and 22 structurally related FGF ligands [8].

A total of 22 genes encode the FGF superfamily. According to phylogenetic analysis, the different FGFs can be divided into different subfamilies: paracrine secreted FGFs (FGF1-10/16-18/20/22), receptor-independent intracrine FGFs (FGF11-14), and hormonal endocrine FGFs (FGF15/19/21/23) [9]. The majority of these FGFs are secreted peptides, although FGF1 and FGF2 are released through an endoplasmic reticulum–Golgi-independent exocytosis pathway or after cell damage in stress conditions [10]. Most of these FGFs are sequestered by matrix glycosaminoglycans within the extracellular space, namely heparin sulfate proteoglycans (HPSGs), or Klotho proteins in the case of hormonal FGFs [11]. The interaction between these HPSGs/Klotho proteins and FGFs serves to protect the ligands from extracellular degradation, generating a local reservoir of FGFs. The spatial pattern of expression of these matrix glycoproteins acts as a tight regulator mechanism of the available FGFs and their half-lives [12].

The four homologous FGFRs are highly conserved tyrosine kinase receptors sharing a common overall structure, comprising an extracellular domain, a single-pass transmembrane domain, and a carboxy-terminal cytoplasmic domain. In the extracellular portion, there are three different immunoglobulin (Ig)-like domains (I–III) that finely modulate the activity of the receptor: (a) the amino-terminal portion containing the acidic box, a serine-enriched portion between the IgI–IgII loops, plays a key role in receptor auto-inhibition; (b) HPSGs and Klotho proteins are docked in the IgII fold, for controlling the release and affinity of FGF ligands; (c) IgII–IgIII domains act as crucial ligand-binding sites for FGFR activation [13]. A fifth related receptor FGFR5 (known as FGFRL1) can bind FGF ligands but has no tyrosine kinase domain. Hence, FGFRL1 is thought to play a role as a ligand trap, also serving as a partner for dimerization with other tyrosine kinase receptors and as a negative feedback loop for limiting the phosphorylation of FGFR1-4 [14].

Under normal conditions, FGFs are engaged with HPSGs/Klotho proteins through electrostatic interactions. Following extracellular protease-mediated release, the binding of these FGFs to the receptors leads to the dimerization of the ternary complex HPSGs:FGFR:FGF. The dimerization triggers a conformational change that induces the autophosphorylation of the tyrosine kinase domain, facilitating the attachment of docking proteins and activation of downstream pathways. The activated FGFR phosphorylates the FGFR substrate 2 (FRS2) on several sites, allowing the recruitment of other adaptor proteins, such as growth factor receptor-bound protein 2 (GRB2) and Son of Sevenless. Further downstream signaling occurs, ultimately leading to an upregulation of PI3K/AKT/mTOR (mammalian target of rapamycin) and RAS/RAF/ERK pathways, phospholipase C gamma (PLCγ), and signal transducer and activator of transcription STAT [15]. After activation, the complex is internalized and transported as an endocytosis vesicle to the lysosomes for its degradation and recycling [9]. Fig. 5.1 illustrates the complexity of this pathway.

5.2.1 FGFR/FGF Deregulation in Human Cancers and Clinical Implications

The FGFR/FGF axis plays a key role in normal cellular processes, such as embryogenesis, cell metabolism and proliferation, migration, and angiogenesis. Hence, it is not surprising that such a crucial pathway may be disrupted in human cancers, precipitating the carcinogenesis process and enabling malignant transformation. In fact, FGFR genes are among the most commonly altered kinase genes in solid tumors. To date, several FGFR/FGF molecular alterations have been described, as depicted in Fig. 5.2.

Cancers may present constitutive activation of the FGFR pathway as a result of gain-of-function missense mutations. Mutations are dichotomized molecular events, meaning that cells are mutated or not. Nevertheless, not all the mutations result in upregulation of the FGFR pathway. FGFR mutations located within the tyrosine kinase domain are rare (~1%), whereas the majority of the mutations considered oncogenic generate novel amino acid residues in the extracellular portion of the receptor [16]. It has been hypothesized that these mutations may activate FGFR independent of ligand binding, by facilitating receptor dimerization through newly formed disulfide or hydrogen bonds [17]. Second, FGFR gene translocations have progressively been described as druggable targets in multiple cancers. A number of different FGFR fusion partners have been identified. The activating mechanism for these partners consists in providing a dimerization domain that facilitates oligomerization of the fusion receptor, leading to constitutive activation of the kinase downstream [18]. Third, FGFR1-4 or FGF gene amplifications have been described [19]. Gene amplification arises from the presence of multiple copies of the particular gene, which in turn translates into an increase in the phenotypic characteristics attributed to that gene [20]. However, the level of FGFR amplification can significantly vary between tumors of different origin, and even more relevant, between tumors within the same histology. Interestingly, higher rates of FGFR amplification seem to correlate with the degree of oncogene addiction of a certain tumor [21].

Recently, other less well-described molecular aberrations have been postulated as potential mechanisms of FGFR upregulation. Certain FGFR isoforms resulting from alternative mRNA splicing processes can result in highly oncogenic proteins [22]. FGFR2-C3 isoforms detected among FGFR2-amplified gastric cancer patients [23] can lead to aberrant receptors that are constitutively FRS2-dependent activated, accompanied by C-terminal modifications that contribute to receptor accumulation [24]. On top of that, extracellular splicing mechanisms of the IgIII portion can modulate the FGFR tissue specificity for the FGFs, even increasing the affinity and number of ligands that can activate the same

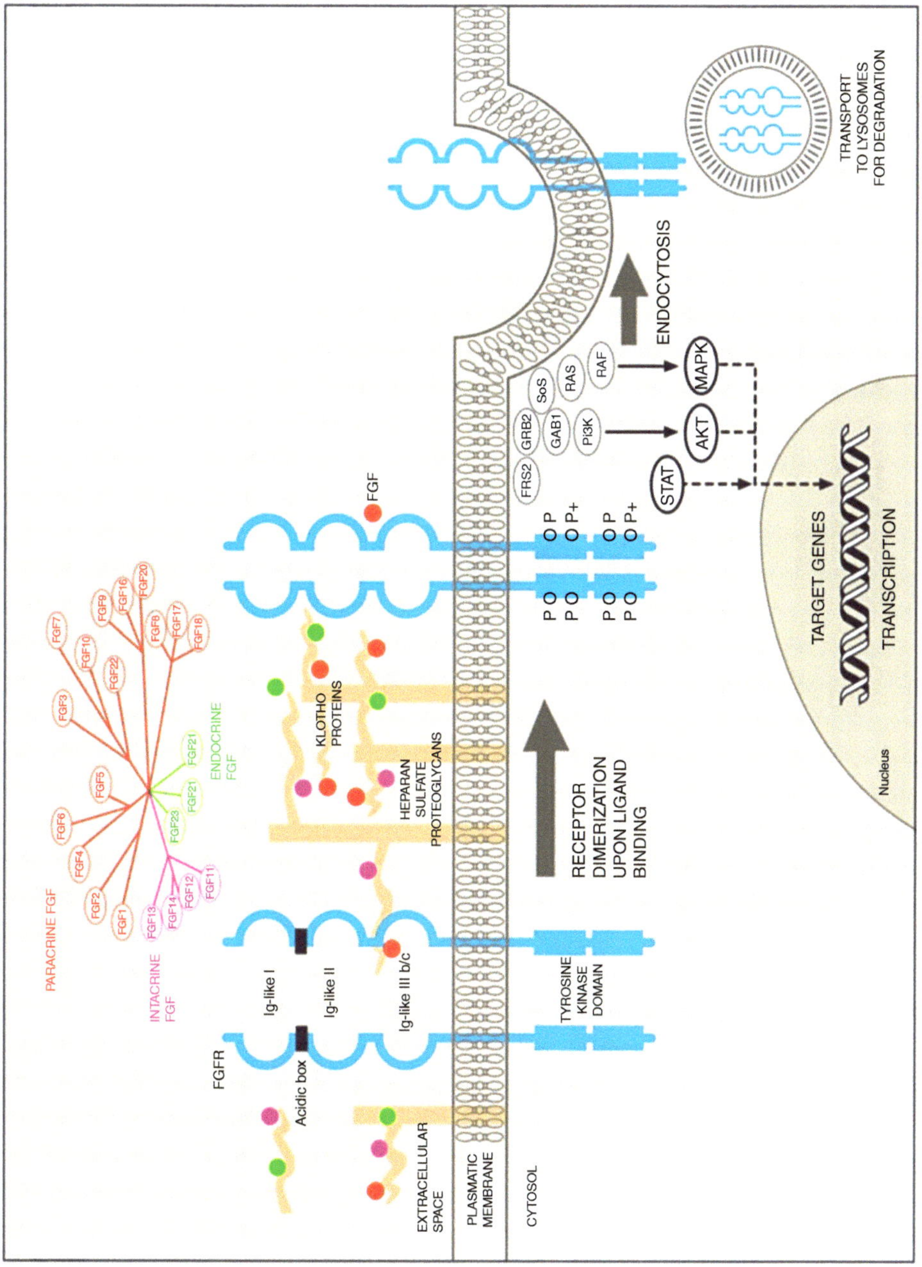

FIG. 5.1 **Illustration of the complexity of the FGFR/FGF pathway.**

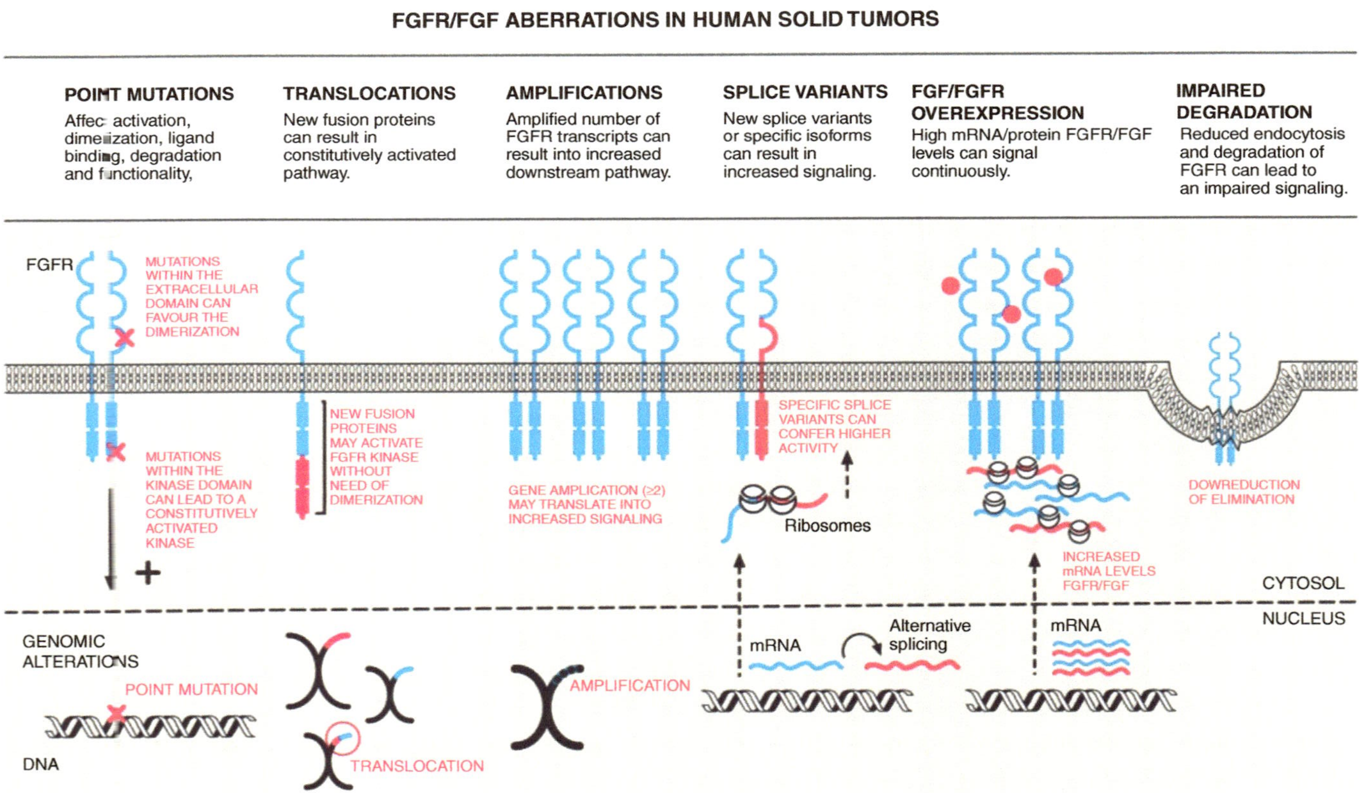

FIG. 5.2 **Heterogeneity of the FGFR/FGF molecular aberrations and associated mechanisms of upregulation of the signaling axis.** FGFR signaling contributes to the carcinogenesis process through several ligand-dependent and -independent mechanisms. In fact, several FGFR/FGF genomic alterations have been described across solid tumors, showing the huge heterogeneity of molecular aberrations that the axis may present: FGFR1-4 point mutations, gene translocations and amplifications, alternative FGFR splice variants that result in activating isoforms, increased mRNA levels that translate into increased protein levels of FGFR/FGF, and altered mechanisms of pre-activated receptor degradation. All these alterations share a common purpose, resulting in an upregulation of FGFR downstream effector signaling.

receptor [25]. Increased levels of mRNA expression of FGFR/FGF and/or protein levels might be produced in an autocrine fashion by cancer cells, establishing a vicious loop that promotes the pathogenic activation. Finally, mechanisms that alter the formation of endocytosis vesicles may lead to an impaired degradation of activated FGFRs, perpetuating the signaling of the pathway.

Interestingly, the FGFRs and their isoforms are distributed unevenly in a tissue-organ-related manner, translating into different roles at certain stages of the development. This tight correlation seems to be maintained throughout the oncogenic process, leading to a highly specific expression of each molecular FGFR/FGF aberration according to each tumor type. Table 5.1 depicts the most common FGFR/FGF alterations described in solid tumors, with the phenotypic characteristics and biological behaviors associated to them.

TABLE 5.1 Common FGFR/FGF Alterations in Solid Tumors and Their Clinical Implications

Tumor type	FGFR/FGF alteration	PREV (%)	Clinical implications	Reference
Squamous NSCLC	FGFR1 amplification	10–20	Poorer prognosis in NSCLC squamous, shorter survival Escape mechanism in NSCLC adenocarcinoma resistant to anti-EGFR chemotherapy resistance in SCLC	[26–34]
	11 amplification	12		
Adenocarcinoma NSCLC	FGFR1 amplification	6		
	11q amplification	4		
SCLC	FGFR1 amplification	6		
Squamous head and neck	FGFR1 amplification	10	Mutually exclusive with HPV infection Poorer overall survival in HPV-negative subtype	[35–36]
Squamous esophageal	FGFR1 amplification	9–12	Poorer prognosis	[37–39]
Gastric adenocarcinoma	FGFR2 amplification	7	Mutually exclusive with HER2/MET amplifications Worst survival with LVI+	[40–42]
Hepatocarcinoma	FGFR4 + βklotho + FGF19 overexpression	15–40	Poorer histological differentiation and higher αfeto-protein	[43–46]
Biliary tract carcinoma	FGFR2 translocation	8–13	Mutually exclusive with KRAS/BRAF mutations	[47,48]

(Continued)

TABLE 5.1 Common FGFR/FGF Alterations in Solid Tumors and Their Clinical Implications (*Cont.*)

Tumor type	FGFR/FGF alteration	PREV (%)	Clinical implications	Reference
Colon adenocarcinoma	FGFR4 amplification + FGF19 overexpression	4	Higher risk for more aggressive tumors Resistance to anti-EGFR if FGF19 upregulation	[49–51]
HR+ breast carcinoma	11q amplification	15	Development of distant metastasis earlier, shorter survival Poorer prognosis in luminal-type, endocrine resistance	[52–55]
	FGFR1 amplification	10		
TBNC	FGFR1 amplification	4		
	FGFR2 amplification	4		
Ovarian carcinoma	FGFR1 amplification	4	Resistance to platinum-based chemotherapies	[56,57]
Endometrial adenocarcinoma	FGFR2 mutation	12	Mutually exclusive with KRAS mutations Shorter disease-free survival interval	[58–61]
	FGFR1 amplification	4		
Squamous cervix	FGFR3 mutation	25		
Urothelial carcinoma (muscle-invasive)	FGFR3 mutation	15	Early recurrence detection in nonmuscle invasive pTa	[62,63]
	FGFR1 amplification	11		
	FGFR3-TACC3 translocation	6		
Prostate adenocarcinoma	FGFR1 amplification	8	Higher risk of resistance to castration in hormone-naïve patients	[64,65]
Glioblastoma	FGFR3-TACC3 translocation	3–7	Mutually exclusive with EGFR/PDGFR/MET amplifications	[66,67]
Melanoma	FGFR2 mutation	9	Unknown significance (passenger mutations?)	[68]
	FGFR1 mutation	4		
	FGFR4 mutation	4		
Rhabdomiosarcoma	FGFR4 mutation	8	Poorer response to chemotherapy in osteosarcomas	[69,70]
Osteosarcoma	FGFR1 amplification	18		

Abbreviations: PREV, prevalence; NSCLC, non-small cell lung cancer; SCLC, small cell lung cancer; HR+ hormone receptor-positive; TBNC, triple breast negative cancer; EGFR, epidermal growth factor receptor; HPV, human papillomavirus; HER2, human epidermal growth factor receptor 2; MET, mesenchymal–epithelial transcription factor; LVI+, lymphovascular invasion-positive.

5.3 FGFR INHIBITORS: TARGETING SOME CRITICAL HALLMARKS OF CANCER

In the last quarter century, researchers have tried to shed some light into the layers of complexity that intrinsically involve cancer. The identification of the biological multistep process that guides the transformation of normal tissue cells into malignant cancers significantly contributed to a better understanding of the process of carcinogenesis in humans [71]. Among these physiological changes acquired by the tumoral cells, there are 10 capabilities known to be involved in the pathogenesis of some, and probably, all cancers. These 10 hallmarks described by Hanahan and Weinberg represent an organizing principle for understanding the neoplastic disease: (1) sustaining proliferative signaling; (2) evading growth suppressors; (3) avoiding immune destruction; (4) enabling replicative immortality; (5) tumor-promoting inflammation; (6) activating invasion and metastasis; (7) inducing angiogenesis; (8) genome instability and mutation; (9) resisting cell death; (10) deregulating cellular energetics [72]. The recognition of these tumoral traits has guided new modern approaches for developing novel anticancer drugs, aiming to overcome some of these hallmarks and/or their pathways involved [73].

The FGFR pathway is involved in normal organ-specific development and differentiation, stimulating cell proliferation and migration by downstream activation of PI3K/AKT/mTOR and RAS/RAF/ERK pathways [74]. Noteworthy, several malformation syndromes that involve skeletal disorders were already associated with mutations of the FGFR genes [75]. As the upregulation of the FGFR signaling is a common event in many cancer types, it has been postulated that this pathway plays a key role in sustaining proliferative signaling. In FGFR-aberrant tumors, the acquired oncogene alteration may induce cell growth and survival, by imitating the stimulatory signals that tightly regulate proliferation in normal conditions. In fact, small molecules blocking FGFR signaling in FGFR-dependent cell lines have translated into cell apoptosis and death [27]. Inhibition of the FGFR pathway decreased tumor growth in xenograft models, clearly reflecting the oncogenic addiction of tumoral cells to the acquired FGFR aberration [76].

Along with organogenesis, angiogenesis was one of the first areas in which the FGFR/FGF axis was demonstrated to be crucial [77]. FGFR1-2 are widely expressed on the surface of endothelial cells, and through FGF1-2 secreted by tumoral or microenvironmental stromal cells, can induce the generation of new blood vessels. Also, FGFR signaling is required for neovascularization after blood vessel injury [78]. Vascular endothelial growth factor (VEGF) and FGF2 may have a synergistic effect with complementary roles, and rising evidence has suggested that upregulation of the FGFR pathway is a resistance mechanism to anti-VEGF therapies [79]. Casanovas et al. [80] have demonstrated that cancer cells are able to develop resistance against anti-angiogenic blockade by the alternative secretion of FGFs, instead of VEGFs. These data have widened the horizon for developing new therapeutic options for overcoming resistance to anti-angiogenic drugs.

The finding of multiple FGFR/FGF aberrations in a wide range of solid tumors has led to a major interest in developing specific targeting agents against this signaling pathway, suggesting the therapeutic benefit of the FGFR blockade. The combined inhibition of two major hallmarks, proliferation and angiogenesis, may represent a convincing reason for developing FGFR inhibitors as anticancer drugs. Fig. 5.3 summarizes the hallmarks of cancer that could be targeted by FGFR inhibitors.

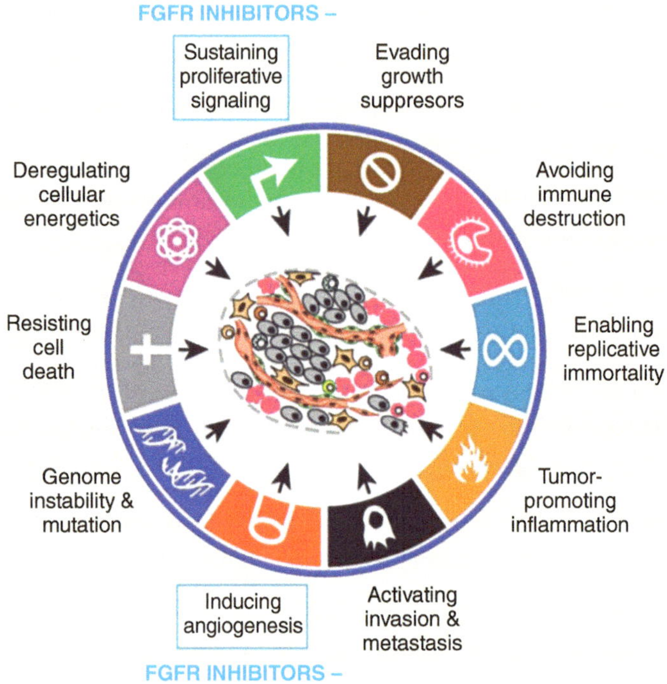

FIG. 5.3 **Hallmarks of cancer that can be targeted by FGFR inhibitors.** Ten different capabilities are known to be involved in the pathogenesis of cancers: (1) sustaining proliferative signaling; (2) evading growth suppressors; (3) avoiding immune destruction; (4) enabling replicative immortality; (5) tumor-promoting inflammation; (6) activating invasion and metastasis; (7) inducing angiogenesis; (8) genome instability and mutation; (9) resisting cell death; (10) deregulating cellular energetics. The FGFR/FGF signaling pathway plays a key role in controlling cellular proliferation and angiogenesis. By combining the dual strategy of blocking these two major hallmarks, FGFR inhibitors arise as a promising targeted option for FGFR-dependent tumors. Adapted from Hanahan and Weinberg [72] with permission from Elsevier.

5.4 FGFR INHIBITORS TO IMPROVE RESPONSE IN SOLID TUMORS

The inhibition of FGFR signaling has shown *in vitro* and *in vivo* activity in many tumor types depending on the FGFR addiction for proliferation and survival [34,81,82]. Therefore, a significant number of novel FGFR inhibitors have entered into the early drug development field in the last decade.

Current FGFR inhibitors can be mainly divided into two groups, according to their mechanism of action: (1) small oral molecules, classified as receptor tyrosine kinase inhibitors (TKIs). TKIs are adenosine triphosphate-competitive molecules that compete for the catalytic binding site of the intracellular kinase domain of FGFRs, diminishing autophosphorylation of the receptor and (2) antibodies. Different antibody strategies have been developed for modulating FGFR activation: (a) anti-FGFR antibodies, which can bind the extracellular portion of the FGFR and reduce the affinity for the FGFs, leading to a reduced FGF:FGFR association and subsequent dimerization and (b) FGF ligand traps, which can sequester multiple extracellular ligands and reduce the activation of their corresponding FGFR partners [83].

The most commonly accepted classification of FGFR inhibitor family is made according to the grade of selectivity that these drugs exert against the different FGFRs and/or other tyrosine kinase receptors: (1) first-generation nonselective FGFR inhibitors include small molecules with a multitargeting kinase activity (MTKI), thereby despite an increased antitumor activity, these can result in increased undesired toxicities due to their off-FGFR target effects; (2) second-generation selective pan-FGFR inhibitors (FGFRinh), which comprise TKIs and FGF ligand traps with increased affinity for only FGFRs ($IC_{50} < 10$ nmol/L); and (3) third-generation ultra-selective FGFRinh, the newest and less developed strategy, which include isoform-specific antibodies aimed to target only one concrete FGFR. Fig. 5.4 illustrates these three generations of FGFR inhibitors and some of the current drugs under development in each one of the subgroups.

Most first-generation MTKIs have been developed as anti-angiogenic therapies, tested in a wide variety of solid tumors irrespective of their FGFR status. That is the case of lenvatinib, subsequently developed in thyroid cancer (NCT01321554). Given the fact that the kinase domains of the FGFR, vascular endothelial growth factor receptor (VEGFR) and platelet-derived growth factor receptor (PDGFR) are phylogenetically close, some of these MTKIs showed FGFR inhibition which could be used in FGFR-aberrant patients. Dovitinib demonstrated *in vivo* and *in vitro* activity in FGFR1-2-amplified breast models, supporting its further development in the clinic [21]. In a phase II trial testing dovitinib in breast cancer patients (NCT01528345), promising efficacy signs were detected: one unconfirmed partial response was reported in the subset of 15% women harboring FGFR1-amplified ER+ tumors. These data raised the question whether the activity was related to VEGFR/PDGFR inhibition, but the fact that no responses were seen among non-FGFR-amplified patients suggested the oncogenic role for FGFR in this subset of patients. Lucitanib, another MTKI that targets FGFR1-3/VEGFR1-4/PDGFRα-β, demonstrated an encouraging antitumor activity in a phase I/II trial (NCT01283945). With a disease control rate of 100%, 6 out of 12 breast cancer patients with FGFR1 amplification achieved long-term partial responses, with progression-free survival intervals of around 10 months [84]. Other MTKIs are currently under development, but the increased nonspecific toxicities commonly seen with these kinase inhibitors have limited the next steps. Hypertension, vomiting, diarrhea, fatigue, and proteinuria represent the most common adverse events seen with these MTKIs that may limit the possibility of an adequate dose-intensity FGFR blockade that translates into efficacy.

Within this context, second-generation selective FGFRinh have been developed, in order to increase the possibility of on-target FGFR inhibition while diminishing the toxicity derived from MTKIs. Most of these FGFRinh have focused on different tumor types but with a high likelihood of FGFR oncogene addiction. Interestingly, the retrospective analysis of the efficacy signs detected in the early clinical trials revealed a significant variability in antitumor activity between genomic alterations. AZD4547, a pan-FGFR1-2-3 inhibitor (NCT00979134), showed partial responses in two FGFR1-amplified squamous NSCLCs and in one FGFR2-amplified gastric cancer patient, on top of other tumor type patients who achieved prolonged stabilizations (one FGFR1-amplified breast cancer and two FGFR3-mutant bladder

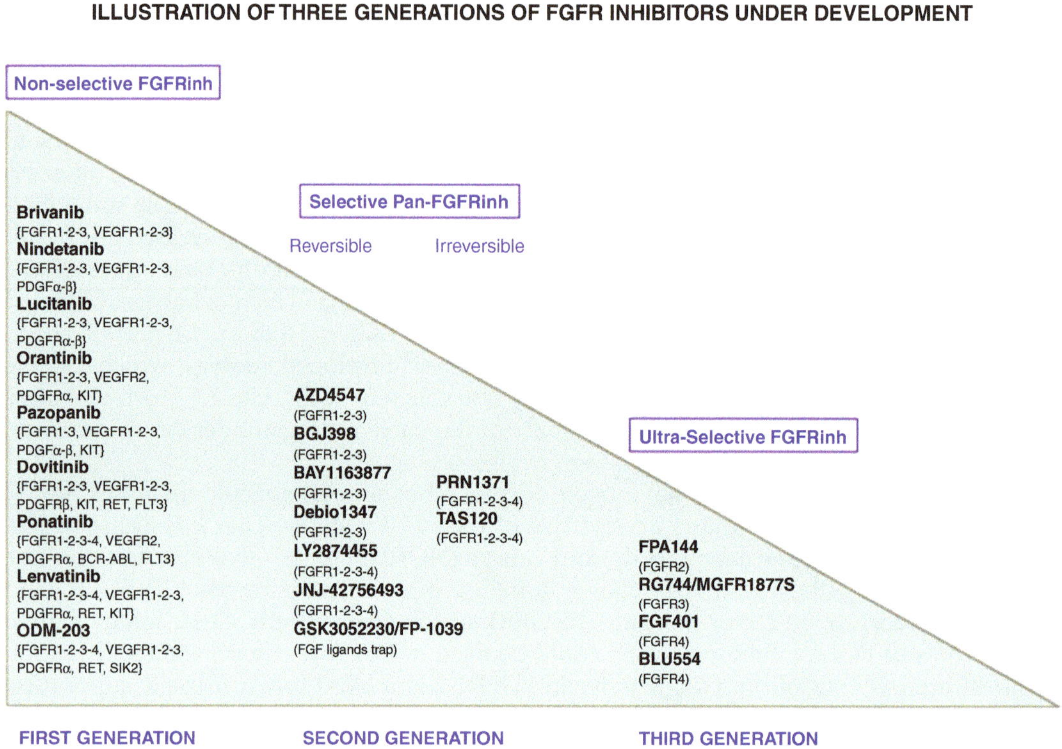

FIG. 5.4 **Clinical development of FGFR inhibitors.** Three major subgroups of FGFR inhibitors have progressively entered different phases of the early drug development field: (1) first generation of nonselective MTKIs include drugs such as brivanib, nindetanib, lucitanib, orantinib, pazopanib, dovitinib, ponatinib, lenvatinib, and ODM-203. Notice below each compound the spectrum of multitargeting kinase activity, with the majority of these inhibitors targeting also angiogenesis by blocking VEGFR and PDGFR; (2) second generation of highly selective pan-FGFR inhibitors (FGFRinh) comprise several molecules with increased affinity for only FGFRs, including both reversible—AZD4547, BGJ398, BAY1163877, Debio1347, LY2874455, JNJ-42756493, and GSK3052230—and irreversible—PRN1371 and TAS120—inhibitors; and (3) third-generation ultraselective FGFRinh, encompassing antibodies specifically developed for targeting a unique FGFR, aiming to treat cancer cells that rely on a particular receptor. That is the case of FPA144, RG744, FGF401, and BLU554. Abbreviation: MTKIs, multityrosine kinase inhibitors.

cancer patients) [85–88]. Another FGFR1-2-3inh, BGJ398, has been tested in a phase I trial enrolling FGFR1/2-amplified or FGFR3-mutant patients (NCT01004224). Partial responses were reported in 14 patients (six FGFR3-mutant bladder cancer and eight FGFR1-amplified squamous NSCLC patients), although minor tumor regressions were also seen among other histologies (FGFR1/2-amplified and FGFR3-mutant breast cancer, FGFR2-mutant and translocated cholangiocarcinoma) [89]. In an effort toward improving patient selection, novel prescreening strategies have been tested, such as assessing FGFR mRNA levels. With this strategy, the BAY1163877 phase I study (NCT01976741) identified new subsets of patients who benefitted from the FGFR blockade: one head and neck squamous carcinoma and one adenoid cystic carcinoma of the tongue [90]. However, most of these FGFRinh inhibit the three receptors to varying degrees, due to similarities in the kinase domain of FGFR1-2-3 [91], but may lack activity against FGFR4 which has a different kinase structure [92]. JNJ-42756493, one of the first developed FGFRinh that also targets FGFR4, showed encouraging antitumor activity, particularly among FGFR-translocated patients (NCT01703481). Partial responses in three FGFR2/3-translocated urothelial carcinomas, one FGFR3-translocated glioblastoma, and one FGFR2-translocated endometrial carcinoma were reported within the escalation phase I [93].

Furthermore, published data have suggested a crucial role for ligand overexpression as an upregulating mechanism of the FGFR pathway [94]. While further studies may clarify whether it is crucial to target FGFs, new FGF ligand traps have emerged as an interesting therapeutic option. By sequestering several extracellular ligands that may signal through a unique FGFR, FP-1039 (NCT00687505) has shown stabilizations in 41% of patients irrespective of their FGFR status, with a significant 20% decrease in tumor size in prostate cancer [95]. Lately, highly potent irreversible FGFRinh have demonstrated *in vivo* and *in vitro* activity against gatekeeper mutations, known to confer resistance to FGFRinh [96]. The fact that new emerging FGFR2 mutations have been progressively detected in tumors treated with reversible FGFRinh has envisaged a new scenario for developing these irreversible FGFRinh [97]. TAS120 (NCT02052778) and PRN1371 (NCT02608125) are currently under development in phase I trials, focused on selected FGFR-aberrant patients and allowing the enrollment of patients previously treated with reversible FGFRinh [98,99].

Finally, it has been noticed that the different FGFR aberrations present a certain preference for organ-specific cancers, such as FGFR4 overexpression in hepatocarcinomas. This finding may be explained by the fact that FGF19/FGFR4 expression plays a key role in the homeostasis of the healthy liver, promoting glycogenesis and suppressing gluconeogenesis/bile acid synthesis [100]. FGFR4 is known to be necessary for FGF19-mediated hepatocarcinoma tumorigenesis *in vivo* [44], and specific treatment with FGFR4-directed strategies may inhibit the proliferation of liver tumors while avoiding unnecessary adverse events derived from other FGFR inhibition. These observations established the basis for developing a third-generation of ultraselective FGFRinh, aimed at specific tumor types that rely on a particular FGFR aberration. Several isoform-specific antibodies are under development in phase I trials. FPA144 is a humanized IgG1 antibody directed against the 2b isoform of the FGFR2 (NCT02318329) and MFGR1877S/RG744 is a human monoclonal antibody targeting FGFR3 (NCT01363024). A couple of specific FGFR4 inhibitors, FGF401 (NCT02325739) and BLU554 (NCT02508467), are currently being tested in phase I trials selectively enrolling hepatocarcinoma patients with high FGFR4/βklotho overexpression. Future development of these selective inhibitors seems a promising strategy for implementing more personalized therapeutic options for FGFR-aberrant tumors.

5.4.1 Combination Strategies of FGFR Inhibitors With Other Anticancer Agents

Despite encouraging efficacy results observed with early studies testing FGFR inhibitors in monotherapy, limited antitumor activity has been reported with these targeted agents so far. Nevertheless, based on the evidence that the combination of these FGFR inhibitors with other anticancer strategies may increase the possibilities of response rate maintaining an adequate safety profile, several combination strategies have been proposed.

FGFR inhibitors have demonstrated promising efficacy signals when combined with different chemotherapy regimens—irinotecan, paclitaxel, and etoposide—by enhancing the apoptosis rates induced by chemotherapeutic agents in drug-resistant cell lines [101]. The combined therapy of 5-fluorouracil with Ki23057 (FGFR2 inhibitor) produced synergistic antitumor effects in FGFR2-aberrant gastric cancer cell lines [102]. However, ponatinib has not shown synergistic effects in combination with gemcitabine plus cisplatin when tested *in vivo* in an FGFR2-CCDC6 fusion intrahepatic cholangiocarcinoma [18]. These contradictory results highlight the need for further research into the synergistic mechanisms that can potentiate certain combination strategies, while avoiding some combinations that may result into severely increased toxicities and detrimental effects. In fact, several combinations with FGFR inhibitors and systemic chemotherapies have been early terminated due to unacceptable toxicity (e.g., phase I trials testing dovitinib plus paclitaxel or platinum/gemcitabine).

Significant cross-talk between FGFR and other oncogenic signaling pathways has been described, as well as the capacity that tumoral cells have for initiating compensatory escape mechanisms through other tyrosine kinase receptors when one pathway is inhibited [14]. The addition of FGFR inhibitors with other targeted agents, such as EGFR, MET, MEK, or PIK3CA inhibitors, may be a feasible option to consider. The BOLERO-2 study testing everolimus and exemestane in breast cancer patients revealed the co-existence of activating mutations in PIK3CA and FGFR1 amplifications [103]. These observations provided the rationale for initiating a phase I combination trial with BGJ398 and BYL719 (PIK3/AKT/mTOR inhibitor) for validating the hypothesis of double blockade in patients harboring co-alterations within the FGFR/PIK3CA pathways (NCT01928459). On top of that, some of the FGFR aberrations have been described as endocrine therapy resistance mechanisms. That is the case of FGFR1 amplification in ER+ breast cancer patients, where increasing data suggest that FGFR inhibitors may be able to overcome and reverse hormone therapy resistance [104]. Several studies have been initiated for assessing the role that the addition of FGFR inhibitors may represent with combined endocrine therapy, such as a phase Ib trial combining lucitanib with fulvestrant in ER+/FGFR1-amplified luminal breast cancer patients (ISRCTN23201971). Table 5.2 depicts some of the clinical trials that have assessed potential combinations of FGFR inhibitors, both MTKIs and FGFRinh, with chemotherapy, endocrine therapy, targeted agents, or immunotherapies.

Despite the strong rationale behind some of these combinations, only nindetanib, a potent first-generation MTKI with triple FGFR/VEGFR/PDGFR antiangiogenic and antitumor activity [105], has been approved in the Europe for a subset of NSCLC. The LUME-Lung 1 trial (NCT00805194) was a phase III trial enrolling both squamous and adenocarcinoma NSCLCs, irrespective of their FGFR status, and patients were randomized to receive nindetanib/placebo plus docetaxel. The experimental combination met the primary endpoint of the study, showing improved progression free survival (median PFS 3.4 months vs 2.7 months; hazard ratio = 0.79 with 95% confidence interval 0.68–0.92, $P = 0.0019$) in all patients. However, nindetanib plus docetaxel only extended overall survival (median OS 12.6 months vs 10.3 months; hazard ratio = 0.83 with 95% confidence interval 0.70–0.99, $P = 0.0359$) in the prespecified adenocarcinoma histology population. In light of these results, nindetanib has been approved in combination with docetaxel for treating metastatic or locally advanced NSCLC with adenocarcinoma histology after receiving one line of platinum-based chemotherapy [106,107]. Unfortunately, no biomarker analysis was performed in this study, and no available

TABLE 5.2 Clinical Trials Testing Some Combinations of FGFR Inhibitors and Other Anticancer Therapies

FGFR inhibitor	Trial identifier	Ph	Status	Combination regimen	Tumor type
1. First-generation multityrosine kinase inhibitors					
Brivanib	NCT00798252	1	Completed	Brivanib + capecitabine or doxorubicin or docetaxel or paclitaxel or ixabelipone	All solid tumors
	NCT00300027	1	Terminated	Brivanib + folfox	GI
	NCT01046864	1	Completed	Brivanib + folfiri or LV + 5FU	GI
	NCT00640471	3	Completed	Brivanib + cetuximab	RAS WT mCRC
Nindetanib	NCT02856425	1	Recruiting	Nindetanib + pembrolizumab	All solid tumors
	NCT02619162	1	Recruiting	Nindetanib + letrozol	ER+ BC
	NCT02393755	1/2	Recruiting	Nindetanib + capecitabine	mCRC
	NCT01684111	1	Completed	Nindetanib + vinorelbina	NSCLC
	NCT01015118	3	Completed	Nindetanib + carboplatin/paclitaxel	Ovarian
	NCT00805194	3	Ongoing	Nindetanib + docetaxel	NSCLC
	NCT00806819	3	Completed	Nindetanib + pemetrexed	NSCLC
Pazopanib	NCT01542047	1	Terminated	Pazopanib + carboplatin	All solid tumors
	NCT02279576	1	Terminated	Pazopanib + paclitaxel	Penile
	NCT01600573	1/2	Recruiting	Pazopanib + topotecan	Ovarian
	NCT01130805	2	Completed	Pazopanib + CAPOX	GOJ/GC
	NCT02331498	1/2	Recruiting	Pazopanib + temozolamida	Glioblastoma
Dovitinib	NCT01548924	1	Terminated	Dovitinib + paclitaxel	All solid tumors
	NCT01496534	1	Terminated	Dovitinib + gemcitabine/cisplatin or carboplatin	All solid tumors
	NCT01921673	1/2	Completed	Dovitinib + docetaxel	GOJ/GC
	NCT01484041	1/2	Terminated	Dovitinib + aromatase inhibitors	BC
Lenvatinib	NCT02640508	1	Recruiting	Lenvatinib + eribulin	All solid tumors
	NCT00832819	1	Completed	Lenvatinib + carboplatin/paclitaxel	NSCLC
	NCT02788708	1	Recruiting	Lenvatinib + paclitaxel	Ovarian
	NCT02501096	1/2	Recruiting	Lenvatinib + pembrolizumab	All solid tumors

(Continued)

TABLE 5.2 Clinical Trials Testing Some Combinations of FGFR Inhibitors and Other Anticancer Therapies (*Cont.*)

FGFR inhibitor	Trial identifier	Ph	Status	Combination regimen	Tumor type
2. Second-generation selective pan-FGFR inhibitors					
AZD4547	NCT01202591	1/2	Completed	AZD4547 + fulvestrant	ER+ BC
	NCT01791985	1/2	Recruiting	AZD4547 + letrozol or anastrozol	ER+ BC
	NCT01824901	1/2	Completed	AZD4547 + docetaxel	NSCLC
GSK3052230	NCT01868022	1	Ongoing	GSK3052230 + carboplatin/paclitaxel or docetaxel or cisplatin/pemetrexed	Nsclc mesothelioma

Abbreviations: FGFR, fibroblast growth factor receptor; Ph, phase; GI, gastrointestinal cancers; WT, wild-type; mCRC, metastatic colorectal cancer; ER+, estrogen receptor-positive; BC, breast cancer; NSCLC, non-small cell lung cancer; GOJ, gastroesophageal junction cancer; GC, gastric cancer.

data regarding FGFR aberrations among enrolled patients is available. This prevents conclusions to be drawn on the potential predictive biomarkers of response to nindetanib and to FGFR inhibitors, in general.

5.5 CONCLUSIONS

Upregulation of FGFR/FGF signaling is a common event in carcinogenesis, and considering the multiple FGFR aberrations detected among cancer patients, FGFR inhibition arises as a promising therapeutic strategy. Despite the preclinical evidence that blocking this pathway leads to a degree of tumor control, the initial results of FGFR inhibitors tested in FGFR-altered patients have not been so successful. This highlights the need for further research into the insights of this signaling pathway and the identification of reliable predictive biomarkers. Only by preselecting the subset population who may achieve greater benefit from FGFR blockade will we be able to avoid unnecessary adverse events to patients with low chances of success and ensure that we are not missing potential candidates that could still benefit from these drugs. In this line, a better delineation of the molecular prescreening strategies in parallel with the performance of these early clinical trials testing new FGFR inhibitors and/or their combinations is crucial [108]. The identification of reliable response biomarkers that define the most suitable patient population that may benefit from receiving FGFRinh will clearly increase the possibilities of success of new therapeutic strategies for FGFR-aberrant patients [109].

In addition, characterization of the mechanism of action of FGFR inhibitors seems mandatory for considering future combination strategies. Taking into consideration the adaptive behavior of the FGFR axis as a resistance mechanism to other anticancer agents, the implementation timing of the FGFR blockade and possibilities of combinations or sequential strategies warrant further consideration. Whether any other hallmarks of cancer apart from proliferation and angiogenesis are modulated by these FGFR inhibitors remains unclear. In the era of emerging immunotherapeutic strategies, epigenetics, and microenvironment modulation, studies are needed to ascertain the role of the FGFR pathway in cancer and the implications of its blockade. Even with these challenges, the preclinical and early clinical data position the FGFR inhibitors as a new promising targeted agent family for enhancing the therapeutic strategies against solid tumors.

Acknowledgments

Dr. Hierro has been granted as sub-investigator by *Instituto de Salud Carlos III* (ISCIII) with the project reference number PI15/00360, which has been selected as part of an operational program co-financed by the European Regional Development Fund (ERDF).

Conflict of Interest: No potential conflicts of interest were disclosed.

References

[1] Druker BJ, Lydon NB. Lessons learned from the development of an abl tyrosine kinase inhibitor for chronic myelogenous leukemia. J Clin Invest 2000;105:3–7.

[2] Joensuu H, Roberts PJ, Sarlomo-Rikala M, et al. Effect of the tyrosine kinase inhibitor STI571 in a patient with a metastatic gastrointestinal stromal tumor. N Engl J Med 2001;344:1052–6.

[3] Chin L, Gray JW. Translating insights from the cancer genome into clinical practice. Nature 2008;452:553–63.

[4] Hauschild A, Grob JJ, Demidov LV, et al. Dabrafenib in BRAF-mutated metastatic melanoma: a multicentre, open-label, phase 3 randomised controlled trial. Lancet 2012;380:358–65.

[5] Chapman PB, Hauschild A, Robert C, et al. Improved survival with vemurafenib in melanoma with BRAF V600E mutation. N Engl J Med 2011;364:2507–16.

[6] Mok TS, Wu YL, Thongprasert S, et al. Gefitinib or carboplatin-paclitaxel in pulmonary adenocarcinoma. N Engl J Med 2009;361:947–57.

[7] Rosell R, Carcereny E, Gervais R, et al. Erlotinib versus standard chemotherapy as first-line treatment for European patients with advanced EGFR mutation-positive non-small-cell lung cancer (EURTAC): a multicentre, open-label, randomised phase 3 trial. Lancet Oncol 2012;13:239–46.

[8] Hierro C, Rodon J, Tabernero J. Fibroblast growth factor (FGF) receptor/FGF inhibitors: novel targets and strategies for optimization of response of solid tumors. Semin Oncol 2015;42:801–19.

[9] Wesche J, Haglund K, Haugsten EM. Fibroblast growth factors and their receptors in cancer. Biochem J 2011;437:199–213.

[10] Raju R, Palapetta SM, Sandhya VK, et al. A network map of FGF-1/FGFR signaling system. J Signal Transduct 2014;2014. 962962.

[11] Kurosu H, Ogawa Y, Miyoshi M, et al. Regulation of fibroblast growth factor-23 signaling by klotho. J Biol Chem 2006;281:6120–3.

[12] Knights V, Cook SJ, De-regulated FGF. De-regulated FGF receptors as therapeutic targets in cancer. Pharmacol Ther 2010;125:105–17.

[13] Hallinan N, Finn S, Cuffe S, et al. Targeting the fibroblast growth factor receptor family in cancer. Cancer Treat Rev 2016;46:51–62.

[14] Chae YK, Ranganath K, Hammerman PS, et al. Inhibition of the fibroblast growth factor receptor (FGFR) pathway: the current landscape and barriers to clinical application. Oncotarget 2017;8(9):16052–74. doi: 10.18632/oncotarget.14109.

[15] Brooks AN, Kilgour E, Smith PD. Molecular pathways: fibroblast growth factor signaling: a new therapeutic opportunity in cancer. Clin Cancer Res 2012;18:1855–62.

[16] di Martino E, Tomlinson DC, Williams SV, Knowles MA. A place for precision medicine in bladder cancer: targeting the FGFRs. Future Oncol 2016;12:2243–63.

[17] Hadari Y, Schlessinger J. FGFR3-targeted mAb therapy for bladder cancer and multiple myeloma. J Clin Invest 2009;119:1077–9.

[18] Wang Y, Ding X, Wang S, et al. Antitumor effect of FGFR inhibitors on a novel cholangiocarcinoma patient derived xenograft mouse model endogenously expressing an FGFR2-CCDC6 fusion protein. Cancer Lett 2016;380:163–73.

[19] Nakatani H, Sakamoto H, Yoshida T, et al. Isolation of an amplified DNA sequence in stomach cancer. Jpn J Cancer Res 1990;81:707–10.

[20] Desai A, Adjei AA. FGFR signaling as a target for lung cancer therapy. J Thorac Oncol 2016;11:9–20.

[21] Andre F, Bachelot T, Campone M, et al. Targeting FGFR with dovitinib (TKI258): preclinical and clinical data in breast cancer. Clin Cancer Res 2013;19:3693–702.

[22] Ueda T, Sasaki H, Kuwahara Y, et al. Deletion of the carboxyl-terminal exons of K-sam/FGFR2 by short homology-mediated recombination, generating preferential expression of specific messenger RNAs. Cancer Res 1999;59:6080–6.

[23] Pearson A, Smyth E, Babina IS, et al. High-level clonal FGFR amplification and response to FGFR inhibition in a translational clinical trial. Cancer Discov 2016;6:838–51.

[24] Cha JY, Maddileti S, Mitin N, et al. Aberrant receptor internalization and enhanced FRS2-dependent signaling contribute to the transforming activity of the fibroblast growth factor receptor 2 IIIb C3 isoform. J Biol Chem 2009;284:6227–40.

[25] Turner N, Grose R. Fibroblast growth factor signalling: from development to cancer. Nat Rev Cancer 2010;10: 116–29.

[26] Heist RS, Mino-Kenudson M, Sequist LV, et al. FGFR1 amplification in squamous cell carcinoma of the lung. J Thorac Oncol 2012;7:1775–80.

[27] Weiss J, Sos ML, Seidel D, et al. Frequent and focal FGFR1 amplification associates with therapeutically tractable FGFR1 dependency in squamous cell lung cancer. Sci Transl Med 2010;2:62ra93.

[28] Cancer Genome Atlas Research Network. Comprehensive genomic characterization of squamous cell lung cancers. Nature 2012;489:519–25.

[29] Cancer Genome Atlas Research Network. Comprehensive molecular profiling of lung adenocarcinoma. Nature 2014;511:543–50.

[30] Cihoric N, Savic S, Schneider S, et al. Prognostic role of FGFR1 amplification in early-stage non-small cell lung cancer. Br J Cancer 2014;110:2914–22.

[31] Kim HR, Kim DJ, Kang DR, et al. Fibroblast growth factor receptor 1 gene amplification is associated with poor survival and cigarette smoking dosage in patients with resected squamous cell lung cancer. J Clin Oncol 2013;31:731–7.

[32] Azuma K, Kawahara A, Sonoda K, et al. FGFR1 activation is an escape mechanism in human lung cancer cells resistant to afatinib, a pan-EGFR family kinase inhibitor. Oncotarget 2014;5:5908–19.

[33] Schultheis AM, Bos M, Schmitz K, et al. Fibroblast growth factor receptor 1 (FGFR1) amplification is a potential therapeutic target in small-cell lung cancer. Mod Pathol 2014;27:214–21.

[34] Pardo OE, Latigo J, Jeffery RE, et al. The fibroblast growth factor receptor inhibitor PD173074 blocks small cell lung cancer growth *in vitro* and *in vivo*. Cancer Res 2009;69:8645–51.

[35] Goke F, Bode M, Franzen A, et al. Fibroblast growth factor receptor 1 amplification is a common event in squamous cell carcinoma of the head and neck. Mod Pathol 2013;26:1298–306.

[36] Koole K, Brunen D, van Kempen PM, et al. FGFR1 is a potential prognostic biomarker and therapeutic target in head and neck squamous cell carcinoma. Clin Cancer Res 2016;22:3884–93.

[37] von Loga K, Kohlhaussen J, Marx AH, et al. FGFR1 amplification is linked to the squamous cell carcinoma subtype in esophageal carcinoma. In: Proceedings of the 104th Annual Meeting of the American Association for Cancer Research (AACR), 6–10 April 2013, Washington, DC; 2013. [Epub ahead of print].

[38] Cancer Genome Atlas Research Network. Integrated genomic characterization of oesophageal carcinoma. Nature 2017;541:169–75.

[39] Sugiura K, Ozawa S, Kitagawa Y, et al. Co-expression of aFGF and FGFR-1 is predictive of a poor prognosis in patients with esophageal squamous cell carcinoma. Oncol Rep 2007;17:557–64.

[40] Cancer Genome Atlas Research Network. Comprehensive molecular characterization of gastric adenocarcinoma. Nature 2014;513:202–9.

[41] Betts G, Valentine H, Pritchard S, et al. FGFR2, HER2 and cMet in gastric adenocarcinoma: detection, prognostic significance and assessment of downstream pathway activation. Virch Arch 2014;464:145–56.

[42] Su X, Zhan P, Gavine PR, et al. FGFR2 amplification has prognostic significance in gastric cancer: results from a large international multicentre study. Br J Cancer 2014;110:967–75.

[43] Sawey ET, Chanrion M, Cai C, et al. Identification of a therapeutic strategy targeting amplified FGF19 in liver cancer by oncogenomic screening. Cancer Cell 2011;19:347–58.

[44] Zhou M, Wang X, Phung V, et al. Separating tumorigenicity from bile acid regulatory activity for endocrine hormone FGF19. Cancer Res 2014;74:3306–16.

[45] Harimoto N, Taguchi K, Shirabe K, et al. The significance of fibroblast growth factor receptor 2 expression in differentiation of hepatocellular carcinoma. Oncology 2010;78:361–8.

[46] Poh W, Wong W, Ong H, et al. Klotho-beta overexpression as a novel target for suppressing proliferation and fibroblast growth factor receptor-4 signaling in hepatocellular carcinoma. Mol Cancer 2012;11:14.
[47] Arai Y, Totoki Y, Hosoda F, et al. Fibroblast growth factor receptor 2 tyrosine kinase fusions define a unique molecular subtype of cholangiocarcinoma. Hepatology 2014;59:1427–34.
[48] Graham RP, Barr Fritcher EG, Pestova E, et al. Fibroblast growth factor receptor 2 translocations in intrahepatic cholangiocarcinoma. Hum Pathol 2014;45:1630–8.
[49] Desnoyers LR, Pai R, Ferrando RE, et al. Targeting FGF19 inhibits tumor growth in colon cancer xenograft and FGF19 transgenic hepatocellular carcinoma models. Oncogene 2008;27:85–97.
[50] Heinzle C, Gsur A, Hunjadi M, et al. Differential effects of polymorphic alleles of FGF receptor 4 on colon cancer growth and metastasis. Cancer Res 2012;72:5767–77.
[51] Mizukami T, Togashi Y, Naruki S, et al. Significance of FGF9 gene in resistance to anti-EGFR therapies targeting colorectal cancer: a subset of colorectal cancer patients with FGF9 upregulation may be resistant to anti-EGFR therapies. Mol Carcinog 2017;56:106–17.
[52] Cancer Genome Atlas Research Network. Comprehensive molecular portraits of human breast tumours. Nature 2012;490:61–70.
[53] Tenhagen M, van Diest PJ, Ivanova IA, et al. Fibroblast growth factor receptors in breast cancer: expression, downstream effects, and possible drug targets. Endocr Relat Cancer 2012;19:R115–29.
[54] Turner N, Pearson A, Sharpe R, et al. FGFR1 amplification drives endocrine therapy resistance and is a therapeutic target in breast cancer. Cancer Res 2010;70:2085–94.
[55] Lee HJ, Seo AN, Park SY, et al. Low prognostic implication of fibroblast growth factor family activation in triple-negative breast cancer subsets. Ann Surg Oncol 2014;21:1561–8.
[56] Cancer Genome Atlas Research Network. Integrated genomic analyses of ovarian carcinoma. Nature 2011;474:609–15.
[57] Roidl A, Berger HJ, Kumar S, et al. Resistance to chemotherapy is associated with fibroblast growth factor receptor 4 up-regulation. Clin Cancer Res 2009;15:2058–66.
[58] Kandoth C, Schultz N, Cherniack AD, et al. Integrated genomic characterization of endometrial carcinoma. Nature 2013;497:67–73.
[59] Krakstad C, Birkeland E, Seidel D, et al. High-throughput mutation profiling of primary and metastatic endometrial cancers identifies KRAS, FGFR2 and PIK3CA to be frequently mutated. PLoS ONE 2012;7:e52795.
[60] Byron SA, Gartside M, Powell MA, et al. FGFR2 point mutations in 466 endometrioid endometrial tumors: relationship with MSI, KRAS, PIK3CA, CTNNB1 mutations and clinicopathological features. PLoS ONE 2012;7:e30801.
[61] Cappellen D, De Oliveira C, Ricol D, et al. Frequent activating mutations of FGFR3 in human bladder and cervix carcinomas. Nat Genet 1999;23:18–20.
[62] Cancer Genome Atlas Research Network. Comprehensive molecular characterization of urothelial bladder carcinoma. Nature 2014;507:315–22.
[63] Billerey C, Chopin D, Aubriot-Lorton MH, et al. Frequent FGFR3 mutations in papillary non-invasive bladder (pTa) tumors. Am J Pathol 2001;158:1955–9.
[64] Carstens JL, Shahi P, Van Tsang S, et al. FGFR1-WNT-TGF-beta signaling in prostate cancer mouse models recapitulates human reactive stroma. Cancer Res 2014;74:609–20.
[65] Armstrong K, Ahmad I, Kalna G, et al. Upregulated FGFR1 expression is associated with the transition of hormone-naive to castrate-resistant prostate cancer. Br J Cancer 2011;105:1362–9.
[66] Singh D, Chan JM, Zoppoli P, et al. Transforming fusions of FGFR and TACC genes in human glioblastoma. Science 2012;337:1231–5.
[67] Parker BC, Annala MJ, Cogdell DE, et al. The tumorigenic FGFR3-TACC3 gene fusion escapes miR-99a regulation in glioblastoma. J Clin Invest 2013;123:855–65.
[68] Gartside MG, Chen H, Ibrahimi OA, et al. Loss-of-function fibroblast growth factor receptor-2 mutations in melanoma. Mol Cancer Res 2009;7:41–54.
[69] Li SQ, Cheuk AT, Shern JF, et al. Targeting wild-type and mutationally activated FGFR4 in rhabdomyosarcoma with the inhibitor ponatinib (AP24534). PLoS ONE 2013;8:e76551.
[70] Fernanda Amary M, Ye H, Berisha F, et al. Fibroblastic growth factor receptor 1 amplification in osteosarcoma is associated with poor response to neo-adjuvant chemotherapy. Cancer Med 2014;3:980–7.
[71] Hanahan D, Weinberg RA. The hallmarks of cancer. Cell 2000;100:57–70.

[72] Hanahan D, Weinberg RA. Hallmarks of cancer: the next generation. Cell 2011;144:646–74.

[73] Bailon-Moscoso N, Romero-Benavides JC, Ostrosky-Wegman P. Development of anticancer drugs based on the hallmarks of tumor cells. Tumour Biol 2014;35:3981–95.

[74] Itoh N, Ornitz DM. Fibroblast growth factors: from molecular evolution to roles in development, metabolism and disease. J Biochem 2011;149:121–30.

[75] Takenouchi T, Sakamoto Y, Miwa T, et al. Severe craniosynostosis with Noonan syndrome phenotype associated with SHOC2 mutation: clinical evidence of crosslink between FGFR and RAS signaling pathways. Am J Med Genet A 2014;164A:2869–72.

[76] Gozgit JM, Wong MJ, Moran L, et al. Ponatinib (AP24534), a multitargeted pan-FGFR inhibitor with activity in multiple FGFR-amplified or mutated cancer models. Mol Cancer Ther 2012;11:690–9.

[77] Lieu C, Heymach J, Overman M, Beyond VEGF, et al. Beyond VEGF: inhibition of the fibroblast growth factor pathway and antiangiogenesis. Clin Cancer Res 2011;17:6130–9.

[78] Oladipupo SS, Smith C, Santeford A, et al. Endothelial cell FGF signaling is required for injury response but not for vascular homeostasis. Proc Natl Acad Sci USA 2014;111:13379–84.

[79] Tran TA, Leong HS, Pavia-Jimenez A, et al. Fibroblast growth factor receptor-dependent and -independent paracrine signaling by sunitinib-resistant renal cell carcinoma. Mol Cell Biol 2016;36:1836–55.

[80] Casanovas O, Hicklin DJ, Bergers G, Hanahan D. Drug resistance by evasion of antiangiogenic targeting of VEGF signaling in late-stage pancreatic islet tumors. Cancer Cell 2005;8:299–309.

[81] Sweeny L, Liu Z, Lancaster W, et al. Inhibition of fibroblasts reduced head and neck cancer growth by targeting fibroblast growth factor receptor. Laryngoscope 2012;122:1539–44.

[82] Xie L, Su X, Zhang L, et al. FGFR2 gene amplification in gastric cancer predicts sensitivity to the selective FGFR inhibitor AZD4547. Clin Cancer Res 2013;19:2572–83.

[83] Herbert C, Lassalle G, Alcouffe C, Bono F. Approaches targeting the FGF-FGFR system: a review of the recent patent literature and associated advanced therapeutic agents. Pharm Pat Anal 2014;3:585–612.

[84] Soria JC, DeBraud F, Bahleda R, et al. Phase I/IIa study evaluating the safety, efficacy, pharmacokinetics, and pharmacodynamics of lucitanib in advanced solid tumors. Ann Oncol 2014;25:2244–51.

[85] Andre F, Ranson M, Dean E, et al. Results of a phase I study of AZD4547, an inhibitor of fibroblast growth factor receptor, in patients with advanced solid tumors. In: Proceedings of the 104th Annual Meeting of the American Association for Cancer Research, Washington, DC, 6–10 April 2013. Philadelphia, PA: AACR; 2013doi: 10.1158/1538-7445.AM2013-LB-145. Cancer Res 73(8 Suppl). [Epub ahead of print].

[86] Kilgour E, Ferry D, Saggese M, et al. Exploratory biomarker analysis of a phase I study of AZD4547, an inhibitor of fibroblast growth factor receptor (FGFR), in patients with advanced solid tumors. J Clin Oncol 2014;325;doi: 10.1158/1538-7445.AM2013-LB-145. (Suppl. abstr. 11010). [Epub ahead of print].

[87] Paik P, Shen R, Ferry D, et al. A phase 1b open-label multicenter study of AZD4547 in patients with advanced squamous cell lung cancers: preliminary antitumor activity and pharmacodynamic data. J Clin Oncol 2014;32(5 Suppl.). [Epub ahead of print]; abstr 8035.

[88] Arkenau H, Saggese M, Hollebecque A, et al. A phase 1 expansion cohort of the fibroblast growth factor receptor (FGFR) inhibitor AZD4547 in patients (pts) with advanced gastric (GC) and gastroesophageal (GOJ) cancer. J Clin Oncol 2014;32(5 Suppl.). [Epub ahead of print]; abstr 2620.

[89] Nogova L, Sequist LV, Perez Garcia JM, et al. Evaluation of BGJ398, a fibroblast growth factor receptor 1-3 kinase inhibitor, in patients with advanced solid tumors harboring genetic alterations in fibroblast growth factor receptors: results of a global phase I, dose-escalation and dose-expansion study. J Clin Oncol 2016;. Jco2016672048, Nov 21 [Epub ahead of print].

[90] Joerger R, Soo B, Cho B, et al. Phase I study of the pan-fibroblast growth factor receptor inhibitor BAY 1163877 with expansion cohorts for patients based on tumour FGFR mRNA expression levels. Ann Oncol 2016;27(6):1–36. doi: 10.1093/annonc/mdw435. [Epub ahead of print].

[91] Johnson DE, Williams LT. Structural and functional diversity in the FGF receptor multigene family. Adv Cancer Res 1993;60:1–41.

[92] Lesca E, Lammens A, Huber R, Augustin M. Structural analysis of the human fibroblast growth factor receptor 4 kinase. J Mol Biol 2014;426:3744–56.

[93] Tabernero J, Bahleda R, Dienstmann R, et al. Phase I dose-escalation study of JNJ-42756493, an oral pan-fibroblast growth factor receptor inhibitor, in patients with advanced solid tumors. J Clin Oncol 2015;33:3401–8.

[94] Bell K, Gaffney D, Martinez Cardona G, et al. Co-amplification of FGF receptors and ligands in FGFR inhibitor-sensitive cell lines. Proceedings of the 106th Annual Meeting of the American Association for Cancer Research, 18–22 April, Philadelphia, PA. Cancer Res 2015;75(15 Suppl.). [Epub ahead of print].

[95] Tolcher AW, Papadopoulos KP, Patnaik A, et al. A Phase 1, First in Human Study of FP-1039 (GSK3052230), a Novel FGF Ligand Trap, in Patients with Advanced Solid Tumors. Ann Oncol 2015;doi: 10.1093/annonc/mdv591. [Epub ahead of print].
[96] Byron SA, Chen H, Wortmann A, et al. The N550K/H mutations in FGFR2 confer differential resistance to PD173074, dovitinib, and ponatinib ATP-competitive inhibitors. Neoplasia 2013;15:975–88.
[97] Goyal L, Saha SK, Liu LY, et al. Polyclonal secondary FGFR2 mutations drive acquired resistance to FGFR inhibition in patients with FGFR2 fusion-positive cholangiocarcinoma. Cancer Discov 2016;doi: 10.1158/2159-8290.cd-16-1000. [Epub ahead of print].
[98] Ochiiwa H, Fujita H, Itoh K, et al. TAS-120, a highly potent and selective irreversible FGFR inhibitor, is effective in tumors harboring various FGFR gene abnormalities. In: Proceedings of the AACR-NCI-EORTC International Conference: Molecular Targets and Cancer Therapeutics, Boston, MA, 19–23 October 2013. Philadelphia, PA: AACR; 2013. Mol Cancer Ther 12(11 Suppl.):2013; doi: 10.1158/1535-7163.TARG-13-A270. [Epub ahead of print].
[99] Piha-Paul S, Hierro C, Funk J, et al. A phase 1, multicenter, dose-escalation study of PRN1371, an irreversible covalent FGFR1-4 kinase inhibitor, in patients with advanced solid tumors, followed by expansion cohorts in patients with FGFR genetic alterations. J Clin Oncol 2016;34. (suppl; abstr TPS2602), [Epub ahead of print].
[100] Owen BM, Mangelsdorf DJ, Kliewer SA. Tissue-specific actions of the metabolic hormones FGF15/19 and FGF21. Trends Endocrinol Metab 2015;26:22–9.
[101] Qiu H, Yashiro M, Zhang X, et al. A FGFR2 inhibitor, Ki23057, enhances the chemosensitivity of drug-resistant gastric cancer cells. Cancer Lett 2011;307:47–52.
[102] Yashiro M, Shinto O, Nakamura K, et al. Synergistic antitumor effects of FGFR2 inhibitor with 5-fluorouracil on scirrhous gastric carcinoma. Int J Cancer 2010;126:1004–16.
[103] Baselga J, Piccart M, Rugo H, et al. Assessment of genetic alterations using next-generation sequencing in postmenopausal women with hormone receptor-positive, HER2-negative advanced breast cancer: results from the BOLERO-2 phase III. In: Proceedings of the 104th Annual Meeting of the American Association for Cancer Research (AACR), 6–10 April, Washington, DC; 2013. Abstr 4564 2013; [Epub ahead of print].
[104] Balko JM, Mayer IA, Sanders ME, et al. Discordant cellular response to presurgical letrozole in bilateral synchronous ER+ breast cancers with a KRAS mutation or FGFR1 gene amplification. Mol Cancer Ther 2012;11:2301–5.
[105] Hilberg F, Roth GJ, Krssak M, et al. BIBF 1120: triple angiokinase inhibitor with sustained receptor blockade and good antitumor efficacy. Cancer Res 2008;68:4774–82.
[106] Reck M, Kaiser R, Mellemgaard A, et al. Docetaxel plus nintedanib versus docetaxel plus placebo in patients with previously treated non-small-cell lung cancer (LUME-Lung 1): a phase 3, double-blind, randomised controlled trial. Lancet Oncol 2014;15:143–55.
[107] Novello S, Kaiser R, Mellemgaard A, et al. Analysis of patient-reported outcomes from the LUME-Lung 1 trial: a randomised, double-blind, placebo-controlled phase III study of second-line nintedanib in patients with advanced non-small cell lung cancer. Eur J Cancer 2015;51:317–26.
[108] Hierro C, Azaro A, Argiles G, et al. Unveiling changes in the landscape of patient populations in cancer early drug development. Oncotarget 2016;doi: 10.18632/oncotarget.13258. [Epub ahead of print].
[109] Hierro C, Alsina M, Sanchez M, et al. Targeting the fibroblast growth factor receptor 2 in gastric cancer: promise or pitfall? Ann Oncol 2017;doi: 10.1093/annonc/mdx081. [Epub ahead of print].

CHAPTER

6

PIK3CA Mutations in Colorectal and Breast Cancer: Impact on Oncogenesis and Response to Nonsteroidal Anti-Inflammatory Drugs

Richard T. Zhu, J. Silvio Gutkind**, Daniel E. Johnson†, Jennifer R. Grandis†*

*The Warren Alpert Medical School of Brown University, Providence, RI, United States
**Moores Cancer Center, University of California at San Diego, La Jolla, CA, United States
†University of California at San Francisco, San Francisco, CA, United States

Abstract

Genetic alterations, including somatic activating mutations of oncogenes and loss of tumor suppressor genes, are initiators of oncogenesis. The phosphatidylinositol-3-kinase (PI3K) signaling pathway regulates a broad spectrum of normal physiological processes, and when altered, contributes to tumor formation, progression, and resistance to therapy. Specifically, mutations in *PIK3CA*, the gene encoding the PI3K catalytic subunit, are found in a variety of human cancers and are associated with more aggressive disease and worse prognosis. Cumulative evidence has demonstrated that regular use of aspirin and other nonsteroidal anti-inflammatory drugs (NSAIDs) is effective in the prevention and treatment of colorectal cancer, breast cancer, and other malignancies. However, the mechanism by which NSAIDs influence cancer risk and survival are poorly understood. Emerging data from observational studies and clinical trials suggest that *PIK3CA* mutational status in conjunction with other biomarkers might help to identify individuals who will benefit most from the NSAID therapy. This chapter reviews the PI3K signaling pathway with a focus on *PIK3CA* mutations and the role of aspirin and other NSAIDs as promising chemopreventive therapies in human cancer.

ABBREVIATIONS

AMPK	5′ Adenosine monophosphate-activated protein kinase
APC	Adenomatous polyposis coli
5 ASA	5-Aminosalicyclic acid
COX	Cyclooxygenase

Targeting Cell Survival Pathways to Enhance Response to Chemotherapy. http://dx.doi.org/10.1016/B978-0-12-813753-6.00006-8
Copyright © 2018 Elsevier Inc. All rights reserved.

COXIBs	COX-2-selective inhibitors
CRC	Colorectal cancer
EGFR	Epidermal growth factor receptor
ER	Estrogen receptor
FAP	Familial adenomatous polyposis
GPCR	G protein-coupled receptor
HER	Human epidermal growth factor receptor
HNPCC	Hereditary nonpolyposis colorectal cancer
HNSCC	Head and neck squamous cell carcinoma
IBD	Inflammatory bowel disease
IGF-1	Insulin-like growth factor-1
MAPK	Mitogen-activated protein kinase
mCRC	Metastatic colorectal cancer
MDSC	Myeloid-derived suppressor cell
MEK	Mitogen-activated protein kinase kinase
mTOR	Mammalian target of rapamycin
NF-kB	Nuclear factor kappa B
NHGRI	National Human Genome Research Institute
NSAIDs	Nonsteroidal anti-inflammatory drugs
PDK1	Phosphoinositide dependent protein kinase-1
PGE_2	Prostaglandin E_2
PGH_2	Prostaglandin H2
PH	Pleckstrin homology
PI	Phosphatidylinositol
PIP	Phosphatidylinositol (3)-phosphate
PIP_2	Phosphatidylinositol (3,4)-biphosphate
PIP_3	Phosphatidylinositol (3,4,5)-triphosphate
PI3K	Phosphatidylinositol-3-kinase
PR	Progesterone receptor
PTEN	Phosphatase and tensin homolog deleted on chromosome ten
RTK	Receptor tyrosine kinase
TCGA	The Cancer Genome Atlas
VEGF	Vascular endothelial growth factor

6.1 INTRODUCTION

There is a growing body of literature supporting the regular use of nonsteroidal anti-inflammatory drugs (NSAIDs), particularly aspirin, in the prevention and treatment of colorectal cancer (CRC) and colorectal adenomas [1]. CRCs are thought to arise from a series of molecular alterations that transform normal colonic epithelial cells into premalignant adenomas and subsequently into carcinomas [2]. Pro-inflammatory prostaglandins, which are produced by the enzyme cyclooxygenase (COX), are believed to support CRC carcinogenesis [3]. Early studies with rodent CRC models showed that treatment with the COX inhibitor, indomethacin, decreased the incidence and size of carcinogen-induced colonic tumors [4–6]. A similar effect in humans has been suggested by small, randomized clinical trials evaluating the efficacy of the NSAIDs sulindac and celecoxib in protecting against colonic polyp formation and reducing the risk of CRC in patients with familial adenomatous polyposis (FAP) [7–9]. These encouraging findings in patients with FAP prompted evaluation of the anticancer properties of aspirin and other NSAIDs in the general population of CRC patients.

Multiple epidemiological studies have examined the chemopreventive properties of NSAIDs in reducing the incidence of CRC and colorectal adenomas [10–12]. In a case–control study of 1326 patients with CRC and 4891 control patients, Rosenberg et al. [10] showed that sustained use of NSAIDs reduced the risk of human colon cancer. A similar investigation from The Nurses' Health Study examined a cohort of 89,446 women and found that regular use of aspirin (two or more times per week), at doses similar to those recommended for prevention of cardiovascular disease, reduced the risk of CRC [12]. However, this clinical benefit was only observed in patients after a decade of aspirin use. Another population-based control study from Australia found a similar reduction in the risk of CRCs among individuals taking aspirin [13]. These findings have been further substantiated in two prospective cohort studies, the Cancer Prevention Study II [14] and the Health Professionals Follow-up Study [11], which demonstrated that regular use of aspirin reduced the risk of fatal CRC and colorectal adenoma recurrence.

Emerging from these cumulative preclinical and observational studies are two landmark, randomized, double-blind clinical trials led by Sandler et al. [15] and Baron et al. [16]. These studies demonstrated that aspirin use was associated with a substantial reduction in incidence of CRCs and colorectal adenomas among patients with a prior history of either CRC or adenomas [15,16]. However, the benefit of aspirin in reducing the risk of CRCs or adenomas was found to disappear when aspirin use was discontinued [16]. A subsequent observational follow-up study reported that patients who discontinued aspirin, and subsequently started using a different NSAID, exhibited significant protection against CRC and adenomas [17]. These findings have encouraged additional clinical investigations evaluating the effects of aspirin on CRC-specific and overall survival [18–20]. In fact, in April 2016, the U.S. Preventive Services Task Force found substantial evidence to recommend that people who are between 50 and 69 years of age, and are not at increased risk for bleeding, consider taking aspirin to prevent CRC and cardiovascular events [21,22].

Although generally well tolerated, use of aspirin daily is not without perils. The most commonly reported adverse side effects include dyspepsia, nausea, vomiting, and abdominal discomfort, while the more threatening complications include hepatotoxicity, gastrointestinal bleeding, and hemorrhagic stroke. Therefore, it is imperative to elucidate molecular biomarkers and mechanisms of aspirin response in the cancer microenvironment in order to better identify which subgroup of patients are most likely to benefit from aspirin's antitumor effect. With the growing population of CRC survivors, there is considerable interest in developing strategies for targeting signaling pathways in both the cancer and the immune system. This is reflected in several *in vitro and in vivo* studies that shed light on the association of specific *PIK3CA* mutations in CRCs with elevated expression of cyclooxygenase (COX-2), a critical enzyme in the synthesis of prostaglandin E_2 (PGE_2), a potent immunosuppressive molecule [23,24]. Notably, COX-2 is inhibited by aspirin (and other NSAIDs) [25]. Thus, it is perhaps not surprising that recent studies have shown that *PIK3CA* mutations are a promising predictive biomarker for successful therapy with aspirin in patients with CRCs. In this chapter, we review the genetic alterations in *PIK3CA* that are found in CRCs and their impact on tumor growth and response to aspirin. Further, we will discuss the potential of aspirin therapy as an emerging chemopreventive strategy in CRCs, breast cancer, and other malignancies.

6.2 ONCOGENIC SIGNALING IN CRCs

CRC is the third most common cancer and second leading cause of cancer-related mortality in the United States. Based on data from 2017, it is estimated that approximately 4.6% of men (1 in 22) and 4.2% of women (1 in 24) will be diagnosed with CRC in their lifetime [26]. The estimated 5-year survival rate varies with the pathological extent of disease (TNM stage), from 90% for localized stage I CRC to 50%–70% for advanced stage III CRC [26]. Major risk factors for CRC development include genetic and epigenetic abnormalities, diet, lifestyle, and chronic inflammation. Since the mid-1980s, enhanced CRC screening and treatment methods have contributed to the rise in CRC survival rates. The detection and removal of preneoplastic polyps is an effective preventive approach to mitigate CRC development. Unfortunately, not all individuals can be identified at the earliest stages of CRCs to reap maximal benefit from surgical resection alone. Thus, chemotherapy and radiation are often combined with surgery to treat patients with more advanced CRC that has spread to surrounding tissues and regional lymph nodes and/or metastasized to distant parts of the body. However, metastatic CRC is considered incurable with a 5-year survival rate of only 11%, and 90% of patients with metastatic disease acquire resistance to therapy [27]. The identification of molecules or pathways that can be targeted to overcome this resistance represents a major avenue of current research.

CRC develops from a cascade of mutations in key cell-cycle regulatory genes, tumor suppressor genes, and proto-oncogenes. The pathogenesis of CRC frequently involves mutations in components of the Wnt signaling pathway in intestinal crypt stem cells, with the *APC* gene being the most frequently altered [28]. This gene encodes the adenomatous polyposis coli (APC) protein, which normally prevents the accumulation of β-catenin, an adaptor protein present in the adherens junctional complex and a transcription regulator. When β-catenin aberrantly accumulates in the cytoplasm, as is the case in CRC with *APC* alterations, it undergoes translocation into the nucleus where it promotes transcription of Wnt target genes, including *c-Myc*, leading to enhanced cellular proliferation and growth [29]. However, it takes more than a "single" hit mutation for malignant transformation in CRC to occur [30]. This is because tumor suppressor genes present in the cell, including *PTEN* and *TP53*, survey dysregulated cellular activity and facilitate either rectification or trigger apoptosis. Eventually, cells acquire additional mutations in tumor suppressor genes leading to "double-hit" (or "multihit") oncogenesis. Abrogation of such tumor suppressor function allows for transformation of the intestinal stem cell into an invasive epithelial carcinoma cell [30].

Mutation or overexpression of proto-oncogenes also promotes CRC development. For example, the epidermal growth factor receptor (EGFR) plays an instrumental role in the transduction of proliferative and anti-apoptotic signals and is commonly overexpressed in metastatic colorectal cancer (mCRC). Accordingly, EGFR has been favorably targeted using two EGFR monoclonal antibodies: cetuximab and panitumumab. These molecular inhibitors selectively block the extracellular binding domain on EGFR and serve to augment conventional chemotherapy regimens in the treatment of mCRC [31,32]. However, dysregulation of signaling components downstream of EGFR, including components of the RAS/RAF, MEK/MAPK, and PI3K/AKT/mTOR pathways, promotes resistance to EGFR inhibitors. Moreover, in chemotherapy-refractory patients with mCRC, activating mutations in *K-Ras* mediate resistance to EGFR monoclonal antibody treatment [33].

6.3 THE PI3K SIGNALING PATHWAY

The phosphatidylinositol-3-kinases (PI3Ks) are intracellular signaling enzymes that translate information from exogenous stimuli including growth factors and cytokines into intracellular second messengers that induce multiple pathways involved in cell-cycle regulation and cellular metabolism, differentiation, migration, and survival [34]. PI3Ks are heterodimeric lipid kinases that phosphorylate 3′-hydroxyl groups on the phosphoinositol ring. By doing so, these kinases generate phospholipid molecules that are involved in critical cell signaling pathways and maintenance of cellular homeostasis. However, aberrant PI3K activity has been observed in numerous pathological conditions including diabetes, chronic inflammation, and cancer. Since their discovery, PI3Ks and the signaling pathways mediated by these kinases have been extensively studied for their promotion of oncogenesis in colorectal, breast, brain, and lung cancers via mechanisms involving increased genomic instability, enhanced cell proliferation, evasion of growth suppressors, resistance to cell death, and induction of angiogenesis and invasion [35–37]. Understanding PI3K signaling pathways will be integral to drug development strategies for these diseases.

Members of the PI3K family can be grouped into classes I, II, and III based on their primary structure, catalytic activity, and substrate specificity [38]. Class I PI3Ks primarily produce phosphatidylinositol (3,4,5)-trisphosphate (PIP_3) from phosphatidylinositol (3,4)-bisphosphate (PIP_2). Class I PI3Ks are further divided into subclasses IA and IB. Class IA PI3Ks consist of a p110 catalytic subunit that heterodimerizes with a p85 regulatory subunit [39]. The genes *PIK3CA*, *PIK3CB*, and *PIK3CD* encode three class IA catalytic isoforms (p110α, β, and δ, respectively) that associate with five different regulatory isoforms: p85α or its two splice variants p55α and p50α encoded by *PIK3R1*, p85β encoded by *PIK3R2*, and p55γ encoded by *PIK3R3* [40]. By contrast, class IB PI3Ks comprise a single catalytic subunit (p110γ) encoded by *PIK3CG* and a regulatory subunit (p101) encoded by *PIK3R5* [40].

Classes II and III PI3Ks are less well characterized. Class II PI3Ks consist of three catalytic isoforms C2α, C2β, and C2γ, but lack regulatory domains and catalyze the phosphorylation of phosphatidylinositol (PI) to produce phosphatidylinositol (3)-phosphate (PIP), with subsequent phosphorylation of PIP to produce PIP_2 [38]. Class II PI3Ks play a critical role in generating the PIP_2 substrate for class I PI3Ks. Class III PI3Ks resemble class I PI3Ks in heterodimeric structure, containing both catalytic (Vps34) and regulatory subunits (Vps15/p150), but differ in function, as class III PI3Ks predominantly produce PIP from PI and are thought to be involved in autophagy, endocytosis, and phagocytosis [41].

The binding of growth factors or cytokines to their cognate transmembrane receptor triggers class I PI3K signaling (Fig. 6.1). Upon activation of G protein-coupled receptors (GPCR) or receptor tyrosine kinases (RTKs), class I PI3Ks are recruited to the inner plasma membrane through interaction with GPCR-associated G protein subunits or tyrosine phosphorylated motifs present on RTKs. Recruitment to the activated receptors relieves the inhibition of p110 imposed by the p85 regulatory subunit and leads to the production of PIP_3 from PIP_2 [40]. PIP_3 is a lipid second messenger that induces AKT-dependent signaling pathways. Specifically, PIP_3 serves as a docking site for proteins containing a pleckstrin homology (PH) domain, including the serine/threonine kinases AKT1, AKT2, and AKT3, as well as phosphoinositide-dependent protein kinase-1 (PDK1). Once localized to the cellular inner membrane, PDK1

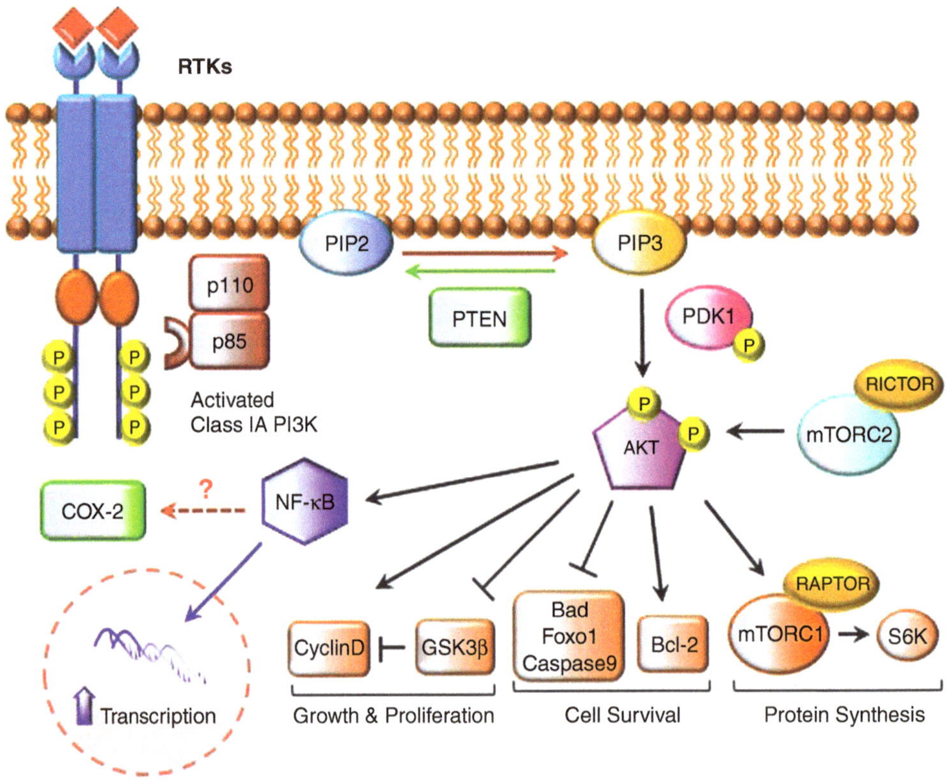

FIG. 6.1 **Class I phosphoinositide 3-kinase (PI3K) signaling pathway.** Upon ligand stimulation and subsequent activation of RTKs, class IA PI3K, consisting of p110/p85, is recruited to the cellular membrane via interaction of the regulatory p85 subunit with the tyrosine-phosphorylated RTK receptor, either directly or through an adaptor protein. The activated p110 catalytic subunit converts phosphatidylinositol-4,5-bisphosphate (PIP2) to phosphatidylinositol-3,4,5-triphosphate (PIP3), which provides docking sites for proteins containing pleckstrin-homology (PH) domains including phosphoinositide-dependent kinase 1 (PKD1). PDK1 phosphorylates and activates the seronine-threonine kinase, AKT, which regulates a broad spectrum of downstream activities that are involved in cell growth, proliferation, and survival. AKT can activate the nuclear transcription factor, NF-κB which is putatively linked to the production of pro-inflammatory prostaglandins by cyclooxygenase-2 (COX-2). Phosphatase and tensin homolog mutated on chromosome ten (PTEN) inhibits PI3K signaling by dephosphorylating PIP3 back to PIP2. Abbreviations: GSK3β, glycogen synthase kinase 3 β; mTORC, mammalian target of rapamycin complex; Bcl-2, B-cell lymphoma-2; S6K, ribosomal protein S6 kinase; Bad, Bcl-2-associated death promoter protein; Foxo1, Forkhead box protein O1; RTK, receptor tyrosine kinase.

phosphorylates AKT on threonine 308 with subsequent phosphorylation of serine 473 by cytoplasmic mTORC2 (complex of mTOR and RICTOR) [42]. Activation of AKT ultimately leads to activation of mTORC1 (complex of mTOR and RAPTOR), a downstream effector kinase that promotes cell growth, enhanced glucose metabolism, and protein and lipid synthesis [43]. Phosphatase and tensin homolog deleted on chromosome ten (PTEN) is a lipid phosphatase that removes the 3′ phosphate moiety from PIP_3, thereby inactivating class I PI3K signaling [40]. The equilibrium between phosphorylation of PIP_2 by kinase and dephosphorylation of PIP_3 by phosphatase modulates the PI3K/AKT/mTOR signaling pathway to orchestrate a diverse array of cellular activities important for growth and proliferation. When aberrantly activated, this pathway is a potent inducer of oncogenesis.

Hyperactivation of the PI3K/AKT/mTOR signaling pathway is one of the most frequent events observed in human cancers. The *PTEN* gene, a critical negative regulator of PI3K signaling, is frequently mutated, lost, or downregulated in human brain, breast, and prostate cancers [44]. The PI3K signaling pathway can also be activated by mutations in the heterodimeric catalytic and regulatory subunits of PI3K. Class I PI3K mutations, particularly alterations in *PIK3CA* that encodes the p110α catalytic subunit, have been implicated as key drivers of CRCs and breast cancer and have emerged as potential drug targets [40]. The oncogenic potential of *PIK3CA* mutation was initially described by studies in the late 1990s, which demonstrated transformation of chicken cells by an avian retroviral oncogene homologous to the p110α catalytic subunit and transformation of human mammary epithelial cells by mutant forms of p110α and p110β [45,46].

Later studies on *PIK3CA* mutations in human cancers identified two hotspots for mutation in the p110α helical domain (E542K, E545K) and one hotspot for mutation in the p110α kinase domain (H1047R) [47]. The E542K, E545K, and H1047R mutations are commonly referred to as canonical PI3K mutations. These canonical mutations are gain-of-function mutations in p110α that promote hyperactivation of the PI3K pathway and have transforming capacity [46]. Mutant p110α proteins harboring the H1047R mutation exhibit increased catalytic activity, resulting in enhanced downstream signaling and oncogenic transforming ability [46]. Similarly, *PIK3CA* helical domain mutations diminish negative regulation by p85, leading to enhanced enzymatic activity [48,49]. Mutations in other genes (*PIK3CB*, *PIK3CD*, and *PIK3CG*) encoding class I PI3K catalytic subunits are more rare [40]. Further, while roles for class I PI3Ks in cancer are well described, the roles of class II and III PI3Ks in human malignancies remain unclear.

6.4 *PIK3CA* MUTATIONS IN CRCs

Advances in tumor tissue harvesting methods and high-throughput sequencing technologies have created new opportunities for identifying somatic PI3K mutations in human colorectal and breast cancers [50]. High-throughput sequencing of all PI3K genes in a panel of 234 CRCs by Samuels et al. [51] revealed that *PIK3CA* was mutated at a frequency of 32% in all CRC cases. All but three of the *PIK3CA* alterations identified in this screen were heterozygous missense mutations and no truncating or nonsense alterations were found. Remarkably, over 80% of the missense mutations in *PIK3CA* localized to the aforementioned hotspots in exon 9 (encoding helicase domain) and 20 (encoding kinase domain), a finding consistent with the presence of hotspot mutations in other CRC oncogenic drivers, including *K-Ras* and *B-Raf* [52,53]. Interestingly, *PIK3CA* mutations were found to be twice as common in tumors with microsatellite instability, indicating an inverse relationship between intact cellular renewal mechanisms and oncogenic *PIK3CA* mutations [51]. This study further examined 76 premalignant colorectal adenomatous polyps and found only two mutations in advanced-stage polyps, suggesting that, perhaps, *PIK3CA* mutations arise late in CRC progression to promote invasion and metastasis.

Other studies have correlated *PIK3CA* mutational status in CRCs to distinct pathologies. For instance, Miyaki et al. [54] found that *PIK3CA* mutations occurred at similar frequencies in sporadic CRCs and inherited CRCs [i.e., FAP and hereditary nonpolyposis colorectal cancer (HNPCC or Lynch syndrome)]. Mutations in patients with inherited predisposition to CRC occurred predominantly in the p110α kinase domain, whereas mutations in sporadic tumors arose primarily in the p110α helical domain [54]. Another study demonstrated a gender bias for *PIK3CA* mutations, with higher frequencies observed in women with CRC [55]. Lastly, several groups showed that *PI3KC*A mutation status was correlated with increased frequency of disease recurrence, resistance to conventional chemotherapy, and poor prognosis, even in patients with curatively resected CRC [56,57]. *PI3KCA* mutations in CRC were further described to confer clinical resistance to anti-EGFR monoclonal antibody treatment [58]. This comes as no particular surprise since inhibition of an upstream signal would not be expected to suppress pathway signaling due to mutation of a downstream signaling protein.

Additional progress in the characterization of *PI3KCA* alterations and other tumor-specific mutations in CRC has emanated from The Cancer Genome Atlas (TCGA) project, a collaborative venture between the National Cancer Institute (NCI) and the National Human Genome Research Institute (NHGRI). The TCGA applies high-throughput genome analysis technologies to explore the full spectrum of genomic changes in human cancer. To characterize genetic alterations in CRCs, the Colorectal Cancer Project conducted genome-wide analysis of 276 tumor samples, of which 224 were sequenced tumors, and found that the 10 most frequently mutated genes in CRCs are *APC*, *TP53*, *K-Ras*, *SYNE1*, *PIK3CA*, *LRP1B*, *FAT4*, *FBXW7*, *DNAH5*, and *LRP2* (Table 6.1) [59,60]. According to TCGA data, over 75% of CRC tumors harbor a mutation in one or more members of the Wnt signaling pathway, predominantly in APC [59].

Additional findings from TCGA have revealed that *PIK3CA* somatic mutation occurs at a frequency of 20.1% (45 of 224 sequenced cases) in CRC tumors [59,60]. Of these *PIK3CA* mutations, the most common are missense mutations that cluster in the p110α helical and kinase domains, a finding consistent with earlier studies. On the other hand, *PIK3CA* amplification, which is observed in 16.5% (83 of 504 sequenced cases) of head and neck squamous cell carcinoma tumors, does not appear to be a frequent event in CRC [61]. Mutations in *PTEN* resulting in loss of PTEN function occur at a frequency of 5.8% (13 of 224 sequenced cases) and are primarily deletion mutations, with truncation and missense mutations occurring at a lower frequency [59,60]. Additional evidence indicates that epigenetic changes may lead to loss of PTEN in a substantial number of cancers. Collectively, these studies suggest that while activation of the Wnt signaling pathway is nearly ubiquitous in CRCs, genetic alterations in the PI3K signaling pathway have also emerged as frequent triggers for oncogenesis and represent promising targets for anticancer strategies.

6.5 PIK3CA MUTATIONS IN BREAST CANCER

Breast cancer is the leading cancer diagnosed in women and the second leading cause of cancer deaths in women (United States), with an estimated 12% lifetime risk (1 in 8 US women). In 2017, an estimated 252,710 new cases of invasive breast cancer are expected to be diagnosed with an average 5-year survival rate of 89% [62]. *PIK3CA* is among the most

TABLE 6.1 Ten Most frequently Mutated Genes in CRCs

Gene	Class	Protein	Pathway affected	Number of mutations	Mutation frequency (cases)
APC	Tumor suppressor	APC	Wnt signaling	247	75% (168)
TP53	Tumor suppressor	Cellular tumor antigen p53	DNA repair, cell cycle, apoptosis, senescence	123	54% (121)
KRAS	Oncogene	GTP3se K-Ras	Signal transduction	94	42% (94)
SYNE1	N/A	Nesprin-1 or nuclear envelope spectrin repeat protein 1	Cerebellar neuronal signaling	101	21% (47)
PIK3CA	Oncogene	p110α catalytic sub-unit of PI3K	Signal transduction	53	20.1% (45)
LRP1B	Tumor suppressor	Low-density lipoprotein receptor-related protein 1B	Negative regulator of cell migration	80	17.9% (40)
FAT4	Tumor suppressor	Protocadherin Fat 4	Cell polarity, Hippo signaling	73	17.4% (39)
FBXW7	Tumor suppressor	F-box/WD repeat-containing protein 7	Ubiquitination	44	16.5% (37)
DNAH5	N/A	Dynein heavy chain 5, axonemal	Microtubule motility	70	15.2% (34)
LRP2	N/A	Low-density lipoprotein receptor-related protein 2	Cell signaling, ligand endocytosis	50	15.2% (34)

Results obtained from TCGA Colorectal Cancer Project [59]. Two hundred and seventy six CRC tumor samples from 276 patients were analyzed and mutation frequency data obtained from 224 sequenced tumors. The *PIK3CA* gene was among the top five most commonly mutated genes in CRC, at a frequency of 20.1% (45 out of 224 cases). Abbreviations: APC, adenomatous polyposis coli; CRC, colorectal cancer.

commonly mutated genes in breast cancer (Table 6.2). Two TCGA studies of primary breast cancers found that somatic *PIK3CA* alterations occur at a frequency between 32.5% (319 out of 982 sequenced cases) in all breast invasive carcinoma and 38% (308 out of 817 sequenced cases) in all invasive lobular breast cancers [63,64]. Specifically, *PIK3CA* mutation occurs in about 28%–47% of estrogen receptor (ER)-positive and progesterone receptor (PR)-positive breast cancers, in approximately 25% of human epidermal growth factor receptor 2 (HER2)-overexpressing breast cancers, and in 8% of triple-negative breast cancers [65]. Similar to the pattern observed in CRCs, the majority of *PIK3CA* mutations in breast cancer are missense mutations and are located in exon 9 (E542K and E545K) and exon 20 (H1047R).

PIK3CA amplification and *PTEN* loss can also augment PI3K activity in invasive breast cancer. These two events occur at higher frequencies in breast cancer compared to CRCs (Table 6.3). The TCGA provisional study on breast invasive carcinoma has observed *PIK-3CA* amplification in 3.1% (30 of 982 sequenced cases) and *PTEN* loss in 5.1% (50 of 982

TABLE 6.2 Ten Most Frequently Mutated Genes in Breast Cancer

Gene	Class	Protein	Pathway affected	Number of mutations	Mutation frequency (cases)
PIK3CA	Oncogene	p110α catalytic subunit of PI3K	Signal transduction	355	32.5% (319)
TP53	Tumor suppressor	Cellular tumor antigen p53	DNA repair, cell cycle, apoptosis, senescence	304	30.7% (301)
CDH1	Tumor suppressor	Cadhem-1	Cell–cell interaction	114	11.4% (112)
GATA3	N/A	GATA Binding Protein 3	Mammary epithelial differentiation. Th2 differentiation	101	9.9% (97)
MAP3K1	Oncogene	Mitogen-activated protein kinase kinase kinase 1	Signal transduction	98	7.2% (71)
KMT2C	N/A	Lysine *N*-methyltransferase	Epigenetic transcriptional activation	83	7.1% (70)
MUC12	N/A	Mucin-12	Mucosal barrier, epithelial renewal	80	5.5% (54)
MUC4	N/A	Mucin-4	Mucosal barrier, epithelial renewal	62	5.4% (53)
FLG	N/A	Filaggrin	Epithelial differentiation	51	4.5% (45)
SYNE1	N/A	Nesprin-1 or nuclear envelope spectrin repeat protein 1	Cerebellar neuronal slanalinc	55	4.4% (43)

Results obtained from TCGA Breast Cancer Project, Provisional. One thousand hundred and five breast cancer tumor samples from 1098 patients were analyzed and mutation frequency data obtained from 982 sequenced tumors. The *PIK3CA* gene was the most frequently mutated gene in breast cancer, at a frequency of 32.5% (319 out of 982 sequenced cases).

sequenced cases) of tumors [63]. While loss-of-function mutations in the *PTEN* gene occur in fewer than 5% of breast cancer, reduced expression of PTEN protein due to promoter methylation or regulation at the RNA or protein level have been reported to occur in 30% of breast cancer cases [66]. *PIK3CA* amplification and/or *PTEN* loss were found to increase AKT phosphorylation and PI3K downstream signaling in 40% of breast invasive lobular carcinoma cases [64]. Inhibitors targeting p110α, mTOR, or AKT as well as dual PI3K/mTOR inhibitors have shown promise in early-phase clinical trials for hormone receptor-positive breast cancer, as well as other cancers, including CRCs and ovarian carcinoma [67,68]. However, as with other targeting agents, clinical response to PI3K pathway inhibitors in breast cancer may be short-lived due to the development of acquired resistance to

TABLE 6.3 *PIK3CA* and *PTEN* Alteration Frequencies in CRC and Breast Cancer

Alteration	Frequency (cases) in CRCs	Frequency (cases) in breast cancer
PIK3CA overall	20.1% (45)	35.6% (350)
PIK3CA mutation	20.1% (45)	32.5% (319)
PIK3CA amplification	N/A	3.1% (30)
PIK3CA deletion	N/A	0.1% (1)
PTEN overall	5.8% (13)	9.0% (88)
PTEN mutation	3.6% (8)	3.6% (35)
PTEN amplification	N/A	0.5% (5)
PTEN deletion	2.2% (5)	5.1% (50)

From TCGA, 276 CRC tumor samples were analyzed that generated 224 sequenced tumors. One thousand hundred and five breast cancer tumor samples were analyzed that generated 982 sequenced tumors. While CRC tumors were characterized by *PIK3CA* mutations, breast cancer tumors also exhibited *PIK3CA* amplification and deletion. *PTEN* deletion occurred at a frequency of 2.2% in CRC (5 of 224 sequenced cases) and 5.1% in breast cancer (50 of 982 sequenced cases). Abbreviation: CRC, colorectal cancer.

therapy. For instance, in the KPL-4PR breast cancer cell line, amplification of an activating mutant *PIK3CA* allele was found to confer resistance to a variety of PI3K inhibitors and PI3K/mTOR dual inhibitors [69]. Likewise, in a phase I clinical trial, loss of *PTEN* was determined to be a frequent mechanism of resistance to treatment with the PI3K inhibitor BYL719 in women with metastatic breast cancer [70]. These preclinical findings suggest that determination of the status of *PIK3CA* amplification and PTEN loss may provide important predictive value in identifying which patients will benefit from treatment with PI3K/AKT/mTOR pathway inhibitors.

Other mechanisms that lead to enhanced PI3K signaling in breast cancer include *HER2* amplification, at a frequency of 13% (107 out of 825 sequenced cases), and *AKT1* mutation, at a frequency of 2.4% (20 out of 825 sequenced cases) [63]. The *HER2* (*ERBB2*) gene encodes a cell-surface receptor tyrosine kinase, HER2, whose expression has been associated with worse prognosis in breast cancer. HER2 lacks a known ligand but heterodimerizes with other members of the HER family of receptors to induce signaling via the PI3K pathway. HER2 preferentially dimerizes with HER3, which contains six docking sites for the p85 subunit of PI3K, thereby allowing the HER2/HER3 heterodimer to trigger PI3K signaling [71]. In fact, HER2/PI3K activation has been shown to play an important role in conferring broad-spectrum chemotherapy resistance on human breast adenocarcinoma cells, suggesting that PI3K activation may prove to be a valuable predictive biomarker and molecular target for therapies aimed at improving outcomes in patients with chemotherapy-resistant breast cancer [72]. *AKT1* gene mutations in breast cancer (<8%) are exclusively restricted to tumors expressing both ER and PR [66].

Several studies on *PIK3CA* mutations in breast cancer have focused on the association of these alterations with clinical and pathological parameters. Saal *et al.* [73] reported that *PIK3CA* mutations correlate with expression of wild-type *PTEN*, positive hormone receptor expression, and node metastasis. Other studies have shown that *PIK3CA* mutations in breast cancer correlate with worse survival and that mutations are more commonly found in larger tumors with ER and PR expression [74]. Moreover, higher frequencies of *PIK3CA* mutations are seen in lobular compared to ductal breast cancers [75]. *PIK3CA* mutational status has also been correlated with resistance to trastuzumab, a monoclonal antibody against HER2 receptor in *HER2*-amplified breast tumors [76].

6.6 *PIK3CA* AND COX-2 EXPRESSION

The mammalian intestinal epithelium undergoes rapid turnover, with cells turning over every 3–5 days [77]. This process is characterized by proliferation of intestinal crypt stem cells, progression of the daughter cells up the crypt-villi axis concurrent with differentiation into mature enterocytes, followed by apoptosis of the differentiated cells [77]. Dysregulation of intestinal epithelial cell proliferation, differentiation, and apoptosis is thought to play an important role in the carcinogenesis of CRC [78]. Prostaglandins are key mediators of coordinated intestinal epithelial cell turnover and homeostasis and are generated from arachidonic acid [79]. COX is the rate-limiting enzyme that catalyzes the conversion of arachidonic acid into prostaglandin-H2 (PGH_2), the precursor to mature prostaglandins. There are two COX isoforms, COX-1 and COX-2, which differ primarily in their pattern of expression and cellular activity. COX-1 is ubiquitously expressed and produces a variety of prostaglandins to coordinate normal metabolic functions. By contrast, COX-2 is primarily an inducible enzyme. Exposure to physiological stresses leads to COX-2 expression and production of prostaglandin-E2 (PGE_2) [80] (Fig. 6.2).

Mounting evidence suggests that COX-2 plays a critical role in CRC progression by enhancing cell survival and metastatic potential [81,82]. In fact, elevated COX-2 expression has been observed in approximately 85% of colorectal adenocarcinomas and 50% of premalignant adenomas and is associated with worse survival among CRC patients [83–85]. Preclinical studies indicate that enhanced COX-2 signaling and PGE_2 synthesis may facilitate CRC progression by augmenting the proliferation and survival of intestinal epithelial cells, increasing the production of pro-angiogenic factors, increasing expression of metalloproteases to promote tumor invasion and metastasis, and suppression of antitumor immunity [86,87]. On the other hand, inactivation of the *COX-2* gene in mice is associated with decreased colorectal tumorigenesis [88]. A similar study in Caco-2 colon carcinoma cells revealed that CRC cell growth may depend on COX-2 expression levels and activity and that COX-2 blockade results in induction of apoptosis [89].

PIK3CA mutations result in the constitutive activation of p110α kinase catalytic activity and appear to stimulate COX-2 expression through downstream AKT signaling. Other alterations in the PI3K signaling pathway, such as *PIK3CA* amplification or *PTEN* loss, may also lead to overexpression of COX-2 and high levels of PGE_2 production [89]. High levels of PGE_2 have been implicated in inflammatory bowel disease (IBD) and colorectal tumorigenesis, where PGE_2 acts to stimulate immunosuppressive myeloid-derived suppressor cells and promote an inflammatory environment [90–92]. Several studies have suggested a link between PI3K signaling and COX-2 expression and activity. For example, PI3K signaling can mediate the proliferative effects of insulin-like growth factor-1 receptor in CRC cells by increasing COX-2 expression and PGE_2 synthesis [89]. Another study on the chemopreventive activity of curcumin in CRC has shown that PI3K/AKT can activate nuclear factor kappa B (NF-κB),

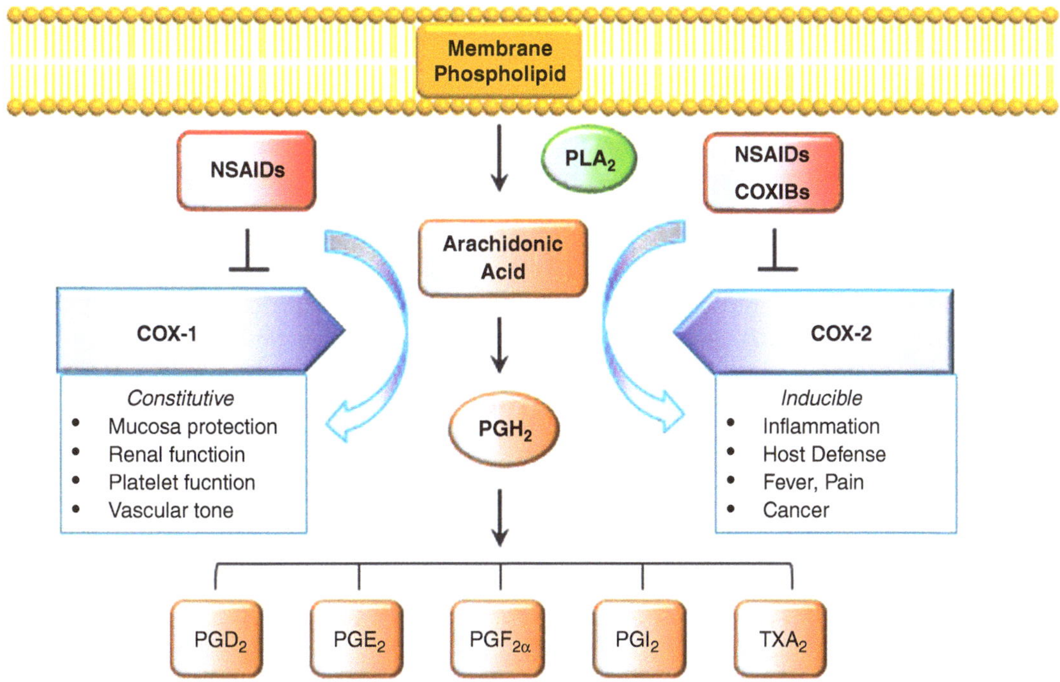

FIG. 6.2 **COX pathway.** Prostaglandins are signaling molecules produced from oxidation of membrane fatty acids. Membrane phospholipid is first converted into arachidonic acid by the enzyme, phospholipase A2 (PLA2). Next, COX catalyzes the rate-limiting step in the conversion of arachidonic acid to the intermediate, prostaglandin H2 (PGH_2) that can then be converted into prostaglandins D_2, E_2, and $F_{2\alpha}$ (PGD_2, PGE_2, $PGF_{2\alpha}$), prostacyclin I_2 (PGI_2), and thromboxane A_2 (TXA_2). There are two known COX isoforms (COX-1 and COX-2). COX-1 is ubiquitously expressed in most cells and plays a role in gastrointestinal mucosa homeostasis, renal function, platelet aggregation, and vascular tone. COX-2 is not normally expressed in most tissues but is induced in response to physiologic stress. COX-2 expression is upregulated by cytokines and is a major source of prostaglandins produced during inflammation and cancer. NSAIDs, including aspirin, inhibit both COX-1 and COX-2. Selective COX-2 inhibitors (COXIBs) preferentially inhibit COX-2 with minimal COX-1 inhibition. The overproduction of PGE_2 is associated with neoplastic effects on tumor growth, apoptosis inhibition, tumor angiogenesis, and immunosuppression. Abbreviation: COX, cyclooxygenase.

a transcription factor involved in the upregulation of pro-inflammatory genes, including inducible *COX-2*, which generates prostaglandins [93].

NF-κB plays a pivotal role in the regulation of cellular proliferation and immune responses, as well as the production of inflammatory mediators that contribute to the pathogenesis of inflammation and cancer [94]. In addition, NF-κB can increase the expression of anti-apoptotic cellular proteins, including members of the Bcl-2 family, which can confer chemotherapy resistance in cancer cells [95]. Several lines of evidence have suggested that aspirin and other NSAIDs, such as sulindac, block NF-κB activity, through inhibition of I-κB phosphorylation, resulting in reduced levels of COX-2 and PGE_2 [96,97]. In fact, sulfasalazine, an anti-inflammatory agent that includes a nonsteroidal anti-inflammatory moiety, 5-aminosalicyclic acid, is a potent inhibitor of NF-κB and has been one of the most commonly prescribed medications for patients with IBD [98].

6.7 ASPIRIN USE IN *PIK3CA*-MUTATED CRCs

A significant effort has been made to understand the role of COX in CRC progression and the clinical benefit of COX inhibition by NSAIDs for CRC prevention and treatment. Although NSAIDs are some of the most commonly used drugs for their analgesic, anti-inflammatory, and potential anticancer properties, the prolonged used of nonselective NSAIDs is associated with adverse events including dyspepsia, gastrointestinal bleeding, and gastroduodenal ulceration. These unwanted side effects are largely attributed to inhibition of COX-1, which plays a key role in the production of prostaglandins protective of gastrointestinal mucosal integrity and induction of platelet aggregation [99]. The development of COX-2-selective inhibitors (COXIBs) such as celecoxib and rofecoxib have made it possible to minimize adverse gastrointestinal complications associated with nonselective NSAIDs while maintaining their therapeutic properties [100].

COX-2, which mediates prostaglandin synthesis during inflammation and is overexpressed in CRC tumors and premalignant polyps, is thought to play an important role in CRC carcinogenesis [83]. COX-2 upregulation in CRCs has been associated with increased tumor growth, angiogenesis, and metastasis [3,87,101]. Emerging animal, epidemiologic, and clinical data have indicated that NSAIDs and COXIBs are promising chemopreventive agents for CRC and other malignancies. Early animal studies from the 1980s demonstrated beneficial effects of NSAIDs on inflammation and tumor regression in carcinogen-induced rat CRC models [5,6,102]. Later, prospective epidemiology studies in humans found an association between regular NSAID use, in particular, aspirin, and decreased CRC risk and cancer-specific mortality. This benefit of aspirin was found to be exclusive in patients with CRC tumors overexpressing COX-2, whereas regular aspirin use did not appear to influence tumors with weak or absent COX-2 [103]. In another study, regular low-dose aspirin use was shown to reduce the risk of fatal CRC [14]. A similar investigation discovered a link between aspirin use and reduction in fatal cancers from other cancers of the gastrointestinal tract including esophageal and gastric adenocarcinoma [104].

Clinical trials evaluating the role of NSAIDs in humans have revealed a notable clinical benefit in the primary and secondary prevention of CRC. Of particular interest, NSAIDs have been shown to diminish the overproduction of colonic polyps in genetically susceptible individuals. For example, treatment with the NSAID sulindac was found to be associated with a significant reduction in the number and size of polyps and overall CRC risk in patients with FAP and HNPCC [7,105]. Mechanisms proposed for the chemopreventive effect of sulindac include induction of the intrinsic apoptosis pathway and inhibition of Wnt/β-catenin pathway [105,106]. Similarly, in nonhereditary CRC, aspirin was found to reduce the recurrence of adenomatous polyps after surgical resection in patients with previous CRC (relative risk: 0.65; 95% CI: 0.46–0.91) [15]. In addition, findings from two large cohorts (Nurses' Health Study and Health Professionals Follow-up Study) showed that following surgical removal of a primary tumor, regular aspirin use significantly reduced the risk of CRC recurrence [103]. Final follow-up of 135,965 health-care professionals (88,084 women and 41,881 men) from these two studies demonstrated that long-term (6 years) regular aspirin use (0.5–1.5 standard aspirin tablets per week) was associated with a modest but significantly reduced risk for overall cancer, particularly CRCs [107]. Furthermore, in a collective analysis of 35,535 patients from six randomized clinical trials, low-dose daily aspirin use resulted in marked reduction in the risk of developing metastatic CRCs [108].

Pivotal studies by Liao et al. [109] and Domingo et al. [110] identified canonical *PIK3CA* mutations as a predictive marker of aspirin therapeutic benefit in CRC. Indeed, regular aspirin use after CRC diagnosis in patients with mutated *PIK3CA* yielded a remarkable improvement in CRC-specific mortality and overall survival (hazards ratio 0.18; 95% CI: 0.06–0.61) [109]. On the other hand, CRC tumors with wild-type *PIK3CA* did not show a clinical benefit with aspirin. Interestingly, the study by Domingo et al. [110] failed to observe a greater benefit of the COX2-selective inhibitor rofecoxib vs placebo in patients with *PIK3CA*-mutated CRC. Nonetheless, the findings from these randomized clinical trials, along with epidemiologic evidence, provide a strong rationale for prospective evaluation of NSAIDs therapy in other malignancies characterized by *PIK3CA* alterations.

6.8 ASPIRIN USE IN *PIK3CA*-MUTATED BREAST CANCER

The production of PGE_2 in breast cancer has been shown to upregulate Wnt signaling and promotes cellular proliferation and survival, angiogenesis, and invasion/metastasis [113]. Elevated levels of COX-2 and PGE_2, which mediate inflammation in the tumor microenvironment, are indicators of poor prognosis in breast cancer [114]. Importantly, COX-2 is upregulated in up to 72% (46 of 64 cases) of breast cancers and associated with larger tumor size, nodal metastasis, and advanced clinical stage [115]. Mounting evidence has described a link between inflammation and breast cancer and stimulated strong interest in targeting COX enzymes, in particular, COX-2, for chemoprevention and breast cancer therapy. Clinical trials and findings from the Women's Health Initiative have reported that women who were diagnosed with breast cancer and were regular aspirin users experienced a decreased risk of distant tumor metastasis and death [116,117]. Furthermore, another study reported aspirin's clinical benefit in decreasing breast cancer-specific mortality and all-cause mortality among women with breast cancer [118].

Since dysregulated COX-2/PGE_2 activity likely contributes to multiple facets of tumorigenesis, targeting this pathway with COX inhibitors offers considerable promise for the prevention and treatment of breast cancer. Preclinical *in vitro* studies have evaluated the cytotoxic mechanism of aspirin and other NSAIDs in breast cancer cells [119,120]. For instance, aspirin was shown to trigger apoptosis in MCF-7 breast cancer cells through increased nuclear translocation and phosphorylation of anti-apoptotic factor, B-cell lymphoma 2 (Bcl-2), leading to its inactivation [119]. Similarly, MDA-MB-453 breast cancer cells treated with aspirin demonstrated increased release of cytochrome c from mitochondria and subsequent activation of caspase-3 [119,120]. Beyond activation of apoptosis, aspirin has also been shown to downregulate NF-κB and inhibit mTOR signaling, leading to cell-cycle arrest and induction of autophagy [121,122]. Likewise, other NSAIDs, such as the selective COX-2 inhibitor celecoxib, have been shown to suppress tumor angiogenesis and growth through inhibition of vascular endothelial growth factor (VEGF) [123], enhance the efficacy of aromatase inhibitors to inhibit mammary carcinoma formation [124], and increase sensitivity to chemotherapy in MCF7 breast cancer cells [125].

Recent preclinical studies have linked *PIK3CA* mutations in breast cancer to enhanced growth suppression by aspirin. Turturro et al. [126] reported that a nontumorigenic breast epithelial cell line harboring canonical *PIK3CA* mutations in either exon 9 or exon 20 demonstrated increased sensitivity to physiologic doses of aspirin. Another study by Henry et al. [127] showed that breast cancer cells expressing mutant *PIK3CA* that were treated daily with aspirin exhibited greater growth inhibition and lower viability than *PIK3CA*-wild-type cancer cells. This effect was attributed to the ability of aspirin to activate the enzyme AMPK, a known inhibitor of mTORC1 signaling. In addition, findings from this study provided a mechanistic rational for the use of aspirin as an adjuvant therapy to augment the efficacy of PI3K inhibitors against breast cancer. Taken together, these data provide evidence that alterations in the PI3K signaling pathway, in particular, *PIK3CA* mutations, sensitize breast cancer cells to aspirin. Prospective clinical trials will be needed to fully evaluate the impact of *PIK3CA* mutations in response to NSAIDs in breast cancer patients.

6.9 CONCLUSIONS

Activating *PIK3CA* mutations is a frequent event in the development and progression of numerous human cancers, including CRCs and breast cancer. Recent epidemiological, preclinical, and clinical studies have propelled NSAIDs, including aspirin, to the forefront of cancer research. The compelling finding that the presence of canonical *PIK3CA* mutations in CRC predicts response to aspirin has given impetus to ongoing clinical trials to validate this observation in other human cancers. Future investigations should also incorporate patients with noncanonical *PIK3CA* mutations, as well as PTEN loss, to potentially broaden the pool of patients who may benefit from NSAID therapy. Moreover, experiments are needed to elucidate the mechanism of COX-2 regulation by PI3K signaling, and the influence of NSAID-mediated COX-2 inhibition on the tumor microenvironment and antitumor immune response. Overall, NSAIDs, including aspirin, have far-reaching clinical benefits beyond their analgesic, anti-inflammatory, and cardiovascular benefits. As they are readily available, low-cost, and relatively safe, NSAIDs and aspirin hold promise for making a major impact on the global burden of human cancers.

Acknowledgments

This work was supported by National Institutes of Health grants R01 DE24728 (DEJ), P50CA097190 (DEJ and JRG), and R01 DE023685 (JRG).

Conflict of Interest: No potential conflicts of interest were disclosed.

References

[1] Jänne P, Mayer R. Chemoprevention of Colorectal Cancer. New Engl J Med 2000;1960–8. https://doi.org/10.1056/NEJM200006293422606.
[2] Brody H. Colorectal cancer. Nature 2015;521(7551):S1. https://doi.org/10.1038/521S1a.
[3] Wang D, DuBois RN. The role of COX-2 in intestinal inflammation and colorectal cancer. Oncogene 2010;29(6):781–8. https://doi.org/10.1038/onc.2009.421.
[4] Pollard M, Luckert PH. Indomethacin treatment of rats with dimethylhydrazine-induced intestinal tumors . Cancer Treat Rep 1980;64(12):1323–7.

[5] Narisawa T, Sato M, Tani M, Kudo T, Takahashi T, Goto A. Inhibition of development of methylnitrosourea-induced rat colon tumors by indomethacin treatment. Cancer Res 1981;41:1954–7.

[6] Kudo T, Narisawa T, Abo S. Antitumor activity of indomethacin on methylazoxymethanol-induced large bowel tumors in rats. Gan 1980;71:260–4.

[7] Giardiello FM, Hamilton SR, Krush AJ, Piantadosi S, Hylind LM, Celano P, et al. Treatment of colonic and rectal adenomas with sulindac in familial adenomatous polyposis. N Engl J Med 1993;328:1313–6. https://doi.org/10.1056/NEJM199305063281805.

[8] Rigau J, Piqué JM, Rubio E, Planas R, Tarrech JM, Bordas JM. Effects of long-term sulindac therapy on colonic polyposis. Ann Intern Med 1991;115(12):952–4. https://doi.org/10.7326/0003-4819-115-12-952.

[9] Steinbach G, Lynch P, Phillips R, Wallace M, Hawk E, Gordon G, et al. The effect of celecoxib, a cyclooxygenase-2 inhibitor, in familial adenomatous polyposis. New Engl J Med 2000;342(26):1946–52. https://doi.org/10.1056/NEJM200006293422603.

[10] Rosenberg L, Palmer JR, Zauber AG, Warshauer ME, Stolley PD, Shapiro S. A hypothesis: nonsteroidal anti-inflammatory drugs reduce the incidence of large-bowel cancer. J Natl Cancer Inst 1991;83:355–8. https://doi.org/10.1093/jnci/83.5.355.

[11] Giovannucci E, Rimm EB, Stampfer MJ, Colditz GA, Ascherio A, Willett WC. Aspirin use and the risk for colorectal cancer and adenoma in male health professionals. Ann Intern Med 1994;121(4):241–6. https://doi.org/10.7326/0003-4819-121-4-199408150-00001.

[12] Giovannucci E, Egan K, Hunter D, Stampfer M, Colditz G, Willett W, et al. Aspirin and the risk of colorectal cancer in women. New Engl J Med 1995;333(10):609–14. https://doi.org/10.1056/NEJM199509073331001.

[13] Kune GA, Kune S, Watson LF. Colorectal cancer risk, chronic illnesses, operations, and medications: case control results from the Melbourne Colorectal Cancer Study. Cancer Res 1988;48(15):4399–404.

[14] Thun MJ, Namboodiri MM, Heath CW. Aspirin use and reduced risk of fatal colon cancer. N Engl J Med 1991;325:1593–6. https://doi.org/10.1056/NEJM199112053252301.

[15] Sandler RS, Halabi S, Baron JA, Budinger S, Paskett E, Keresztes R, et al. A randomized trial of aspirin to prevent colorectal adenomas in patients with previous colorectal cancer. N Engl J Med 2003;348:883–90. https://doi.org/10.1056/NEJMoa021633.

[16] Baron J, Cole B, Sandler R, Haile R, Ahnen D, Bresalier R, et al. A randomized trial of aspirin to prevent colorectal adenomas. New Engl J Med 2003;348(10):891–9. https://doi.org/10.1056/NEJMoa021735.

[17] Grau M, Sandler R, McKeown-Eyssen G, Bresalier R, Haile R, Barry E, et al. Nonsteroidal anti-inflammatory drug use after 3 years of aspirin use and colorectal adenoma risk: observational follow-up of a randomized study. J Natl Cancer Inst 2009;101(4):267–76. https://doi.org/10.1093/jnci/djn484.

[18] Drew D, Chin S, Gilpin K, Parziale M, Pond E, Schuck M, et al. ASPirin intervention for the REDuction of colorectal cancer risk (ASPIRED): a study protocol for a randomized controlled trial. Trials 2017;18(1):50. https://doi.org/10.1186/s13063-016-1744-z.

[19] Coyle C, Cafferty FH, Rowley S, MacKenzie M, Berkman L, Gupta S, et al. ADD-ASPIRIN: a phase III, double-blind, placebo controlled, randomised trial assessing the effects of aspirin on disease recurrence and survival after primary therapy in common non-metastatic solid tumours. Contemp Clin Trials 2016;51:56–64. https://doi. org/10.1016/j.cct.2016.10.004.

[20] Zell JA. Clinical trials update: tertiary prevention of colorectal cancer. J Carcinog 2011;10:8. https://doi.org/10.4103/1477-3163.78271.

[21] Chubak J, Kamineni A, Buist DS, Anderson ML, Whitlock EP. Aspirin use for the prevention of colorectal cancer: an updated systematic evidence review for the US Preventive Services Task Force. Agency for Healthcare Research and Quality (US) 2015;.

[22] Dehmer SP, Maciosek MV, Flottemesch TJ. Aspirin Use to Prevent Cardiovascular Disease and Colorectal Cancer: A Decision Analysis: Technical Report. Agency for Healthcare Research and Quality (US) 2015;.

[23] Nosho K, Yamamoto H, Takahashi T, Mikami M, Taniguchi H, Miyamoto N, et al. Genetic and epigenetic profiling in early colorectal tumors and prediction of invasive potential in pT1 (early invasive) colorectal cancers. Carcinogenesis 2007;28:1364–70. https://doi.org/10.1093/carcin/bgl246.

[24] Nosho K, Kawasaki T, Ohnishi M, Suemoto Y, Kirkner GJ, Zepf D, et al. PIK3CA mutation in colorectal cancer: relationship with genetic and epigenetic alterations. Neoplasia 2008;10:534–41.

[25] Vane JR, Botting RM. The mechanism of action of aspirin. Thromb Res 2003;110:255–8.

[26] Siegel RL, Miller KD, Fedewa SA, Ahnen DJ, Meester RGSG, Barzi A, et al. Colorectal cancer statistics, 2017. CA Cancer J Clin 2017;67:177–93. https://doi.org/10.3322/caac.21395.

[27] Longley DB, Johnston PG. Molecular mechanisms of drug resistance. J Pathol 2005;205:275–92. https://doi.org/10.1002/path.1706.
[28] Abdul Khalek FJ, Gallicano GI, Mishra L. Colon cancer stem cells. Gastrointest Cancer Res 2010;(Suppl. 1):S16–23.
[29] Fodde R. The APC gene in colorectal cancer. Eur J Cancer 2002;38:867–71.
[30] Knudson AG. Two genetic hits (more or less) to cancer. Nat Rev Cancer 2001;1:157–62. https://doi.org/10.1038/35101031.
[31] Jean GW, Shah SR. Epidermal growth factor receptor monoclonal antibodies for the treatment of metastatic colorectal cancer. Pharmacotherapy 2008;28:742–54. https://doi.org/10.1592/phco.28.6.742.
[32] Smith D, Bosacki C, Merrouche Y. Use of anti-EGFR antibodies (cetuximab and panitumumab) in the treatment of metastatic colorectal cancer in KRAS wild type patients. Bull Cancer 2009;96:S31–40. https://doi.org/10.1684/bdc.2009.0995.
[33] Amado RG, Wolf M, Peeters M, Van Cutsem E, Siena S, Freeman DJ, et al. Wild-type KRAS is required for panitumumab efficacy in patients with metastatic colorectal cancer. J Clin Oncol 2008;26:1626–34. https://doi.org/10.1200/JCO.14. 7116.
[34] Engelman JA, Luo J, Cantley LC. The evolution of phosphatidylinositol 3-kinases as regulators of growth and metabolism. Nat Rev Genet 2006;7:606–19. https://doi.org/10.1038/nrg1879.
[35] Hanahan D, Weinberg RA. Hallmarks of cancer: the next generation. Cell 2011;144:646–74. https://doi.org/10.1016/j.cell.2011.02.013.
[36] Zhao JJ, Roberts TM. PI3 kinases in cancer: from oncogene artifact to leading cancer target. Sci STKE 2006;2006:pe52. https://doi.org/10.1126/stke.3652006pe52.
[37] Beagle B, Fruman DA. A lipid kinase cousin cooperates to promote cancer. Cancer Cell 2011;19:693–5. https://doi. org/10.1016/j.ccr.2011.05.020.
[38] Leevers SJ, Vanhaesebroeck B, Waterfield MD. Signalling through phosphoinositide 3-kinases: the lipids take centre stage. Curr Opin Cell Biol 1999;11:219–25.
[39] Carpenter CL, Duckworth BC, Auger KR, Cohen B, Schaffhausen BS, Cantley LC. Purification and characterization of phosphoinositide 3-kinase from rat liver. J Biol Chem 1990;265:19704–11.
[40] Thorpe LM, Yuzugullu H, Zhao JJ. PI3K in cancer: divergent roles of isoforms, modes of activation and therapeutic targeting. Nat Rev Cancer 2015;15:7–24. https://doi.org/10.1038/nrc3860.
[41] Jaber N, Dou Z, Chen J-SS, Catanzaro J, Jiang Y-PP, Ballou LM, et al. Class III PI3K Vps34 plays an essential role in autophagy and in heart and liver function. Proc Natl Acad Sci USA 2012;109:2003–8. https://doi.org/10.1073/pnas.1112848109.
[42] Vivanco I, Sawyers CL. The phosphatidylinositol 3-Kinase AKT pathway in human cancer. Nat Rev Cancer 2002;2:489–501. https://doi.org/10.1038/nrc839.
[43] Dowling RJ, Topisirovic I, Fonseca BD, Sonenberg N. Dissecting the role of mTOR: lessons from mTOR inhibitors. Biochim Biophys Acta 2010;1804:433–9. https://doi.org/10.1016/j.bbapap.2009.12.001.
[44] Li J, Yen C, Liaw D, Podsypanina K, Bose S, Wang SI, et al. PTEN, a putative protein tyrosine phosphatase gene mutated in human brain, breast, and prostate cancer. Science 1997;275:1943–7.
[45] Chang HW, Aoki M, Fruman D, Auger KR, Bellacosa A, Tsichlis PN, et al. Transformation of chicken cells by the gene encoding the catalytic subunit of PI 3-kinase. Science 1997;276:1848–50.
[46] Zhao JJ, Liu Z, Wang L, Shin E, Loda MF, Roberts TM. The oncogenic properties of mutant p110alpha and p110beta phosphatidylinositol 3-kinases in human mammary epithelial cells. Proc Natl Acad Sci USA 2005;102:18443–8. https://doi.org/10.1073/pnas.0508988102.
[47] Kang S, Bader AG, Vogt PK. Phosphatidylinositol 3-kinase mutations identified in human cancer are oncogenic. Proc Natl Acad Sci USA 2005;102:802–7. https://doi.org/10.1073/pnas.0408864102.
[48] Zhao L, Vogt PK. Hot-spot mutations in p110alpha of phosphatidylinositol 3-kinase (pI3K): differential interactions with the regulatory subunit p85 and with RAS. Cell Cycle 2010;9:596–600. https://doi.org/10.4161/cc.9.3.10599.
[49] Mangone FR, Bobrovnitchaia IG, Salaorni S, Manuli E, Nagai MA. PIK3CA exon 20 mutations are associated with poor prognosis in breast cancer patients. Clinics (Sao Paulo) 2012;67:1285–90.
[50] Wood LD, Parsons DW, Jones S, Lin J, Sjöblom T, Leary RJ, et al. The genomic landscapes of human breast and colorectal cancers. Science 2007;318:1108–13. https://doi.org/10.1126/science.1145720.
[51] Samuels Y, Wang Z, Bardelli A, Silliman N, Ptak J, Szabo S, et al. High frequency of mutations of the PIK3CA gene in human cancers. Science 2004;304:554. https://doi.org/10.1126/science.1096502.

[52] Bos JL, Fearon ER, Hamilton SR, Verlaan-de Vries M, van Boom JH, van der Eb AJ, et al. Prevalence of ras gene mutations in human colorectal cancers. Nature 1987;327:293–7. https://doi.org/10.1038/327293a0.
[53] Davies H, Bignell GR, Cox C, Stephens P, Edkins S, Clegg S, et al. Mutations of the BRAF gene in human cancer. Nature 2002;417:949–54. https://doi.org/10.1038/nature00766.
[54] Miyaki M, Iijima T, Yamaguchi T, Takahashi K, Matsumoto H, Yasutome M, et al. Mutations of the PIK3CA gene in hereditary colorectal cancers. Int J Cancer 2007;121:1627–30. https://doi.org/10.1002/ijc.22829.
[55] Benvenuti S, Frattini M, Arena S, Zanon C, Cappelletti V, Coradini D, et al. PIK3CA cancer mutations display gender and tissue specificity patterns. Hum Mutat 2008;29:284–8. https://doi.org/10.1002/humu.20648.
[56] Kato S, Iida S, Higuchi T, Ishikawa T, Takagi Y, Yasuno M, et al. PIK3CA mutation is predictive of poor survival in patients with colorectal cancer. Int J Cancer 2007;121:1771–8. https://doi.org/10.1002/ijc.22890.
[57] Ogino S, Nosho K, Kirkner GJ, Shima K, Irahara N, Kure S, et al. PIK3CA mutation is associated with poor prognosis among patients with curatively resected colon cancer. J Clin Oncol 2009;27:1477–84. https://doi.org/10.1200/JCO.2008.18.6544.
[58] Sartore-Bianchi A, Martini M, Molinari F, Veronese S, Nichelatti M, Artale S, et al. PIK3CA mutations in colorectal cancer are associated with clinical resistance to EGFR-targeted monoclonal antibodies. Cancer Res 2009;69:1851–7. https://doi.org/10.1158/0008-5472.CAN-08-2466.
[59] Cancer Genome Atlas Network. Comprehensive molecular characterization of human colon and rectal cancer. Nature 2012;487:330–7. https://doi.org/10.1038/nature11252.
[60] The Cancer Genome Atlas Network. Comprehensive molecular characterization of human colon and rectal cancer. Nature 2012;487(7407):330-337. doi:10.1038/nature11252.
[61] Lawrence M, Sougnez C, Lichtenstein L, Cibulskis K, Lander E, Gabriel S, et al. Comprehensive genomic characterization of head and neck squamous cell carcinomas. Nature 2015;517(7536):576–82. https://doi.org/10.1038/nature14129.
[62] Siegel R, Miller K, Jemal A. Cancer statistics. Cancer J Clin 2017;7–30. https://doi.org/10.3322/caac.21387.
[63] Comprehensive molecular portraits of human breast tumours. Nature 2012;490:61–70. https://doi.org/10.1038/nature11412.
[64] Ciriello G, Gatza ML, Beck AH, Wilkerson MD, Rhie SK, Pastore A, et al. Comprehensive molecular portraits of invasive lobular breast cancer. Cell 2015;163:506–19. https://doi.org/10.1016/j.cell.2015.09.033.
[65] Lee JJ, Loh K, Yap Y-SS. PI3K/Akt/mTOR inhibitors in breast cancer. Cancer Biol Med 2015;12:342–54. https://doi.org/10.7497/j.issn.2095-3941.2015.0089.
[66] Stemke-Hale K, Gonzalez-Angulo AM, Lluch A, Neve RM, Kuo W-LL, Davies M, et al. An integrative genomic and proteomic analysis of PIK3CA, PTEN, and AKT mutations in breast cancer. Cancer Res 2008;68:6084–91. https://doi.org/10.1158/0008-5472.CAN-07-6854.
[67] Ihle NT, Powis G. Take your PIK: phosphatidylinositol 3-kinase inhibitors race through the clinic and toward cancer therapy. Mol Cancer Ther 2009;8:1–9. https://doi.org/10.1158/1535-7163.MCT-08-0801.
[68] Janku F, Wheler JJ, Naing A, Falchook GS, Hong DS, Stepanek VM, et al. PIK3CA mutation H1047R is associated with response to PI3K/AKT/mTOR signaling pathway inhibitors in early-phase clinical trials. Cancer Res 2013;73:276–84. https://doi.org/10.1158/0008-5472.CAN-12-1726.
[69] Huw L-YY, O'Brien C, Pandita A, Mohan S, Spoerke JM, Lu S, et al. Acquired PIK3CA amplification causes resistance to selective phosphoinositide 3-kinase inhibitors in breast cancer. Oncogenesis 2013;2:e83. https://doi.org/10.1038/oncsis.2013.46.
[70] Castel P, Juric D, Won H, Ainscough B, Ellis H, Ebbesen S, et al. Abstract LB-327: loss of PTEN leads to clinical resistance to the PI3Kα inhibitor BYL719 and provides evidence of convergent evolution under selective therapeutic pressure. Cancer Res 2014;LB-327. https://doi.org/10.1158/1538-7445.AM2014-LB-327.
[71] Hynes NE, Lane HA. ERBB receptors and cancer: the complexity of targeted inhibitors. Nat Rev Cancer 2005;5:341–54. https://doi.org/10.1038/nrc1609.
[72] Knuefermann C, Lu Y, Liu B, Jin W, Liang K, Wu L, et al. HER2/PI-3K/Akt activation leads to a multidrug resistance in human breast adenocarcinoma cells. Oncogene 2003;22:3205–12. https://doi.org/10.1038/sj.onc. 1206394.
[73] Saal LH, Holm K, Maurer M, Memeo L, Su T, Wang X, et al. PIK3CA mutations correlate with hormone receptors, node metastasis, and ERBB2, and are mutually exclusive with PTEN loss in human breast carcinoma. Cancer Res 2005;65:2554–9. https://doi.org/10.1158/0008-5472-CAN-04-3913.
[74] Li SY, Rong M, Grieu F, Iacopetta B. PIK3CA mutations in breast cancer are associated with poor outcome. Breast Cancer Res Treat 2006;96:91–5. https://doi.org/10.1007/s10549-005-9048-0.

[75] Buttitta F, Felicioni L, Barassi F, Martella C, Paolizzi D, Fresu G, et al. PIK3CA mutation and histological type in breast carcinoma: high frequency of mutations in lobular carcinoma. J Pathol 2006;208:350–5. https://doi.org/10.1002/path.1908.
[76] Berns K, Horlings HM, Hennessy BT, Madiredjo M, Hijmans EM, Beelen K, et al. A functional genetic approach identifies the PI3K pathway as a major determinant of trastuzumab resistance in breast cancer. Cancer Cell 2007;12:395–402. https://doi.org/10.1016/j.ccr.2007.08.030.
[77] Barker N. Adult intestinal stem cells: critical drivers of epithelial homeostasis and regeneration. Nat Rev Mol Cell Biol 2014;15:19–33. https://doi.org/10 1038/nrm3721.
[78] Kinzler KW, Vogelstein B. Lessons from hereditary colorectal cancer. Cell 1996;87:159–70.
[79] Eberhart CE, Dubois RN. Eicosanoids and the gastrointestinal tract. Gastroenterology 1995;109:285–301.
[80] Williams CS, Mann M, DuBois RN. The role of cyclooxygenases in inflammation, cancer, and development. Oncogene 1999;18:7908–16. https://doi.org/10.1038/sj.onc.1203286.
[81] Williams CS, Smalley W, DuBois RN. Aspirin use and potential mechanisms for colorectal cancer prevention. J Clin Invest 1997;100:1325–9. https://doi.org/10.1172/JCI119651.
[82] Prescott SM, Fitzpatrick FA. Cyclooxygenase-2 and carcinogenesis. Biochim Biophys Acta 2000;1470: M69–78.
[83] Eberhart CE, Coffey RJ, Radhika A, Giardiello FM, Ferrenbach S, DuBois RN. Up-regulation of cyclooxygenase 2 gene expression in human colorectal adenomas and adenocarcinomas. Gastroenterology 1994;107:1183–8. https://doi.org/10.1016/0016-5085(94)90246-1.
[84] Gupta RA, Dubois RN. Colorectal cancer prevention and treatment by inhibition of cyclooxygenase-2. Nat Rev Cancer 2001;1:11–21. https://doi.org/10.1038/35094017.
[85] Ogino S, Kirkner GJ, Nosho K, Irahara N, Kure S, Shima K, et al. Cyclooxygenase-2 expression is an independent predictor of poor prognosis in colon cancer. Clin Cancer Res 2008;14:8221–7. https://doi.org/10.1158/1078-0432.CCR-08-1841.
[86] Tsujii M, DuBois RN. Alterations in cellular adhesion and apoptosis in epithelial cells overexpressing prostaglandin endoperoxide synthase 2. Cell 1995;83:493–501.
[87] Tsujii M, Kawano S, DuBois RN. Cyclooxygenase-2 expression in human colon cancer cells increases metastatic potential. Proc Natl Acad Sci USA 1997;94:3336–40. https://doi.org/10.1073/pnas.94.7.3336.
[88] Oshima M, Dinchuk JE, Kargman SL, Oshima H, Hancock B, Kwong E, et al. Suppression of intestinal polyposis in Apc delta716 knockout mice by inhibition of cyclooxygenase 2 (COX-2). Cell 1996;87:803–9.
[89] Di Popolo A, Memoli A, Apicella A, Tuccillo C, di Palma A, Ricchi P, et al. IGF-II/IGF-I receptor pathway up-regulates COX-2 mRNA expression and PGE2 synthesis in Caco-2 human colon carcinoma cells. Oncogene 2000;19:5517–24. https://doi.org/10.1038/sj.onc.1203952.
[90] Nakanishi M, Menoret A, Tanaka T, Miyamoto S, Montrose D, Vella A, et al. Selective PGE(2) suppression inhibits colon carcinogenesis and modifies local mucosal immunity. Cancer Prev Res 2011;4(8):1198–208. https://doi.org/10.1158/1940-6207.capr-11-0188.
[91] McCartney SA, Mitchell JA, Fairclough PD, Farthing MJ, Warner TD. Selective COX-2 inhibitors and human inflammatory bowel disease. Aliment Pharmacol Ther 1999;13(8):1115–7. https://doi.org/10.1046/j.1365-2036.1999.00585.x.
[92] Sheibanie A, Yen J-H, Khayrullina T, Emig F, Zhang M, Tuma R, et al. The proinflammatory effect of prostaglandin E2 in experimental inflammatory bowel disease is mediated through the IL-23→IL-17 axis. J Immunol 2007;178(12):8138–47. https://doi.org/10.4049/jimmunol.178.12.8138.
[93] Karin M, Greten F. NF-kappaB: linking inflammation and immunity to cancer development and progression. Nat Rev Immunol 2005;5(10):749–59.
[94] Yamamoto Y, Gaynor R. Therapeutic potential of inhibition of the NF-κB pathway in the treatment of inflammation and cancer. J Clin Invest 2001;107(2):135–42. https://doi.org/10.1172/JCI11914.
[95] Wang CY, Guttridge DC, Mayo MW, Baldwin ASFJr. NF-kappaB induces expression of the Bcl-2 homologue A1/Bfl-1 to preferentially suppress chemotherapy-induced apoptosis. Mol Cell Biol 1999;19(9):5923–9.
[96] Pierce JW, Read MA, Ding H, Luscinskas FW, Collins T. Salicylates inhibit I kappa B-alpha phosphorylation, endothelial-leukocyte adhesion molecule expression, and neutrophil transmigration. J Immunol 1996;156(10):3961–9.
[97] Yamamoto Y, Yin MJ, Lin KM, Gaynor RB. Sulindac inhibits activation of the NF-kappaB pathway. J Biol Chem 1999;274(38):27307–14.

[98] Wahl C, Liptay S, Adler G, Schmid RM. Sulfasalazine: a potent and specific inhibitor of nuclear factor kappa B. J Clin Invest 1998;101(5):1163–74. https://doi.org/10.1172/JCI992.

[99] Hawkey CJ. COX-1 and COX-2 inhibitors. Best Pract Res Clin Gastroenterol 2001;15:801–20. https://doi.org/10.1053/bega.2001.0236.

[100] FitzGerald GA, Patrono C. The coxibs, selective inhibitors of cyclooxygenase-2. N Engl J Med 2001;345:433–42. https://doi.org/10.1056/NEJM200108093450607.

[101] Nie D, Honn KV, Cyclooxygenase, lipoxygenase and tumor angiogenesis. . Cell Mol Life Sci 2002;59(5):799–807. https://doi.org/10.1007/s00018-002-8468-9.

[102] Pollard M, Luckert PH. Effect of indomethacin on intestinal tumors induced in rats by the acetate derivative of dimethylnitrosamine. Science 1981;214:558–9.

[103] Chan AT, Ogino S, Fuchs CS. Aspirin and the risk of colorectal cancer in relation to the expression of COX-2. N Engl J Med 2007;356:2131–42. https://doi.org/10.1056/NEJMoa067208.

[104] Thun MJ, Namboodiri MM, Calle EE, Flanders WD, Heath CW. Aspirin use and risk of fatal cancer. Cancer Res 1993;53:1322–7.

[105] Koornstra JJ, Rijcken FE, Oldenhuis CN, Zwart N, van der Sluis T, Hollema H, et al. Sulindac inhibits beta-catenin expression in normal-appearing colon of hereditary nonpolyposis colorectal cancer and familial adenomatous polyposis patients. Cancer Epidemiol Biomarkers Prev 2005;14:1608–12. https://doi.org/10.1158/1055-9965.EPI-05-0112.

[106] Chan TA. Nonsteroidal anti-inflammatory drugs, apoptosis, and colon-cancer chemoprevention. Lancet Oncol 2002;3:166–74.

[107] Cao Y, Nishihara R, Wu K, Wang M, Ogino S, Willett W, et al. Population-wide impact of long-term use of aspirin and the risk for cancer. JAMA Oncol 2016;2(6):762–9. https://doi.org/10.1001/jamaoncol.2015.6396.

[108] Rothwell PM, Price JF, Fowkes FG, Zanchetti A, Roncaglioni MC, Tognoni G, et al. Short-term effects of daily aspirin on cancer incidence, mortality, and non-vascular death: analysis of the time course of risks and benefits in 51 randomised controlled trials. Lancet 2012;379:1602–12. https://doi.org/10.1016/ S0140-6736(11)61720-0.

[109] Liao X, Lochhead P, Nishihara R, Morikawa T, Kuchiba A, Yamauchi M, et al. Aspirin use, tumor PIK3CA mutation, and colorectal-cancer survival. N Engl J Med 2012;367:1596–606. https://doi.org/10.1056/ NEJMoa1207756.

[110] Domingo E, Church DN, Sieber O, Ramamoorthy R, Yanagisawa Y, Johnstone E, et al. Evaluation of PIK3CA mutation as a predictor of benefit from nonsteroidal anti-inflammatory drug therapy in colorectal cancer. J Clin Oncol 2013;31:4297–305. https://doi.org/10.1200/JCO.2013.50.0322.

[111] Kothari N, Kim R, Jorissen R, Desai J, Tie J, Wong H-L, et al. Impact of regular aspirin use on overall and cancer-specific survival in patients with colorectal cancer harboring a PIK3CA mutation. Acta Oncol 2015;54(4):487–92. https://doi.org/10.3109/0284186X.2014.990158.

[112] Murphy C, Turner N, Wong H, Sinnathamby M, Tie J, Lee B, et al. Examining the impact of regular aspirin use and PIK3CA mutations on survival in stage 2 colon cancer. Intern Med J 2017;47(1):88–98. https://doi.org/10.1111/imj.13312.

[113] Howe L. Inflammation and breat cancer: cyclooxygenase/prostaglandin signaling and breast cancer. Breast Cancer Res 2007;9(4):210. https://doi.org/10.1186/bcr1678.

[114] Kochel T, Goloubeva O, Fulton A. Upregulation of cyclooxygenase-2/prostaglandin E2 (COX-2/PGE2) pathway member multiple drug resistance-associated protein 4 (MRP4) and downregulation of prostaglandin transporter (PGT) and 15-prostaglandin dehydrogenase (15-PGDH) in triple-negative breast cancer. Breast Cancer 2016;10:61–70. https://doi.org/10.4137/BCBCR.S38529.

[115] Shim J, An H, Lee Y, Kim S, Lee K, Lee K. Overexpression of cyclooxygenase-2 is associated with breast carcinoma and its poor prognostic factors. Modern Pathol 2003;16(12):1199–204. https://doi.org/10.1097/01.MP.0000097372.73582.CB.

[116] Harris R, Chlebowski R, Jackson R, Frid D, Ascenseo J, Anderson G, et al. Breast cancer and nonsteroidal anti-inflammatory drugs: prospective results from the Women's Health Initiative. Cancer Res 2003;63(18): 6096–101.

[117] Rothwell P, Wilson M, Price J, Belch J, Meade T, Mehta Z. Effect of daily aspirin on risk of cancer metastasis: a study of incident cancers during randomised controlled trials. Lancet 2004;379(9826):1591–601. https://doi.org/10.1016/S0140-6736(12)60209-8.

[118] Fraser DM, Sullivan FM, Thompson AM, McCowan C. Aspirin use and survival after the diagnosis of breast cancer: a population-based cohort study. Br J Cancer 2014;111(3):623–7. https://doi.org/10.1038/bjc.2014.264.

[119] Choi B-H, Chakraborty G, Baek K, Yoon H. Aspirin-induced Bcl-2 translocation and its phosphorylation in the nucleus trigger apoptosis in breast cancer cells. Exp Mol Med 2013;45:e47. https://doi.org/10.1038/emm.2013.91.

[120] Yan F, He Q, Hu X, Li W, Wei K, Li L, et al. Direct regulation of caspase-3 by the transcription factor AP-2α is involved in aspirin-induced apoptosis in MDA-MB-453 breast cancer cells. Mol Med Rep 2013;7(3):909–14. https://doi.org/10.3892/mmr.2013.1257.

[121] Agrawal A, Fentiman IS. NSAIDs and breast cancer: a possible prevention and treatment strategy. Int J Clin Practice 2008;62(3):444–9. https://doi.org/10.1111/j.1742-1241.2007.01668.x.

[122] Din F, Valanciute A, Houde V, Zibrova D, Green K, Sakamoto K, et al. Aspirin inhibits mTOR signaling, activates AMP-activated protein kinase, and induces autophagy in colorectal cancer cells. Gastroenterology 2012;e3:1504–15. https://doi.org/10.1053/j.gastro.2012.02.050.

[123] Xu L, Stevens J, Hilton M, Seaman S, Conrads T, Veenstra T, et al. COX-2 inhibition potentiates antiangiogenic cancer therapy and prevents metastasis in preclinical models. Sci Transl Med 2014;6(242):242ra84. https://doi.org/10.1126/scitranslmed.3008455.

[124] Mazhar D, Ang R, Waxman J. COX inhibitors and breast cancer. Br J Cancer 2006;94(3):346–50. https://doi.org/10.1038/sj.bjc.6602942.

[125] Zatelli M, Luchin A, Tagliati F, Leoni S, Piccin D, Bondanelli M, et al. Cyclooxygenase-2 inhibitors prevent the development of chemoresistance phenotype in a breast cancer cell line by inhibiting glycoprotein p-170 expression. Endocr Relat Cancer 2007;14(4):1029–38. https://doi.org/10.1677/ERC-07-0114.

[126] Turturro S, Najor M, Ruby C, Cobleigh M, Abukhdeir A. Mutations in PIK3CA sensitize breast cancer cells to physiologic levels of aspirin. Breast Cancer Res Treat 2016;156(1):33–43. https://doi.org/10.1007/s10549-016-3729-8.

[127] Henry W, Laszewski T, Tsang T, Beca F, Beck A, McAllister S, et al. Aspirin suppresses growth in PI3K-mutant breast cancer by activating AMPK and inhibiting mTORC1 signaling. Cancer Res 2016;77(3):790–801. https://doi.org/10.1158/0008-5472.CAN-16-2400.

CHAPTER

7

STAT3 as a Major Contributor to Chemoresistance

David S. Lee, Jennifer R. Grandis, Daniel E. Johnson

University of California at San Francisco, San Francisco, CA, USA

Abstract

Signal transducer and activator of transcription 3 (STAT3) is a well-validated anticancer target that promotes tumorigenesis. Although it has historically been considered "undruggable", multiple classes of direct STAT3 inhibitors have recently been discovered that successfully abrogate STAT3 activation or DNA-binding activity, ultimately generating selective cytotoxicity against cancerous cells. Considering the growing body of evidence implicating STAT3 in the development of treatment resistance, direct STAT3 inhibitors have promising potential to augment the antitumor effects of conventional chemotherapy drugs, radiation, and molecular targeting agents, leading to improved patient outcomes. This chapter focuses on the progress made in developing direct STAT3 inhibitors and their evaluation in preclinical models and clinical trials.

ABBREVIATIONS

DBDI	DNA binding domain inhibitor
DN	Dominant-negative
EGFR	Epidermal growth factor receptor
FDA	Food and Drug Administration
FGFR	Fibroblast growth factor receptor
HNSCC	Head and neck squamous cell carcinoma
hSIE	High affinity serum-inducible element
IL-6	Interleukin-6
IL-6R	Interleukin-6 receptor
JAK	Janus kinase
mts	Membrane-translocation sequence
NDI	N-terminal domain inhibitor
NMR	Nuclear magnetic resonance
NSCLC	Non-small cell lung cancer
NTD	N-terminal domain
pY	Phosphotyrosine
SDI	SH2 domain inhibitor
SH2	SRC-homology 2
STAT	Signal transducer and activator of transcription
VEGF	Vascular endothelial growth factor

Targeting Cell Survival Pathways to Enhance Response to Chemotherapy. http://dx.doi.org/10.1016/B978-0-12-813753-6.00007-X
Copyright © 2018 Elsevier Inc. All rights reserved.

7.1 INTRODUCTION

The family of signal transducer and activator of transcription (STAT) proteins facilitates a wide variety of physiological functions essential to cell survival, including proliferation, differentiation, and inflammation [1,2]. In total, there are seven members of the STAT family (STAT1-4, STAT5a, STAT5b, and STAT6) [3,4]. Their structures consist of an N-terminal domain, coiled-coil region, DNA-binding sequence, and SRC-homology 2 (SH2) domain. The N-terminal domain facilitates STAT3 oligomerization, a poorly understood process that is important for enhanceosome stability and association with regulatory proteins. Similarly, the coiled-coil region serves as a loading site for regulatory proteins. The DNA-binding domain recognizes a specific DNA sequence motif, enabling direct interactions with response elements in the promoter regions of STAT target genes. The SH2 domain binds to phosphotyrosine (pY) residues present on activated tyrosine kinases, leading to subsequent phosphorylation of the STAT protein by the bound kinase. The SH2 domain also facilitates STAT dimerization via reciprocal associations with a key pY residue (pY705 in STAT3) of another monomer [1]. This dimerization plays an essential role in the binding of the STAT protein to DNA. Transient and highly regulated activation of STATs leads to expression of a variety of target genes, encoding proteins such as Bcl-X_L, c-Myc, cyclin D1, and VEGF [1,5].

STAT activation can be achieved through multiple mechanisms (Fig. 7.1). In response to stimulation by growth factors and cytokines, cell-surface receptors dimerize, leading to activation of an intrinsic receptor tyrosine kinase activity or the activation of receptor-associated kinases like Janus kinases (JAKs). These activated kinases then facilitate recruitment of inactive STAT monomers by phosphorylating tyrosine residues in the cytoplasmic region of the receptor protein. The receptor pY residues serve as docking sites for the SH2 domain of STAT monomers within the cytoplasm. Once recruited, STAT monomers undergo receptor- or JAK-mediated tyrosine phosphorylation, followed by dimerization via interactions between the SH2 domain of one STAT monomer with a pY residue on another monomer. The STAT dimers translocate into the nucleus where they act to modulate the expression of STAT target genes [1]. Well-studied signaling pathways, including those activated by interferon-gamma, interleukin-6 (IL-6), epidermal growth factor, and platelet-derived growth factor, employ STAT activation as described [6–9]. In addition, nonreceptor tyrosine kinases, such as Abelson leukemia protein and SRC-related kinases, have been shown to phosphorylate/activate STATs in the absence of an extracellular stimulus [10–12]. Cell confluence, leading to cadherin–cadherin stimulation, has been proposed as another mechanism of nonreceptor-mediated STAT activation, although the role of cellular microenvironment in STAT activation is less well characterized [13–15].

Consistent with the multifaceted role of STAT in cellular proliferation and survival, a growing body of evidence supports the idea that dysregulated STAT activity—STAT3, in particular—plays a direct role in oncogenesis [16,17]. Constitutive STAT3 activation has been observed in multiple cancers and is associated with poor clinical prognoses [18–20]. *In vitro* studies have demonstrated that inhibition of STAT3 in cancer cell lines with aberrantly activated STAT3 results in decreased proliferation and increased apoptosis [21–23]. However, inhibition of STAT3 does not appear to result in cytotoxicity in models of normal adult tissue, suggesting pathway redundancy under physiological circumstances and oncogenic dependency in the context of cancer [1,24]. With numerous studies establishing additional roles for

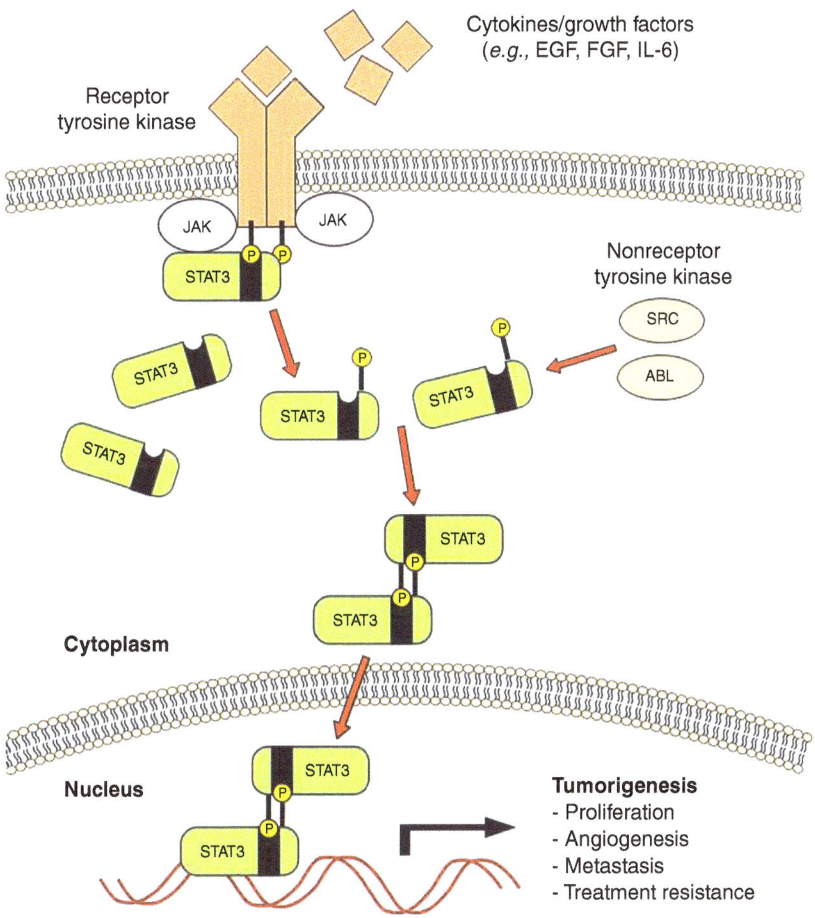

FIG. 7.1 **STAT3 activation and signal transduction pathway.** Ligand-activated receptor tyrosine kinases and activated nonreceptor tyrosine kinases phosphorylate STAT3 on pY705. The SH2 domain, indicated by the black region on STAT3, serves as a binding site for the pY705 residue of another activated monomer, facilitating reciprocal interactions essential to STAT3 dimerization. STAT3 dimers translocate to the nucleus for subsequent target gene expression.

STAT3 in metastasis, angiogenesis, treatment resistance, and suppression of antitumor immunity, a large body of evidence identifies STAT3 inhibition as a promising therapeutic strategy with a low toxicity profile [1,25,26].

Four main classes of drugs have been developed to directly inhibit the activity of STAT3: SH2 domain inhibitors (SDI), DNA-binding domain inhibitors (DBDI), N-terminal domain inhibitors (NDIs), and STAT3 antisense (Figs. 7.2 and 7.3) [27]. Each of the first three classes utilizes a different aspect of the STAT3 domain structure to abrogate dimerization, protein–DNA interactions, or formation of protein complexes necessary for transcription, while antisense inhibitors act to prevent expression of the protein. Collectively, direct inhibitors of

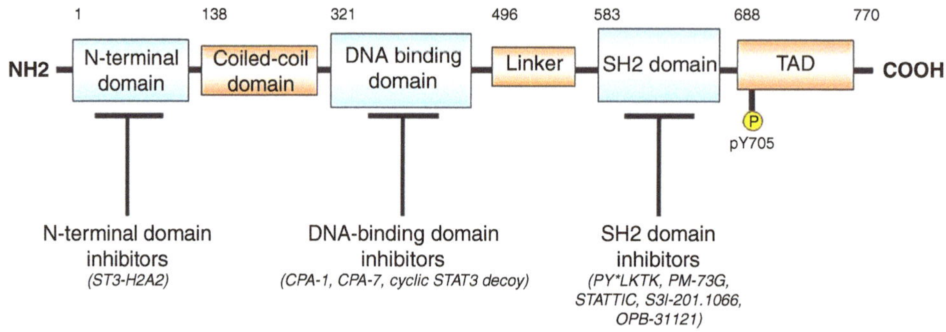

FIG. 7.2 **Schematic of STAT3 domain structure and corresponding target sites for direct STAT3 inhibitors.**

STAT3 face the multiple challenges of intracellular drug delivery, including cell membrane permeability, selectivity, and systemic toxicities. Although a fifth class of indirect STAT3 inhibitors exists in the form of upstream receptor/receptor-associated tyrosine kinase inhibitors, this chapter will focus on depicting current progress related to direct STAT3 inhibitors and their potential use as standalone or adjuvant therapies for cancer.

7.2 ROLE OF STAT3 IN TREATMENT RESISTANCE

A growing body of literature has identified a connection between STAT3 activation and treatment resistance through experiments employing STAT3 dominant-negative (DN) mutants, RNA interference, or treatment with STAT3 inhibitors. For example, exogenous expression of DN-STAT3(Y705F) markedly enhances radiation sensitivity in treatment-resistant glioblastoma cancer cells (U87) that harbor constitutively activated STAT3 [28]. Similarly, siRNA-mediated silencing of STAT3 in chemoradioresistant colorectal carcinoma results in increased sensitivity to 5-fluorouracil and radiation [29,30]. Indeed, a broad number of investigations have reported sensitization to chemotherapy and/or radiation in multiple solid tumors, including breast, prostate, and esophageal cancers, following STAT3 inhibition or downregulation [26,31–33].

Mechanistic studies have characterized the role of STAT3 in generating resistance to radiation, chemotherapy, and immunotherapy. Investigation of non-small-cell lung cancer (NSCLC), melanoma, and colon cancer models resistant to epidermal growth factor receptor (EGFR) inhibitor have suggested that development of resistance to small-molecule inhibitors is associated with autocrine stimulation of interleukin-6 receptor (IL-6R) and fibroblast growth factor receptor (FGFR), ultimately leading to downstream STAT3 activation [15,34,35]. The link between disruption of EGFR signaling pathways and activation of STAT3 has also been shown to be conserved across multiple cancer types, indicating that inhibition of STAT3 may provide a general avenue for overcoming treatment resistance [36]. Similar autocrine feedback loops have been proposed to explain STAT3-mediated resistance to conventional chemotherapy drugs and radiation [26,37–39].

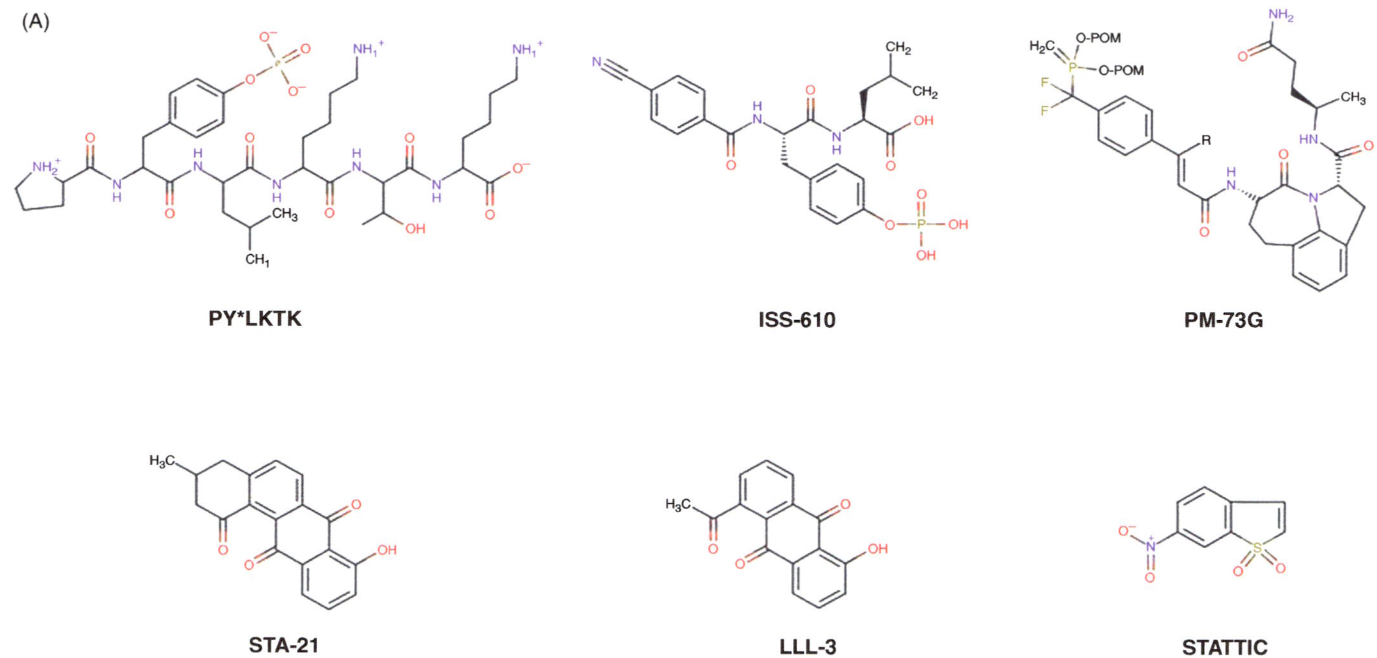

FIG. 7.3 **Chemical structures of STAT3 inhibitors. (A) SDIs. (B) DBDIs.** (C) Antisense inhibitor. Lowercase nucleotides in the STAT3 decoy molecules indicate phosphorothioate bases. Lowercase nucleotides in AZD9150 indicate bases modified with cET. Chemical structures were drawn using Marvin Sketch. Marvin 17.11.0, 2017, ChemAxon (http://www.chemaxon.com).

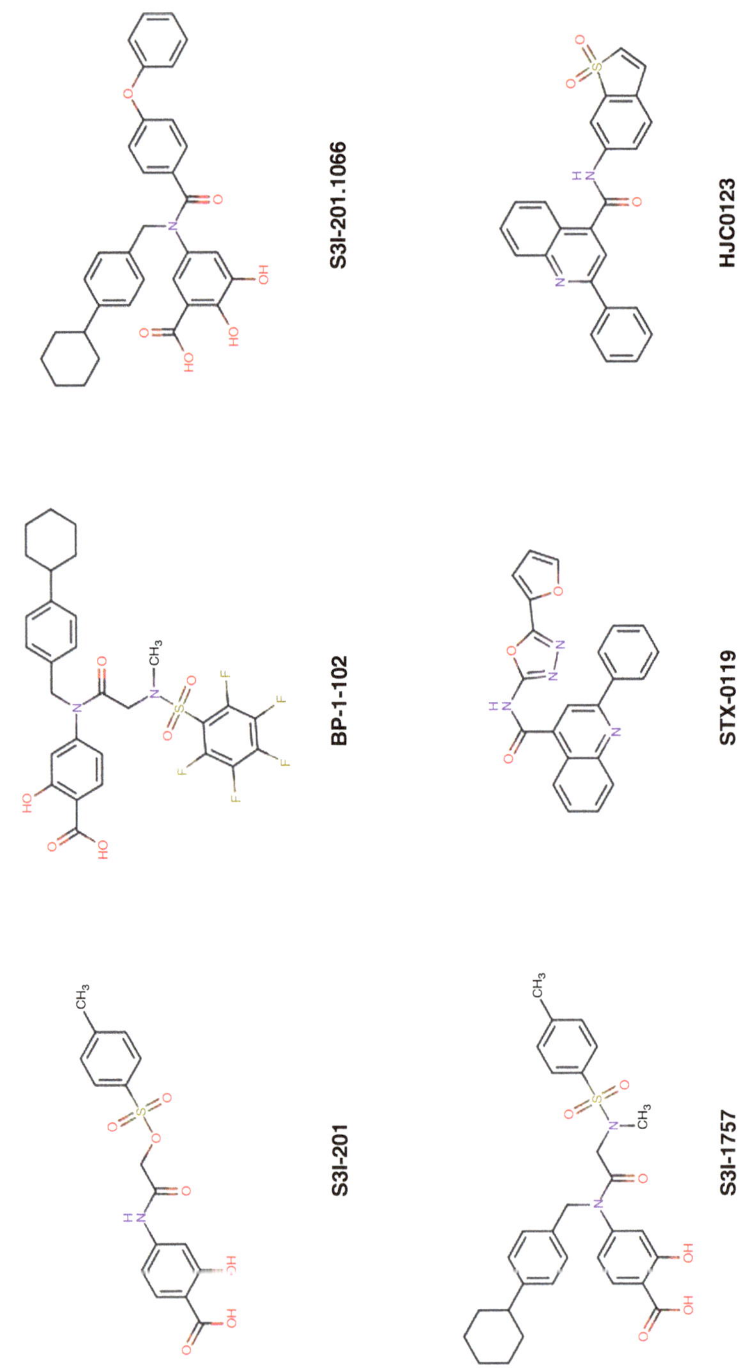

FIG. 7.3 (cont.)

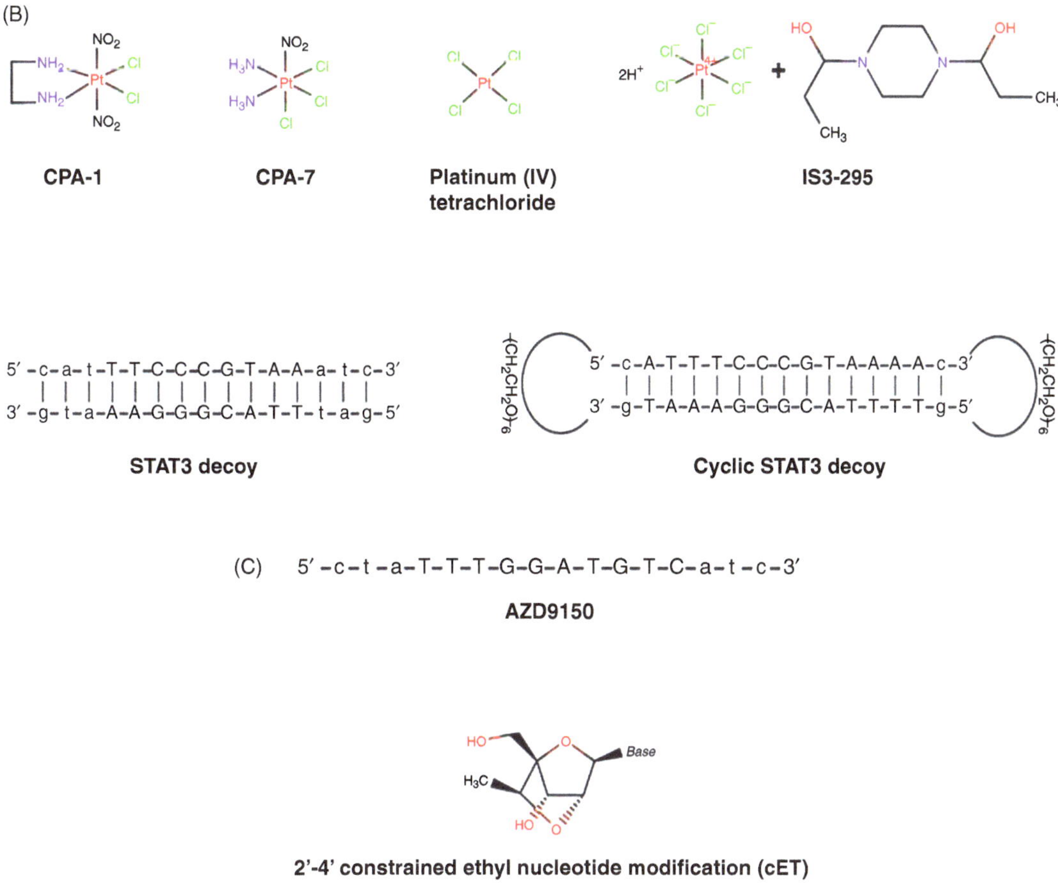

FIG. 7.3 *(cont.)*

Considering that multimodal therapy has become the standard approach to managing patients with solid tumors, targeting STAT3 may provide a promising method for enhancing the antitumor effects of conventional therapies, as well as novel molecular targeting and immunomodulatory agents. Although direct STAT3 inhibitors have not yet been translated to the clinic for standard of care use, concerted efforts to discover and evaluate drugs leading to STAT3 abrogation have generated multiple promising leads.

7.3 DIRECT INHIBITORS OF STAT3

7.3.1 SH2 Domain Inhibitors

The SH2 domain is essential to STAT3 activation. It facilitates association with receptor kinases at pY residues in the receptor cytoplasmic region. Ultimately, this physical association leads to STAT3 phosphorylation by either the receptor or receptor-associated kinases (e.g.,

JAKs). The SH2 domain also stabilizes reciprocal pY/SH2 interactions necessary for homo- or heterodimerization with other STAT monomers. Disruption of SH2-mediated interactions can abrogate STAT3 function and prevent the downstream expression of target genes, leading to inhibition of STAT3-mediated tumor growth. SH2 domains typically span approximately 100 residues and recognize and bind pY residues that occur within specific sequence contexts [40]. The consensus SH2 recognition site for STAT3 has been reported as PY*LKTK (Y* indicates the pY residue) [41]. While nuclear magnetic resonance (NMR) and X-ray crystallography studies of phosphopeptide/SH2 domain complexes have been reported, including the SH2 domains of Src, Lck, and Grb2, the conformational structure of STAT3 binding to various receptors remains elusive [42]. Nonetheless, several different approaches, including phosphopeptide library screening, computer-assisted virtual screening, modification of naturally occurring STAT3 inhibitors, and fragment-based drug design have led to the discovery of two subcategories of SDIs: peptides/peptidomimetics and nonpeptidic small-molecule inhibitors (Table 7.1, Fig. 7.3A) [43,44].

The first proof-of-concept study for phosphopeptide and peptidomimetic STAT3 inhibitors was reported in 2001 [41]. In these studies, the phosphopeptide PY*LKTK demonstrated an *in vitro* dose-dependent inhibition of STAT3 DNA-binding activity in nuclear extracts. It was proposed that PY*LKTK binds to the SH2 domain of inactive STAT3 monomers and competitively inhibits their homodimerization, preventing activation and target gene expression. However, the suppression of STAT3 activity by PY*LKTK was only observed at high concentrations, with an IC_{50} of 235 μM. With respect to selectivity, DNA-binding assays illustrated that PY*LKTK did not appreciably affect STAT5 activity, but did modestly inhibit STAT1 activity [41]. To overcome the obstacle of transporting a phosphorylated peptide across the cell membrane, a hydrophobic membrane-translocating sequence (mts) was linked to PY*LKTK peptide at its C-terminus. Treatment of NIH 3T3/v-Src fibroblasts exhibiting hyperactivation of STAT3 with the PY*LKTK-mts peptide decreased STAT3 DNA-binding activity relative to treatment with an unphosphorylated control peptide (PYLKTK-mts). The decreased STAT3 DNA-binding activity was accompanied by increased apoptosis and reduced malignant transformation [41]. The concentration required to abrogate STAT3-mediated gene transcription was 1 mM, highlighting a need for greater potency to achieve clinical translation [41,89].

Although specificity, stability, and intracellular transport will continue to challenge the clinical translation of peptide inhibitors, some peptidic agents have yielded noteworthy preclinical results. ISS-610, a tripeptide phosphopeptidomimetic based on the original PY*LKTK peptide, demonstrated increased potency with an *in vitro* IC_{50} of 42 μM for suppression of STAT3 DNA-binding activity [45]. It also suppressed the growth and survival of multiple cell lines with constitutively active STAT3, including Src-transformed fibroblasts and NSCLC and breast carcinoma cells.

In an additional attempt to overcome the poor cell permeability and stability associated with phosphopeptides, Mandal et al. [47] designed PM-73G, a novel pro-drug. PM-73G contains a phosphonodifluoromethyl group to prevent dephosphorylation via phosphatases and esterase-labile pivaloyloxymethyl groups to hide the negative phosphate charges. In breast cancer cells, PM-73G disrupted both constitutively activated and IL-6-induced STAT3 DNA-binding activity with an IC_{50} of 100–500 nM [47]. *In vivo* PM-73G also abrogated tumor growth and angiogenesis in mice harboring MDA-MB-468 breast cancer xenografts, although no

TABLE 7.1 Direct STAT3 Inhibitors

	Drug candidate	Target domain	Proposed mechanism of action	Preclinical/Clinical evidence	References
Short peptides	PY*LKTK	SH2	Inhibits homodimerization	*In vitro*	[41]
	ISS-610	SH2	Inhibits homodimerization	*In vitro*	[45]
	PM-73G	SH2	Inhibits STAT3 phosphorylation	*In vitro, in vivo*	[46,47]
Peptidomimetics	STA-21	SH2	Inhibits homodimerization	*In vitro*	[48,49]
	LLL-3	SH2	Inhibits homodimerization	*In vitro, in vivo*	[50,51]
	STATTIC	SH2	Inhibits STAT3 phosphorylation	*In vitro, in vivo*	[52–54]
Nonpeptidic small-molecule inhibitors	BP-1-102	SH2	Inhibits homodimerization	*In vitro, in vivo*	[55]
	S3I-201.1066	SH2	Inhibits homodimerization	*In vitro, in vivo*	[56]
	S3I-1757	SH2	Inhibits homodimerization	*In vitro*	[23]
	STX-0119	SH2	Inhibits homodimerization	*In vitro, in vivo*	[57,58]
	HJC0123	SH2	Inhibits homodimerization	*In vitro, in vivo*	[59]
	OPB-31121	SH2	Inhibits homodimerization	*In vitro, in vivo,* Phase I (advanced solid tumors)	[60–64]
	OPB-51602	SH2	Inhibits homodimerization	*In vitro, in vivo,* Phase I (advanced solid tumors)	[62,65,66]
Platinum-based compounds	CPA-1	DNA binding domain	Inhibits STAT3 DNA binding	*In vitro*	[67]
	CPA-7	DNA binding domain	Inhibits STAT3 DNA binding	*In vitro, in vivo*	[67–69]
	Platinum (IV) tetrachloride	DNA binding domain	Inhibits STAT3 DNA binding	*In vitro*	[67]
	IS3 295	DNA binding domain	Inhibits STAT3 DNA binding domain	*In vitro*	[70]

(Continued)

TABLE 7.1 Direct STAT3 inhibitors (*cont.*)

	Drug candidate	Target domain	Proposed mechanism of action	Preclinical/Clinical evidence	References
Double-stranded oligonucleotides	STAT3 decoy	DNA binding domain	Competitively inhibits STAT3 DNA binding	*In vitro, in vivo,* Phase 0 (HNSCC)	[71–79]
	Cyclic STAT3 decoy	DNA binding domain	Competitively inhibits STAT3 DNA binding	*In vitro, in vivo*	[5,74]
Antisense oligonucleotides	2′-*O*-methylethyl antisense oligonucleotide	STAT3 mRNA	Binds STAT3 mRNA for degradation	*In vitro, in vivo*	[80–82]
	AZD9150	STAT3 mRNA	Binds STAT3 mRNA for degradation	*In vitro, in vivo,* Phase 1 (lymphoma and NSCLC)	[83–86]
N-terminal helices	ST3-H2A2	N-terminal domain	Inhibition of downstream suppressors of pro-apoptosis	*In vitro*	[87,88]

impact on apoptosis or cyclin D1 was detected [46]. Subsequent studies of PM-73G in other tumor models have not been reported.

Collectively, peptide and peptidomimetic inhibitors of the SH2 domain have sufficiently overcome the challenges of poor intracellular transport and stability in preclinical studies to warrant their evaluation in clinical trials. To date, PY*LKTK has primarily been employed as a research tool compound to identify and characterize STAT3-dependent processes, including activation of STAT3 by cell confluence, innate and adaptive immune responses of bone marrow progenitor cells, and STAT3-mediated resistance to apoptosis in polyamine-depleted cells [89–91]. Despite their current lack of clinical utility, peptide and peptidomimetic inhibitors that bind to the STAT3 SH2 domain have validated the concept that phosphopeptides can successfully disrupt or prevent STAT3 homodimerization. Further optimization may lead to the discovery of clinically viable agents in this class of compounds.

The search for STAT3 inhibitors has primarily shifted away from peptides to nonpeptidic small molecules because of the latters increased potency, stability, and cell membrane penetration. Computational docking studies and molecular modeling have led to the rational design of multiple promising compounds (Fig. 7.3A) [44,49,54,55]. STA-21 is a small-molecule inhibitor derived from a naturally occurring deoxytetrangomycin that binds to pY*705 residues of STAT3 and prevents dimerization [49]. In human breast carcinoma, rhabdomyosarcoma, and osteosarcoma cell lines that overexpress active STAT3, inhibition of STAT3 DNA binding and induction of apoptosis was observed with STA-21 concentrations of 20–30 μM [48,49]. Furthermore, STA-21 exhibited minimal cytotoxicity against human skin fibroblasts and breast carcinoma cells that lack overexpression of active STAT3 [49]. LLL-3 is an analog of STA-21 and was found to induce apoptosis of glioblastoma cell lines with hyperactivation of STAT3 at concentrations in the range 10–40 μM [50]. *In vivo* administration of LLL-3 to mice with intracranial glioblastoma tumors resulted in suppression of tumor growth and increased survival, relative to control treatment [50]. In view of evidence suggesting that disruption of STAT3 activity improves therapy response in malignant glioma cells, LLL-3 or related analogs hold potential for clinical application as a chemoradiotherapy adjunct in this devastatingly lethal malignancy [36,50,92,93]. Additional studies have shown that LLL-3 enhances the activity of imatinib against the chronic myelogenous leukemia cell line K562 [51]. However, despite promising preclinical data, neither LLL-3 nor any of its derivatives have been evaluated in human trials.

STAT3 inhibitory compound (STATTIC) is another SDI of STAT3. STATTIC was identified via fluorescence polarization-based assays that screened more than 17,000 compounds for their ability to dissociate phosphopeptide binding to STAT3 SH2 domain; subsequent electrophoretic mobility shift assays confirmed that STATTIC inhibited STAT3 DNA-binding activity [53,54]. *In vivo* studies have shown that STATTIC inhibits tumor growth and promotes apoptosis in breast cancer and head and neck squamous cell carcinoma (HNSCC) xenograft tumors [54]. Interestingly, STATTIC was demonstrated to sensitize nasopharyngeal carcinoma cell lines to treatment with cisplatin and radiation [52]. Nasopharyngeal carcinomas are frequently characterized by constitutively activated STAT3, modest response to chemoradiation, and poor clinical prognosis [94]. However, no further investigations have been performed with STATTIC in multimodal treatment for this carcinoma, leaving its clinical utility largely unexplored.

A few nonpeptidic SDIs have demonstrated efficacy as orally bioavailable compounds. BP-1-102, a derivative of the lead compound S3I-201, has been shown to inhibit STAT3 DNA-binding activity at an IC_{50} of 6.8 μM [55]. Oral administration of BP-1-102 to mice harboring breast cancer and NSCLC xenografts resulted in inhibition of tumor cell proliferation and invasiveness [55]. These data are consistent with results obtained with other nonoral analogs of S3I-201, including S3I-201.1066 and S3I-1757 [23,56]. Pharmacokinetic profiling in tumor-bearing mice showed that steady-state BP-1-102 plasma levels can be achieved that exceed the *in vitro* IC_{50} for STAT3 inhibition. Uptake and accumulation within tumor tissues also reached therapeutic levels, supporting the clinical potential of this orally bioavailable drug [55].

STX-0119 is an orally bioavailable inhibitor that has been shown to inhibit the growth of human lymphomas in mouse models. In these studies, STX-0119 treatment was not associated with any observable nonspecific toxicities to the mice [57,58]. HJC0123, another orally bioavailable STAT3 inhibitor, was generated by fusing together STATTIC and STX-0119 [59]. HJC0123 has been reported to inhibit the growth of breast and pancreatic cancer cell lines with IC_{50} values (0.1–1.25 μM) lower than that can be achieved by either STATTIC or STX-0119 components alone. Oral administration of HJC0123 inhibited the growth of estrogen receptor-negative breast cancer xenograft tumors in mice without evidence of adverse toxicities [59].

OPB-31121 and OPB-51602 are the only two orally administered nonpeptidic SH2 domain antagonists to reach clinical trials to date. Computational docking and molecular dynamics simulation suggest that these antagonists bind with high affinity (K_d 10 nM) to the STAT3 SH2 domain at a binding pocket distinct from the binding sites of other STAT3 SH2 domain antagonists [61,62]. Preclinical data for OPB-31121 demonstrate inhibition of STAT3 DNA binding across multiple cell lines and against xenograft tumors expressing constitutively active STAT3 [61,63,64]. Furthermore, OPB-31121 has been shown to induce apoptosis in models of drug-resistant leukemia, while exhibiting synergistic antitumor effects in gastric cancer models when combined with 5-fluorouracil [63,64]. These findings indicate the potential of OPB-31121 to act as a monotherapy and to sensitize cancer cells to conventional chemotherapy. However, the first-in-human phase I clinical trial of OPB-31121 in patients with advanced solid tumors demonstrated dose-limiting toxicities (nausea/vomiting, diarrhea, and fatigue) at drug concentrations significantly lower than needed to achieve target inhibition, ultimately resulting in failure to halt or reverse tumor progression [60]. In preclinical studies, OPB-51602 demonstrated similar anti-tumor effects as OPB-31121 [65]. A phase I dose-escalation trial in patients with treatment-refractory solid tumors found that OPB-51602 initially exhibited tolerable toxicities (nausea/vomiting, diarrhea, peripheral neuropathy, and lactic acidosis) at therapeutic concentrations, although side effects increased after multiple treatment cycles and led to eventual treatment discontinuation [66]. Following administration of OPB-51602, phosphorylated STAT3 levels were decreased in peripheral monocytes, and tumor regression was observed in two NSCLC patients (5% of overall cohort; one patient had complete resolution of target lesions after 10 cycles of OPB-51602 and a 6.9-month progression-free interval; the other patient had a 41% reduction of measurable tumor burden, but stopped OPB-51602 treatment after five cycles due to intolerable peripheral neuropathy) [66]. Both patients with tumor regression had known treatment resistance against EGFR tyrosine kinase inhibitors and had received at least one round of chemotherapy. However, post-treatment biopsies were not performed to establish changes in activated STAT3 concentrations [66]. A subsequent phase I trial has examined OPB-51602 in refractory hematologic cancers [65].

As a class, the landscape of STAT3 SDIs has steadily improved since their initial conception in 2001. Optimization of lead molecules through innovative methods of structure-based design has led to improved potency as evidenced by IC_{50} levels in the nanomolar range, increased intracellular uptake, enhanced antitumor effects *in vivo*, and greater specificity for STAT3. In spite of this progress, systemic toxicities and modest bioavailabilities continue to limit their translation to the clinic.

7.3.2 DNA-Binding Domain Inhibitors

As a transcription factor, STAT3 contains a unique DNA-binding region that enables recognition and association with corresponding response elements in the promoters of target genes. Targeting this STAT3:DNA association offers another approach to mitigating aberrant STAT3-mediated cellular processes that facilitate tumor growth and survival. STAT3 DBDIs have largely been derived from two distinct classes of compounds: platinum-based molecules and decoy oligonucleotides (Table 7.1, Fig. 7.3B). Although the DBDIs have not received as much research focus as SDIs, a few stand out as potentially viable therapeutic agents.

In 2004, a screen of platinum-based compounds by Turkson et al. [67] identified CPA-1, CPA-7, and platinum (IV) tetrachloride, as molecules capable of disrupting STAT3:DNA associations (Fig. 7.3B). Although platinum (II)-based compounds such as cisplatin are known to produce antitumor effects by cross-linking to purines, platinum (IV)-based compounds were found to exert cytotoxicities by downmodulating expression of STAT3-dependent genes [67,95]. The effects of CPA-1, CPA-7, and platinum (IV) tetrachloride have been studied in multiple cancer cell lines with constitutively activated STAT3, including v-Src-transformed mouse fibroblasts (NIH 3T3/v-Src), human breast carcinoma (MDA-MB-231, MDA-MB-435, and MDA-MB-468), colon cancer (CT26), and melanoma (M2 and C110) cells [67]. Treatment with these platinum-based compounds resulted in lower levels of activated STAT3, decreased STAT3 DNA binding activity, and induction of apoptosis (IC_{50} values ≤5.0 μM). Additionally, these novel compounds exhibited 2.5- to 4-fold greater selectivity for STAT3 over STAT1 [67]. Overall, CPA-1, CPA-7, and platinum (IV) tetrachloride demonstrate potency and selectivity comparable to the previously mentioned SDIs.

Despite promising *in vitro* findings with the platinum-based compounds, *in vivo* studies have only been performed with CPA-7. Intravenous treatment of mice harboring colon carcinoma xenografts with CPA-7 (5 mg/kg) resulted in decreased STAT3 activity and induction of complete or partial tumor regression relative to treatment with vehicle control [67]. Additional investigations on human melanoma xenograft tumors confirmed inhibition of angiogenesis and induction of tumor regression following intravenous administration of CPA-7 [69]. However, CPA-7 failed to promote tumor regression in mice harboring intracranial melanoma and glioblastoma xenograft tumors, suggesting poor central nervous system penetration [68].

Another platinum (IV)-based compound, IS3-295, exhibits similar antitumor effects as CPA-7 [70]. IS3-295 was originally discovered by screening the NCI 2000 diversity set for compounds that decrease STAT3 DNA binding activity. Although its exact mechanism has not fully been elucidated, IS3-295 likely binds to the STAT3 DNA-binding domain [70]. Treatment of breast carcinoma cell lines with IS3-295 resulted in G_0/G_1-phase cell-cycle arrest. Furthermore, TUNEL staining suggested that malignant cell lines harboring constitutively activated STAT3, including human breast cancer (MDA-MD-435), NSCLC (A549), prostate carcinoma (DU145), multiple myeloma (U266), and pancreatic carcinoma (Panc1), exhibited greater apoptosis after treatment with IS3-295 compared to control groups. Although IS3-295 exhibits potent inhibitory activity against STAT3 and reduces STAT3-mediated expression of cyclin D1 and Bcl-X_L *in vitro*, no *in vivo* studies have been reported to date [70]. As a subclass of STAT3 DBDIs, platinum (IV)-based compounds have demonstrated intriguing therapeutic potential. However, their translation is currently limited by lack of toxicity and pharmacology studies and they remain untested in treatment-resistant models.

Double-stranded "decoy" oligonucleotides offer an alternative approach to disrupting the STAT3:DNA-binding interaction. Decoy oligonucleotides contain double-stranded DNA sequences that mimic the binding site of transcription factors, leading to competitive inhibition of authentic interactions between the transcription factors and their target genes [96]. Previous studies on arterial wall gene therapy and STAT6-mediated T helper 2 cell activity demonstrated that adequate intracellular concentrations of decoy lead to successful inhibition of transcription factor DNA-binding activity and decreased expression of target genes [97,98]. Coupled with evidence that antisense oligonucleotides and dominant-negative STAT3 mutants induce apoptosis in multiple cancer cell lines [99–102], these studies provided the conceptual framework to pursue the design of a double-stranded decoy oligonucleotide that modulates STAT3-mediated gene expression.

In 2003, Leong et al. [72] synthesized and evaluated the first STAT3 double-stranded decoy oligonucleotide (Fig. 7.3B). The decoy consisted of the sequence 5′-CATTTCCCGTAAATC-3′ according to the STAT3 DNA-binding sequence located in the *c-fos* gene promoter (high-affinity serum-inducible element; hSIE) [41,103]. Confocal microscopy demonstrated cellular uptake and nuclear and cytoplasmic localization of the STAT3 decoy following treatment of cancerous and noncancerous cells for 6 h. Dose-dependent inhibition of the proliferation of HNSCC cell lines was seen using STAT3 decoy concentrations ranging from 0.0125 to 25 μM. By contrast, a point mutant control decoy exerted only minimal effects on proliferation. Treatment of HNSCC cells harboring an hSIE-luciferase construct with the STAT3 decoy resulted in decreased luciferase activity relative to mutant control decoy; similarly, reduced expression was also seen for antiapoptotic Bcl-X_L—the product of a STAT3 target gene. Importantly, normal human oral keratinocytes demonstrated no cytotoxic effects after incubation with STAT3 decoy, maintaining growth at rates similar to keratinocytes incubated with mutant control decoy or no treatment. Altogether, *in vitro* studies suggest that STAT3 decoy achieves intracellular transport, leads to enhanced apoptosis, and decreases expression of STAT3 target genes, highlighting its potential as a therapy [72].

Follow-up *in vivo* studies on the antitumor efficacy of STAT3 decoy provided promising results in terms of both efficacy and specificity. In one study, athymic nude mice were xenografted with two HNSCC tumors each [77]. One tumor received daily intratumoral injections of STAT3 decoy and the other received mutant control decoy. A 3.25-fold increase in cell death was observed in tumors treated with STAT3 decoy vs mutant control decoy. Decreased tumor cell expression of cyclin D1 and Bcl-X_L was also seen [77]. STAT5 activation was unaffected by treatment with STAT3 decoy, highlighting the specificity of the decoy. Similar inhibition of tumor growth and reduction in STAT3 activity has been observed in other tumor xenograft models following intratumoral injection of STAT3 decoy, including human glioblastoma multiforme (U251), lung carcinoma (A549), and ovarian epithelial cancer (SKOV3) models [76,78,79]. The *in vivo* evidence suggests that STAT3 decoy facilitates tumor growth inhibition via STAT3 blockade and is at least partially due to increased apoptosis.

STAT3 decoy therapy may be particularly useful in the case of cancers resistant to EGFR inhibitors [104–107]. Administration of the STAT3 decoy to multiple HNSCC cell lines with resistance to erlotinib, a Food and Drug Administration (FDA)-approved EGFR inhibitor, downmodulated STAT3 target gene expression and decreased cell proliferation compared to treatment with mutant control decoy (EC_{50} values ranged from 5.2 to 11.1 nM) [73]. Similar *in vitro* results were observed in bladder cancers resistant to cetuximab, an FDA-approved EGFR monoclonal antibody [73]. Studies of cetuximab-sensitive and cetuximab-resistant HNSCC xenograft tumors illustrated that STAT3 decoy augmented the antitumor effects of cetuximab in both the sensitive and resistant models compared to treatment with cetuximab plus mutant control decoy [73]. Since treatment with cetuximab and radiation has been shown to modestly improve survival in HNSCC, relative to radiation alone, the addition of STAT3 decoy to this regimen may yield a more efficacious approach [108].

Synergistic activity of STAT3 decoy in combination with bortezomib has been observed in HNSCC cell lines [71]. These data are particularly encouraging in light of clinical evidence that show sustained decreases in tumor burden and disease stabilization among patients with recurrent HNSCC after treatment with bortezomib and radiation, suggesting that further optimization of bortezomib/radiation therapy has the potential to improve patient outcomes [109,110].

Treatment with STAT3 decoy has also demonstrated success in sensitizing tumors to chemotherapy. Combination of STAT3 decoy and the chemotherapy drug cisplatin resulted in a

2.6-fold increase in apoptosis when compared to STAT3 decoy monotherapy in HNSCC cell lines [77,111]. *In vivo* studies using an HNSCC xenograft model confirmed that STAT3 decoy exhibited an additive effects with cisplatin, as evidenced by decreased tumor cell expression of VEGF, Bcl-X_L, cyclin D1, and PCNA in comparison to treatment with the decoy alone [77]. In ovarian cancer cells (SKOV3 and OVCAR3), treatment with STAT3 decoy was able to render the cells more sensitive to paclitaxel [112]. Collectively, these preclinical data support further investigation of STAT3 decoy in combination with conventional chemotherapy drugs in multiple tumor models.

Toxicology studies with the STAT3 decoy have revealed no evidence of systemic or local toxicities [75]. Based on these favorable results, a phase 0 clinical trial was conducted to assess the safety profile and pharmacodynamic effects of STAT3 decoy in patients with resectable HNSCC [74]. STAT3 decoy and normal saline (control) were injected directly into the tumors at doses ranging from 250 μg to 1 mg per injection. No toxicities were reported in either study arm and the maximum tolerated dose remains unknown. Downmodulation of cyclin D1 and Bcl-X_L was observed in tumor biopsies treated with STAT3 decoy compared to those treated with saline injections, demonstrating pharmacodynamic activity of the decoy in the target tissue [74].

Additional studies have assessed the antitumor effects of STAT3 decoy after intravenous administration. In an HNSCC xenograft model, systemic administration of STAT3 decoy failed to abrogate tumor growth, presumably due to nucleolytic degradation and/or thermal denaturation of the oligonucleotide free ends [74]. To overcome this barrier, Sen et al. [74] generated a cyclic version of the decoy with hexaethylene glycol linkages replacing the free ends (Fig. 7.3B). The cyclic STAT3 decoy demonstrated markedly enhanced thermal stability and resistance to nucleases. Importantly, intravenous administration of the cyclic STAT3 decoy inhibited the growth of HNSCC xenograft tumors, relative to treatment with cyclic mutant STAT3 decoy [74]. Subsequent toxicity studies in mice demonstrated that systemically administered cyclic STAT3 decoy did not lead to organ damage, significant changes in weight, or abnormal serum chemistries over a 2-week study period (no-observable-adverse-effect-level was 100 mg/kg) [5]. Thus, the cyclic STAT3 decoy exhibits antitumor activity following systemic administration has a favorable safety profile and warrants further evaluation in clinical trials. Overall, STAT3 DBDIs demonstrate promising potential for translation to clinical settings. In particular, these inhibitors of STAT3 may be well suited for enhancing the antitumor effects of conventional chemotherapy drugs, radiation, and molecular targeting agents. Cyclic STAT3 decoy exhibits notable potential due to a high degree of specificity for STAT3, lack of systemic toxicities, capacity for systemic delivery, and demonstrated efficacy in cancers resistant to multimodal therapy.

7.3.3 N-Terminal Domain Inhibitors

The N-terminal domain (NTD) of STAT3 provides another target site for inhibition of STAT3 activity. NMR and X-ray crystallography data suggest that the STAT3 NTD is comprised of 130 amino acids and 8 alpha helices and reflects a well-conserved structural region across STAT family members [87,113]. The STAT3 NTD facilitates tetramerization of activated STAT3 dimers; this tetramerization enables interactions with proteins that stabilize enhanceosomes [114,115]. Presumably, these interactions increase the range of STAT3-mediated gene expression to include suppressors of pro-apoptotic genes, although further studies need to be performed to define the complete set of target genes regulated by the STAT3 NTD [88,113]. Other functions of the NTD include a role in chromatin remodeling and transnuclear shuttling of both phosphorylated and unphosphorylated STAT3, suggesting additional ancillary roles to gene regulation [88,116,117].

In 2007, Timofeeva et al. [87] designed and evaluated a library of short peptide NTD inhibitors. The library was designed based on the second and eighth helices of the STAT3 NTD. Diffusion of the peptides across the cell membrane was augmented by covalently attaching Penetratin, a peptide transduction motif, to the C-termini [87]. *In vitro* studies have shown that ST3-H2A2, an optimized NTD inhibitor, exhibits IC_{50} values in the low micromolar range against breast cancer cells harboring constitutively active STAT3 (MDA-MB-231, MDA-MB-435, T47D, AND MCF-7) [87]. Similar effects of ST3-H2A2 on growth suppression and induction of apoptosis were observed in prostate cancer cells with high basal levels of phosphorylated STAT3 (LNCaP, PC3, and DU145) [88]. Although the precise mechanism of action is not completely understood, data from ChIP assays and human promoter arrays suggest that ST3-H2A2 decreases the number of genes bound by STAT3, possibly disrupting expression of pro-apoptotic suppressors [88,113]. Importantly, the cytotoxic effects of ST3-H2A2 were not observed in normal mouse embryonic fibroblasts, suggesting the possibility of acceptable treatment toxicities [87].

Although preclinical data suggest therapeutic potential for NTD inhibitors, it is likely that the same issues of stability, specificity, and intracellular transportation that have challenged other peptide inhibitors of STAT3 will be encountered. In particular, cell membrane penetration will pose an obstacle for *in vivo* studies of ST3-H2A2 since the Penetratin motif exhibits low serum stability, limiting drug delivery to its site of action [87]. Moreover, peptide inhibitors are likely to generate an immune response, necessitating the need to develop peptidomimetic anologs. Nonetheless, NTD inhibitors provide a novel alternative approach for abrogating STAT3-mediated cancer proliferation. Considering the nascent history of NTD inhibitors, future areas of research may include characterizing the mechanism of NTD inhibitor action, defining the full complement of response genes modulated by the STAT3 NTD and NTD inhibitors and development of additional NTD inhibitor molecules.

7.3.4 STAT3 Antisense

Antisense oligonucleotides designed to bind and promote the destruction of STAT3 mRNA molecules provide another option for abrogating STAT3-mediated cell proliferation and tumor growth. Studies have been performed using a STAT3 antisense oligonucleotide comprised of the sequence 5′-GCTCCAGCATCTGCTGCTTC-3′, capped with 2′-*O*-methylethyl groups to prevent degradation of free ends (Table 7.1) [80,82]. Treatment with STAT3 antisense reduced the expression of STAT3 response genes in human melanoma, breast carcinoma, and prostate cancer cell lines containing elevated levels of phosphorylated STAT3 [80,118,119]. Treatment with antisense STAT3 led to significant reductions in angiogenesis and tumor size in mice harboring prostate and hepatocellular carcinoma xenograft tumors, including androgen-resistant prostate cancer models and models of hepatocellular lung metastases [80,81]. Considering the high recurrence rate of treatment-resistant prostate cancers and the aggressive nature of hepatocellular carcinomas, these results raise the prospect of integrating STAT3 antisense oligonucleotides into current treatment regimens for these malignancies [120–122].

Although preclinical studies suggest the potential for clinical application of antisense oligonucleotides, AZD9150 (ISIS 481464) is the only STAT3 antisense that has been evaluated in human clinical trials (Fig. 7.3C). This second-generation STAT3 antisense oligonucleotide was optimized by replacing the 2′-*O*-methylethyl groups of previous iterations with 2′-4′ constrained ethyl-modified residues to increase stability [84]. *In vitro* experiments performed in lymphoma (KARPAS299 and SUP-M2) and neuroblastoma (IMR 32) cell lines showed that AZD9150 treatment led to reductions in total STAT3 protein and downstream targets [84,86]. Moreover, systemic administration of AZD9150 promoted antitumor effects against xenograft models of lymphoma, lung carcinoma, and neuroblastoma [84,86]. Interestingly, pretreatment with AZD9150 sensitized neuroblastoma tumors to cisplatin [86]. Toxicity studies with systemically administered AZD9150 have revealed a lack of end-organ damage and adverse toxicities [83]. These promising preclinical data provided the impetus for a phase I clinical trial (NCT01563302) in patients with advanced lymphoma and treatment-refractory solid tumors [85]. Intravenous administration of AZD9150 was well tolerated at doses of 2 and 4 mg/kg; notable toxicities included a 70% reduction in platelet counts from baseline and one instance of treatment-related thrombotic microangiopathy at 4 mg/kg [85]. One patient in the 15-patient cohort experienced more than a 55% reduction in tumor size and another patient achieved sufficient partial response to undergo successful autologous stem-cell transplantation [85]. The favorable safety profile and promising initial results of AZD9150 offer compelling evidence for further optimization and clinical trials.

7.4 CONCLUSIONS

In the last 30 years, improved understanding of the role of STAT3 in cancer biology has led to great strides in the discovery, optimization, and evaluation of anti-STAT3 agents. Collectively, preclinical experiments have progressed well beyond proof-of-concept studies and should encourage optimism for the eventual realization of novel therapies that directly inhibit STAT3, although adequate specificity, potency, stability, and drug delivery persist as challenging issues. Despite these advances, patient response to treatment will ultimately determine the clinical utility of STAT3 inhibitors, and given the paucity of randomized controlled trials that have been conducted, research efforts should continue to pursue clinical validation. Possible future directions after clinical validation include identifying tumor biomarkers that stratify patients according to STAT3 inhibitor sensitivity and investigating other disease processes that depend upon aberrant expression of targetable transcription factors. Overall, therapeutic inhibition of STAT3 has the potential to be another crucial turning point in the management of cancer, and the rapidly expanding base of knowledge surrounding STAT3 suggests that this promise may reach fruition within the upcoming decade.

Acknowledgments

This work was supported by the National Institutes of Health grants R01 DE24728 (D.E.J.), P50CA097190 (D.E.J. and J.R.G.), and R01 DE023685 (J.R.G.), and American Cancer Society grant CRP-13-308-06-COUN (J.R.G.).

Conflict of Interest: No potential conflicts of interest were disclosed.

References

[1] Yu H, Jove R. The STATs of cancer—new molecular targets come of age. Nat Rev Cancer 2004;4:97–105.

[2] Yu H, Pardoll D, Jove R. STATs in cancer inflammation and immunity: a leading role for STAT3. Nat Rev Cancer 2009;9:798–809. Available: http://www.ncbi.nlm.nih.gov/pubmed/19851315.

[3] Darnell JE. STATs and gene regulation. Science 1997;277:1630–5.

[4] Darnell JE, Kerr IM, Stark GR. Jak-STAT pathways and transcriptional activation in response to IFNs and other extracellular signaling proteins. Science 1994;264:1415–21.

[5] Sen M, Paul K, Freilino ML, Li H, Li C, Johnson DE, et al. Systemic administration of a cyclic signal transducer and activator of transcription 3 (STAT3) decoy oligonucleotide inhibits tumor growth without inducing toxicological effects. Mol Med 2014;20:46–56. Available: http://www.pubmedcentral.nih.gov/articlerender.fcgi?artid=3951464&tool=pmcentrez&rendertype=abstract.

[6] Hirano T, Ishihara K, Hibi M. Roles of STAT3 in mediating the cell growth, differentiation and survival signals relayed through the IL-6 family of cytokine receptors. Oncogene 2000;19:2548–56. Available: http://www.nature.com/onc/journal/v19/n21/full/1203551a.html.

[7] Silvennoinen O, Schindler C, Schlessinger J, Levy DE. Ras-independent growth factor signaling by transcription factor tyrosine phosphorylation. Science 1993;261:1736–9. Available: http://www.ncbi.nlm.nih.gov/pubmed/8378775.

[8] Stark GR, Kerr IM, Williams BR, Silverman RH, Schreiber RD. How cells respond to interferons. Annu Rev Biochem 1998;67:227–64. Available: http://www.ncbi.nlm.nih.gov/pubmed/9759489.

[9] Zhong Z, Wen Z, Darnell JE. Stat3: a STAT family member activated by tyrosine phosphorylation in response to epidermal growth factor and interleukin-6. Science 1994;264:95–8. Available: http://www.ncbi.nlm.nih.gov/pubmed/8140422.

[10] Bowman T, Garcia R, Turkson J, Jove R. STATs in oncogenesis. Oncogene 2000;19:2474–88.

[11] Danial NN, Rothman P. JAK-STAT signaling activated by Abl oncogenes. Oncogene 2000;19:2523–31.

[12] Irby RB, Yeatman TJ. Role of Src expression and activation in human cancer. Oncogene 2000;19:5636–42.

[13] Arulanandam R, Vultur A, Cao J, Carefoot E, Elliott BE, Truesdell PF, et al. Cadherin-cadherin engagement promotes cell survival via Rac1/Cdc42 and signal transducer and activator of transcription-3. Mol Cancer Res 2009;7:1310–27.

[14] Stairs DB, Bayne LJ, Rhoades B, Vega ME, Waldron TJ, Kalabis J, et al. Deletion of p120-catenin results in a tumor microenvironment with inflammation and cancer that establishes it as a tumor suppressor gene. Cancer Cell 2011;19:470–83.

[15] Vultur A, Villanueva J, Krepler C, Rajan G, Chen Q, Xiao M, et al. MEK inhibition affects STAT3 signaling and invasion in human melanoma cell lines. Oncogene 2014;33:1850–61. Available: http://www.pubmedcentral.nih.gov/articlerender.fcgi?artid=3769503&tool=pmcentrez&rendertype=abstract.

[16] Darnell JE. Validating Stat3 in cancer therapy. Nat Med 2005;11:595–6.

[17] Germain D, Frank D a. Targeting the cytoplasmic and nuclear functions of signal transducers and activators of transcription 3 for cancer therapy. Clin Cancer Res 2007;13:5665–9. Available: http://www.ncbi.nlm.nih.gov/pubmed/17908954.

[18] Chen CL, Cen L, Kohout J, Hutzen B, Chan C, Hsieh FC, et al. Signal transducer and activator of transcription 3 activation is associated with bladder cancer cell growth and survival. Mol Cancer 2008;7:78.

[19] Takemoto S, Ushijima K, Kawano K, Yamaguchi T, Terada A, Fujiyoshi N, et al. Expression of activated signal transducer and activator of transcription-3 predicts poor prognosis in cervical squamous-cell carcinoma. Br J Cancer 2009;101:967–72.

[20] Wen W, Wu J, Liu L, Tian Y, Buettner R, Hsieh M-Y, et al. Synergistic anti-tumor effect of combined inhibition of EGFR and JAK/STAT3 pathways in human ovarian cancer. Mol Cancer 2015;14:100. Available: http://www.pubmedcentral.nih.gov/articlerender.fcgi?artid=4437681&tool=pmcentrez&rendertype=abstract.

[21] Alas S, Bonavida B. Inhibition of constitutive STAT3 activity sensitizes resistant non-Hodgkin's lymphoma and multiple myeloma to chemotherapeutic drug-mediated apoptosis. Clin Cancer Res 2003;9:316–26.

[22] Rivat C, De Wever O, Bruyneel E, Mareel M, Gespach C, Attoub S. Disruption of STAT3 signaling leads to tumor cell invasion through alterations of homotypic cell-cell adhesion complexes. Oncogene 2004;23:3317–27. Available: http://www.ncbi.nlm.nih.gov/pubmed/15077188.

[23] Zhang X, Sun Y, Pireddu R, Yang H, Urlam MK, Lawrence HR, et al. A novel inhibitor of STAT3 homodimerization selectively suppresses STAT3 activity and malignant transformation. Cancer Res 2013;73:1922–33.

[24] Schlessinger K, Levy DE. Malignant transformation but not normal cell growth depends on signal transducer and activator of transcription 3. Cancer Res 2005;65:5828–34.
[25] Fukuda A, Wang SC, Morris JP, Folias AE, Liou A, Kim GE, et al. Stat3 and MMP7 contribute to pancreatic ductal adenocarcinoma initiation and progression. Cancer Cell 2011;19:441–55.
[26] Spitzner M, Ebner R, Wolff HA, Michael Ghadimi B, Wienands J, Grade M. STAT3: a novel molecular mediator of resistance to chemoradiotherapy. Cancers (Basel) 2014;6:1986–2011.
[27] Yue P, Turkson J. Targeting STAT3 in cancer: how successful are we? Expert Opin Invest Drugs 2009;18:45–56. Available: http://www.pubmedcentral.nih.gov/articlerender.fcgi?artid=2610472&tool=pmcentrez&rendertype=abstract.
[28] Zhou H, Miki R, Eeva M, Fike FM, Seligson D, Yang L, et al. Reciprocal regulation of SOCS 1 and SOCS3 enhances resistance to ionizing radiation in glioblastoma multiforme. Clin Cancer Res 2007;13:2344–53. Available: http://www.ncbi.nlm.nih.gov/pubmed/17438093.
[29] Spitzner M, Emons G, Kramer F, Gaedcke J, Rave-Fränk M, Scharf JG, et al. A gene expression signature for chemoradiosensitivity of colorectal cancer cells. Int J Radiat Oncol Biol Phys 2010;78:1184–92.
[30] Spitzner M, Roesler B, Bielfeld C, Emons G, Gaedcke J, Wolff HA, et al. STAT3 inhibition sensitizes colorectal cancer to chemoradiotherapy *in vitro* and *in vivo*. Int J Cancer 2014;134:997–1007.
[31] Chen M-F, Chen P-T, Lu MS, Lin PY, Chen W-C, Lee K-D. IL-6 expression predicts treatment response and outcome in squamous cell carcinoma of the esophagus. Mol Cancer 2013;12:26. Available: http://www.ncbi.nlm.nih.gov/pubmed/23561329%0Ahttp://www.pubmedcentral.nih.gov/articlerender.fcgi?artid=PMC3667147.
[32] Kim KW, Mutter RW, Cao C, Albert JM, Shinohara ET, Sekhar KR, et al. Inhibition of signal transducer and activator of transcription 3 activity results in down-regulation of Survivin following irradiation. Mol Cancer Ther 2006;5:2659–65.
[33] Wu C-T, Chen M-F, Chen W-C, Hsieh C-C. The role of IL-6 in the radiation response of prostate cancer. Radiat Oncol 2013;8:159. Available: http://ro-journal.biomedcentral.com/articles/10.1186/1748-717X-8-159.
[34] Kim SM, Kwon O-J, Hong YK, Kim JH, Solca F, Ha S-J, et al. Activation of IL-6R/JAK1/STAT3 signaling induces de novo resistance to irreversible EGFR inhibitors in non-small cell lung cancer with T790M resistance mutation. Mol Cancer Ther 2012;11:2254–64. Available: http://www.ncbi.nlm.nih.gov/pubmed/22891040.
[35] Prahallad A, Sun C, Huang S, Di Nicolantonio F, Salazar R, Zecchin D, et al. Unresponsiveness of colon cancer to BRAF(V600E) inhibition through feedback activation of EGFR. Nature 2012;483:100–3. Available: http://www.nature.com/doifinder/10.1038/nature10868%5Cnhttp://www.ncbi.nlm.nih.gov/pubmed/22281684.
[36] Lee HJ, Zhuang G, Cao Y, Du P, Kim HJ, Settleman J. Drug resistance via feedback activation of stat3 in oncogene-addicted cancer cells. Cancer Cell 2014;26:207–21.
[37] Kong J, Kong F, Gao J, Zhang Q, Dong S, Gu F, et al. YC-1 enhances the anti-tumor activity of sorafenib through inhibition of signal transducer and activator of transcription 3 (STAT3) in hepatocellular carcinoma. Mol Cancer 2014;13:7. Available: http://www.pubmedcentral.nih.gov/articlerender.fcgi?artid=3895679&tool=pmcentrez&rendertype=abstract.
[38] Yin ZJ, Jin FG, Liu TG, Fu EQ, Xie YH, Sun RL. Overexpression of STAT3 potentiates growth, survival, and radioresistance of non-small-cell lung cancer (NSCLC) cells. J Surg Res 2011;171:675–83.
[39] You S, Li R, Park D, Xie M, Sica GL, Cao Y, et al. Disruption of STAT3 by niclosamide reverses radioresistance of human lung cancer. Mol Cancer Ther 2014;13:606–16.
[40] Pawson T. Protein modules and signalling networks. Nature 1995;373:573–80.
[41] Turkson J, Ryan D, Kim JS, Zhang Y, Chen Z, Haura E, et al. Phosphotyrosyl peptides block Stat3-mediated DNA binding activity, gene regulation, and cell transformation. J Biol Chem 2001;276:45443–55. Available: http://www.ncbi.nlm.nih.gov/pubmed/11579100.
[42] Makriyannis A, Biegel D. Drug discovery strategies methods. New York: Marcel Dekker; 2004.
[43] Li H, Liu A, Zhao Z, Xu Y, Lin J, Jou D, et al. Fragment-based drug design and drug repositioning using multiple ligang simultaneous docking (MLSD): identifying celecoxib and template compounds as novel inhibitors of signal transducer and activator of transcription 3 (STAT3). J Med Chem 2011;54:5592–6.
[44] Siddiquee K, Zhang S, Guida WC, Blaskovich MA, Greedy B, Lawrence HR, et al. Selective chemical probe inhibitor of Stat3, identified through structure-based virtual screening, induces antitumor activity. Proc Natl Acad Sci USA, 104. p. 7391–6.
[45] Turkson J, Kim JS, Zhang S, Yuan J, Huang M, Glenn M, et al. Novel peptidomimetic inhibitors of signal transducer and activator of transcription 3 dimerization and biological activity. Mol Cancer Ther 2004;3:261–9.

[46] Auzenne EJ, Klostergaard J, Mandal PK, Liao WS, Lu Z, Gao F, et al. A phosphopeptide mimetic prodrug targeting the SH2 domain of Stat3 inhibits tumor growth and angiogenesis. J Exp Ther Oncol 2012;10:155–62. Available: http://www.ncbi.nlm.nih.gov/pubmed/23350355.
[47] Mandal PK, Gao F, Lu Z, Ren Z, Ramesh R, Birtwistle JS, et al. Potent and selective phosphopeptide mimetic prodrugs targeted to the Src homology 2 (SH2) domain of signal transducer and activator of transcription 3. J Med Chem 2011;54:3549–63.
[48] Chen CL, Loy A, Cen L, Chan C, Hsieh FC, Cheng G, et al. Signal transducer and activator of transcription 3 is involved in cell growth and survival of human rhabdomyosarcoma and osteosarcoma cells. BMC Cancer 2007;7:111. Available: http://www.ncbi.nlm.nih.gov/entrez/query.fcgi?cmd=Retrieve&db=PubMed&dopt=Citation&list_uids=17598902.
[49] Song H, Wang R, Wang S, Lin J. A low-molecular-weight compound discovered through virtual database screening inhibits Stat3 function in breast cancer cells. Proc Natl Acad Sci USA, 102. p. 4700–5. Available: http://www.ncbi.nlm.nih.gov/pubmed/15781862%5Cnhttp://www.ncbi.nlm.nih.gov/pmc/articles/PMC555708/pdf/pnas-0409894102.pdf.
[50] Fuh B, Sobo M, Cen L, Josiah D, Hutzen B, Cisek K, et al. LLL-3 inhibits STAT3 activity, suppresses glioblastoma cell growth and prolongs survival in a mouse glioblastoma model. Br J Cancer 2009;100:106–12.
[51] Mencalha AL, Rocher BD, Salles D, Binato R, Abdelhay E. LLL-3, a STAT3 inhibitor, represses BCR-ABL-positive cell proliferation, activates apoptosis and improves the effects of Imatinib mesylate. Cancer Chemother Pharmacol 2010;65:1039–46.
[52] Pan Y, Zhou F, Zhang R, Claret FX. Stat3 inhibitor static exhibits potent antitumor activity and induces chemo- and radio-sensitivity in nasopharyngeal carcinoma. PLoS ONE 2013;8. e54565.
[53] Schust J, Berg T. A high-throughput fluorescence polarization assay for signal transducer and activator of transcription 3. Anal Biochem 2004;330:114–8. Available: http://www.ncbi.nlm.nih.gov/pubmed/15183768.
[54] Schust J, Sperl B, Hollis A, Mayer TU, Berg T. Stattic: a small-molecule inhibitor of STAT3 activation and dimerization. Chem Biol 2006;13:1235–42.
[55] Zhang X, Yue P, Page BDG, Li T, Zhao W, Namanja AT, et al. Orally bioavailable small-molecule inhibitor of transcription factor Stat3 regresses human breast and lung cancer xenografts. Proc Natl Acad Sci, 109. p. 9623–8.
[56] Zhang X, Yue P, Fletcher S, Zhao W, Gunning PT, Turkson J. A novel small-molecule disrupts Stat3 SH2 domain-phosphotyrosine interactions and Stat3-dependent tumor processes. Biochem Pharmacol 2010;79:1398–409.
[57] Ashizawa T, Miyata H, Ishii H, Oshita C, Matsuno K, Masuda Y, et al. Antitumor activity of a novel small molecule STAT3 inhibitor against a human lymphoma cell line with high STAT3 activation. Int J Oncol 2011;38:1245–52.
[58] Matsuno K, Masuda Y, Uehara Y, Sato H, Muroya A, Takahashi O, et al. Identification of a new series of STAT3 inhibitors by virtual screening. ACS Med Chem Lett 2010;1:371–5.
[59] Chen H, Yang Z, Ding C, Chu L, Zhang Y, Terry K, et al. Fragment-based drug design and identification of HJC0123, a novel orally bioavailable STAT3 inhibitor for cancer therapy. Eur J Med Chem 2013;62:498–507.
[60] Bendell JC, Hong DS, Burris HA, Naing A, Jones SF, Falchook G, et al. Phase 1, open-label, dose-escalation, and pharmacokinetic study of STAT3 inhibitor OPB-31121 in subjects with advanced solid tumors. Cancer Chemother Pharmacol 2014;74:125–30.
[61] Brambilla L, Genini D, Laurini E, Merulla J, Perez L, Fermeglia M, et al. Hitting the right spot: mechanism of action of OPB-31121, a novel and potent inhibitor of the signal transducer and activator of transcription 3 (STAT3). Mol Oncol 2015;9:1194–206.
[62] Genini D, Brambilla L, Laurini E, Civenni G, Pinton S, Sarti M, et al. CVC: novel inhibitors of signal transducer and activator of transcription 3 (STAT3) show potent activity in cell cultures and tumor xenografts. 105 Annu Meet Am Assoc Cancer Res 2014;.
[63] Hayakawa F, Sugimoto K, Harada Y, Hashimoto N, Ohi N, Kurahashi S, et al. A novel STAT inhibitor, OPB-31121, has a significant antitumor effect on leukemia with STAT-addictive oncokinases. Blood Cancer J 2013;3:e166. Available: http://www.mendeley.com/research/hallmarks-cancer-next-generation-3/.
[64] Kim MJ, Nam HJ, Kim HP, Han SW, Im SA, Kim TY, et al. OPB-31121, a novel small molecular inhibitor, disrupts the JAK2/STAT3 pathway and exhibits an antitumor activity in gastric cancer cells. Cancer Lett 2013;335:145–52.
[65] Ogura M, Uchida T, Terui Y, Hayakawa F, Kobayashi Y, Taniwaki M, et al. Phase I study of OPB-51602, an oral inhibitor of STAT3, in patients with relapsed/refractory hematological malignancies. Cancer Sci 2015;106:896–901. Available: http://www.ncbi.nlm.nih.gov/pubmed/25912076.

[66] Wong AL, Soo RA, Tan DS, Lee SC, Lim JS, Marban PC, et al. Phase I and biomarker study of OPB-51602, a novel signal transducer and activator of transcription (STAT) 3 inhibitor, in patients with refractory solid malignancies. Ann Oncol 2015;26:998 1005.
[67] Turkson J, Zhang S, Palmer J, Kay H, Stanko J, Mora LB, et al. Inhibition of constitutive signal transducer and activator of transcription 3 activation by novel platinum complexes with potent antitumor activity. Mol Cancer Ther 2004;3:1533–42. Available: http://www.ncbi.nlm.nih.gov/pubmed/15634646.
[68] Assi HH, Paran C, VanderVeen N, Savakus J, Doherty R, Petruzzella E, et al. Preclinical characterization of signal transducer and activator of transcription 3 small molecule inhibitors for primary and metastatic brain cancer therapy. J Pharmacol Exp Ther 2014;349:458–69. Available: http://www.ncbi.nlm.nih.gov/pubmed/24696041.
[69] Xu Q, Briggs J, Park S, Niu G, Kortylewski M, Zhang S, et al. Targeting Stat3 blocks both HIF-1 and VEGF expression induced by multiple oncogenic growth signaling pathways. Oncogene 2005;24:5552–60. Available: http://dx.doi.org/10.1038/sj.onc.1208719.
[70] Turkson J, Zhang S, Mora LB, Burns A, Sebti S, Jove R. A novel platinum compound inhibits constitutive Stat3 signaling and induces cell cycle arrest and apoptosis of malignant cells. J Biol Chem 2005;280:32979–88.
[71] Changyou L, Zang Y, Sen M, Leeman-Neill RJ, Man DS, Grandis JR, et al. Bortezomib up-regulates activated signal transducer and activator of transcription-3 and synergizes with inhibitors of signal transducer and activator of transcription-3 to promote head and neck squamous cell carcinoma cell death. Mol Cancer Ther 2009;8:2211–20.
[72] Leong PL, Andrews GA, Johnson DE, Dyer KF, Xi S, Mai JC, et al. Targeted inhibition of Stat3 with a decoy oligonucleotide abrogates head and neck cancer cell growth. Proc Natl Acad Sci USA, 100. p. 4138–43.
[73] Sen M, Joyce S, Panahandeh M, Li C, Thomas SM, Maxwell J, et al. Targeting Stat3 abrogates EGFR inhibitor resistance in cancer. Clin Cancer Res 2012;18:4986–96.
[74] Sen M, Thomas SM, Kim S, Yeh JI, Ferris RL, Johnson JT, et al. First-in-human trial of a STAT3 decoy oligonucleotide in head and neck tumors: Implications for cancer therapy. Cancer Discov 2012;2:694–705.
[75] Sen M, Tosca PJ, Zwayer C, Ryan MJ, Johnson JD, Knostman KAB, et al. Lack of toxicity of a STAT3 decoy oligonucleotide. Cancer Chemother Pharmacol 2009;63:983–95.
[76] Shen J, Li R, Li G. Inhibitory effects of decoy-ODN targeting activated STAT3 on human glioma growth *in vivo*. In Vivo (Brooklyn) 2009;23:237–44.
[77] Xi S, Gooding WE, Grandis JR. *In vivo* antitumor efficacy of STAT3 blockade using a transcription factor decoy approach: implications for cancer therapy. Oncogene 2005;24:970–9.
[78] Zhang X, Liu P, Zhang B, Mao H, Shen L, Ma Y. Inhibitory effects of STAT3 decoy oligodeoxynucleotides on human epithelial ovarian cancer cell growth *in vivo*. Int J Mol Med 2013;32:623–8.
[79] Zhang X, Zhang J, Wang L, Wei H, Tian Z. Therapeutic effects of STAT3 decoy oligodeoxynucleotide on human lung cancer in xenograft mice. BMC Cancer 2007;7:149. Available: http://www.pubmedcentral.nih.gov/articlerender.fcgi?artid=1988829&tool=pmcentrez&rendertype=abstract.
[80] Barton BE, Murphy TF, Shu P, Huang HF, Meyenhofer M, Barton A. Novel single-stranded oligonucleotides that inhibit signal transducer and activator of transcription 3 induce apoptosis in vitro and in vivo in prostate cancer cell lines. Mol Cancer Ther 2004;3:1183–91.
[81] Li W-C, Ye S-L, Sun R-X, Liu Y-K, Tang Z-Y, Kim Y, et al. Inhibition of growth and metastasis of human hepatocellular carcinoma by antisense oligonucleotide targeting signal transducer and activator of transcription 3. Clin Cancer Res 2006;12:7140–8.
[82] Mora LB, Buettner R, Seigne J, Diaz J, Ahmad N, Garcia R, et al. Constitutive activation of Stat3 in human prostate tumors and cell lines: Direct inhibition of Stat3 signaling induces apoptosis of prostate cancer cells. Cancer Res 2002;62:6659–66.
[83] Burel SA, Han S-R, Lee H-S, Norris DA, Lee B-S, Machemer T, et al. Preclinical evaluation of the toxicological effects of a novel constrained ethyl modified antisense compound targeting signal transducer and activator of transcription 3 in mice and cynomolgus monkeys. Nucleic Acid Ther 2013;23:213–27. Available: http://www.ncbi.nlm.nih.gov/pubmed/23692080.
[84] Hong D, Kurzrock R, Kim Y, Woessner R, Younes A, Nemunaitis J, et al. AZD9150, a next-generation antisense oligonucleotide inhibitor of STAT3 with early evidence of clinical activity in lymphoma and lung cancer. Sci Transl Med 2015;7:314ra185. Available: http://stm.sciencemag.org/cgi/doi/10.1126/scitranslmed.aac5272.
[85] Hong DS, et al. A phase I study of ISIS 481464 (AZD9150), a first-in-human, first in-class, antisense oligonucleotide inhibitor of STAT3, in patients with advanced cancers. ASCO 2013;. Abstract No. 8523. Available: http://abstracts2.asco.org/AbstView_132_112636.html.

[86] Odate S, Veschi V, Yan S, Lam N, Woessner R, Thiele C. Inhibition of STAT3 with the generation 2.5 antisense oligonucleotide, AZD9150, decreases neuroblastoma tumorigenicity and increases chemosensitivity. Clin Cancer Res 2017;23:1771–84.
[87] Timofeeva OA, Gaponenko V, Lockett SJ, Tarasov SG, Jiang S, Michejda CJ, et al. Rationally designed inhibitors identify STAT3 N-domain as a promising anticancer drug target. ACS Chem Biol 2007;2:799–809.
[88] Timofeeva OA, Tarasova NI, Zhang X, Chasovskikh S, Cheema AK, Wang H, et al. STAT3 suppresses transcription of proapoptotic genes in cancer cells with the involvement of its N-terminal domain. Proc Natl Acad Sci, 110. p. 1267–72. Available: http://www.pnas.org/lookup/doi/10.1073/pnas.1211805110.
[89] Vultur A, Cao J, Arulanandam R, Turkson J, Jove R, Greer P, et al. Cell-to-cell adhesion modulates Stat3 activity in normal and breast carcinoma cells. Oncogene 2004;23:2600–16. Available: http://www.ncbi.nlm.nih.gov/pubmed/15007380.
[90] Bhattacharya S, Ray RM, Johnson LR. STAT3-mediated transcription of Bcl-2, Mcl-1 and c-IAP2 prevents apoptosis in polyamine-depleted cells. Biochem J 2005;392:335–44.
[91] Wang T, Niu G, Kortylewski M, Burdelya L, Shain K, Zhang S, et al. Regulation of the innate and adaptive immune responses by Stat-3 signaling in tumor cells. Nat Med 2004;10:48–54.
[92] Hussain SF, Kong L-Y, Jordan J, Conrad C, Madden T, Fokt I, et al. A novel small molecule inhibitor of signal transducers and activators of transcription 3 reverses immune tolerance in malignant glioma patients. Cancer Res 2007;67:9630–6.
[93] Rosenberg S a, Yang JC, Restifo NP. Cancer immunotherapy: moving beyond current vaccines. Nat Med 2004;10:909–15.
[94] Lui VWY, Wong EYL, Ho Y, Hong B, Wong SCC, Tao Q, et al. STAT3 activation contributes directly to Epstein–Barr virus-mediated invasiveness of nasopharyngeal cancer cells *in vitro*. Int J Cancer 2009;125:1884–93.
[95] Siddik ZH. Cisplatin: mode of cytotoxic action and molecular basis of resistance. Oncogene 2003;22:7265–79. Available: http://www.ncbi.nlm.nih.gov/pubmed/14576837.
[96] Bielinska A, Shivdasani RA, Zhang LQ, Nabel GJ. Regulation of gene expression with double-stranded phosphorothioate oligonucleotides. Science 1990;250:997–1000. Available: http://www.ncbi.nlm.nih.gov/pubmed/2237444.
[97] Nabel EG, Plautz G, Nabel GJ. Site-specific gene expression *in vivo* by direct gene transfer into the arterial wall. Science 1990;249:1285–8.
[98] Wang LH, Yang XY, Kirken RA, Resau JH, Farrar WL. Targeted disruption of stat6 DNA binding activity by an oligonucleotide decoy blocks IL-4-driven T(H)2 cell response. Blood 2000;95:1249–57. Available: http://www.ncbi.nlm.nih.gov/pubmed/10666197.
[99] Catlett-Falcone R, Landowski TH, Oshiro MM, Turkson J, Levitzki A, Savino R, et al. Constitutive activation of Stat3 signaling confers resistance to apoptosis in human U266 myeloma cells. Immunity 1999;10:105–15. Available: http://www.ncbi.nlm.nih.gov/pubmed/10023775.
[100] Grandis JR, Drenning SD, Zeng Q, Watkins SC, Melhem MF, Endo S, et al. Constitutive activation of Stat3 signaling abrogates apoptosis in squamous cell carcinogenesis *in vivo*. Proc Natl Acad Sci USA 2000;97:4227–32.
[101] Nielsen M, Kstel C, Eriksen K, Woetmann A, Stokkedal T, Kaltoft K, et al. Inhibition of constitutively activated Stat3 correlates with altered Bcl-2/Bax expression and induction of apoptosis in mycosis fungoides tumor cells. Leukemia 1999;13:735–8. Available: http://www.stockton-press.co.uk/leu.
[102] Niu G, Heller R, Catlett-Falcone R, Coppola D, Jaroszeski M, Dalton W, et al. Gene therapy with dominant-negative Stat3 suppresses growth of the murine melanoma B16 tumor *in vivo*. Cancer Res 1999;59:5059–63.
[103] Sadowski HB, Shuai K, Darnell JE, Gilman MZ. A common nuclear signal transduction pathway activated by growth factor and cytokine receptors. Science 1993;261:1739–44. Available: http://www.ncbi.nlm.nih.gov/pubmed/8397445.
[104] Ciardiello F, Tortora G, Magrassi SF, Lanzara A. EGFR antagonists in cancer treatment. N Engl J Med 2008;358:1160–74.
[105] Harandi A, Zaidi AS, Stocker AM, Laber DA. Clinical efficacy and toxicity of anti-EGFR therapy in common cancers. J Oncol 2009;2009. 567486.
[106] Thatcher N, Chang A, Parikh P, Rodrigues Pereira J, Ciuleanu T, von Pawel J, et al. Gefitinib plus best supportive care in previously treated patients with refractory advanced non-small-cell lung cancer: results from a randomised, placebo-controlled, multicentre study (Iressa Survival Evaluation in Lung Cancer). Lancet 2005;366:1527–37. Available: http://www.sciencedirect.com/science?_ob=MImg&_imagekey=B6T1B-4HDY-FM3-R-7&_cdi=4886&_user=1029205&_pii=S0140673605676258&_origin=&_coverDate=11/04/2005&_sk=996330503&view=c&wchp=dGLbVlz-zSkWB&md5=d67f3c6c2d4256106eb755f18ef91a18&ie=/sdarticle.pdf.

[107] Yamatodani T, Ekblad L, Kjellen E, Johnsson A, Mineta H, Wennerberg J. Epidermal growth factor receptor status and persistent activation of Akt and p44/42 MAPK pathways correlate with the effect of cetuximab in head and neck and colon cancer cell lines. J Cancer Res Clin Oncol 2009;135:395–402.
[108] Bonner JA, Harari PM, Giralt J, Azarnia N, Shin DM, Cohen RB, et al. Radiotherapy plus cetuximab for squamous-cell carcinoma of the head and neck. N Engl J Med 2006;354:567–78. Available: http://www.ncbi.nlm.nih.gov/pubmed/19567214.
[109] Allen C, Saigal K, Nottingham L, Arun P, Chen Z, Van Waes C. Bortezomib-induced apoptosis with limited clinical response is accompanied by inhibition of canonical but not alternative nuclear factor-kB subunits in head and neck cancer. Clin Cancer Res 2008;14:4175–85.
[110] Van Waes C, Chang A a, Lebowitz PF, Druzgal CH, Chen Z, Elsayed YA, et al. Inhibition of nuclear factor-kappaB and target genes during combined therapy with proteasome inhibitor bortezomib and reirradiation in patients with recurrent head-and-neck squamous cell carcinoma. Int J Radiat Oncol Biol Phys 2005;63:1400–12.
[111] Mayer F, Honecker F, Looijenga LHJ, Bokemeyer C. Towards an understanding of the biological basis of response to cisplatin-based chemotherapy in germ-cell tumors. Ann Oncol 2003;14:825–32.
[112] Zhang X, Liu P, Zhang B, Wang A, Yang M. Role of STAT3 decoy oligodeoxynucleotides on cell invasion and chemosensitivity in human epithelial ovarian cancer cells. Cancer Genet Cytogenet 2010;197:46–53.
[113] Hu T, Yeh JE, Pinello L, Jacob J, Chakravarthy S, Yuan G-C, et al. Impact of the N-terminal domain of STAT3 in STAT3-dependent transcriptional activity. Mol Cell Biol 2015;35:3284–300. Available: http://mcb.asm.org/lookup/doi/10.1128/MCB.00060-15.
[114] Lerner L, Henriksen MA, Zhang X, Darnell JE. STAT3-dependent enhanceosome assembly and disassembly: synergy with GR for full transcriptional increase of the a2-macroglobulin gene. Genes Dev 2003;17:2564–77.
[115] Zhang X, Darnell JE. Functional importance of Stat3 tetramerization in activation of the a2-macroglobulin gene. J Biol Chem 2001;276:33576–81.
[116] Pranada AL, Metz S, Herrmann A, Heinrich PC, Mu G, Müller-Newen G. Real time analysis of STAT3 nucleocytoplasmic shuttling. J Biol Chem 2004;279:15114–23. Available: http://www.ncbi.nlm.nih.gov/pubmed/14701810.
[117] Vogt M, Domoszlai T, Kleshchanok D, Lehmann S, Schmitt A, Poli V, et al. The role of the N-terminal domain in dimerization and nucleocytoplasmic shuttling of latent STAT3. J Cell Sci 2011;124:900–9. Available: http://jcs.biologists.org/cgi/doi/10.1242/jcs.072520.
[118] Barton BE, Karras JG, Murphy TF, Barton A, Huang HF-S. Signal transducer and activator of transcription 3 (STAT3) activation in prostate cancer: Direct STAT3 inhibition induces apoptosis in prostate cancer lines. Mol Cancer Ther 2004;3:11–20.
[119] Niu G, Wright KL, Huang M, Song L, Haura E, Turkson J, et al. Constitutive Stat3 activity up-regulates VEGF expression and tumor angiogenesis. Oncogene 2002;21:2000–8. Available: http://www.nature.com/doifinder/10.1038/sj.onc.1205260.
[120] Brawer MK. Prostate cancer: where we have been, where we are, where we are going. World J Surg 2000;24:1200–4.
[121] Denmeade SR, Lin XS, Isaacs JT. Role of programmed (apoptotic) cell death during the progression and therapy for prostate cancer. Prostate 1996;28:251–65.
[122] Tang ZY, Ye SL, Liu YK, Qin LX, Sun HC, Ye QH, et al. A decade's studies on metastasis of hepatocellular carcinoma. J Cancer Res Clin Oncol 2004;130:187–96.

CHAPTER

8

Targeting the Hippo Pathway to Improve Response to Chemotherapy

Yulei Zhao, Xiaolong Yang
Queen's University, Kingston, ON, Canada

Abstract

The Hippo pathway plays an essential role in tumorigenesis, stem-cell self-renewal and differentiation, organ size control, and many other biological processes. Currently, increasing studies also suggest that the components of the Hippo pathway are involved in the sensitivities of different cancer types to various chemotherapies. As a major approach for cancer treatments, chemotherapeutic therapies can sometimes effectively suppress tumor growth in cancer patients. However, a significant proportion of patients are either intrinsically resistant or later develop acquired resistance to primary chemotherapy, leading to disease relapse and patient mortality. The best way to conquer this resistance is through a better understanding of the molecular networks that are activated in cancer cells in response to drugs. Therefore, identification of signaling pathways and molecules involved in drug resistance is essential for successful treatment of cancers. Here, we will discuss the specific roles of the Hippo pathway in chemotherapy, potential applications for studying this network in response to drugs, and the future strategies to increase chemotherapy efficiency through targeting the Hippo pathway.

ABBREVIATIONS

AC	Adenylate cyclase
AMPK	5' AMP-activated protein kinase
BC	Breast cancer
CDK	Cyclin-dependent kinase
CSC	Cancer stem cell
ER	Estrogen receptor
Ex	Expanded
5-FU	5-Fluorouracil
GPCR	G protein-coupled receptor
HCC	Hepatocellular carcinoma
Hpo	Hippo
LIFR	Leukemia inhibitory factor receptor
LKB1	Liver kinase B1
Lats	Large tumor suppressor
Mats	Mob as tumor suppressor

Targeting Cell Survival Pathways to Enhance Response to Chemotherapy. http://dx.doi.org/10.1016/B978-0-12-813753-6.00008-1
Copyright © 2018 Elsevier Inc. All rights reserved.

MDR	Multi-drug resistance
MST1/2	Mammalian Ste-20 like kinase ½
NF2	Neurofibromatosis 2
NSCLC	Non-small cell lung cancer
PKA	Protein kinase A
PKB	Protein kinase B
PKC	Protein kinase C
RASSF1A	Ras-associated domain family member 1A
SMDs	Small molecular drugs
TEAD	Transcriptional enhancer associated domain
UPS	Ubiquitin proteasome system
VP	Verteporfin
YKI	Yorkie

8.1 INTRODUCTION

8.1.1 The Hippo Signaling Pathway

8.1.1.1 Canonical Core Cascade of the Hippo Pathway

The Hippo pathway, which was first identified in *Drosophila* and later in mammals [1], is a recently discovered signaling pathway that plays essential roles in both physiological and pathological processes, such as organ size control, tissue regeneration, stem-cell renewal and differentiation, and tumorigenesis [1–9]. The core components of the Hippo pathway were first identified in *Drosophila* (Fig. 8.1) with the development of a new technique called the mosaic genetic screen [10]. The first player of the Hippo pathway detected through this screening was the tumor suppressor and serine/threonine (S/T) kinase named *lats* (large tumor suppressor) or *wts* (Warts) [11,12]. Loss of *lats/wts* causes cell overproliferation and tissue overgrowth [11,12]. Using a similar genetic screen, the S/T kinase *Hpo* (Hippo) was identified with similar functions and as an upstream regulator of *lats/wts* [13–15]. At the same time, scaffold protein Sav (Salvador) was found to promote the functions of Hpo [13] and 2 years later, Mats (Mob as tumor suppressor) was identified as another scaffold protein interacting and potentiating the functions of lats/Wts [16,17]. All these proteins were found to work together as a size-control phosphorylation cascade to regulate the downstream effector Yki (Yorkie), which is a transcriptional co-activator [18,19] (Fig. 8.1).

With increasing studies later on, the Hippo pathway was identified and found to be well conserved in mammals. In the mammalian Hippo pathway (Fig. 8.1), the core components are as follows: kinases MST1/2 (Mammalian Ste-20-like kinase 1/2; homolog of *Drosophila* Hpo; MST will be used instead of MST1/2 in the following text) and LATS1/2 (homolog of *Drosophila* lats/Warts; LATS will be used for LATS1/2 in the following text), scaffold proteins WW45/Sav (homolog of Sav in *Drosophila*) and Mob1 (homolog of *Drosophila* Mats), as well as the transcriptional co-activators YAP (Yes-associated protein; homolog of *Drosophila* Yki) and its paralog TAZ (transcriptional co-activator with PDZ binding motif) [6,20]. Similar to *Drosophila* Hippo signaling, the mammalian Hippo pathway functions through phosphorylation. When the Hippo pathway is activated in certain conditions such as increased cell–cell contact caused by high cell density, the upstream core component MST first gets phosphorylated and activated, which further phosphorylates and activates the downstream kinase

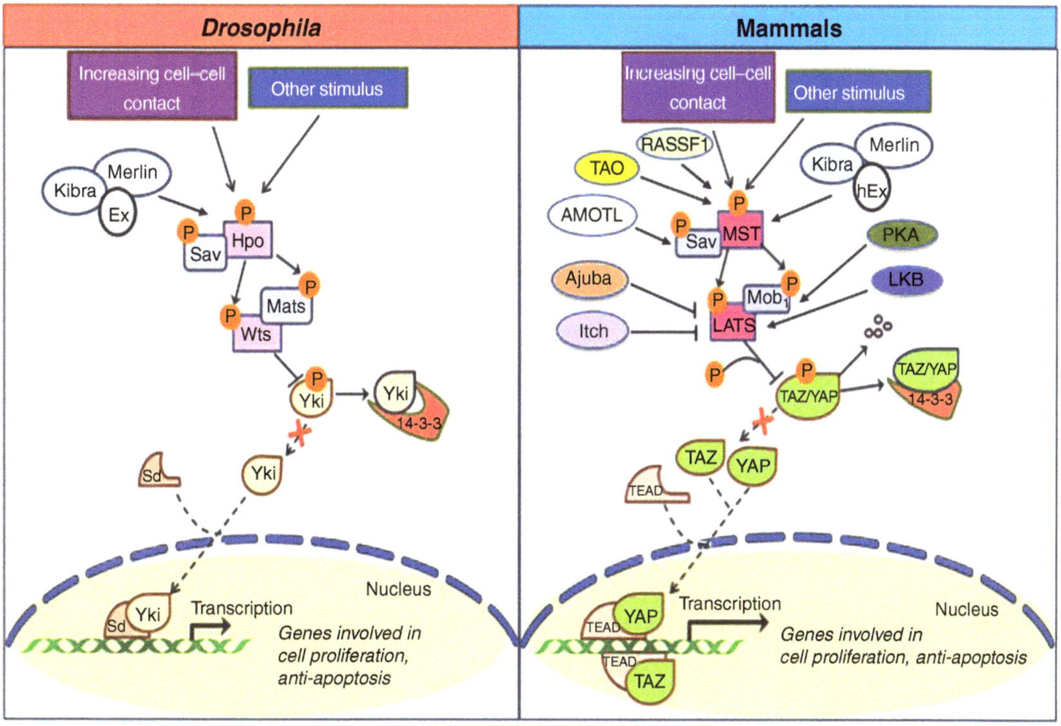

FIG. 8.1 **Hippo pathway in *Drosophila* and mammals.** Abbreviations: Ex, expanded; Sav, Salvador; Hpo, Hippo; Wts, Warts; Yki, Yorki; MST, mammalian Ste-20 like kinase; LATS, large tumor suppressor; hEx, human expanded; RASSF1, Ras-association domain family member 1; PKA, protein kinase A; LKB, liver kinase B.

LATS [1,6]. The phosphorylated LATS can inhibit the transactivating functions of YAP and TAZ through phosphorylation of specific S sites in the motif of "HxH/R/KxxS/T" (H, histidine; R, arginine; K, lysine) on YAP and TAZ [21–23]. As a result, TAZ and YAP are anchored in the cytoplasm [21–23] and/or degraded [24,25], which interrupt their interactions with the transcription factor TEAD (transcriptional enhancer associate domain) family to block their roles in assisting the transcription of downstream target genes involved in cell-cycle progression, cell proliferation, and anti-apoptosis (Fig. 8.1) [21,23–27]. Therefore, dysregulation of the Hippo pathway is involved in tumorigenesis and metastasis [28–31], which mainly depend on cell proliferation and cell survival.

8.1.1.2 Upstream Regulators of the Hippo Pathway

With increasing studies on the Hippo pathway, many regulators of the Hippo pathway have been identified in addition to the core components. Studies on *Drosophila* determined that a protein complex consisting of the FERM domain proteins Merlin (NF2, neurofibromatosis 2) and Ex (Expanded) as well as Kibra (kidney and brain expressed protein) activate Hippo/MST [32–34]. A conserved association of Merlin and Kibra exists in human cells to trigger activation of the Hippo pathway via activation of MST [34] (Fig. 8.1). However, unlike *Ex* in *Drosophila*, *hEx* (*human Expanded*), which is also named *FRDM6* and *Willin*, functions as a tumor suppressor gene (TSG) in both Hippo pathway-dependent and -independent mechanisms [35,36]. In addition to forming a complex with Kibra, Merlin is also shown to recruit MST and LATS to the plasma membrane to activate the pathway [37]. In addition, AMOTL2, which is a member of AMOT (angiomotin) family, functions as an adapter protein to interact with MST, LATS2, and YAP to promote the activation of LATS2 by MST [38]. Besides, the apicobasal polarity of epithelial cells also plays a role in the activation of the Hippo pathway through basolateral membrane-located protein Scribble, which is required for recruiting MST to the LATS-TAZ/YAP complex [39]. Furthermore, leukemia-inhibitory factor receptor inhibits YAP-promoted metastasis through Scribble involved Hippo pathway activation [40]. In addition, as a tumor suppressor, RASSF1A (Ras-association domain family member 1A) interacts with MST1 to induce apoptosis [41]. Further studies indicate that this protein can protect activated MST from dephosphorylation/inactivation to trigger the Hippo pathway [42,43]. Furthermore, the STE-20 family kinase TAO1, PKA (protein kinase A), and LKB1 (liver kinase B1) can activate the Hippo pathway through phosphorylation of MST and/or LATS (Fig. 8.1) [44–48].

In addition to the positive regulators, kinase Akt [also known as PKB (protein kinase B)] has been shown to inhibit the Hippo pathway through phosphorylation of MST [49,50]. In addition, the proteins of Ajuba Lim family can interact with and inhibit LATS activity to block the Hippo pathway [51–53]. Moreover, LATS can be specifically targeted by Itch ubiquitin ligase to be degraded through ubiquitin proteasome system [54], which also suppresses the activity of the Hippo pathway.

Besides, as a cell-surface glycoprotein and a marker for cancer stem cells (CSCs), CD44 is found to attenuate the activity of the Hippo pathway through association with Merlin in glioblastoma cells [55–57]. Further, CD44 can also directly activate YAP independent of the Hippo pathway [58]. In addition to CD44, several recent studies propose that some extracellular ligands can regulate the Hippo pathway through different G protein-coupled receptors (GPCRs) [2,59–61]. Moreover, a variety of stimuli (e.g., increasing cell density, DNA damage, energy stress, mechanotransduction, etc.) activate or inhibit the Hippo pathway [2,62–67].

8.1.2 Hippo Signaling in Human Cancer

The Hippo pathway is a tumor suppressor pathway. Dysregulation of the Hippo pathway is involved in tumorigenesis and metastasis [1,2,40]. Evidence of the involvement of the Hippo pathway in tumorigenesis comes from the mouse models with gene knockout of upstream components in the Hippo pathway. These mice develop variable types of tumors. For example, LATS1 knockout mice develop soft tissue sarcoma and ovarian carcinoma [68], whereas MST1/2 knockout mice exhibit several types of tumors including hepatocellular carcinoma (HCC) [69], which is the same as mice with YAP overexpression in livers [18]. Moreover, the core components in the Hippo pathway are clinically involved in diverse human cancers. For example, YAP and/or TAZ overexpressions are identified in tumor tissues from patients with breast cancer (BC) [31,70,71], non-small-cell lung cancer (NSCLC) [72,73], HCC [74], colorectal cancer [75,76], head and neck cancer [77,78], oral squamous cell carcinoma [79,80], ovarian cancer [81–85], prostate cancer [86], and so on. On the other hand, downregulation of LATS is detected in BC [87] and NSCLC [88,89].

8.2 HIPPO PATHWAY IN CHEMOTHERAPEUTIC DRUG RESISTANCE

Abnormal expression levels and dysfunctions of components in the Hippo pathway have been implicated in the resistance of cells to different chemotherapeutic drugs and we will discuss how each component of the core Hippo pathway is involved in chemoresistance during cancer treatment in the following text.

MST

MST1 and its homolog MST2 are located upstream of the core Hippo pathway to control the activity of this pathway (Fig. 8.1). Downregulation of MST causes resistance of prostate cancer cells to the DNA-damaging reagent cisplatin, whereas increasing levels of MST sensitize these cells to cisplatin [90]. However, how reduced MST contributes to cisplatin resistance is not clear. One possibility is that decreased MST may cause drug resistance by inactivating the Hippo pathway, which subsequently activates its downstream targets YAP and/or TAZ (see below).

LATS

As the core kinases in the Hippo pathway, *LATS1* and its homolog *LATS2* are well-known TSGs and loss of their functions is found in various human cancers [91]. We have previously shown that knockdown of LATS1 by small interference RNAs can dramatically decrease the sensitivity of HeLa cervical cancer cells to Taxol, an antimicrotubule drug commonly used for treatment of breast and lung cancers [92]. In addition, *LATS1* has been identified as a gene causing Taxol resistance through a functional genomic screen using short-hairpin RNA (shRNA) library targeting TSGs in the lung cancer cell line A549 [93]. In addition, knockdown of ITCH, which is a ubiquitin ligase causing LATS1 degradation [54], increases cell sensitivity to doxorubicin [94]. This suggests that ITCH may play a role in the chemosensitivity of cells through regulation of LATS1. However, the molecular mechanism of how LATS1 is involved in chemoresistance remains unknown. Similar to LATS1, LATS2 is found to be negatively regulated in leukemia and the low level of LATS2 contributes to the resistance of leukemic cells to the DNA-damaging agents doxorubicin and etoposide (standard drugs for leukemia treatment) [95]. In addition, silencing of LATS2 can upregulate the transcription of ERα (estrogen receptor alpha)-regulated genes, which may render patients with ER+ BC resistant to Tamoxifen and other ER antagonists [96]. In contrast to the above studies, by examination of *LATS2* mRNA levels in tissues from chemotherapy-treated patients, one study found that BC patients with low levels of LATS2 mRNA are more sensitive to epirubicin plus cyclophophamide (EC). A potential explanation for these conflicting findings is that EC treatment functions most effectively in cancer cells that are in S phase of replication. LATS2 can inhibit CDK2 function, indirectly preventing cells from entering S phase and being targeted by EC efficiently [97]. These results suggest that a gene may function differently in response to different types of drugs. Therefore, it is necessary to consider the mechanisms of action for different drug treatments when studying the role of LATS2 in chemotherapeutic drug resistance.

YAP

YAP was identified in 1994 as a protein interacting with Yes and Src tyrosine kinases [98]. Later studies indicated that YAP functions as a transcriptional co-activator and an oncoprotein by interacting with many transcription factors and initiating the expression of downstream oncogenic genes [27,28,99]. In 2007, YAP was identified as a major downstream component of the Hippo pathway. The regulation of YAP by LATS in the Hippo pathway determines its subcellular location, which further affects its oncogenic functions through transactivation of a variety of genes involved in anti-apoptosis and cell proliferation [21,22,24].

The expression of YAP is associated with resistance to various chemotherapeutic drugs confirmed in several studies. Upregulation of YAP causes resistance of mammary and ovarian cancer cells to the chemotherapeutic drugs Taxol and cisplatin [30,81,82,85]. Besides, YAP is found de-acetylated by SIRT1 and translocated into the nucleus in HCC during cisplatin treatment, which leads to cisplatin resistance in HCC [100]. YAP nuclear translocation induced by hypoxia in HCC also causes cells to become less sensitive to sorafenib, a first-line treatment drug for HCC [101]. HCC with high levels of YAP expression are also resistant to doxorubicin [102]. Besides, overexpression of YAP mediates the resistance of senescent cells to doxorubicin through upregulation of the anti-apoptotic protein survivin [103]. Additionally, high levels of nuclear-localized YAP has been found in colon cancer cells that are resistant to antimetabolite 5-flurouracil as well as castration-resistant prostate tumor samples [104,105]. Moreover, YAP also mediates the resistance of BRAF mutant cancer cells to either BRAF inhibitor vemurafenib or MEK inhibitor trametinib through upregulation of the downstream anti-apoptotic protein BCL-XL [106]. However, YAP also transcriptionally upregulates *RAS* genes through TEAD in NF2-loss thyroid cancers and sensitizes cells to MEK inhibitor AZD6244 [107]. In addition, enhanced YAP is associated with the resistance of lung cancer cells to EGFR tyrosine kinase inhibitors through upregulation of the downstream target AXL and the activation of ERK [108]. Consistently, reduced levels of YAP in various cancer cells sensitize these cells to cisplatin and the EGFR inhibitors erlotinib and cetuximab [85,109,110].

The activation of YAP may play a critical role in the resistance of cancer cells to various therapeutic drugs. For example, we have recently demonstrated that in response to antitubulin drug treatments, activated CDK1 can negatively regulate YAP by directly phosphorylating YAP on five sites with an SP (S, Serine; P, Proline) motif. This phosphorylation can dramatically decrease the interaction between YAP and TEAD transcription factor and sensitize cancer cells to antimicrotubule drug treatments [111]. Therefore, the identification of the upstream regulators and downstream mediators of YAP during chemotherapeutic drug response may be critical to understanding how activation of YAP causes resistance to different drug treatments.

TAZ

TAZ was identified as a 14-3-3 binding protein in 2000 [112]. Further studies have indicated that TAZ is a paralog of YAP in mammals and regulated by the Hippo pathway [23]. Therefore, TAZ shares many functions with YAP. TAZ has been identified as an oncogene and is involved in the development and metastasis of various types of cancers [23,113]. High levels of TAZ are detected in basal-like BC cells and patient tissues [31,70,114].

Overexpression of TAZ causes resistance of mammary cells to Taxol, whereas knockdown of TAZ by shRNA in BC cells sensitizes these cells to Taxol [114]. Besides, we also found that TAZ can be directly phosphorylated by CDK1, which further causes TAZ degradation and therefore reverses TAZ-related cell resistance to antitubulin drugs [115]. Moreover, TAZ overexpression in Ras-transformed MCF10A-T1K cells increases multidrug resistance protein levels and results in cellular resistance to paclitaxel and doxorubicin [39]. Besides, TAZ is required for maintaining the properties of breast CSCs and causes resistance of these cells to Taxol and doxorubicin [31,39].

Our laboratory provided first evidence that overexpression of TAZ causes resistance of mammary cells to Taxol treatment by upregulating its downstream target genes, CTGF and Cyr61, through activation of transcription factor TEAD [114]. It has previously been shown that Cyr61 reduces chemosensitivity to Taxol by activating an integrin-MAPK pathway [116]. Therefore, it is possible that TAZ induces Taxol resistance through a novel TEAD-Cyr61/CTGF-integrin-MAPK pathway. However, it remains to be explored whether this pathway is also important in mediating TAZ-mediated resistance to other drugs.

8.3 TARGETING THE HIPPO PATHWAY FOR CANCER CHEMOTHERAPY

As stated above, dysregulation of the Hippo pathway plays important roles in the development and progression of a wide variety of human cancers and is critical for their response to chemotherapeutic drug treatments. Therefore, targeting the Hippo pathway will be a novel strategy for cancer therapy.

8.3.1 Drugs Targeting the Hippo Pathway

Currently, most of the drugs that target the Hippo pathway act by either activating the TSGs MST/LATS or inactivating the oncogenes YAP/TAZ directly or indirectly (Table 8.1 and Fig. 8.2).

8.3.1.1 MST and LATS Activation

Activation of MST or LATS can inhibit cancer cell growth by promoting the phosphorylation and inactivation of YAP and TAZ (Fig. 8.1). Although no drug has been developed so far to directly activate MST/LATS, many drugs have been used to indirectly activate MST/LATS by modulating its upstream signaling (Table 8.1; Fig. 8.2). For example, since F-actin has been shown to inhibit MST/LATS [117,118], inhibition of F-actin polymerization directly [Cytochalasin (Cyt D) and Latrunculin (Lat) A/B] or indirectly through inhibition of upstream regulators such as Rho (Botulinum toxin C3) and ROCK (Y27632 or HA1077) can activate the Hippo pathway by activating MST/LATS [60,119] (Fig. 8.2). In addition, statins and simvastatin can also activate LATS through Rho inhibition by suppressing HMG-CoA reductase activity in the mevalonate pathway [120,121]. Moreover, dobutamine or epinephrine, agonists of GPCRs (coupled with Gs), can also activate LATS and the Hippo pathway through activation of PKA, whereas QLT0267, an ILK inhibitor, reduces BC cell growth by activating MST [122,123] (Fig. 8.2).

TABLE 8.1 Drugs Targeting the Hippo Pathway

Target	Drugs	Working mechanism	References
AMPK (5' AMP-activated protein kinase)	Metformin, C19	Inhibiting YAP activity by activating AMPK	[62–64,134,135]
Cdk1	Taxol	Inhibiting YAP/TAZ activity by activating Cdk1	[111,115]
F-actin	Latrunculin A/B	Activating LATS through regulating F-actin polymerization	[117]
	Cytochalasin D	Activating LATS through regulating F-actin polymerization	[117,118]
GPCR	Dobutamine, epnephrine	Activating LATS by regulating its upstream regulator GPCR	[59,123]
HMG-CoA reductase, Rho	Statins, Simvastatin	Activating MST/LATS activity through Rho GTPases	[60]
		by inhibiting HMG-CoA in the mevalonate pathway	[120,121]
ILK	QLT0267	Activating MST by inhibiting ILK activity	[122]
MST	C19	Activating MST	[136]
PKA	Ibudilast, rolipram	Activating LATS by increasing PKA activity	[47,59,60]
PKC	Auranofin	Inhibiting YAP through AMOT by inhibiting PKC	[135]
Rho	Botulinum toxin C3	Activating LATS through inhibition of Rho GTP activity	[66,118]
ROCK	Y27632,HA1077	Activating LATS though inhibition of ROCK	[66,119]
Tankyrase	XAV939	Inhibiting YAP activity by activating its inhibitor angiomotin	[110]
YAP	Super-TDU	Peptides inhibiting YAP-TEAD interaction	[127]
	Cyclic peptides	Peptides disrupting YAP-TEAD interaction	[125,126]
	C108	Promoting YAP degradation	Guan Patent
	Flufenamic acid	Disrupting YAP-TEAD interacting	[124]
	Verteporfin	Disrupt YAP-TEAD interaction	[128–130]
TAZ	Rottlerin	TAZ inhibitor	[136]
VEGFR PDGFR	Pazopanib	Inhibit YAP/TAZ nuclear localization by inhibiting VEGFR and PDGFR	[142]
Yes	Dasatinib	Activating kinase activity of Yes (YAP/TAZ activator)	[143]

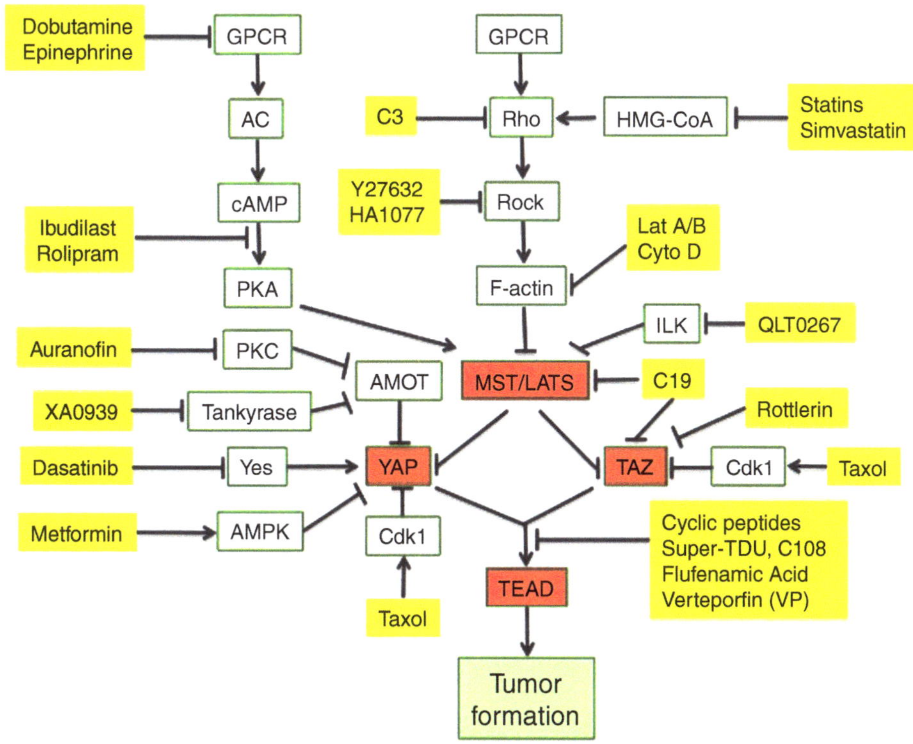

FIG. 8.2 **Regulators of the Hippo pathway and relevant drugs targeting the Hippo pathway.** Abbreviations: GPCR, G protein-coupled receptor; ILK, integrin-linked kinase; PKC, protein kinase C; AC, adenylate cyclase; cAMP, cyclic adenosine monophosphate; AMPK, 5′ AMP-activated protein kinase.

8.3.1.2 Inhibition of YAP/TAZ-TEAD Interaction

The YAP/TAZ oncogene is a major output of the Hippo pathway (Fig. 8.1). Since YAP/TAZ is a transcriptional co-activator with no enzymatic activity, directly inhibiting YAP/TAZ is challenging. However, it has been shown that YAP/TAZ causes tumorigenic phenotypes through interaction and activation of transcription factor TEAD [27]. Therefore, disruption of YAP/TAZ interaction with TEAD to inhibit YAP/TAZ tumorigenic function represents an attractive strategy to target the Hippo pathway for cancer therapy [124]. Based on the crystallographic structure of the YAP-TEAD4 complex and using a combination of engineering approaches, a series of cyclic peptide mimics have been designed and shown to be able to effectively block YAP-TEAD interaction [125,126]. In addition, a 48-mer peptide called "Super-TDU", which is derived from the sequence of VGLL4 that is critical for inhibition of YAP, was also designed and found to specifically disrupt YAP-TEAD interaction [127]. Significantly, these peptides were shown to suppress tumor growth both *in vitro* in cell lines and *in vivo* in a xenograft mouse model. However, since the cost of manufacturing peptide-based compounds is high and peptide-based compounds can be quickly degraded *in vivo*, it is still challenging to administer these peptide drugs to cancer patients.

Small molecular drugs (SMDs) are easy to synthesize and relatively stable *in vivo*. Therefore, they are the cost-effective drugs commonly used for targeted therapies. Currently, Verteporfin (VP) is the only SMD that has been shown to directly disrupt the YAP–TEAD interaction. In a search for SMDs blocking YAP–TEAD interaction, Pan and colleagues used a YAP–TEAD transcriptional activating luciferase reporter to screen a library of 3300 FDA-approved SMDs. VP was identified from the screen [128]. VP can directly disrupt YAP-TEAD interaction *in vitro* and *in vivo* in a dose-dependent manner and inhibit YAP-TEAD transcriptional activation function. Most significantly, this and later studies showed that VP treatment dramatically suppressed the growth of various cancers caused by YAP overexpression in xenograft mouse models *in vivo* [128–130].

8.3.1.3 Targeting Proteins Regulating YAP/TAZ

(1) Kinases: Since direct targeting of YAP/TAZ is difficult, targeting the positive or negative regulators of YAP/TAZ becomes one of the important strategies targeting the Hippo pathway. Since kinases are druggable by SMDs, they are attractive targets for inhibiting YAP/TAZ activity. Recently, Yes proto-oncogene was found to be crucial for the formation of a transcriptional complex including YAP and β-catenin [131]. Interestingly, inhibition of Yes by Dasatinib can suppress cancer cell growth by inhibiting YAP–catenin complex function. In addition, 5' AMP-activated protein kinase (AMPK) has also been shown to phosphorylate YAP and inhibit its function under energy (e.g., glucose) deprivation. Metformin, a drug used for the treatment of diabetes, can suppress YAP through activation of AMPK [62–64]. Moreover, we have recently shown that antitubulin drugs such as Taxol can activate Cdk1, which subsequently phosphorylates and inactivates YAP/TAZ to induce cancer cell death [111,115]. In this case, studies on drugs modulating the activity of Cdk1 could also be used to identify drugs that regulate YAP/TAZ functions.

(2) Proteins regulating YAP/TAZ nuclear localization and stability: Since YAP/TAZ functions as a transcriptional co-activator in the nucleus, inhibition of their nuclear localization is another strategy for inhibiting YAP/TAZ function. The AMOT family of proteins has recently been shown to negatively regulate the oncogenic properties of YAP/TAZ by inhibiting its nuclear localization through direct protein–protein interactions [132–134]. Therefore, activation of AMOT could be another strategy to inhibit YAP/TAZ in cancer (Fig. 8.2). It has been recently reported that XA0939, an inhibitor of tankyrase that can cause AMOT degradation through E3 ligase RNF146, can inhibit YAP nuclear localization through AMOT [110]. On the other hand, atypical protein kinase C (PKCι) can directly phosphorylate AMOT, whose phosphorylation inhibits YAP1 binding and increases YAP nuclear localization. Therefore, treatment of tumor cells with the PKCι inhibitor Auranofin can decrease YAP nuclear localization and tumor growth *in vitro* and *in vivo* through activation of AMOT [135].

Another way to inhibit YAP/TAZ function is to induce YAP/TAZ protein degradation [136]. Recently, C19, a newly identified SMD, was shown to cause TAZ degradation by activation of GSK3β, a negative regulator of TAZ [137,138]. In addition, in a screen for YAP/TAZ inhibitors, the SMD C108 was shown to inhibit YAP/TAZ by promoting their degradation (Guan Patent). The molecular mechanism underlying C108-induced YAP/TAZ degradation is unclear. However, it is suggested that C108 may be targeting the de-ubiquitinating enzymes critical for regulating YAP/TAZ protein stability [139].

8.3.2 Targeting Hippo Pathways to Improve Response to Chemotherapy

As discussed above, mounting evidence suggests that inactivation of the Hippo pathway by upregulating oncoproteins such as YAP/TAZ or downregulating TSGs such as LATS1/2 or MST1/2 may represent a major event leading to resistance of cancer cells to various chemotherapeutic drugs. Therefore, activation of the Hippo pathway by activating MST/LATS or inactivating YAP/TAZ may sensitize drug-resistant cancer cells to chemotherapeutic drug treatments. Since MST/LATS is not druggable, YAP and TAZ are the potential therapeutic targets to improve response to chemotherapy.

Many studies have shown that knockdown of YAP or TAZ genetically by RNA interference (RNAi) in various cancer cells leads to improved response to a wide variety of chemotherapeutic drugs including antitubulin drugs (Taxol, vinblastine, etc.), cisplatin, EGFR inhibitors (e.g., erlotinib, gefitinib, and cetuximab), c-Abl inhibitor (e.g., imatinib), and RAF and MEK inhibitors [106,108,114,140]. One good example of these findings is the manipulation of YAP in coping with RAF and MEK inhibitor drug resistance. Although RAF- and MEK-targeted therapy has been widely used for the treatment of melanoma and NSCLC, resistance to this therapy is a major obstacle for successful treatment. Through a genetic screen, YAP was identified as a gene-promoting resistance of cancer cells to RAF and MEK inhibitors [106]. Most significantly, knockdown of YAP by RNAi in drug-resistant melanoma, colon, and thyroid cells caused synthetically lethal with RAF/MEK inhibitor treatment, suggesting that inactivation of YAP is an attractive strategy to treat cancers resistant to RAF and MEK inhibitors.

Other studies demonstrate that targeting YAP/TAZ using SMDs may sensitize cancers to various chemotherapies. For example, the levels of YAP and TAZ are negatively correlated with the sensitivity of cancer cells to Taxol [111,114]. Most significantly, it has recently been shown that treatment of Taxol-resistant colon cancer cells with the YAP inhibitor VP reverses the resistance to Taxol [141]. In addition, combined treatment of BC cells with the YAP/TAZ inhibitors Dasatinib and pazopanib also sensitizes BC cells to both doxorubicin and Taxol [142]. Moreover, activation of YAP was found to be responsible for hypoxia-induced resistance of HCC to Sorafenib, a multikinase inhibitor [101]. Treatment of drug-resistant HCC cells with VP sensitizes these cells to Sorafenib treatment. Similarly, treatment of Sorafenib-resistant HCC cells with statins, which inactivate YAP through activation of the Hippo pathway (Fig. 8.2), overcomes Sorafenib resistance, leading to improved response to Sorafenib therapy [101].

Interestingly, a screen for genetic markers associated with sensitivity or resistance to SMDs revealed a combination of two FDA-approved SMDs, statins and Dasatinib, as an effective cotreatment strategy to potently inhibit YAP/TAZ in cancer cells [143]. Compared to single drug treatment, combined treatment of BC cells with these two drugs significantly sensitized cancer cells to Taxol treatment. Therefore, combined treatment of cancer cells with inhibitors targeting the Hippo pathway may result in better outcomes in the treatment of drug-resistant cancers.

8.4 CONCLUSION

In conclusion, dysregulation of the Hippo pathway plays important roles not only in tumorigenesis, but also in drug resistance. Newly published data strongly suggest that targeting the Hippo pathway is a novel and attractive strategy to improve the response of drug-resistant cancer cells to chemotherapy. Therefore, the development of more SMDs directly targeting the core Hippo components will significantly enhance our ability to treat drug-resistant cancer patients in the future.

Acknowledgments

We would like to thank Canadian Institute of Health Research (#119325, 148629) (XY), the Canadian Breast Cancer Foundation/Canadian Cancer Society (XY) and the Ontario Trillium Scholarship (OTS, Canada) and the CIHR/Terry Fox Foundation Training Program in Transdisciplinary Cancer Research (YZ) for their financial support.

Conflict of Interest: No potential conflicts of interest were disclosed.

References

[1] Pan D. The hippo signaling pathway in development and cancer. Dev Cell 2010;19:491–505.
[2] Yu F, Guan K. The Hippo pathway: regulators and regulations. Genes Dev 2013;27:355–71.
[3] Chan SW, Lim CJ, Huang C, Chong YF, Gunaratne HJ, Hogue KA, et al. WW domain-mediated interaction with Wbp2 is important for the oncogenic property of TAZ. Oncogene 2011;30:600–10.
[4] Halder G, Johnson RL. Hippo signaling: growth control and beyond. Development 2011;138(1):9–22.
[5] Harvey KF, Zhang X, Thomas DM. The Hippo pathway and human cancer. Nat Rev Cancer 2013;13:246–57.
[6] Harvey K, Tapon N. The Salvador–Warts–Hippo pathway—an emerging tumour-suppressor network. Nat Rev Cancer 2007;7:182–91.
[7] Hiemer SE, Varelas X. Stem cell regulation by the Hippo pathway. Biochim Biophys Acta 2013;1830:2323–34.
[8] Zhao B, Lei Q, Guan K. The Hippo–YAP pathway: new connections between regulation of organ size and cancer. Curr Opin Cell Biol 2008;20:638–46.
[9] Zhao B, Tumaneng K, Guan K-L. The Hippo pathway in organ size control, tissue regeneration and stem cell self-renewal. Nat Cell Biol 2011;13:877–83.
[10] Xu T, Rubin GM. Analysis of genetic mosaics in developing and adult Drosophila tissues. Development 1993;117(4):1223–37.
[11] Xu T, Wang W, Zhang S, Stewart RA, Yu W. Identifying tumor suppressors in genetic mosaics: the *Drosophila* lats gene encodes a putative protein kinase. Development 1995;121(4):1053–63.
[12] Justice RW, Zilian O, Woods DF, Noll M, Bryant PJ. The *Drosophila* tumor suppressor gene warts encodes a homolog of human myotonic dystrophy kinase and is required for the control of cell shape and proliferation. Genes Dev 1995;9(5):534–46.
[13] Wu S, Huang J, Dong J, Pan D. Hippo encodes a Ste-20 family protein kinase that restricts cell proliferation and promotes apoptosis in conjunction with salvador and warts. Cell 2003;114(4):445–56.
[14] Harvey KF, Pfleger CM, Hariharan IK. The *Drosophila* Mst ortholog, hippo, restricts growth and cell proliferation and promotes apoptosis. Cell 2003;114:457–67.
[15] Udan RS, Kango-Singh M, Nolo R, Tao C, Halder G. Hippo promotes proliferation arrest and apoptosis in the Salvador/Warts pathway. Nat Cell Biol 2003;5(10):914–20.
[16] Chan EHY, Nousiainen M, Chalamalasetty RB, Schäfer A, Nigg EA, Sillјé HHW. The Ste20-like kinase Mst2 activates the human large tumor suppressor kinase Lats1. Oncogene 2005;24:2076–86.
[17] Lai Z-C, Wei X, Shimizu T, Ramos E, Rohrbaugh M, Nikolaidis N, et al. Control of cell proliferation and apoptosis by Mob as tumor suppressor Mats. Cell 2005;120(5):675–85.
[18] Dong J, Feldmann G, Huang J, Wu S, Zhang N, Comerford SA, et al. Elucidation of a universal size-control mechanism in *Drosophila* and mammals. Cell 2007;130:1120–33.
[19] Huang J, Wu S, Barrera J, Matthews K, Pan D. The Hippo signaling pathway coordinately regulates cell proliferation and apoptosis by inactivating Yorkie, the *Drosophila* homolog of YAP. Cell 2005;122:421–34.
[20] Saucedo LJ, Edgar BA. Filling out the Hippo pathway. Nat Rev Mol Cell Biol 2007;8(8):613–21.
[21] Zhao B, Wei X, Li W, Udan RS, Yang Q, Kim J, et al. Inactivation of YAP oncoprotein by the Hippo pathway is involved in cell contact inhibition and tissue growth control. Genes Dev 2007;21:2747–61.
[22] Hao Y, Chun A, Cheung K, Rashidi B, Yang X. Tumor suppressor LATS1 is a negative regulator of oncogene YAP. J Biol Chem 2008;283:5496–509.
[23] Lei Q-Y, Zhang H, Zhao B, Zha Z-Y, Bai F, Pei X-H, et al. TAZ promotes cell proliferation and epithelial–mesenchymal transition and is inhibited by the hippo pathway. Mol Cell Biol 2008;28:2426–36.

[24] Zhao B, Li L, Tumaneng K, Wang C, Guan K. A coordinated phosphorylation by Lats and CK1 regulates YAP stability through. Genes Dev 2010;24:72–85.
[25] Liu C-Y, Zha Z-Y, Zhou X, Zhang H, Huang W, Zhao D, et al. The hippo tumor pathway promotes TAZ degradation by phosphorylating a phosphodegron and recruiting the SCF{beta}-TrCP E3 ligase. J Biol Chem 2010;285:37159–69.
[26] Thompson BJ, Cohen SM. The Hippo pathway regulates the bantam microRNA to control cell proliferation and apoptosis in *Drosophila*. Cell 2006;126(4):767–74.
[27] Zhao B, Ye X, Yu J, Li L, Li W, Li S, et al. TEAD mediates YAP-dependent gene induction and growth control. Genes Dev 2008;22:1962–71.
[28] Lamar JM, Stern P, Liu H, Schindler JW, Jiang Z-G, Hynes RO. The Hippo pathway target, YAP, promotes metastasis through its TEAD-interaction domain. Proc Natl Acad Sci USA 2012;109:E2441–50.
[29] Yang N, Morrison CD, Liu P, Miecznikowski J, Bshara W, Han S, et al. TAZ induces growth factor-independent proliferation through activation of EGFR ligand amphiregulin. Cell Cycle 2012;11(15):2922–30.
[30] Overholtzer M, Zhang J, Smolen GA, Muir B, Li W, Sgroi DC, et al. Transforming properties of YAP, a candidate oncogene on the chromosome 11q22 amplicon. Proc Natl Acad Sci USA 2006;103:12405–10.
[31] Bartucci M, Dattilo R, Moriconi C, Pagliuca A, Mottolese M, Federici G, et al. TAZ is required for metastatic activity and chemoresistance of breast cancer stem cells. Oncogene 2014;34:681–90.
[32] Yu J, Zheng Y, Dong J, Klusza S, Deng W-M, Pan D. Kibra functions as a tumor suppressor protein that regulates Hippo signaling in conjunction with Merlin and Expanded. Dev Cell 2010;18:288–99.
[33] Baumgartner R, Poernbacher I, Buser N, Hafen E, Stocker H. The WW domain protein Kibra acts upstream of Hippo in *Drosophila*. Dev Cell 2010;18:309–16.
[34] Genevet A, Wehr MC, Brain R, Thompson BJ, Tapon N. Kibra is a regulator of the Salvador/Warts/Hippo signaling network. Dev Cell 2010;18:300–8.
[35] Visser-Grieve S, Hao Y, Yang X. Human homolog of *Drosophila* expanded, hEx, functions as a putative tumor suppressor in human cancer cell lines independently of the Hippo pathway. Oncogene 2011;31: 1189–95.
[36] Angus L, Moleirinho S, Herron L, Sinha A, Zhang X, Niestrata M, et al. Willin/FRMD6 expression activates the Hippo signaling pathway kinases in mammals and antagonizes oncogenic YAP. Oncogene 2012;31: 238–50.
[37] Yin F, Yu J, Zheng Y, Chen Q, Zhang N, Pan D. Spatial organization of Hippo signaling at the plasma membrane mediated by the tumor suppressor Merlin/NF2. Cell 2013;154:1342–55.
[38] Murugan P, Sarkeshik A, Yates JR, McCollum D. Angiomotin family proteins are novel activators of the LATS2 kinase tumor suppressor. Mol Biol Cell 2011;22:3725–33.
[39] Cordenonsi M, Zanconato F, Azzolin L, Forcato M, Rosato A, Frasson C, et al. The Hippo transducer TAZ confers cancer stem cell-related traits on breast cancer cells. Cell 2011;147:759–72.
[40] Chen D, Sun Y, Wei Y, Zhang P, Rezaeian AH, Teruya-Feldstein J, et al. LIFR is a breast cancer metastasis suppressor upstream of the Hippo–YAP pathway and a prognostic marker. Nat Med 2012;18:1511–7.
[41] Khokhlatchev A, Rabizadeh S, Xavier R, Nedwidek M, Chen T, Zhang X, et al. Identification of a novel Ras-regulated proapoptotic pathway. Curr Biol 2002;12:253–65.
[42] Guo C, Zhang X, Pfeifer GP. The tumor suppressor RASSF1A prevents dephosphorylation of the mammalian STE20-like kinases MST1 and MST2. J Biol Chem 2011;286:6253–61.
[43] Guo C, Tommasi S, Liu L, Yee J-K, Dammann R, Pfeifer GP. RASSF1A is part of a complex similar to the *Drosophila* Hippo/Salvador/Lats tumor-suppressor network. Curr Biol 2007;17:700–5.
[44] Poon CLC, Lin JI, Zhang X, Harvey KF. The sterile 20-like kinase Tao-1 controls tissue growth by regulating the Salvador–Warts–Hippo pathway. Dev Cell 2011;21:896–906.
[45] Boggiano JC, Vanderzalm PJ, Fehon RG. Tao-1 phosphorylates Hippo/MST kinases to regulate the Hippo–Salvador–Warts tumor suppressor pathway. Dev Cell 2011;21:888–95.
[46] Mohseni M, Sun J, Lau A, Curtis S, Goldsmith J, Fox VL, et al. A genetic screen identifies an LKB1-MARK signalling axis controlling the Hippo–YAP pathway. Nat Cell Biol 2014;16:108–17.
[47] Yu F-X, Zhang Y, Park HW, Jewell JL, Chen Q, Deng Y, et al. Protein kinase A activates the Hippo pathway to modulate cell proliferation and differentiation. Genes Dev 2013;27:1223–32.
[48] Kim M, Kim M, Lee S, Kuninaka S, Saya H, Lee H, et al. cAMP/PKA signalling reinforces the LATS–YAP pathway to fully suppress YAP in response to actin cytoskeletal changes. EMBO J 2013;32:1543–55.
[49] Romano D, Matallanas D, Weitsman G, Preisinger C, Ng T, Kolch W. Proapoptotic kinase MST2 coordinates signaling crosstalk between RASSF1A, Raf-1, and Akt. Cancer Res 2010;70:1195–203.

[50] Romano D, Nguyen LK, Matallanas D, Halasz M, Doherty C, Kholodenko BN, et al. Protein interaction switches coordinate Raf-1 and MST2/Hippo signalling. Nat Cell Biol 2014;16:673–84.
[51] Reddy B, Irvine KD. Regulation of Hippo signaling by EGFR-MAPK signaling through Ajuba family proteins. Dev Cell 2013;24:459–71.
[52] Das Thakur M, Feng Y, Jagannathan R, Seppa MJ, Skeath JB, Longmore GD. Ajuba LIM proteins are negative regulators of the Hippo signaling pathway. Curr Biol 2010;20:657–62.
[53] Rauskolb C, Sun S, Sun G, Pan Y, Irvine KD. Cytoskeletal tension inhibits Hippo signaling through an Ajuba–Warts complex. Cell 2014;158:143–56.
[54] Ho K, Zhou Z, She Y, Chun A, Cyr TD, Yang X. Itch E3 ubiquitin ligase regulates large tumor suppressor 1 tumor-suppressor stability. PNAS 2011;108:4870–5.
[55] Xu Y, Stamenkovic I, Yu Q. CD44 attenuates activation of the hippo signaling pathway and is a prime therapeutic target for glioblastoma. Cancer Res 2010;70:2455–64.
[56] Sainio M, Zhao F, Heiska L, Turunen O, den Bakker M, Zwarthoff E, et al. Neurofibromatosis 2 tumor suppressor protein colocalizes with ezrin and CD44 and associates with actin-containing cytoskeleton. J Cell Sci 1997;110:2249–60.
[57] Yu S, Cai X, Wu C, Wu L, Wang Y, Liu Y, et al. Adhesion glycoprotein CD44 functions as an upstream regulator of a network connecting ERK AKT and Hippo–YAP pathways in cancer progression. Oncotarget 2015;6:2951–65.
[58] Zhang Y, Xia H, Ge X, Chen Q, Yuan D, Chen Q, et al. CD44 acts through RhoA to regulate YAP signaling. Cell Signal 2014;26:2504–13.
[59] Yu F-X, Zhao B, Panupinthu N, Jewell JL, Lian I, Wang LH, et al. Regulation of the Hippo–YAP pathway by G-protein-coupled receptor signaling. Cell 2012;150(4):780–91.
[60] Mo J-S, Yu F-X, Gong R, Brown JH, Guan K-L. Regulation of the Hippo–YAP pathway by protease-activated receptors (PARs). Genes Dev 2012;26:2138–43.
[61] Zhou X, Wang S, Wang Z, Feng X, Liu P, et al. Estrogen regulates Hippo signaling via GPER in breast cancer. J Clin Invest 2015;125:2123–35.
[62] DeRan M, Yang J, Shen C-H, Peters EC, Fitamant J, Chan P, et al. Energy stress regulates Hippo–YAP signaling involving AMPK-mediated regulation of angiomotin-like 1 protein. Cell Rep 2014;9:495–503.
[63] Wang W, Xiao Z, Li X, Aziz KE, Gan B, Johnson RL, et al. AMPK modulates Hippo pathway activity to regulate energy homeostasis. Nat Cell Biol 2015;17:490–9.
[64] Mo J-S, Meng Z, Kim YC, Park HW, Hansen CG, Kim S, et al. Cellular energy stress induces AMPK-mediated regulation of YAP and the Hippo pathway. Nat Cell Biol 2015;17:500–10.
[65] Calvo F, Ege N, Grande-Garcia A, Hooper S, Jenkins RP, Chaudhry SI, et al. Mechanotransduction and YAP-dependent matrix remodelling is required for the generation and maintenance of cancer-associated fibroblasts. Nat Cell Biol 2013;15(6):637–46.
[66] Dupont S, Morsut L, Aragona M, Enzo E, Giulitti S, Cordenonsi M, et al. Role of YAP/TAZ in mechanotransduction. Nature 2011;474:179–83.
[67] Wrighton KH. Mechanotransduction: YAP and TAZ feel the force. Nat Rev Mol Cell Biol 2011;12(7):404.
[68] St John MA, Tao W, Fei X, Fukumoto R, Carcangiu ML, Brownstein DG. Mice deficient of Lats1 develop soft-tissue sarcomas, ovarian tumours and pituitary dysfunction. Nat Genet 1999;21:182–6.
[69] Zhou D, Conrad C, Xia F, Park J-S, Payer B, Yin Y, et al. Mst1 and Mst2 maintain hepatocyte quiescence and suppress hepatocellular carcinoma development through inactivation of the Yap1 oncogene. Cancer Cell 2009;16:425–38.
[70] Chan SW, Lim CJ, Guo K, Ng CP, Lee I, Hunziker W, et al. A role for TAZ in migration, invasion, and tumorigenesis of breast cancer cells. Cancer Res 2008;68:2592–8.
[71] Wang X, Su L, Ou Q. Yes-associated protein promotes tumour development in luminal epithelial derived breast cancer. Eur J Cancer 2012;48:1227–34.
[72] Su LL, Ma WX, Yuan JF, Shao Y. XW and JS expression of Yes-associated protein in non-small cell lung cancer and its relationship with clinical pathological factors. Chin Med J (Engl) 2012;125:4003–8.
[73] Noguchi S, Saito A, Horie M, Mikami Y, Suzuki HI, Morishita Y, et al. An integrative analysis of the tumorigenic role of TAZ in human non-small cell lung cancer. Clin cancer Res 2014;20:4660–72.
[74] Han S, Bai E, Jin G, He C, Guo X, Wang L, et al. Expression and clinical significance of YAP, TAZ, and AREG in hepatocellular carcinoma. J Immunol Res 2014;2014:261365.

[75] Wang L, Shi S, Guo Z, Zhang X, Han S, Yang A, et al. Overexpression of YAP and TAZ is an independent predictor of prognosis in colorectal cancer and related to the proliferation and metastasis of colon cancer cells. PLoS ONE 2013;8:e65539.

[76] Yuen H-F, McCrudden CM, Huang Y-H, Tham JM, Zhang X, Zeng Q, et al. TAZ expression as a prognostic indicator in colorectal cancer. PLoS ONE 2013;8:e54211.

[77] Ehsanian R, Brown M, Lu H, Yang XP, Pattatheyil A, Yan B, et al. YAP dysregulation by phosphorylation or ΔNp63-mediated gene repression promotes proliferation, survival and migration in head and neck cancer subsets. Oncogene 2010;29(46):6160–71.

[78] Ge L, Smail M, Meng W, Shyr Y, Ye F, Fan K-H, et al. Yes-associated protein expression in head and neck squamous cell carcinoma nodal metastasis. PLoS ONE 2011;6(11). e27529.

[79] Wei Z, Wang Y, Li Z, Yuan C, Zhang W, Wang D, et al. Overexpression of Hippo pathway effector TAZ in tongue squamous cell carcinoma: correlation with clinicopathological features and patients' prognosis. J Oral Pathol Med 2013;42(10):747–54.

[80] Hiemer SE, Zhang L, Kartha VK, Packer TS, Almershed M, Noonan V, et al. A YAP/TAZ-regulated molecular signature is associated with oral squamous cell carcinoma. Mol Cancer Res 2015;13(6):957–68.

[81] Jeong W, Kim S-B, Sohn BH, Park Y-Y, Park ES, Kim SC, et al. Activation of YAP1 is associated with poor prognosis and response to taxanes in ovarian cancer. Anticancer Res 2014;34:811–7.

[82] Zhang X, George J, Deb S, Degoutin JL, Takano EA, Fox SB, et al. The Hippo pathway transcriptional coactivator, YAP, is an ovarian cancer oncogene. Oncogene 2011;30:2810–22.

[83] Hall CA, Wang R, Miao J, Oliva E, Shen X, Wheeler T, et al. Hippo pathway effector Yap is an ovarian cancer oncogene. Cancer Res 2010;70:8517–25.

[84] He C, Lv X, Hua G, Lele SM, Remmenga S, Dong J, et al. YAP forms autocrine loops with the ERBB pathway to regulate ovarian cancer initiation and progression. Oncogene 2015;34(50):6040–54.

[85] Xia Y, Chang T, Wang Y, Liu Y, Li W, Li M, et al. YAP promotes ovarian cancer cell tumorigenesis and is indicative of a poor prognosis for ovarian cancer patients. PLoS ONE 2014;9(3). e91770.

[86] Zhang L, Yang S, Chen X, Stauffer S, Yu F, Lele SM, et al. The Hippo pathway effector, YAP, regulates motility, invasion and castration-resistant growth of prostate cancer cells. Cell Biol 2015;35:00102–15.

[87] Takahashi Y, Miyoshi Y, Takahata C, Irahara N, Taguchi T. Down-regulation of LATS1 and LATS2 mRNA expression by promoter hypermethylation and its association with biologically aggressive phenotype in human breast cancers. Clin Cancer Res 2005;11:1380–5.

[88] Lin X-Y, Zhang X-P, Wu J-H, Qiu X-S, Wang E-H. Expression of LATS1 contributes to good prognosis and can negatively regulate YAP oncoprotein in non-small-cell lung cancer. Tumour Biol 2014;35:6435–43.

[89] Sasaki H, Hikosaka Y, Kawano O, Yano M, Fujii Y. Hypermethylation of the large tumor suppressor genes in Japanese lung cancer. Oncol Lett 2010;1:303–7.

[90] Ren A, Yan G, You B, Sun J. Down-regulation of mammalian sterile 20-like kinase 1 by heat shock protein 70 mediates cisplatin resistance in prostate cancer cells. Cancer Res 2008;68:2266–74.

[91] Visser S, Yang X. LATS tumor suppressor: a new governor of cellular homeostasis. Cell Cycle 2010;9: 3892–903.

[92] Visser S, Yang X. Identification of LATS transcriptional targets in HeLa cells using whole human genome oligonucleotide microarray. Gene 2010;449:22–9.

[93] Ji D, Deeds SL, Weinstein EJ. A screen of shRNAs targeting tumor suppressor genes to identify factors involved in A549 paclitaxel sensitivity. Oncol Rep 2007;18(6):1499–505.

[94] Hansen TM, Rossi M, Roperch JP, Ansell K, Simpson K, Taylor D, et al. Itch inhibition regulates chemosensitivity *in vitro*. Biochem Biophys Res Commun 2007;361(1):33–6.

[95] Kawahara M, Hori T, Chonabayashi K, Oka T, Sudol M, Uchiyama T. Kpm/Lats2 is linked to chemosensitivity of leukemic cells through the stabilization of p73. Blood 2008;112(9):3856–66.

[96] Lit LC, Scott S, Zhang H, Stebbing J, Photiou A, Giamas G. LATS2 is a modulator of estrogen receptor alpha. Anticancer Res 2013;33:53–63.

[97] Takahashi Y, Miyoshi Y, Morimoto K, Taguchi T, Tamaki Y, Noguchi S. Low LATS2 mRNA level can predict favorable response to epirubicin plus cyclophosphamide, but not to docetaxel, in breast cancers. J Cancer Res Clin Oncol 2007;133(8):501–9.

[98] Sudol M. Yes-associated protein (YAP65) is a proline-rich phosphoprotein that binds to the SH3 domain of the Yes proto-oncogene product. Oncogene 1994;9:2145–52.

[99] Yagi R, Chen LF, Shigesada K, Murakami Y, Ito Y. A WW domain-containing yes-associated protein (YAP) is a novel transcriptional co-activator. EMBO J 1999;18:2551–62.
[100] Mao B, Hu F, Cheng J, Wang P, Xu M, Yuan F, et al. SIRT1 regulates YAP2-mediated cell proliferation and chemoresistance in hepatocellular carcinoma. Oncogene 2014;33:1468–74.
[101] Zhou T, Zhuang L, Hu Y, Zhou Y, Lin W, Wang D, et al. Inactivation of hypoxia-induced YAP by statins overcomes hypoxic resistance tosorafenib in hepatocellular carcinoma cells. Sci Rep 2016;6:30483.
[102] Huo X, Zhang Q, Liu AM, Tang C, Gong Y, Bian J, et al. Overexpression of Yes-associated protein confers doxorubicin resistance in hepatocellullar carcinoma. Oncol Rep 2013;29:840–6.
[103] Ma K, Xu Q, Wang S, Zhang W, Liu M, Liang S, et al. Nuclear accumulation of Yes-associated protein (YAP) maintains the survival of doxorubicin-induced senescent cells by promoting survivin expression. Cancer Lett 2016;375(1):84–91.
[104] Touil Y, Igoudjil W, Corvaisier M, Dessein A-F, Vandomme J, Monté D, et al. Colon cancer cells escape 5FU chemotherapy-induced cell death by entering stemness and quiescence associated with the c-Yes/YAP axis. Clin Cancer Res 2014;20:837–46.
[105] Jiang N, Hjorth-Jensen K, Hekmat O, Iglesias-Gato D, Kruse T, Wang C, et al. *In vivo* quantitative phosphoproteomic profiling identifies novel regulators of castration-resistant prostate cancer growth. Oncogene 2015;34:2764–76.
[106] Lin L, Sabnis AJ, Chan E, Olivas V, Cade L, Pazarentzos E, et al. The Hippo effector YAP promotes resistance to RAF- and MEK-targeted cancer therapies. Nat Genet 2015;47:250–6.
[107] Garcia-Rendueles MER, Ricarte-Filho JC, Untch BR, Landa I, Knauf JA, Voza F, et al. NF2 loss promotes oncogenic RAS-induced thyroid cancers via YAP-dependent transactivation of RAS proteins and sensitizes them to MEK inhibition. Cancer Discov 2015;5(11):1178–93.
[108] Lee JE, Park HS, Lee D, Yoo G, Kim T, Jeon H, et al. Hippo pathway effector YAP inhibition restores the sensitivity of EGFR-TKI in lung adenocarcinoma having primary or acquired EGFR-TKI resistance. Biochem Biophys Res Commun 2016;474(1):154–60.
[109] Huang J, Nagatomo I, Suzuki E, Mizuno T, Kumagai T, Berezov A, et al. YAP modifies cancer cell sensitivity to EGFR and survivin inhibitors and is negatively regulated by the non-receptor type protein tyrosine phosphatase 14. Oncogene 2013;32:2220–9.
[110] Wang H, Castillo J, Zhang Y, Yang Z, McAllister G, Lindeman A, et al. Tankyrase inhibitor sensitizes lung cancer cells to endothelial growth factor receptor (EGFR) inhibition via stabilizing angiomotins and inhibiting yap signaling. J Biol Chem 2016;291(29):15256–66.
[111] Zhao Y, Khanal P, Savage P, She Y, Cyr TD, Yang X. YAP-induced resistance of cancer cells to antitubulin drugs is modulated by a Hippo-independent pathway. Cancer Res 2014;74:4493–503.
[112] Kanai FA, Marignani P, Sarbassova D, Yagi R, Hall RA, Donowitz M, et al. TAZ: a novel transcriptional co-activator regulated by interactions with 14-3-3 and PDZ domain proteins. EMBO J 2000;19:6778–91.
[113] Zhao Y, Yang X. WWTR1 (WW domain containing transcription regulator 1). Atlas Genet Cytogenet Oncol Haematol 2014;18(11):849–52.
[114] Lai D, Ho KC, Hao Y, Yang X. Taxol resistance in breast cancer cells is mediated by the hippo pathway component TAZ and its downstream transcriptional targets Cyr61 and CTGF. Cancer Res 2011;71:2728–38.
[115] Zhao Y, Yang X. Regulation of sensitivity of tumor cells to antitubulin drugs by Cdk1-TAZ signalling. Oncotarget 2015;6:21906–17.
[116] Menendez JA, Vellon L, Mehmi I, Teng PK, Griggs DW, Lupu R. A novel CYR61-triggered "CYR61- alphavbeta3 integrin loop" regulates breast cancer cell survival and chemosensitivity through activation of ERK1/ERK2 MAPK signaling pathway. Oncogene 2005;24:761–79.
[117] Sun S, Irvine KD. Cellular organization and cytoskeletal regulation of the Hippo signaling network. Trends Cell Biol 2016;26(9):694–704.
[118] Zhao B, Li L, Wang L, Wang C-Y, Yu J, Guan K-L. Cell detachment activates the Hippo pathway via cytoskeleton reorganization to induce anoikis. Genes Dev 2012;26:54–68.
[119] Wada K-I, Itoga K, Okano T, Yonemura S, Sasaki H. Hippo pathway regulation by cell morphology and stress fibers. Development 2011;138(18):3907–14.
[120] Sorrentino G, Ruggeri N, Specchia V, Cordenonsi M, Mano M, Dupont S, et al. Metabolic control of YAP and TAZ by the mevalonate pathway. Nat Cell Biol 2014;16(4):357–66.
[121] Wang Z, Wu Y, Wang H, Zhang Y, Mei L, Fang X, et al. Interplay of mevalonate and Hippo pathways regulates RHAMM transcription via YAP to modulate breast cancer cell motility. Proc Natl Acad Sci USA 2014;111(1):E89–98.

[122] Serrano I, McDonald PC, Lock F, Muller WJ, Dedhar S. Inactivation of the Hippo tumour suppressor pathway by integrin-linked kinase. Nat Commun 2013;4:2976.
[123] Bao Y, Nakagawa K, Yang Z, Ikeda M, Withanage K, Ishigami-Yuasa M. A cell-based assay to screen stimulators of the Hippo pathway reveals the inhibitory effect of dobutamine on the YAP-dependent gene transcription. J Biochem 2011;150(2):199–208.
[124] Pobbati AV, Han X, Hung AW, Weiguang S, Huda N, Chen G-Y, et al. Targeting the central pocket in human transcription factor TEAD as a potential cancer therapeutic strategy. Structure 2015;23(11):2076–86.
[125] Zhang Z, Lin Z, Zhou Z, Shen HC, Yan SF, Mayweg AV, et al. Structure-based design and synthesis of potent cyclic peptides inhibiting the YAP-TEAD protein–protein interaction. ACS Med Chem Lett 2014;5(9):993–8.
[126] Zhou Z, Hu T, Xu Z, Lin Z, Zhang Z, Feng T, et al. Targeting Hippo pathway by specific interruption of YAP-TEAD interaction using cyclic YAP-like peptides. FASEB J 2015;29(2):724–32.
[127] Jiao S, Wang H, Shi Z, Dong A, Zhang W, Song X, et al. A peptide mimicking VGLL4 function acts as a YAP antagonist therapy against gastric cancer. Cancer Cell 2014;25(2):166–80.
[128] Liu-Chittenden Y, Huang B, Shim JS, Chen Q, Lee S-J, Anders RA, et al. Genetic and pharmacological disruption of the TEAD-YAP complex suppresses the oncogenic activity of YAP. Genes Dev 2012;26(12):1300–5.
[129] Feng X, Degese MS, Iglesias-Bartolome R, Vaque JP, Molinolo AA, Rodrigues M, et al. Hippo-independent activation of YAP by the GNAQ uveal melanoma oncogene through a trio-regulated rho GTPase signaling circuitry. Cancer Cell 2014;25(6):831–45.
[130] Yu F-X, Luo J, Mo J-S, Liu G, Kim YC, Meng Z, et al. Mutant Gq/11 promote uveal melanoma tumorigenesis by activating YAP. Cancer Cell 2014;25(6):822–30.
[131] Rosenbluh J, Nijhawan D, Cox AG, Li X, Neal JT, Schafer EJ, et al. β-Catenin-driven cancers require a YAP1 transcriptional complex for survival and tumorigenesis. Cell 2012;151(7):1457–73.
[132] Chan SW, Lim CJ, Chong YF, Pobbati AV, Huang C, Hong W. Hippo pathway-independent restriction of TAZ and YAP by angiomotin. J Biol Chem 2011;286:7018–26.
[133] Zhao B, Li L, Lu Q, Wang LH, Liu C, Lei Q, et al. Angiomotin is a novel Hippo pathway component that inhibits YAP oncoprotein. Genes Dev 2011;25:51–63.
[134] Wang W, Huang J, Chen J. Angiomotin-like proteins associate with and negatively regulate YAP1. J Biol Chem 2011;286(6):4364–70.
[135] Wang Y, Justilien V, Brennan KI, Jamieson L, Murray NR, Fields AP. PKCβ regulates nuclear YAP1 localization and ovarian cancer tumorigenesis. Oncogene 2017;36(4):534–45.
[136] Zhao Z, Zheng N, Wang L, Hou Y, Zhou X, Wang Z, et al. Rottlerin exhibits antitumor activity via down-regulation of TAZ in non-small cell lung cancer. Oncotarget 2017;8(5):7827–38. Available from: http://www.oncotarget.com/abstract/13974.
[137] Basu D, Lettan R, Damodaran K, Strellec S, Reyes-Mugica M, Rebbaa A. Identification, mechanism of action, and antitumor activity of a small molecule inhibitor of Hippo, TGF-, and Wnt signaling pathways. Mol Cancer Ther 2014;13(6):1457–67.
[138] Huang W, Lv X, Liu C, Zha Z, Zhang H, Jiang Y, et al. The N-terminal phosphodegron targets TAZ/WWTR1 protein for SCFβ-TrCP-dependent degradation in response to phosphatidylinositol 3-kinase inhibition. J Biol Chem 2012;287:26245–53.
[139] Gong R, Yu F-X. Targeting the Hippo pathway for anti-cancer therapies. Curr Med Chem 2015;22(35):4104–17.
[140] Zhao Y, Yang X. The Hippo pathway in chemotherapeutic drug resistance. Int J Cancer 2015;137:2767–73.
[141] Pan W, Wang Q, Zhang Y, Zhang N, Qin J, Li W, et al. Verteporfin can reverse the paclitaxel resistance induced by YAP over-expression in HCT-8/T cells without photoactivation through inhibiting YAP expression. Cell Physiol Biochem 2016;39(2):481–90.
[142] Oku Y, Nishiya N, Shito T, Yamamoto R, Yamamoto Y, Oyama C, et al. Small molecules inhibiting the nuclear localization of YAP/TAZ for chemotherapeutics and chemosensitizers against breast cancers. FEBS Open Biol 2015;5:542–9.
[143] Taccioli C, Sorrentino G, Zannini A, Caroli J, Beneventano D, Anderlucci L, et al. MDP, a database linking drug response data to genomic information, identifies dasatinib and statins as a combinatorial strategy to inhibit YAP/TAZ in cancer cells. Oncotarget 2015;6(36):38854–65.

CHAPTER

9

Modulation of the Epigenome (Methylome) to Improve Chemotherapeutic Efficacy

William B. Coleman

Department of Pathology and Laboratory Medicine, Curriculum in Toxicology, UNC Program in Translational Medicine, UNC Lineberger Comprehensive Cancer Center, University of North Carolina School of Medicine, Chapel Hill, NC, United States

Abstract

Current clinical management of cancer typically includes cytotoxic chemotherapy—in the neo-adjuvant setting prior to surgery, in the adjuvant setting following surgery, or as a first-line treatment in patients that present with advanced disease. However, cancer cells often display de novo resistance to chemotherapeutic drugs or develop resistance with treatment. Once a cancer becomes chemoresistant, progression and demise of the patient become inevitable. New strategies that combine epigenetic therapies and cytotoxic chemotherapy show promise for more effective treatments which may lead to improvements in long-term patient outcomes. Demethylating drugs can sensitize cancer cells to chemotherapy, resulting in enhanced cell killing. New studies have shown that this effect is related to re-expression of genes encoding proteins of pro-apoptotic pathways, rendering cancer cells susceptible to cell killing by chemotherapeutic drugs. With further investigation of epigenetic mechanisms of chemoresistance and sensitization in cancer cells, new clinical strategies for management of cancer patients will emerge.

ABBREVIATIONS

5-aza	5-Aza-2′-deoxycytidine
BRCA1	Breast cancer 1
DNMT3b	DNA methyltransferase 3 beta
ERa	Estrogen receptor alpha
HDAC	Histone deacetylase
miRs	MicroRNAs
PR	Progesterone receptor
RARb2	Retinoic acid receptor beta 2
RNAi	RNA interference
TIMP-3	Tissue inhibitor of metalloproteinases 3
TNF	Tumor necrosis factor
TNFR	Tumor necrosis factor receptor

Targeting Cell Survival Pathways to Enhance Response to Chemotherapy. http://dx.doi.org/10.1016/B978-0-12-813753-6.00009-3
Copyright © 2018 Elsevier Inc. All rights reserved.

9.1 INTRODUCTION

Chemotherapy has long been a mainstay of systemic cancer treatment [1]. Even today, many cancer patients present with advanced disease that has spread from the primary site of occurrence into the vasculature, the lymphatics, or to distant tissue sites [2]. Once cancer becomes a systemic disease (metastatic cancer), surgical interventions are ineffective and/or incapable of cure. In fact, patients that present with early-stage (localized) invasive disease often have micrometastases that become evident following surgical removal of the primary cancer. Hence, cytotoxic chemotherapy has utility for treatment of patients with advanced disease (where the whole body is treated), as well as patients with early-stage disease to eliminate micrometastatic spread (adjuvant chemotherapy) (Fig. 9.1). In current cancer management, patients with localized disease may be treated with neo-adjuvant chemotherapy to shrink the primary cancer prior to surgery, resulting in more effective removal of the neoplastic lesion (Fig. 9.1).

Management of patients with chemotherapy is complex as there are numerous morbidities associated with treatments using cytotoxic drugs. The effectiveness of chemotherapeutic drugs tends to be directly associated with the dose of drug utilized. Hence, patients are typically treated with the maximum tolerated dose. In addition, patients typically receive combination chemotherapy which targets multiple pathways in the cancer cells. Durable responses to chemotherapy result in subsets of patients, although the underlying mechanism for chemosensitivity is not always evident. In some cases, the apparent success of adjuvant chemotherapy may actually reflect the effectiveness of surgical removal of the primary cancer (Fig. 9.1). Despite the use of high-dose combination chemotherapy regimens, many patients fail therapy. Some cancers display a natural resistance to specific classes of chemotherapeutic agents, rendering them largely ineffective at cell killing, and others evolve to develop resistance, resulting in cancer cell clones that cannot be killed by the chemotherapeutic drug combination. The development of chemoresistance is associated with progression of neoplastic disease and represents a harbinger of the demise of the patient as treatment fails and the cancer spreads.

Numerous mechanisms have been implicated in the development of chemoresistance (Fig. 9.2). In general, resistant cancer cells avoid contact with the drug by decreasing drug uptake and/or increasing drug efflux. With less intracellular drug, DNA damage (or interaction with other drug targets) that might lead to cell death is minimized. Cancer cells can also become resistant to chemotherapy by increasing DNA repair mechanisms to address the DNA damage that results from drug exposure with genotoxic agents (Fig. 9.2). With effective DNA repair, cancer cells avoid death despite drug exposure. Furthermore, alterations in mechanisms of apoptosis can result in chemoresistance. Increased expression of anti-apoptotic proteins (and pathways) protects cancer cells from drug-induced programmed cell death. Likewise, decreased expression of pro-apoptotic proteins (and pathways) protects cancer cells from programmed cell death (Fig. 9.2).

Several genetic mechanisms leading to chemoresistance in cancer cells have been elucidated. In this chapter, epigenetic mechanisms of chemoresistance and the use of epigenetic drugs to sensitize cancer cells to chemotherapy will be discussed. This discussion will focus on breast cancer as a cancer model system, DNA hypermethylation as a prominent epigenetic mechanism, and demethylating drugs that demonstrate potential strategies for overcoming chemoresistance. In particular, this chapter will present evidence that epigenetic

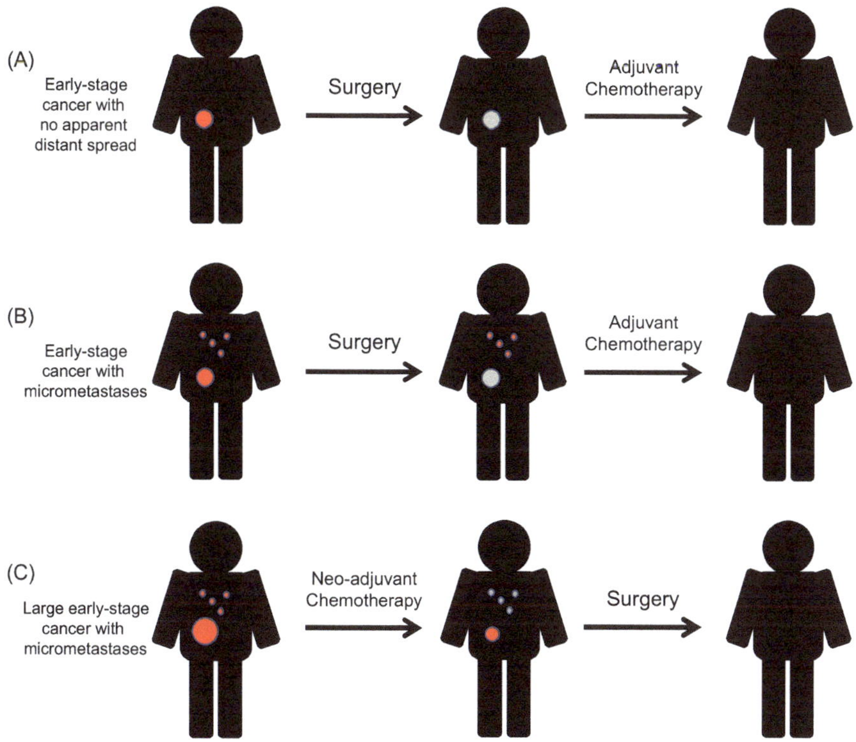

FIG. 9.1 **Utilization of cancer chemotherapy strategies and ideal/potential patient outcomes.** (A) In patients with an early-stage cancer that is small and is not associated with distant spread (micrometastases or other), surgery can often provide a cure. However, adjuvant chemotherapy is typically employed to treat any cancer that might be present, but is clinically undetectable. (B) Many patients present with early-stage cancer that is associated with microscopic spread (micrometastases). These micrometastases are not realized until after surgery when they expand in size. Adjuvant chemotherapy can eliminate these micrometastases in some cases, resulting in cure when combined with surgical removal of the primary cancer. (C) Patients that present with a large primary cancer (with or without micrometastases) may be treated with neoadjuvant therapy to shrink the primary cancer prior to surgery. Neoadjuvant therapy might also eliminate micrometastases, resulting in cure once surgery is performed.

loss of pro-apoptotic proteins and pathways accounts for drug resistance in breast cancer cell lines that display an aberrant hypermethylation phenotype and that treatment with demethylating drugs results in re-expression of the associated apoptotic genes concurrent with increased chemosensitivity.

9.2 DNA METHYLATION IN CANCER

DNA methylation (chemical alteration of cytosine bases) is the most studied and best understood mechanism of epigenetic regulation [3,4] and is regarded as the hallmark of epigenetic modification. DNA methylation occurs almost exclusively on cytosines within CpG dinucleotides, which are found in the genome at ~20% of the predicted frequency, and the

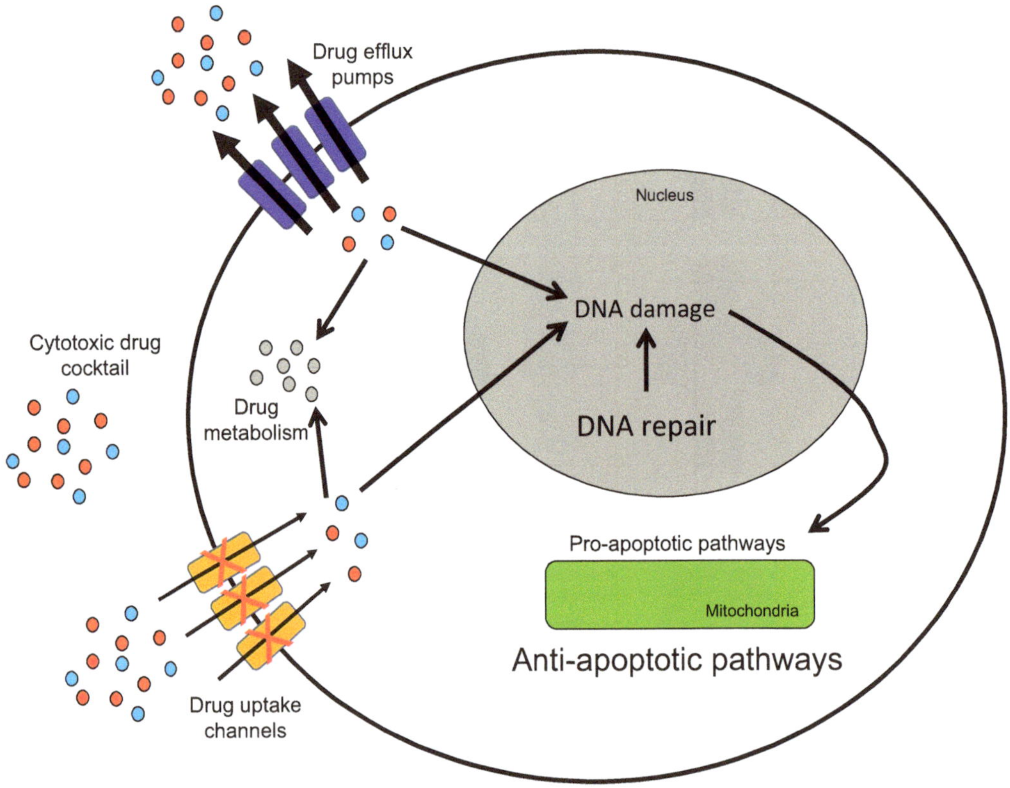

FIG. 9.2 **Cellular mechanisms of chemotherapy resistance.** Most cancer patients are treated with a cocktail of chemotherapeutic drugs (represented here by red and blue circles) which attack different pathways in the cancer cells. Cancer cells can avoid death in response to chemotherapy using various mechanisms. Shown here are decreased drug uptake, increased drug efflux, increased drug metabolism, increased DNA repair, and aberrant apoptotic mechanisms (decreased pro-apoptotic activity and increased anti-apoptotic activity).

majority (>70%) of CpG dinucleotides are typically methylated in any given cell type [5]. However, regions of CpG density, termed CpG islands [6,7], occur in the promoter sequences of many genes, proximal to their transcription start site [8]. CpG islands are conventionally defined as $\geq$200 bp with $\geq$50% G + C and $\geq$0.6 CpG observed/CpG expected [9], although a more rigorous definition ($\geq$200 bp with $\geq$60% G + C and $\geq$0.7 CpG observed/CpG expected) has been proposed [7]. In reality, many investigators report detection of CpG islands in gene promoters that do not contain regions of CpG density conforming to these commonly accepted guidelines. However, good evidence has emerged that gene promoters with intermediate CpG features, or that lack CpG density, may be sensitive to and regulated by DNA methylation [10–12]. Gene promoters and other regulatory regions of housekeeping genes are largely unmethylated in most cell types, and tissue-specific genes are not methylated only in the cell types where these genes are transcriptionally active [13]. Most promoter CpG islands are unmethylated in normal tissues, but may become methylated in cancers leading to transcriptional silencing of tumor suppressor genes (and other negative mediators of neoplastic growth) resulting in selective growth advantages for emergent neoplastic cells.

Neoplastic transformation is associated with alterations in DNA methylation, including both global hypomethylation and gene-specific hypermethylation [14–16]. DNA hypomethylation in cancer was first recognized in rodent models of liver cancer [17] and subsequently was reported in human cancer [18–20]. Hypomethylation of cancer cell genomes reflect loss of methylation in CpG-depleted regions where most CpG dinucleotides would be expected to be methylated [18,20,21]. The loss of methylation in these regions of the genome may be associated with aberrant or inappropriate expression of some genes that could contribute to neoplastic transformation, tumorigenesis, or cancer progression [19]. In addition, genome-wide demethylation has been shown to contribute to chromosomal instability by destabilizing pericentromeric regions of certain chromosomes [22–24]. Gains in DNA methylation in cancer cells typically reflect hypermethylation of CpG islands or other regions of CpG density in gene promoter regions, which can lead to gene silencing [14]. Methylation-dependent gene silencing is a normal mechanism for regulation of gene expression [25], but in cancer cells methylation-dependent epigenetic gene silencing represents a mutation-independent mechanism for inactivation of tumor suppressor genes [26]. A significant number of cancer-related genes have been identified that are subject to methylation-dependent silencing [5,27] and many of these genes contribute to the hallmarks of cancer [28,29]. These observations combine to strongly suggest that epigenetic events, and particularly those involving DNA methylation, represent fundamental aspects of cancer and play key roles in neoplastic transformation and progression.

9.3 ABNORMAL DNA METHYLATION IN BREAST CANCER

Abnormal DNA methylation is a well-recognized hallmark of cancer [4,15,16,30]. DNA methylation-dependent epigenetic silencing of tumor suppressor genes and other genes required for neoplastic transformation, tumorigenesis, and tumor progression have been described in a number of studies. Recent evidence suggests that epigenetic mechanisms play a major role in breast carcinogenesis [10,31–37]. Epigenetic alterations differ from genetic alterations in that they occur more frequently, are reversible, and occur at defined regions of specific genes. DNA methylation is a well-known epigenetic mechanism, and a number of different genes have been shown to be inactivated in breast cancer through methylation-dependent gene silencing. Some of these genes are silenced through a direct effect of DNA methylation, whereas others are affected through indirect mechanisms.

Numerous methylation-sensitive genes have been identified in breast cancer cell lines through experimental studies [10,38], but not all have been functionally characterized. However, a number of important cancer-related genes are known to be epigenetically regulated in breast cancer. Among the genes that have been determined to be directly silenced by DNA methylation in breast cancer are cell-cycle control genes ($p16^{INK4a}$), steroid receptor genes (*ERα, PR, RARβ2*), tumor suppressor genes (*BRCA1*), genes associated with cancer metastasis (*E-cadherin, TIMP-3*), and others [39–43]. The $\mathrm{p16^{INK4a}}$ cyclin-dependent kinase inhibitor is inactivated through methylation in several human cancers. In breast cancer, $p16^{INK4a}$ is methylated in 20%–30% of tumors and cell lines, with a concomitant loss of expression [44,45]. Loss of $p16^{INK4a}$ expression in this subset of breast cancers may contribute to unregulated cell proliferation and tumorigenesis. A significant percentage of breast cancers lack expression of the estrogen receptor (and other steroid receptors), but loss of *ER* gene expression is not associated with gene deletion or somatic mutation [46]. Rather, methylation-dependent silencing of the *ER* gene is responsible for the loss of expression in these tumors [47,48]. Somatic mutations of the *BRCA1* gene do not frequently occur in nonhereditary breast cancers [49]. Therefore, an alternative mechanism for *BRCA1* inactivation involving DNA methylation was proposed [50–52]. Subsequently, several studies documented methylation-dependent epigenetic silencing of *BRCA1* in sporadic breast cancer [53–57]. Loss of *E-cadherin* gene expression in breast cancer is associated with an aggressive tumor phenotype and decreased patient survival [58]. Methylation-dependent loss of *E-cadherin* gene expression has been shown in 30% of primary breast cancers and up to 60% of metastatic tumors [59]. Loss of *TIMP-3* expression in breast tumors potentially results in increased proteolytic activity from matrix metalloproteinase enzymes [60]. The *TIMP-3* promoter is methylated in ~30% of primary breast cancers and breast cancer cell lines [61]. Both these methylation-related losses of gene expression are likely to contribute to tumor progression and spread.

9.4 EPIGENETIC THERAPY IN BREAST CANCER TREATMENT

Given that aberrant DNA methylation and epigenetic silencing of gene expression are now well-recognized hallmarks of cancer [4,15,16,30], numerous investigators have suggested that cancer should be treated with "epigenetic therapy" [62–64]. In contrast to gene therapy, epigenetic therapy (demethylating treatment) alters gene expression patterns in breast cancer without complications from enhanced immune response to therapeutic DNA. The reversibility of epigenetic alterations makes them excellent targets for improving breast cancer outcomes. The goal of such therapy would be to effect changes in gene expression, including re-expression of silenced genes (like tumor suppressor genes), that alter the clinical behavior of the tumor or the response of the tumor to other therapeutic modalities (such as chemotherapy). This concept has been tested in a breast cancer cell model system based on the MCF7 cell line using known demethylating drugs [65,66]. These studies provide strong evidence for enhancement of chemotherapeutic effect in MCF7 cells following demethylation of genomic DNA. These investigators have also initiated clinical trials with combination therapy using demethylating and cytotoxic drugs [67–69]. These studies strongly suggest that this epigenetic therapy will benefit breast cancer patients.

The majority of triple-negative breast cancers (basal-like and claudin-low breast cancers) overexpress DNMT3b, leading to aberrant DNA hypermethylation and concurrent loss of numerous methylation-sensitive genes [11,38]. It is likely that the methylation-dependent silencing of critical genes contributes significantly to the lack of sensitivity to standard chemotherapeutic agents by triple-negative breast cancers. The goal of epigenetic therapy in triple-negative breast cancers is to render cancer cells sensitive to chemotherapy through re-expression of methylation-sensitive genes secondary to normalization of DNA methylation patterns in response to inhibition of DNMT3b. The observation that primary breast cancers can be classified for aberrant DNA hypermethylation based upon methylation-sensitive gene expression signatures presents the opportunity to exploit gene-silencing events as biomarkers to identify which triple-negative breast cancer patients are most likely to benefit from epigenetic therapy [11].

Various forms of epigenetic therapy have been investigated experimentally. We explored pharmacologic epigenetic therapy and targeted epigenetic therapy in combination with cytotoxic chemotherapy in a breast cancer cell line model where the cell lines exhibit aberrant DNA hypermethylation related to overexpression of DNMT3b [70]. The goal of these studies was to evaluate the effectiveness of targeting the DNA methylation machinery to modify the sensitivity of breast cancer cells to cytotoxic drugs. Epigenetic treatment was accomplished through pharmacologic inhibition of DNA methyltransferase activity using 5-aza-2′-deoxycytidine (5-aza) and targeted inhibition of overexpressed DNMT3b using RNAi-mediated *DNMT3b* knockdown [70]. The results show that 5-aza pretreatment sensitizes breast cancer cells with aberrant DNA hypermethylation to cell killing by cytotoxic drugs and that the improved chemotherapeutic efficacy is a function of dose and duration of exposure to 5-aza [70]. Treatment of breast cancer cells for 7 days with low-dose 5-aza resulted in substantial decreases in the IC50 for doxorubicin (60%), paclitaxel (37%), and 5-fluorouracil (93%) [70].

Although we used 5-aza for these investigations, several other epigenetic drugs have been studied in a similar fashion [64,71–73]. In addition to DNMT inhibitors, histone deacetylase inhibitors may also prove to be useful in the epigenetic sensitization of breast cancer cells to chemotherapy, used alone or in conjunction with drugs like 5-aza. We also observed an increase in the effectiveness of chemotherapeutic drugs after targeted inhibition of DNMT3b using RNAi [70]. Although RNAi may not currently represent a viable approach for clinical treatment of breast cancer, model systems using RNAi in vitro demonstrate how targeted agents like small-molecule inhibitors might function therapeutically. Several small-molecule inhibitors of DNMT3b have been described [74]. Our results strongly suggest that DNMT3b is an excellent target for development of rational therapeutic approaches for breast cancers that exhibit aberrant DNA hypermethylation (such as triple-negative breast cancers). Our studies in cell lines provide proof-of-concept that targeting DNMT3b in breast cancer cells with aberrant DNA hypermethylation sensitizes them to cell killing by cytotoxic drugs and that this strategy can be exploited to improve patient outcomes that involve standard therapeutic approaches. Increasing the efficacy of chemotherapy through epigenetic therapy might benefit patients in at least two different ways. First, increasing the efficacy of a certain fixed dose of drug may increase the benefits of chemotherapy without associated increases in toxic side effects. Second, a lower dose of chemotherapeutic drug may be used to achieve a certain fixed therapeutic effect, but with diminished side effects.

Another approach to epigenetic therapy has emerged from our recent elucidation of the molecular basis of aberrant DNA hypermethylation in triple-negative breast cancers [75,76]. We have termed this approach biologic epigenetic therapy. Biologic epigenetic therapy is based upon the concept that normal regulation of DNMT3b can be restored by replacement of lost regulatory miRs [12]. Our experimental studies in cell lines showed that restoration of post-transcriptional regulation of DNMT3b mRNA can be accomplished using specific pre-miRs as single agents [75]. With recent development of nanoparticles for delivery of drugs and other active agents specifically to cancer cells, it becomes possible that pre-miRs might have practical utility as therapeutic agents. We would suggest that pre-miRs might be used in combination to target *DNMT3b* regulation in triple-negative breast cancers to normalize the methylome of these cancers and sensitize them to standard chemotherapies.

9.5 MECHANISMS GOVERNING IMPROVED RESPONSES TO CHEMOTHERAPY AFTER EPIGENETIC SENSITIZATION

Having made the observation that cell killing efficacy by chemotherapeutic drugs was enhanced following epigenetic drug exposure in breast cancer cell lines, the logical follow-up question revolves around the mechanism of sensitization. We speculated that exposure of cell lines to demethylating drugs restored expression of genes that are critical to pathways that produce cell death in response to cytotoxic drugs. Hence, pro-apoptotic and anti-apoptotic genes emerged as targets of interest. It is known that cancer cells avoid cell death through modification of apoptotic pathways.

We surveyed expression of 328 genes associated with apoptosis and survival in Hs578T breast cancer cells (which express the aberrant DNA hypermethylator phenotype) and identified several pro-apoptotic genes that are not expressed (undetected) or expressed at negligible levels in Hs578T cells, including *FASLG*, *IGF1*, *CD27*, and *BLK* (Table 9.1). Epigenetic silencing of these genes (directly or indirectly) would convey a survival advantage to the cancer cells through elimination of pro-apoptotic signaling (Fig. 9.3A). In addition, several members of the tumor necrosis factor (TNF) and TNF receptor (TNFR) superfamilies were identified that are not expressed (undetected) or expressed at negligible levels in Hs578T cells, including *TNFSF8*, *TNFSF10*, *TNFSF11*, *TNFSF14*, *TNFSF15*, *TNFRSF13B*, *TNFRSF17*, and *TNFRSF18* (Table 9.1). Among these TNF/TNFR superfamily genes, *TNFSF10*, *TNFSF14*, *TNFSF15*, *TNFRSF1B*, and *TNFRSF8* have been directly implicated as activators of apoptosis (Fig. 9.3B). Several of these pro-apoptotic genes and/or members of associated pathways have been shown to be subject to methylation-dependent silencing in cancer cells. In contrast, several anti-apoptotic genes were found to be abundantly expressed in Hs578T breast cancer cells, including *HSPB1* and *NPM1*. It is not known whether epigenetic mechanisms contribute to overexpression of anti-apoptotic genes in breast cancer. The contributions of aberrant DNA hypermethylation to the silencing of pro-apoptotic genes (and/or activation of anti-apoptotic genes) as a mechanism of chemotherapeutic resistance was explored further using a panel of breast cancer cell lines. Among the genes examined that regulate pro-apoptotic pathways, lack of expression was a consistent feature among breast cancer cell lines propagated in control medium (Fig. 9.4). However, significant induction of gene expression was observed for each of these genes in the index Hs578T breast cancer cell line following 5-aza treatment. In a similar fashion, substantial induction was seen in specific cell lines for specific genes in response to 5-aza treatment (Fig. 9.4). BLK is known to be methylation-sensitive, consistent with these responses to 5-aza treatment. These results suggest that FASLG, CD27, and IGF1 may also be subject to methylation-dependent silencing in breast cancer. Lack of expression of the genes that regulate pro-apoptotic pathways, either as a TNF signaling molecule (*TNFSF10*, *TNFSF11*, *TNFSF14*, *TNFSF15*) or as a TNFR molecule (*TNFRSF1B*, *TNFRSF8*, *TNFRSF13B*, *TNFRSF17*), was a consistent feature among breast cancer cell lines examined (Fig. 9.5). However, significant induction of gene expression was observed for each of these genes in the index Hs578T breast cancer cell line in response to 5-aza treatment. Likewise, substantial induction of specific TNF/TNFR superfamily genes was observed in response to 5-aza treatment of some breast cancer cell lines (Fig. 9.5). *TNFSF10*, *TNFRSF1B*, and *TNFRSF8* are known to be methylation-sensitive, consistent with the observed responses to 5-aza

TABLE 9.1 Discovery of Candidate Genes Related to Apoptosis and Survival in Human Hs578T Breast Cancer Cells

Genes not expressed		
ADCY1	Hs.192215	Adenylate cyclase 1
BCL2A1	Hs.227817	BCL2-related protein A1
BLK	Hs.146591	B lymphoid tyrosine kinase
CCL4	Hs.75703	Chemokine (C–C motif) ligand 4
CCL19	Hs.50002	Chemokine (C–C motif) ligand 29
CD27	Hs.355307	CD27 molecule
CR2	Hs.445757	Complement component receptor 2
CXCL13	Hs.100431	Chemokine (C–X–C motif) ligand 13
FASLG	Hs.2007	Fas ligand (TNF superfamily, member 6)
FCER2	Hs.465778	Fc fragment of IgE receptor, low-affinity II (CD23)
FLT3	Hs.507590	Fms-related tyrosine kinase 3
IGF1	Hs.160562	Insulin-like growth factor 1
IL-10	Hs.193717	Interleukin 10
IL-2	Hs.89679	Interleukin 2
INPP5D	Hs.262886	Inositol polyphosphate-5-phosphatase
NOS1	Hs.654410	Nitric oxide synthase 1
NOS3	Hs.647092	Nitric oxide synthase 3
PRKACG	Hs.158029	Protein kinase, cAMP-dependent, catalytic, gamma
TNFSF10	Hs.478275	Tumor necrosis factor superfamily, member 10
TNFSF11	Hs.333791	Tumor necrosis factor superfamily, member 11
TNFSF14	Hs.129708	Tumor necrosis factor superfamily, member 14
TNFSF15	Hs.23349	Tumor necrosis factor superfamily, member 15
TNFRSF1B	Hs.256278	Tumor necrosis factor receptor superfamily, member 1B
TNFRSF8	Hs.1314	Tumor necrosis factor receptor superfamily, member 8
TNFRSF13B	Hs.158341	Tumor necrosis factor receptor superfamily, member 13B
TNFRSF17	Hs.2556	Tumor necrosis factor receptor superfamily, member 17
VAV1	Hs.116237	Vav 1 guanine nucleotide exchange factor
Genes expressed at low levels		
CCL21	Hs.57907	Chemokine (C–C motif) ligand 21
CD40LG	Hs.592244	CD40 ligand
PRKCZ	Hs.496255	Protein kinase C, zeta
TNFSF8	Hs.494901	Tumor necrosis factor superfamily, member 8
Genes expressed at high levels		
GNAS	Hs.125898	GNAS complex locus
HSPB1	Hs.520973	Heat shock 27 kDa protein 1
NPM1	Hs.557550	Nucleophosmin
PPIB	Hs.434937	Peptidylprolyl isomerase B
SQSTM1	Hs.724025	Sequestosome 1

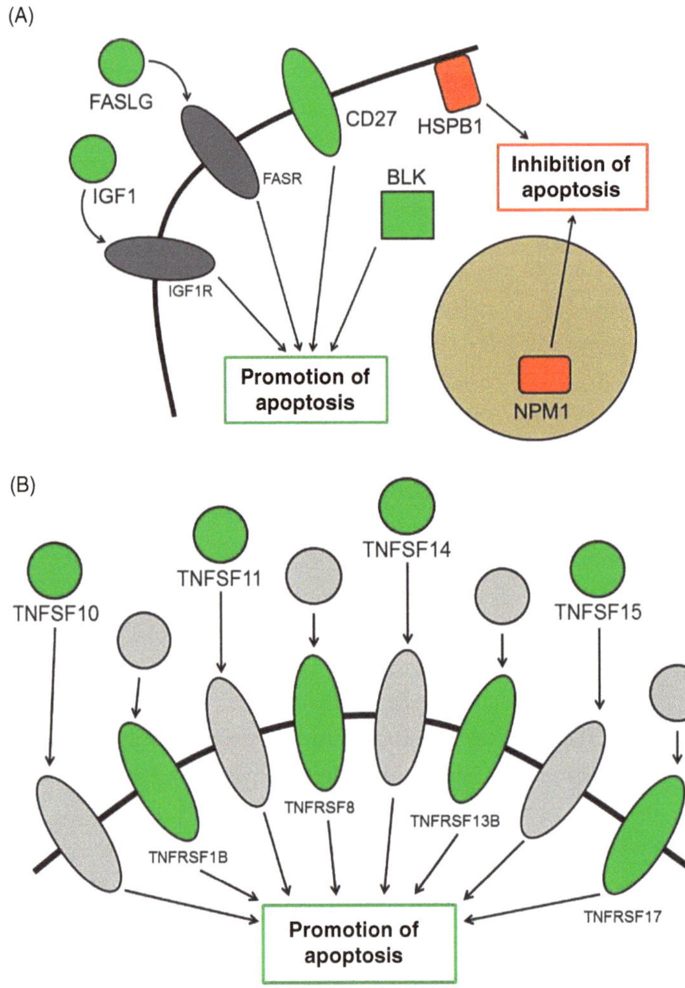

FIG. 9.3 **Pro-apoptotic and TNF superfamily genes.** (A) The response of four genes directly involved in pro-apoptotic pathways to demethylating drug treatment was examined, including cell-surface receptor CD27 (TNFRSF7), ligands for FASR (FASLG) and IGF1R (IGF1), and cytoplasmic tyrosine kinase (BLK). The protein products for these genes are shown in green in the schematic. Two genes that inhibit apoptosis, including HSPB1 (HSP27) and NPM1, are described in the text. The protein products for these genes are shown in red in the schematic. (B) The response of eight TNF/TNFR superfamily genes to demethylating drug treatment was examined. The protein products of these genes are directly involved in pro-apoptotic pathways, including TNF ligands (TNFSF10, TNFSF11, TNFSF14, TNFSF15), and cell-surface TNF receptors (TNFRSF1B, TNFRSF8, TNFRSF13B, TNFRSF17). The protein products for these genes are shown in green in the schematic. Abbreviation: TNF, tumor necrosis factor.

treatment. These results suggest that *TNFSF11*, *TNFSF14*, *TNFSF15*, *TNFRSF13B*, and *TNFRSF17* may also be subject to methylation-dependent silencing in breast cancer. Further studies are needed to examine promoter methylation events in these pro-apoptotic and TNF/TNFR superfamily genes in breast cancer cells that display the aberrant DNA hypermethylation phenotype and in response to demethylating drugs like 5-aza.

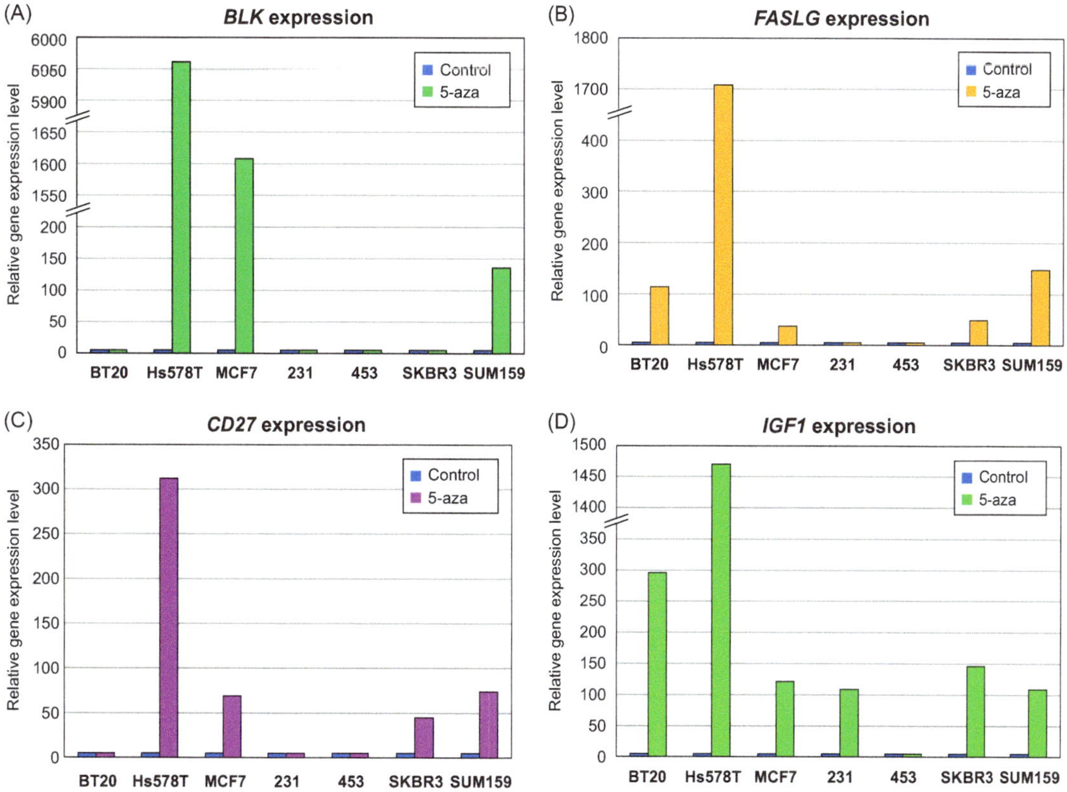

FIG. 9.4 **Re-expression of pro-apoptotic genes in breast cancer cell lines following treatment with demethylating drugs.** A panel of human breast cancer cell lines consisting of BT20, Hs578T, MCF7, MDA-MB-231 (represented as 231), MDA-MB-453 (represented as 453), SKBR3, and SUM159 were propagated in DMEM/F12 mix medium containing 10% fetal calf serum with or without 500 nM 5-aza-2′-deoxycytidine for 7 days [70]. At the end of the culture period, cells were harvested, RNA prepared, and real-time PCR was conducted for the genes shown: (A) *BLK*; (B) *FASLG*; (C) *CD27*; (D) *IGF1*.

9.6 EPIGENETICS OF CHEMORESISTANCE

The studies reviewed here suggest strongly that re-expression of genes associated with pro-apoptotic pathways in response to demethylating drugs represents one mechanism for sensitizing breast cancer cells to cytotoxic chemotherapy. However, additional investigations are warranted to determine whether re-expression of pro-apoptotic genes is the direct result of demethylating events in their associated promoters, or whether indirect mechanisms account for some of these responses. With ever-broadening studies of the human epigenome (and various cancer epigenomes), important information related to mechanisms of chemotherapy resistance and sensitization will be gained. Likewise, as greater understanding of these phenomena is gained, appropriate strategies can be developed to address epigenetic therapy of human cancer in the clinical setting. Possibilities include the use of broad-spectrum demethylating drugs (like 5-aza) or novel targeted drugs (like

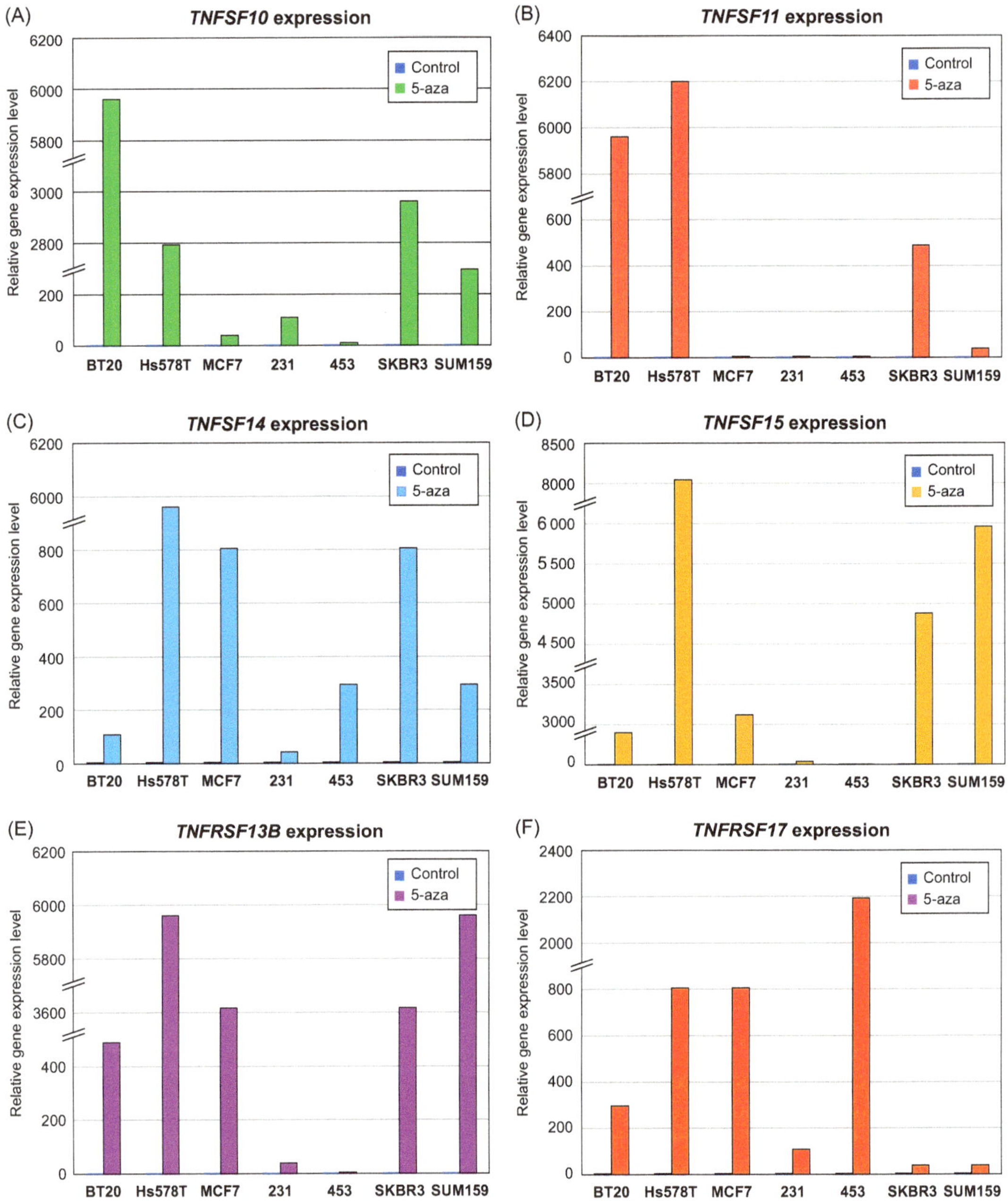

FIG. 9.5 **Re-expression of TNF and TNFR genes in breast cancer cell lines following treatment with demethylating drugs.** A panel of human breast cancer cell lines consisting of BT20, Hs578T, MCF7, MDA-MB-231 (represented as 231), MDA-MB-453 (represented as 453), SKBR3, and SUM159 were propagated in DMEM/F12 mix medium containing 10% fetal calf serum with or without 500 nM 5-aza-2′-deoxycytidine for 7 days [70]. At the end of the culture period, cells were harvested, RNA prepared, and real-time PCR was conducted for the genes shown: (A) *TNFSF10*; (B) *TNFSF11*; (C) *TNFSF14*; (D) *TNFSF15*; (E) *TNFRSF13B*; (F) *TNFRSF17*.

small-molecule inhibitors) that correct the aberrant methylome of cancer cells resulting in sensitization to cytotoxic drugs (or other therapeutic modalities such as immunotherapy or other targeted therapies).

9.7 EPIGENETIC CANCER THERAPY

The ultimate goal of epigenetic therapy in cancer is to reverse the biology associated with the malignant phenotype associated with aberrant DNA hypermethylation events (and/or changes in histone modifications), which results in changes in gene expression patterns. The changes in gene expression patterns observed with demethylating treatment are complex and extend beyond upregulation of a subset of genes due to promoter demethylation [10,12]. Of importance, some genes show downregulation in response to demethylating drug treatment, which might be due to direct mechanisms or indirect mechanisms. The response to demethylating drugs will require much more study before all of these molecular outcomes are known.

There are several possible beneficial outcomes of epigenetic therapy in cancer. One of these is the sensitization of the cancer cells to traditional cytotoxic chemotherapy, to enhance cell killing or to overcome resistance. This has been shown experimentally to be a plausible goal with respect to the use of demethylating drugs. In some cancers, re-expression of critical genes may result in a slowing of cell proliferation or loss of viability of the tumor-initiating cells (perhaps related to re-expression of a tumor suppressor gene or pathway). Loss of proliferative ability of the cancer in response to tumor suppressor gene expression might reflect an exit from the cell cycle (movement toward differentiation) or activation of some SOS system and cell death. All of these possibilities could improve outcomes for cancer patients. However, there are a number of unanswered questions. For instance, will epigenetic sensitization of cancer cells result in a sensitization of the host? This could result in a greater magnitude of cytotoxic effects on the bone marrow and other critical systems in the body and would be counterproductive to therapy (especially if the cytotoxic drug dose needed to be reduced). Experimental studies in animals will address some of these questions effectively if properly designed and controlled. Likewise, carefully planned human studies could also answer some of these critical questions prior to wide application of epigenetic drugs in the clinical setting in routine oncology practice.

Conflict of Interest: No potential conflicts of interest were disclosed.

References

[1] DeVita VT Jr, Chu E. A history of cancer chemotherapy. Cancer Res 2008;68(21):8643–53.
[2] Siegel RL, Miller KD, Jemal A. Cancer statistics. Cancer J Clin 2017;67(1):7–30.
[3] Feinberg AP. The epigenetics of cancer etiology. Semin Cancer Biol 2004;14(6):427–32.
[4] Baylin SB. DNA methylation and gene silencing in cancer. Nat Clin Pract Oncol 2005;2(Suppl. 1):S4–S11.
[5] Widschwendter M, Jones PA. DNA methylation and breast carcinogenesis. Oncogene 2002;21(35):5462–82.
[6] Jones PA, Takai D. The role of DNA methylation in mammalian epigenetics. Science 2001;293:1068–70.
[7] Takai D, Jones PA. Comprehensive analysis of CpG islands in human chromosomes 21 and 22. Proc Natl Acad Sci USA 2002;99:3740–5.
[8] Bird AP. CpG-rich islands and the function of DNA methylation. Nature 1986;321:209–13.
[9] Gardiner-Garden M, Frommer M. CpG islands in vertebrate genomes. J Mol Biol 1987;196(2):261–82.
[10] Rivenbark AG, Jones WD, Risher JD, Coleman WB. DNA methylation-dependent epigenetic regulation of gene expression in MCF-7 breast cancer cells. Epigenetics 2006;1(1):32–44.

[11] Roll JD, Rivenbark AG, Sandhu R, Parker JS, Jones WD, Carey LA, et al. Dysregulation of the epigenome in triple-negative breast cancers: basal-like and claudin-low breast cancers express aberrant DNA hypermethylation. Exp Mol Pathol 2013;95(3):276–87.
[12] Sandhu R, Roll JD, Rivenbark AG, Coleman WB. Dysregulation of the epigenome in human breast cancer: contributions of gene-specific DNA hypermethylation to breast cancer pathobiology and targeting the breast cancer methylome for improved therapy. Am J Pathol 2015;185(2):282–92.
[13] Paulsen M, Tierling S, Walter J. DNA methylation and the mammalian genome. In: Tost J, editor. Epigenetics. Norfolk, UK: Caister Academic Press; 2008. p. 1–21.
[14] Herman JG, Baylin SB. Gene silencing in cancer in association with promoter hypermethylation. N Engl J Med 2003;349(21):2042–54.
[15] Baylin S. DNA methylation and epigenetic mechanisms of carcinogenesis. Dev Biol (Basel) 2001;106:85–7.
[16] Baylin SB, Herman JG, Graff JR, Vertino PM, Issa JP. Alterations in DNA methylation: a fundamental aspect of neoplasia. Adv Cancer Res 1998;72:141–96.
[17] Lapeyre JN, Becker FF. 5-Methylcytosine content of nuclear DNA during chemical hepatocarcinogenesis and in carcinomas which result. Biochem Biophys Res Commun 1979;87(3):698–705.
[18] Feinberg AP, Vogelstein B. Hypomethylation distinguishes genes of some human cancers from their normal counterparts. Nature 1983;301(5895):89–92.
[19] Feinberg AP, Vogelstein B. Hypomethylation of ras oncogenes in primary human cancers. Biochem Biophys Res Commun 1983;111(1):47–54.
[20] Goelz SE, Vogelstein B, Hamilton SR, Feinberg AP. Hypomethylation of DNA from benign and malignant human colon neoplasms. Science 1985;228(4696):187–90.
[21] Feinberg AP, Vogelstein B. Alterations in DNA methylation in human colon neoplasia. Semin Surg Oncol 1987;3(3):149–51.
[22] Narayan A, Ji W, Zhang XY, Marrogi A, Graff JR, Baylin SB, et al. Hypomethylation of pericentromeric DNA in breast adenocarcinomas. Int J Cancer 1998;77(6):833–8.
[23] Eden A, Gaudet F, Waghmare A, Jaenisch R. Chromosomal instability and tumors promoted by DNA hypomethylation. Science 2003;300(5618):455.
[24] Gaudet F, Hodgson JG, Eden A, Jackson-Grusby L, Dausman J, Gray JW, et al. Induction of tumors in mice by genomic hypomethylation. Science 2003;300(5618):489–92.
[25] Momparler RL. Cancer epigenetics. Oncogene 2003;22(42):6479–83.
[26] Jones PA, Laird PW. Cancer epigenetics comes of age. Nat Genet 1999;21(2):163–7.
[27] Tsou JA, Hagen JA, Carpenter CL, Laird-Offringa IA. DNA methylation analysis: a powerful new tool for lung cancer diagnosis. Oncogene 2002;21(35):5450–61.
[28] Hanahan D, Weinberg RA. The hallmarks of cancer. Cell 2000;100(1):57–70.
[29] Hanahan D, Weinberg RA. Hallmarks of cancer: the next generation. Cell 2011;144(5):646–74.
[30] Baylin SB, Esteller M, Rountree MR, Bachman KE, Schuebel K, Herman JG. Aberrant patterns of DNA methylation, chromatin formation and gene expression in cancer. Hum Mol Genet 2001;10(7):687–92.
[31] Bae YK, Brown A, Garrett E, Bornman D, Fackler MJ, Sukumar S, et al. Hypermethylation in histologically distinct classes of breast cancer. Clin Cancer Res 2004;10(18 Pt 1):5998–6005.
[32] Domann FE, Rice JC, Hendrix MJ, Futscher BW. Epigenetic silencing of maspin gene expression in human breast cancers. Int J Cancer 2000;85(6):805–10.
[33] Fackler MJ, McVeigh M, Evron E, Garrett E, Mehrotra J, Polyak K, et al. DNA methylation of RASSF1A, HIN-1, RAR-beta, Cyclin D2 and Twist in *in situ* and invasive lobular breast carcinoma. Int J Cancer 2003;107(6):970–5.
[34] Huang TH, Perry MR, Laux DE. Methylation profiling of CpG islands in human breast cancer cells. Hum Mol Genet 1999;8(3):459–70.
[35] Jhaveri MS, Morrow CS. Methylation-mediated regulation of the glutathione S-transferase P1 gene in human breast cancer cells. Gene 1998;210(1):1–7.
[36] Rivenbark AG, Jones WD, Coleman WB. DNA methylation-dependent silencing of CST6 in human breast cancer cell lines. Lab Invest 2006;86(12):1233–42.
[37] Rivenbark AG, Livasy CA, Boyd CE, Keppler D, Coleman WB. Methylation-dependent silencing of CST6 in primary human breast tumors and metastatic lesions. Exp Mol Pathol 2007;83(2):188–97.
[38] Roll JD, Rivenbark AG, Jones WD, Coleman WB. DNMT3b overexpression contributes to a hypermethylator phenotype in human breast cancer cell lines. Mol Cancer 2008;7:15.
[39] Yang X, Yan L, Davidson NE. DNA methylation in breast cancer. Endocr Relat Cancer 2001;8(2):115–27.

[40] Yan PS, Shi H, Rahmatpanah F, Hsiau TH, Hsiau AH, Leu YW, et al. Differential distribution of DNA methylation within the RASSF1A CpG island in breast cancer. Cancer Res 2003;63(19):6178–86.

[41] Dammann R, Yang G, Pfeifer GP. Hypermethylation of the cpG island of Ras association domain family 1A (RASSF1A), a putative tumor suppressor gene from the 3p21.3 locus, occurs in a large percentage of human breast cancers. Cancer Res 2001;61(7):3105–9.

[42] Widschwendter M, Berger J, Muller HM, Zeimet AG, Marth C. Epigenetic downregulation of the retinoic acid receptor-beta2 gene in breast cancer. J Mammary Gland Biol Neoplasia 2001;6(2):193–201.

[43] Widschwendter M, Berger J, Hermann M, Muller HM, Amberger A, Zeschnigk M, et al. Methylation and silencing of the retinoic acid receptor-beta2 gene in breast cancer. J Natl Cancer Inst 2000;92(10):826–32.

[44] Herman JG, Merlo A, Mao L, Lapidus RG, Issa JP, Davidson NE, et al. Inactivation of the CDKN2/p16/MTS1 gene is frequently associated with aberrant DNA methylation in all common human cancers. Cancer Res 1995;55(20):4525–30.

[45] Woodcock DM, Linsenmeyer ME, Doherty JP, Warren WD. DNA methylation in the promoter region of the p16 (CDKN2/MTS-1/INK4A) gene in human breast tumours. Br J Cancer 1999;79(2):251–6.

[46] Lapidus RG, Nass SJ, Davidson NE. The loss of estrogen and progesterone receptor gene expression in human breast cancer. J Mammary Gland Biol Neoplasia 1998;3(1):85–94.

[47] Ottaviano YL, Issa JP, Parl FF, Smith HS, Baylin SB, Davidson NE. Methylation of the estrogen receptor gene CpG island marks loss of estrogen receptor expression in human breast cancer cells. Cancer Res 1994;54(10):2552–5.

[48] Lapidus RG, Ferguson AT, Ottaviano YL, Parl FF, Smith HS, Weitzman SA, et al. Methylation of estrogen and progesterone receptor gene 5′ CpG islands correlates with lack of estrogen and progesterone receptor gene expression in breast tumors. Clin Cancer Res 1996;2(5):805–10.

[49] Merajver SD, Pham TM, Caduff RF, Chen M, Poy EL, Cooney KA, et al. Somatic mutations in the BRCA1 gene in sporadic ovarian tumours. Nat Genet 1995;9(4):439–43.

[50] Dobrovic A, Simpfendorfer D. Methylation of the BRCA1 gene in sporadic breast cancer. Cancer Res 1997;57(16):3347–50.

[51] Magdinier F, Ribieras S, Lenoir GM, Frappart L, Dante R. Down-regulation of BRCA1 in human sporadic breast cancer; analysis of DNA methylation patterns of the putative promoter region. Oncogene 1998;17(24):3169–76.

[52] Rice JC, Massey-Brown KS, Futscher BW. Aberrant methylation of the BRCA1 CpG island promoter is associated with decreased BRCA1 mRNA in sporadic breast cancer cells. Oncogene 1998;17(14):1807–12.

[53] Bianco T, Chenevix-Trench G, Walsh DC, Cooper JE, Dobrovic A. Tumour-specific distribution of BRCA1 promoter region methylation supports a pathogenetic role in breast and ovarian cancer. Carcinogenesis 2000;21(2):147–51.

[54] Catteau A, Harris WH, Xu CF, Solomon E. Methylation of the BRCA1 promoter region in sporadic breast and ovarian cancer: correlation with disease characteristics. Oncogene 1999;18(11):1957–65.

[55] Hedenfalk I, Duggan D, Chen Y, Radmacher M, Bittner M, Simon R, et al. Gene-expression profiles in hereditary breast cancer. N Engl J Med 2001;344(8):539–48.

[56] Magdinier F, Billard LM, Wittmann G, Frappart L, Benchaib M, Lenoir GM, et al. Regional methylation of the 5′ end CpG island of BRCA1 is associated with reduced gene expression in human somatic cells. FASEB J 2000;14(11):1585–94.

[57] Miyamoto K, Fukutomi T, Asada K, Wakazono K, Tsuda H, Asahara T, et al. Promoter hypermethylation and post-transcriptional mechanisms for reduced BRCA1 immunoreactivity in sporadic human breast cancers. Jpn J Clin Oncol 2002;32(3):79–84.

[58] Bringuier PP, Umbas R, Schaafsma HE, Karthaus HF, Debruyne FM, Schalken JA. Decreased E-cadherin immunoreactivity correlates with poor survival in patients with bladder tumors. Cancer Res 1993;53(14):3241–5.

[59] Nass SJ, Herman JG, Gabrielson E, Iversen PW, Parl FF, Davidson NE, et al. Aberrant methylation of the estrogen receptor and E-cadherin 5′ CpG islands increases with malignant progression in human breast cancer. Cancer Res 2000;60(16):4346–8.

[60] Uria JA, Ferrando AA, Velasco G, Freije JM, Lopez-Otin C. Structure and expression in breast tumors of human TIMP-3, a new member of the metalloproteinase inhibitor family. Cancer Res 1994;54(8):2091–4.

[61] Bachman KE, Herman JG, Corn PG, Merlo A, Costello JF, Cavenee WK, et al. Methylation-associated silencing of the tissue inhibitor of metalloproteinase-3 gene suggest a suppressor role in kidney, brain, and other human cancers. Cancer Res 1999;59(4):798–802.

[62] Brueckner B, Kuck D, Lyko F. DNA methyltransferase inhibitors for cancer therapy. Cancer J 2007;13(1):17–22.

[63] Brueckner B, Lyko F. DNA methyltransferase inhibitors: old and new drugs for an epigenetic cancer therapy. Trends Pharmacol Sci 2004;25(11):551–4.
[64] Yoo CB, Cheng JC, Jones PA. Zebularine: a new drug for epigenetic therapy. Biochem Soc Trans 2004;32(Pt 6): 910–2.
[65] Segura-Pacheco B, Perez-Cardenas E, Taja-Chayeb L, Chavez-Blanco A, Revilla-Vazquez A, Benitez-Bribiesca L, et al. Global DNA hypermethylation-associated cancer chemotherapy resistance and its reversion with the demethylating agent hydralazine. J Transl Med 2006;4:32.
[66] Chavez-Blanco A, Perez-Plasencia C, Perez-Cardenas E, Carrasco-Legleu C, Rangel-Lopez E, Segura-Pacheco B, et al. Antineoplastic effects of the DNA methylation inhibitor hydralazine and the histone deacetylase inhibitor valproic acid in cancer cell lines. Cancer Cell Int 2006;6:2.
[67] Zambrano P, Segura-Pacheco B, Perez-Cardenas E, Cetina L, Revilla-Vazquez A, Taja-Chayeb L, et al. A phase I study of hydralazine to demethylate and reactivate the expression of tumor suppressor genes. BMC Cancer 2005;5(1):44.
[68] Candelaria M, Gallardo-Rincon D, Arce C, Cetina L, Aguilar-Ponce J, Arrieta O, et al. A phase II study of epigenetic therapy with hydralazine and magnesium valproate to overcome chemotherapy resistance in refractory solid tumors. Ann Oncol 2007;18(9):1529–38.
[69] Arce C, Perez-Plasencia C, Gonzalez-Fierro A, de la Cruz-Hernandez E, Revilla-Vazquez A, Chavez-Blanco A, et al. A proof-of-principle study of epigenetic therapy added to neoadjuvant doxorubicin cyclophosphamide for locally advanced breast cancer. PLoS ONE 2006;1:e98.
[70] Sandhu R, Rivenbark AG, Coleman WB. Enhancement of chemotherapeutic efficacy in hypermethylator breast cancer cells through targeted and pharmacologic inhibition of DNMT3b. Breast Cancer Res Treat 2012;131(2):385–99.
[71] Cheng JC, Matsen CB, Gonzales FA, Ye W, Greer S, Marquez VE, et al. Inhibition of DNA methylation and reactivation of silenced genes by zebulariwne. J Natl Cancer Inst 2003;95(5):399–409.
[72] Chuang JC, Yoo CB, Kwan JM, Li TW, Liang G, Yang AS, et al. Comparison of biological effects of non-nucleoside DNA methylation inhibitors versus 5-aza-2′-deoxycytidine. Mol Cancer Ther 2005;4(10):1515–20.
[73] Marquez VE, Kelley JA, Agbaria R, Ben-Kasus T, Cheng JC, Yoo CB, et al. Zebularine: a unique molecule for an epigenetically based strategy in cancer chemotherapy. Ann NY Acad Sci 2005;1058:246–54.
[74] Lyko F, Brown R. DNA methyltransferase inhibitors and the development of epigenetic cancer therapies. J Natl Cancer Inst 2005;97(20):1498–506.
[75] Sandhu R, Rivenbark AG, Coleman WB. Loss of post-transcriptional regulation of DNMT3b by microRNAs: a possible molecular mechanism for the hypermethylation defect observed in a subset of breast cancer cell lines. Int J Oncol 2012;41(2):721–32.
[76] Sandhu R, Rivenbark AG, Mackler RM, Livasy CA, Coleman WB. Dysregulation of microRNA expression drives aberrant DNA hypermethylation in basal-like breast cancer. Int J Oncol 2014;44(2):563–72.

CHAPTER

10

Targeting the Ataxia Telangiectasia and Rad3 Signaling Pathway to Overcome Chemoresistance in Cancer

Marwan Kwok,**, Tatjana Stankovic**

*Institute of Cancer and Genomic Sciences, University of Birmingham, United Kingdom

**Centre for Clinical Haematology, Queen Elizabeth Hospital Birmingham, United Kingdom

Abstract

The ataxia telangiectasia and Rad3-related (ATR) signaling pathway controls DNA replication initiation and cell-cycle progression and coordinates the repair of damaged replication forks. It is, therefore, an important regulator of the cellular response to DNA replication stress. Inhibition of the ATR signaling pathway results in dysregulation of replication origin initiation, leading to exacerbation of replication stress manifested by the slowing and stalling of replication forks, and enhanced susceptibility to fork breakage. Accordingly, inhibition of ATR or its downstream effector Chk1 is potently cytotoxic to tumor cells with heightened levels of endogenous replication stress, particularly in tumors with DNA damage response or cell-cycle checkpoint defects. This results in a functional addiction to the ATR pathway. Small-molecule inhibitors of ATR and Chk1 are currently under investigation. Preclinical studies of these inhibitors have demonstrated single-agent efficacy across a range of malignancies, and results from clinical trials are emerging. In addition, markers of sensitivity to ATR and Chk1 inhibitors are being identified, and novel chemotherapeutic combinations involving these agents are being developed. A number of unresolved issues notwithstanding, the ATR signaling pathway represent a promising target for cancer therapeutics.

ABBREVIATIONS

ALT	Alternative lengtheningof telomeres
AML	Acute myeloid leukemia
ARID1A	AT-rich interaction domain1A
ATM	Ataxia telangiectacia mutated
ATR	Ataxia telangiectacia andRad3-related
ATRIP	ATR-interacting protein
BLM	Bloom syndrome protein
BRCA1	Breast cancer type 1 susceptibility protein
cdk	Cyclin-dependent kinase
chk1	Checkpoint kinase 1
CLL	Chronic lymphocytic leukemia

Targeting Cell Survival Pathways to Enhance Response to Chemotherapy. http://dx.doi.org/10.1016/B978-0-12-813753-6.00010-X
Copyright © 2018 Elsevier Inc. All rights reserved.

DDR	DNA damageresponse
DLBCL	Diffuse large B cell lymphoma
DSBs	Double-stranded breaks
FANCM	Fanconi anemia complementation group M
GCB	Germinal center B-cell like
HRR	Homologousrecombination repair
NHEJ	Nonhomologous end joining
PARP	Poly-(ADP-ribose)polymerase
PDX	Patient-derived xenotransplantation
PIKK	Phosphoinositide 3-kinase-relatedkinase
RPA	Replication protein A
RS	Replication stress
ssDNA	Single-strandedDNA
T-ALL	T-cell acute lymphoblastic leukemia
TOPB1	Topoisomerase 2-bindingprotein 1
WRN	Werner syndrome ATP-dependent helicase

10.1 INTRODUCTION

The DNA damage response (DDR) is essential for the maintenance of genome integrity. This is accomplished through a network of intricately connected pathways, resulting in a dichotomous outcome: either resolution of the incipient DNA damage or apoptotic cell death, depending on the nature and severity of the afflicting lesion. Tumorigenesis is associated with a pervasive corruption of one or more of these pathways, often through mutation or deletion of key DNA repair or regulatory proteins, conferring on malignant cells a survival advantage, avoidance of apoptosis, and chemoresistance [1]. However, the subversion of normal DDR mechanisms can render tumor cells susceptible to further DNA insults, the accumulation of which may become incompatible with survival and results in cell death during mitosis, a process termed mitotic catastrophe [2,3]. Moreover, the disruption of a DNA repair pathway constrains tumor cells to rely on collateral repair pathways to maintain genome stability, thereby creating a vulnerability that can be amenable to therapeutic targeting [4].

Ataxia telangiectasia and Rad3-related (ATR) is a phosphoinositide 3-kinase-related kinase (PIKK) that is the master regulator of one such DDR pathway, with wide-ranging roles in DNA replication [5]. Studies in the last decade have shed light into its function and interaction with other DDR pathways, as well as the effects of inhibiting this pathway. We herein review our current understanding of the mechanistic basis of ATR signaling and illustrate how some tumor cells are distinctively reliant on this pathway and how targeting this pathway leads to tumor cell death. We also highlight important preclinical and clinical studies on this therapeutic target, as a single agent and in combination with other chemotherapeutic agents, and discuss unresolved questions and directions for future research.

10.2 FUNCTIONAL ROLES OF THE ATR SIGNALING CASCADE

10.2.1 The Role of ATR in the Regulation of DNA Replication Initiation

Initiation of DNA replication is a stochastic event, occurring at replication origins from which replication forks arise. Replication origins are clustered and replicated together within replication factories. An important function of ATR is its ability to protect replication forks by acting as a negative regulator of replication origin initiation. This limits the activation of new replication factories, thereby directing replication toward already active factories [6,7]. Such negative regulation of replication initiation is of paramount importance in preventing the depletion of cellular pools of nucleotides and replication proteins, which would otherwise result in genomic instability. Indeed, only a small proportion of licensed origins is active in an unperturbed cell cycle, with the rest remaining dormant [6].

10.2.2 The Role of the ATR Signaling Pathway in the Maintenance of Replication Fork Stability

During DNA replication when the double-stranded DNA helix unwinds, replication fork progression can slow or stall due to replication obstacles such as DNA damage. Continued unwinding of the DNA helix despite stalled DNA polymerase leads to exposure of long stretches of single-stranded DNA (ssDNA) that are prone to damage. This results in replication fork breakage. The ATR pathway functions to prevent this outcome by protecting exposed ssDNA and repairing DNA lesions at these sites. Exposed ssDNA is rapidly coated with replication protein A (RPA), which stabilizes the replication fork and protects ssDNA from deleterious degradation by nucleases [8].

The ATR signaling cascade is initiated by the independent recruitment of two complexes, namely the ATR and ATR-interacting protein (ATRIP) and the RAD9–RAD1–HUS1 (9–1–1) complex, to RPA-coated ssDNA where RPA binds ATRIP and engages with the 9–1–1 complex. Further recruitment of topoisomerase 2-binding protein 1 (TOPBP1) activates ATR by stimulating ATR kinase activity [9]. Recently, the RPA-binding protein Ewing's tumor-associated antigen 1 was also reported to have ATR stimulatory activity and ability to activate ATR independent of TOPBP1 in a parallel pathway [10–12]. Functionally active ATR in turn phosphorylates the downstream effector molecule Chk1 and other substrates such as Werner syndrome ATP-dependent helicase (WRN), Bloom syndrome protein (BLM), and SMARCAL1. A recent study using iPOND method confirmed the recruitment of these proteins to DNA replisomes upon replication fork stalling and highlighted their functional importance in preventing fork collapse and breakage during DNA replication [13].

Through these proteins, ATR promotes replication fork stability and mediates replication restart in the event of fork stalling. This is achieved through fork reversal, whereby stalled forks reverse their course and through fork remodeling resulting in the formation of four-way junctions [14,15]. Helicases such as WRN and BLM, as well as translocases such as SMARCAL1 and the Fanconi anemia complementation group M protein (FANCM) are thought to play a major role in this process [16–21]. Fork reversal, remodeling, and restart together ensure completion of DNA replication and prevent fork breakage.

Proteins downstream of ATR also serve additional functions in facilitating fork repair, through suppression of new origin firing (BLM and FANCM) and protection of replication forks from unscheduled DNA degradation (WRN) [22,23]. Furthermore, Chk1 diffuses throughout the nucleus to suppress new origin firing. Thus, the regulatory activities of ATR signaling are amplified and propagated globally in the event of replication fork stalling [6,24]. This is crucial in curtailing further fork stalls to conserve the supply of RPA for the hitherto perturbed sites and to ensure that an excess of RPA over ssDNA is maintained [25].

10.2.3 The Role of the ATR Signaling Pathway in Cell-Cycle Regulation

Finally, through the downstream kinase Chk1, ATR also plays a major role in controlling cell-cycle progression. Chk1 regulates cellular entry into mitosis through the G2/M checkpoint. Mechanistically, this involves phosphorylation of Cdc25 phosphatases to prevent its binding to the cyclin B1–cyclin-dependent kinase (Cdk1) complex. Cdc25 phosphatases remove inhibitory phosphorylation from Cdk1 that is a prerequisite for the activation of cyclin B1–Cdk1, which initiates mitotic entry. In the presence of stalled replication forks, abolition of cyclin B1–Cdk1 activity through Chk1-mediated phosphorylation of Cdc25 phosphatases prevents premature entry of replicatively perturbed cells into mitosis until replication stalling is rectified and genomic duplication complete [26,27].

10.3 THE ATR PATHWAY AS A CANCER THERAPEUTIC TARGET

10.3.1 Induction of Replication Stress as a Consequence of ATR Pathway Inhibition

When replication fork progression is disrupted, slowing or stalling of replication forks ensues. This results in replication stress (RS), with cells continuing to progress through cell cycle despite the presence of unreplicated DNA [28]. Low levels of RS are ubiquitous in cycling cells and are often tolerated [29]. On the other hand, excessive amounts of RS arising, for instance, from the accumulation of unrepaired DNA damage, misincorporated ribonucleotides, or from difficult-to-replicate segments of DNA within fragile sites, inevitably lead to cell death [28]. The ATR signaling pathway ameliorates RS through its role in the regulation of DNA replication initiation, replication fork stability, and cell-cycle progression, as outlined above. Conversely, inhibition of the ATR pathway is a potent inducer of RS.

A consequence of ATR pathway inhibition is excessive replication origin initiation in an unscheduled and uncontrolled manner. This occurs independent of any additional sources of RS, either endogenous or exogenous. Indeed, DNA fiber analysis showed that the number of active replication forks in cancer cells was increased when ATR is inhibited, even in the absence of DNA damaging agents or other sources of RS [18,30]. Excessive and uncontrolled origin initiation as a result of ATR inhibition depletes essential replication factors such as nucleotides and replication proteins. This leads to a decrease in the rate of replication fork elongation and stalling of replication forks [24] and, as evidenced in cancer cells, to rapid reduction in inter-origin distance indicating increased replication origins [18,31]. These observations are consistent with the induction of RS.

Excessive and uncontrolled initiation of dormant replication origins also generates increased ssDNA. Exposed ssDNA is normally coated with RPA. However, when ATR is inhibited, the excess of ssDNA exhausts the nuclear pool of RPA. Unprotected ssDNA is susceptible to fork breakage, thereby precipitating DNA damage [30]. Furthermore, ATR inhibition prevents proper regulation of SMARCAL1, the deregulated activities of which promote excessive replication fork regression and fork collapse [18]. Together, these mechanisms underpin the cytotoxic effect of ATR pathway inhibition.

10.3.2 Replication Stress as a Therapeutic Vulnerability in Cancer Cells

An elevated level of RS is a defining feature of malignant cells. Although the precise mechanisms underlying oncogene-induced RS remain unclear, it is plausible that the accelerated consumption of RPA resulting from unchecked proliferation mediated by oncogenes could be a contributing factor [30]. Other potential mechanisms, such as the deregulation of replication initiation as well as interference between replication and transcription, arising either directly or indirectly from oncogenic activity, have also been postulated [32,33]. Given that the high constitutive levels of cellular RS is inherent in many tumors, it follows that they are likely to be particularly sensitive to RS overload instigated by inhibition of the ATR signaling pathway.

Until recently, it has been assumed that ATR is physiologically indispensable and not amenable to therapeutic targeting [34]. Indeed, human ATR mutations are uncommon, even in cancer. Among the few cancer types where *ATR* mutations have been reported, the acquisition of these mutations has been associated with tumor progression and poor clinical outcome [35,36]. Moreover, abrogation of the *ATR* gene is embryonically lethal, whereas its deletion in adult mice results in rapid aging and stem-cell loss [37,38]. In patients with Seckel syndrome in whom ATR signaling is defective due to hypomorphic germline mutation of the *ATR* gene, growth retardation, dwarfism, microcephaly, and mental impairment are typical manifestations [39,40].

Nonetheless, although complete abolition of ATR activity likely results in toxicity to healthy tissues, suppression of ATR activity was shown to be tolerable in healthy cells but not in tumor cells. In an elegant study by Schoppy et al. [41], conditional reduction of ATR in mice to 10% of normal levels using Cre–lox recombination resulted in minimal adverse effect on healthy hematopoietic and intestinal tissues. In contrast, suppression of ATR to this level was sufficient to severely restrict the growth of fibrosarcomas driven by H-Ras and p53 loss, as well as acute myeloid anemia driven by K-Ras and the *MLL-ENL* translocation [41]. Such reduction in ATR levels likewise prevented the development of Myc-driven lymphomas and pancreatic tumors featuring high levels of RS, as demonstrated by Murga et al. [42]. Corroborating these reports, we recently showed that therapeutic doses of an ATR kinase inhibitor were well tolerated in mice, while achieving effective tumor reduction [31]. Therefore, healthy and tumor cells demonstrate differential sensitivity to ATR pathway inhibition, providing a therapeutic window for the targeting of tumor cells.

10.4 PREDICTIVE BIOMARKERS OF SENSITIVITY TO ATR PATHWAY INHIBITION

10.4.1 ATR Pathway Addiction and the Concept of Synthetic Lethality

The substantial functional redundancy within cellular DDR pathways is widely recognized, wherein more than one pathway is capable of performing the same role. Physiologically, this provides protection against disruption of normal DDR mechanisms, particularly in cancer where DDR genes are mutated with high frequency. However, when one pathway is disrupted, cells become greatly reliant on collateral pathways. Thus, where two independent pathways regulate an essential DDR process, the absence of one pathway is compatible with cell survival, whereas the absence of both results in cell death. This gives rise to an emerging therapeutic strategy known as synthetic lethality, in which collaborating pathways are abolished to induce cytotoxicity [43,44].

From the studies highlighted above, it is apparent that tumors with heightened levels of RS, such as those driven by Myc or Ras, are exquisitely sensitive to ATR pathway inhibition. Understanding other circumstances under which tumor cells become addicted to the ATR signaling pathway is therefore essential to identifying further predictive biomarkers of sensitivity to ATR inhibition. In the following sections, we summarize current evidence for synthetic lethality of ATR pathway inhibition with various DDR defects prevalent in cancers.

10.4.2 Synthetic Lethality With G1/S Cell-Cycle Checkpoint Defects

The G1/S cell-cycle checkpoint is regulated by p53, a key regulatory transcription factor. In response to DNA damage, p53 transcriptionally activates certain genes and represses others to promote G1/S cell-cycle arrest [45]. The *TP53* tumor suppressor gene encoding p53 is among the most frequently mutated in cancer [46]. The loss of the G1/S checkpoint due to *TP53* loss imposes upon tumor cells a dependence on the G2/M checkpoint controlled primarily through ATR/Chk1. Consequently, ATR pathway inhibition has synthetically lethal properties in *TP53*-defective tumors, the evidence of which can be gleaned from several studies.

Nghiem et al. [47] provided one of the earliest evidence for this concept demonstrating hypersensitivity of G1/S checkpoint-deficient cells to ATR loss. In this study, the authors observed that the osteosarcoma cell line U2OS in the absence of ATR function displayed premature chromatin condensation, a feature associated with mitotic catastrophe, in response to hydroxyurea or ultraviolet radiation-induced RS. Moreover, repression of p53 function by overexpression of MDM2 markedly potentiated the lethal effect of ATR inhibition in these cells [47]. These findings were recapitulated in a later study showing that a p53-deficient colorectal cell line rendered ATR deficient by knock-in of a Seckel gene ($ATR^{s/s}$) exhibited marked sensitivity to hydroxyurea- and cisplatin-induced RS, whereas restoration of p53 function reduced its sensitivity to these agents [48].

These *in vitro* studies were complemented by experiments in murine models by Ruzankina et al. [49], who generated *ATR*-mosaic knockout models from $TP53^{-/-}$ mice. Mice with concomitant loss of both *ATR* and *TP53* displayed markedly reduced survival compared to those harboring isolated loss of *ATR* and exhibited high levels of DNA damage as evidenced by the accumulation of γH2AX-positive cells [49]. In a further study using xenotransplantation models of both p53-wild-type and p53-deficient triple-negative breast cancer, Chk1 inhibition was shown to potentiate chemotherapy-induced apoptosis in p53-mutant xenografts but not in p53-wild-type counterparts. Combining Chk1 inhibition with the topoisomerase inhibitor irinotecan resulted in suppression of tumor growth and prolonged survival in xenografts with p53 deficiency but not in wild-type xenografts [50].

In addition to *TP53* loss, cyclin E overexpression has also been reported to exacerbate the effects of ATR pathway inhibition [51]. Cyclin E promotes G1/S cell-cycle progression and hence contributes to a heightened level of cellular replication stress.

10.4.3 Synthetic Lethality With Double-Stranded DNA Repair Defects

Ataxia telangiectasia mutated (ATM) is a kinase that acts upstream of p53 and controls a DDR pathway critical to resolving double-stranded DNA breaks (DSBs). ATM has considerable functional redundancy with the ATR pathway [5,52]. Therefore, a second scenario in which tumor cells can become addicted to the ATR pathway is through ATM loss. Like the *TP53* gene, deletions or mutations of the *ATM* gene are also frequent in various malignancies [53]. Moreover, loss of ATM promotes tumor progression and chemoresistance similar to p53 loss [54–56].

When ATR function is inhibited, stalled replication forks with persistent ssDNA collapse, resulting in the formation of partially replicated sister chromatid fragments with DNA double-stranded ends [5,13,28]. This is a form of DSBs that requires homologous recombination repair (HRR) through the ATM pathway. When HRR is defective, these chromatid fragments accumulate, and while they can be ligated through nonhomologous end joining (NHEJ), the latter is a low-fidelity repair process that potentially gives rise to aberrant sequence deletions or chromosomal translocations. In addition, ATM also plays an important role in cell-cycle regulation through both p53-dependent and -independent mechanisms [57]. Thus, when ATR is inhibited in ATM-deficient cells, accumulation of DNA damage ensues and defective cell-cycle checkpoints due to combined loss of ATR and ATM permit unrestricted entry into mitosis, leading to mitotic catastrophe.

Consistent with this, Reaper et al. [58] showed that ATR inhibition was invariably cytotoxic to ATM or p53-defective tumor cells. Moreover, combination of ATR inhibition with RS-inducing genotoxic agents such as cisplatin and carboplatin was more profoundly synergistic in tumor cells with constitutive ATM or p53 defects, and in cells subjected to ATM pharmacological inhibition or p53 siRNA knockdown [58]. In our recently published study, we extended these observations to primary tumor samples and patient-derived xenotransplantation (PDX) models. ATR inhibition was selectively cytotoxic to primary chronic lymphocytic leukemia (CLL) cells harboring deletions or mutations of *ATM* or *TP53*, in comparison with a panel of CLL samples without these aberrations and with healthy donor peripheral blood lymphocytes. Remarkably, ATR inhibitor treatment in murine xenografts derived from CLL patients with biallelic *ATM* loss led to a reduction not only in tumor load, but also in the proportion of CLL cells with *ATM* defects, indicating specificity of ATR inhibition for these defects [31].

Several other studies have demonstrated synthetic lethality between ATR inhibition and ATM or other HRR defects. ATR inhibition was lethal in cell lines derived from patients suffering from gastric cancer with ATM deficiency, in ovarian cancer with Breast cancer type 1 susceptibility protein 1 (BRCA1) deficiency, and in Rad51-depleted cells [59–62]. Taken together, these studies demonstrate that defects in HRR-dependent DSB repair are markers of sensitivity to ATR pathway inhibition.

Fig. 10.1 summarizes the synthetically lethal interaction between ATR pathway inhibition and ATM or p53 loss.

10.4.4 Other Markers of Sensitivity

A number of additional DNA repair defects have been shown to confer enhanced sensitivity to ATR pathway inhibition. These defects include deficiency in the DNA repair proteins XRCC1, ERCC1, POLD1, or PRIM1 [63–65]. XRCC1 is a protein involved in base excision

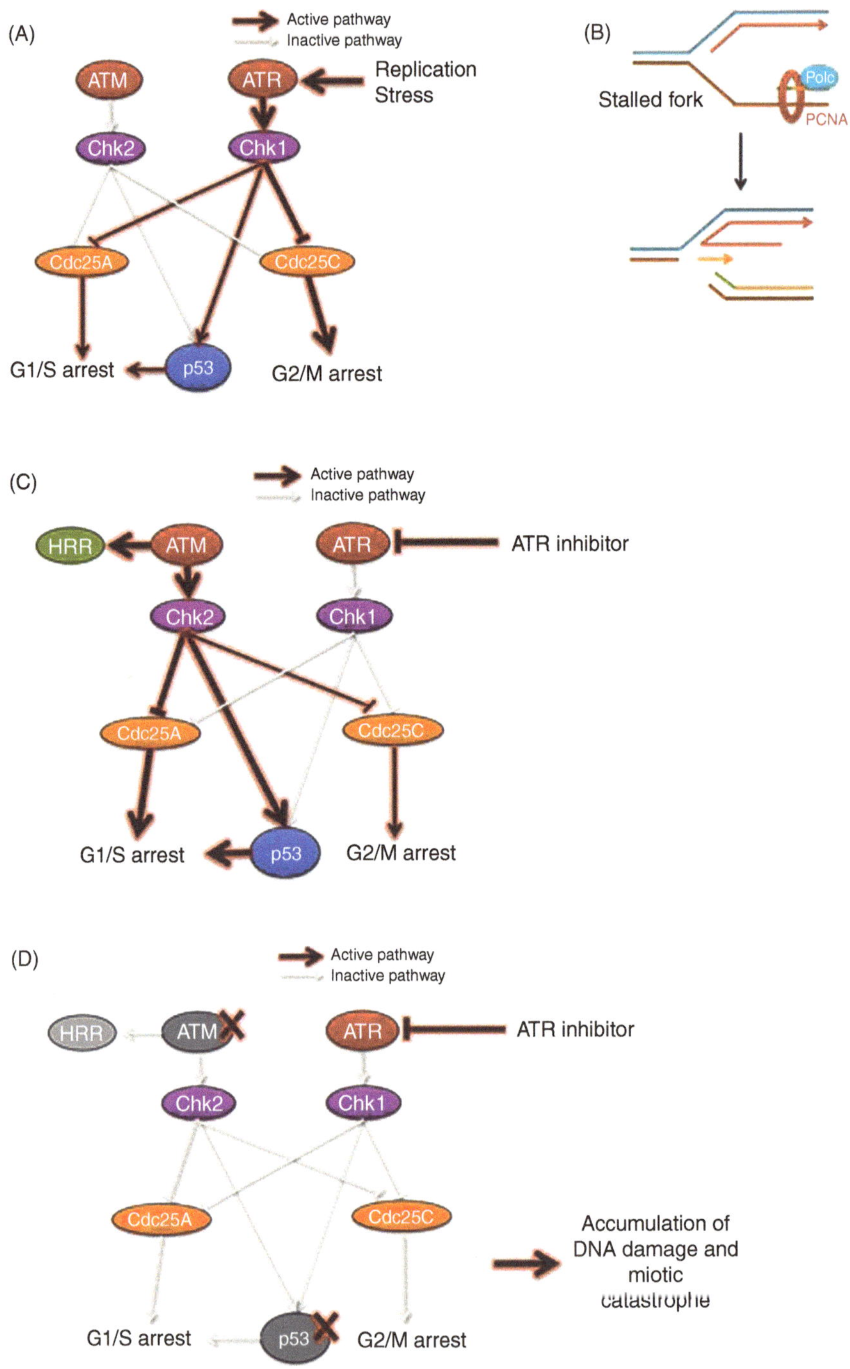

FIG. 10.1 **A model for synthetic lethality in CLL cells with ATM or p53 deficiency by inhibition of ATR.** ATM and ATR are master regulators of DDR, with ATM being activated in response to DNA double-strand breaks, and ATR in response to replication stress. (A) Activation of the ATR pathway leads to cell-cycle arrest mediated primarily

repair and the single-strand break repair pathway, whereas ERCC1 mediates the repair of several types of DNA lesions including bulky adducts, DSBs, and inter-strand cross-links and facilitates the separation of sister chromatids at fragile sites. POLD1 and PRIM1, on the other hand, are involved in DNA replication synthesis. Recently, an RNAi screen identified AT-rich interaction domain 1A (ARID1A) deficiency as an additional marker of sensitivity to ATR inhibition. Mutations in ARID1A are common in cancers and lead to disrupted topoisomerase localization and cell-cycle progression, thus imposing on cancer cells a dependence on the ATR signaling pathway [66].

Finally, a study by Flynn et al. [67] has alluded to the hypersensitivity to ATR inhibition in cancer cells reliant on a mechanism of telomere maintenance known as alternative lengthening of telomeres (ALT), whereby telomeres are elongated through recombination. The absence of ATR leads to abrogation of ALT, compromising telomere stability in ALT-dependent cancer cells, thus resulting in DNA damage, telomere loss, and selective lethality of these cells.

10.5 PRECLINICAL EVALUATION OF SMALL-MOLECULE INHIBITORS TARGETING THE ATR PATHWAY

10.5.1 Development and Evaluation of ATR Pathway Inhibitors

The design and development of selective ATR kinase inhibitors has been coupled with difficulty because of the atypical nature of PIKKs relative to other more conventional kinases. Early inhibitors of ATR were not specific, but more recently two drug companies have manufactured potent and highly specific ATR inhibitors. They are VE-821 and its analog VX-970 produced by Vertex Pharmaceuticals, as well as AZ20 and its analog AZD6738 developed by AstraZeneca [68,69]. Both compounds are ATP-competitive ATR kinase inhibitors, with AZD6738 additionally being available in an oral formulation.

In contrast, Chk1 inhibitors were accessible much earlier, and to date a number of Chk1 inhibitors have been developed. The earlier Chk1 inhibitors, such as UCN-01, however, lacked target selectivity and had poor pharmacokinetic properties, thus limiting their utility [70]. Next-generation Chk1 inhibitors, including AZD7762, MK-8776 (Sch900776), LY2603618, LY2606368, PF-00477736, V158411, and SAR020106, had considerably less off-target effects with an improved pharmacokinetic profile [71].

Numerous preclinical studies on ATR and Chk1 kinase inhibitors have been conducted on an array of solid and hematological malignancies, using a variety of experimental models. Although cell line models of human malignancies are easy to manipulate and provide an excellent basis for the initial evaluation of novel therapeutic agents, effects observed on cell lines

through the G2/M checkpoint, and repair of stalled replication forks. (B–C) Inhibition of ATR leads to collapse of stalled replication forks into partially replicated DNA fragments with double-stranded ends that are repaired through the ATM/p53 pathway. This involves cell-cycle arrest mediated primarily through the G1/S checkpoint and HRR. (D) In cells with defective ATM or p53, inhibition of ATR results in intolerable accumulation of unrepaired DNA damage. This arises from impaired HRR due to defective ATM and/or impaired cell-cycle regulation resulting from combined loss of functional ATR and ATM/p53. Abbreviations: CLL, chronic lymphocytic leukemia; ATM, ataxia telangiectasia mutated; ATR, ataxia telangiectasia and Rad3; HRR, homologous recombination repair. Adapted from Kwok et al., Blood, 2016, 127:582–95) [31].

are not always reflective of their effects *in vivo*. Therefore, results from studies on primary tumor material, as well as animal models, especially PDX, are invaluable additions to *in vitro* experimentation in informing the potential drug utility and effects, beneficial or otherwise, prior to their clinical deployment. In the following sections, we examine currently available *in vitro* and *in vivo* preclinical data on ATR and Chk1 inhibitors.

10.5.2 Single-Agent Efficacy of ATR Inhibitors in Cancer

In vitro, exposure of *TP53*-mutant multiple myeloma cell lines to 1 μM of VE-821 for 3 days resulted in a reduction in cell viability by >50% [72]. VE-821 monotherapy was shown to be cytotoxic also to glioma and lymphoma cell lines, to rectal carcinoma cell lines under hypoxic conditions, and to breast cancer and non-small-cell lung cancer lines with ERCC1 defect [60,63,73,74]. In colon cancer, HCT166 cells, particularly those deficient in ARID1A, responded to VE-821, VX-970, and AZ20, as reflected by marked reduction (by >90%) in viability and clonogenic survival upon treatment with ATR inhibitor doses of ≤1 μM [66]. In acute myeloid leukemia (AML), AZ20 doses of 350 nM to 1.4 μM were sufficient to reduce viability of AML cell lines by 50% (EC_{50}), whereas in primary AML samples this was achieved with AZ20 doses of 800 nM to 27 μM [75]. With AZD6738, single-agent efficacy was seen in cell lines of HER2-positive breast cancer and ATM-deficient gastric cancer [61,76]. Furthermore, in CLL, we demonstrated that AZD6738 treatment for 4 days led to lethality in *TP53* or *ATM* defective CLL cell lines and proliferating primary CLL samples, with an average EC_{50} of 1.4 μM in cell lines and 8.5 μM in primary CLL cells [31].

In vivo, VX-970 suppressed tumor growth and prevented tumor establishment in mice xenografted with AR1D1A-deficient HCT116 or ovarian cancer TOV-21G cells, whereas AZ20 showed single-agent efficacy in reducing tumor infiltration and improving survival of murine models transplanted with *MLL*-rearranged AML cells [66,77]. Moreover, AZD6738 monotherapy, administered orally for 2 weeks, led to marked reduction in tumor load in PDX of biallelic *ATM*- or *TP53*-inactivated CLL, accompanied by a reduction in the proportion of CLL cells with these defects [31]. AZD6738 also countered tumor growth and induced apoptosis in a murine xenograft model of ATM-deficient gastric cancer [61].

A summary of the major preclinical studies involving ATR inhibitors in the various cancer types is presented in Table 10.1.

10.5.3 Single-Agent Efficacy of Chk1 Inhibitors in Cancer

In vitro, single-agent Chk1 inhibition with ≤1 μM AZD7762 led to >90% reduction in the viability of radioresistant breast cancer cell lines as well as viability reductions of varying magnitudes in metastatic melanoma cell lines [78,79]. With MK-8776 (Sch900776), single-agent cytotoxic effects were seen in neuroblastoma lines (median EC_{50} 900 nM), in cell lines of myeloid leukemia and in BRCA-mutant ovarian cancer [80,81]. Cell lines that were sensitive to MK-8776 monotherapy were found to display aberrant Cdk2 activation in S phase upon Chk1 inhibition, leading to DSBs, possibly because these cells depend highly on constitutive suppression of Cdc25 phosphatases by Chk1 [82]. Moreover, LY2603618 exerted single-agent cytotoxicity on AML cell lines (EC_{50} 0.1–1.6 μM) and patient samples (EC_{50} <9 μM) as well as in osteosarcoma cell lines [83,84]. Furthermore, single-agent PF-00477736 activity was evident

TABLE 10.1 Preclinical Studies involving ATR Inhibitors

Malignancy	Study	Inhibitor	Experimental Model	Biomarker of Sensitivity	Treatments Potentiated by ATR Inhibitor
AML	Ma et al., 2017 [75]	AZ20	Cell lines, primary cells	—	Cytarabine
	Chauduri et al., 2014 [93]	VE-821	Cell lines	—	Wee1 inhibitor
Breast	Yazinski et al., 2017 [62]	AZ20, VE-821	Cell lines, primary cells	BRCA1 deficiency	PARP1 inhibitor
	Kim et al., 2017 [76]	AZD6738	Cell lines	—	Cisplatin
Cervical	Teng et al., 2015 [115]	ETP-46464	Cell lines	—	Cisplatin, radiotherapy
CLL	Kwok et al., 2016 [31]	AZD6738	Cell lines, primary cells, PDX	ATM, p53 deficiency	Chlorambucil, fludarabine, cyclophosphamide, bendamustine, ibrutinib
Colorectal	Hocke et al., 2016 [65]	VX-970, NU-6027	Cell lines	POLD1 deficiency, PRIM1 deficiency	—
Endometrial	Teng et al., 2015 [115]	ETP-46464	Cell lines	—	Cisplatin, radiotherapy
Esophageal	Leszczynska et al., 2016 [116]	VX-970	Cell lines, cell-line xenografts	—	Cisplatin, carboplatin, radiotherapy
Gastric	Min et al., 2017 [61]	AZD6738	Cell lines, cell-line xenografts	ATM deficiency	—
Lung	Vendetti et al., 2015 [90]	AZD6738	Cell lines, cell-line xenografts	—	Cisplatin, gemcitabine, radiotherapy
	Hall et al., 2014 [89]	VX-970	Cell lines, primary cells, PDX	—	Cisplatin
Lymphoma	Muralidharan et al., 2016 [73]	AZ20, VE-821	Cell lines, cell-line xenografts	—	BET inhibitor
	Menezes et al., 2015 [117]	WO2010/073034	Cell lines, cell-line xenografts	ATM deficiency	—

(Continued)

TABLE 10.1 Preclinical Studies Involving ATR Inhibitors (*cont.*)

Malignancy	Study	Inhibitor	Experimental Model	Biomarker of Sensitivity	Treatments Potentiated by ATR Inhibitor
Myeloma	Cottini et al., 2015 [72]	VE-821	Cell lines, primary cells	Myc expression, p53 deficiency	Piperlongumine
Ovarian	Kim et al., 2016 [81]	AZD6738	Cell lines, PDX	—	PARP1 inhibitor
	Huntoon et al., 2013 [118]	VE-821	Cell lines	—	Cisplatin, gemcitabine, PARP inhibitor
Pancreatic	Prevo et al., 2012 [119]	VE-821	Cell lines, primary cells	—	Gemcitabine, radiotherapy
	Fokas et al., 2012 [120]	VX-970	Cell lines, cell-line xenografts	—	Gemcitabine, radiotherapy
Various	Toledo et al., 2011 [51]	ETP-46464	Cell lines	p53 deficiency, cyclin E overexpression	Hydroxyurea, Chk1 inhibitor
	Sultana et al., 2013 [64]	NU-6027	Cell lines	XRCC1 deficiency	Cisplatin
	Repear et al., 2011 [58]	VE-821	Cell lines	ATM, p53 deficiency	Cisplatin
	Pires et al., 2012 [74]		Cell lines	—	Radiotherapy
	Mohni et al., 2015 [91]		Cell lines	TLS polymerase, 53BP1	Cisplatin
	Krajewska et al., 2015 [59]		Cell lines	HRR deficient, Rad51 deficiency	—
	Middelton et al., 2015 [60]		Cell lines	HR/BER defects	—
	Josse et al., 2014 [92]	VE-821, VX-970	Cell lines, cell-line xenografts	—	Topoisomerase inhibitor
	Williamson et al., 2016 [66]		Cell lines, cell-line xenografts	ARID1A deficiency	—
	Mohni et al., 2014 [63]	VE-821	Cell lines	ERCC1 deficiency	—

Abbreviations: AML, acute myeloid leukemia; CLL, chronic lymphocytic leukemia; PDX, patient-derived xenografts; TLS, translesion synthesis.

in T-cell acute lymphoblastic leukemia (T-ALL) cell lines (EC_{50} 20–109 nM) and primary cells (EC_{50} <200 nM) which overexpress Chk1 [85]. PF-00477736 also exerted single-agent activity in cell lines of mantle cell lymphoma (EC_{50} 0.68 nM), germinal center B-cell like (GCB) diffuse large B cell lymphoma (DLBCL; EC_{50} 10.2 nM) and activated B-cell (ABC) DLBCL (EC_{50} 87.3 nM) [86]. Finally, V158411 inhibited cell proliferation and induced caspase activation in breast and ovarian cancer cell lines [87].

In vivo, AZD6722 treatment for 3 days resulted in delay in tumor growth in MCF-7/C6 radioresistant breast cancer xenografts [78]. MK-8776 (Sch900776), on the other hand, reduced tumor volume in MDA-MB-231 breast cancer xenografts [88]. Furthermore, in a xenotransplantation model of T-ALL, 30 days of treatment with PF-00477736 yielded a significant reduction in tumor growth compared to vehicle-treated controls [85].

A selection of important preclinical studies in different cancers involving Chk1 inhibitors is presented in Table 10.2.

10.5.4 Chemotherapeutic Combinations Involving ATR Pathway Inhibitors

ATR and Chk1 inhibitors can be readily combined with conventional chemotherapeutic agents. Although the effectiveness of ATR/Chk1 inhibitor monotherapy is often dependent on specific tumor phenotypes, such as p53 deficiency or Myc overexpression, ATR and Chk1 inhibitors, when tested in preclinical models, synergistically enhance the effect of chemotherapy across a broad range of cancer phenotypes, DDR defective or otherwise. This can be ascribed to the potentiation of chemotherapy-induced RS by ATR/Chk1 inhibitors. The ability of ATR pathway inhibitors to resensitize chemoresistant tumor cells to conventional chemotherapeutic agents is of particular clinical significance.

Among the numerous chemotherapeutic agents that synergize with ATR/Chk1 inhibitors, cisplatin and gemcitabine are the most frequently studied. Cisplatin causes DNA breaks and crosslinks, whereas gemcitabine induces RS by decreasing the deoxyribonucleotide triphosphate required for DNA replication. ATR inhibitors broadly sensitized a panel of lung cancer cell lines, cell line xenografts, and PDX to cisplatin or gemcitabine, and ovarian cancer cell lines to cisplatin [89,90]. The sensitization to cisplatin is especially profound in tumor cells with deficiency in ATM, p53, or XRCC1, or with loss of 53BP1 or polymerase ζ involved in translesion synthesis [48,58,64,91]. Other DNA damaging agents that synergize with ATR inhibitors include cytarabine, as observed in AML cell lines and primary cells, as well as chlorambucil, bendamustine, fludarabine, and cyclophosphamide, as observed in CLL cell lines and primary cells. ATR inhibitors also sensitized tumor cells to topoisomerase inhibitors through disruption of DNA replication initiation and fork elongation processes [92].

Furthermore, ATR pathway inhibitors display synergy across a range of novel agents. Two studies, for instance, have highlighted the synergistic interaction between poly-(ADP-ribose) polymerase (PARP) inhibitors and ATR or Chk1 inhibitors in ovarian and breast cancer, respectively, using cell lines complemented by primary tumor samples or PDX [62,81]. PARP inhibitor treatment led to increased accumulation of cells in G2 phase, thus intensifying their reliance on the ATR pathway for checkpoint control and the maintenance of genome stability. ATR and/or Chk1 inhibitors have also shown synergy with inhibitors of other DNA repair molecules, such as Wee1, which contributes to the regulation of the G2/M checkpoint, and MK2, which is critical for prolonged checkpoint maintenance [79,86,93–96]. Finally,

TABLE 10.2 Preclinical Studies Involving Chk1 Inhibitors

Malignancy	Study	Inhibitor	Experimental Model	Biomarker of Sensitivity	Treatments Potentiated by Chk1 Inhibitor
ALL	Sarmento et al., 2015 [85]	PF477736	Cell lines, cell-line xenografts, primary cells	—	—
AML	Zhao et al., 2016 [83]	LY2603618	Cell lines, primary cells	—	Bcl-2 inhibitor
	Chauduri et al., 2014 [93]	MK-8776	Cell lines	—	Wee1 inhibitor
	Yuan et al., 2014 [80]	SCH900766, AZD7762	Cell lines	—	—
Bladder	Wang et al., 2015 [121]	Gö6976	Cell lines	—	Gemcitabine
Breast	Zhang et al., 2016 [78]	AZD7762	Cell lines, cell-line xenografts	—	—
	Zhou et al., 2017 [88]	MK-8776	Cell lines, cell-line xenografts	—	Radiotherapy
	Ma et al., 2012 [50]	UCN-01, AZD7762	PDX	p53 deficiency	Topoisomerase inhibitor
	Tang et al., 2012 [122]		Cell lines, cell-line xenografts	—	PARP inhibitor, radiotherapy
	Bryant et al., 2014 [87]	V158411, PF477736, AZD7762	Cell lines	—	Gemcitabine, cisplatin
Colon	Martino-Echarri et al., 2014 [123]	MK-8776, AZD7762	Cell lines	—	5-Fluorouracil
	Origanti et al., 2013 [124]	UCN-01	Murine models, orthotropic models	p53 deficiency, p21 deficiency	Topoisomerase inhibitor
Head & neck	Gadhikar et al., 2013 [125]	AZD7762	Cell lines	p53 deficiency	Cisplatin
	Barker et al., 2016 [126]	CCT244747	Cell lines, cell-line xenografts, primary cells	—	Paclitaxel
Lung	Bartucci et al., 2012 [127]	SB218078, AZD7762	Primary cells, PDX	—	Gemcitabine, cisplatin, paclitaxel

TABLE 10.2 Preclinical Studies Involving Chk1 Inhibitors (*cont.*)

Malignancy	Study	Inhibitor	Experimental Model	Biomarker of Sensitivity	Treatments Potentiated by Chk1 Inhibitor
Lymphoma	Chila et al., 2015 [86]	PF477736	Cell lines, cell-line xenografts	—	Wee1 inhibitor
	Zemanova et al., 2016 [128]	SCH900776	Cell lines, primary cells, murine models	p53 deficiency	Fludarabine, cytarabine, gemcitabine
	Murga et al., 2011 [42]	UCN-01	Murine models	Myc expression	—
Melanoma	Magnussen et al., 2015 [79]	AZD7762	Cell lines, cell-line xenografts	—	Wee1 inhibitor
Myeloma	Pei et al., 2014 [129]	CEP3891	Cell lines, primary cells	—	MEK1/2 inhibitor
	Dai et al., 2011 [130]	UCN-01	Cell lines, primary cells, PDX	—	SRC inhibitor
Nasopharyngeal	Mak et al., 2015 [95]	AZD7762	Cell lines, cell-line xenografts	—	Wee1 inhibitor
Neuroblastoma	Russell et al., 2013 [94]	MK-8776	Cell lines, cell-line xenografts	—	Wee1 inhibitor
	Cole et al., 2011 [131]	SB218078, TCS2312	Cell lines, primary cells	Myc expression	—
Oral SCC	Sankunny et al., 2014 [132]	PF477736	Cell lines	ATM deficiency	Radiotherapy
Osteosarcoma	Duan et al., 2014 [84]	LY2603618	Cell lines	—	Cisplatin
Ovarian	Kim et al., 2016 [81]	MK-8776	Cell lines, PDX	—	PARP inhibitor
	Bryant et al., 2014 [87]	V158411, PF477736, AZD7762	Cell lines	—	Gemcitabine, cisplatin
Pancreatic	Morgan et al., 2010 [92]	AZD7762	Cell lines, cell-line xenografts, PDX	—	Radiotherapy
	Vance et al., 2011 [133]		Cell lines	p53 deficiency	PARP inhibitor
	Engelke et al., 2013 [134]	MK-8776	Cell lines, cell-line xenografts	HRR deficiency	Gemcitabine, radiotherapy

(*Continued*)

TABLE 10.2 Preclinical Studies Involving Chk1 Inhibitors (*cont.*)

Malignancy	Study	Inhibitor	Experimental Model	Biomarker of Sensitivity	Treatments Potentiated by Chk1 Inhibitor
Various	McNeely et al., 2010 [135]	AZD7762	Cell lines	BRCA2 deficiency, XRCC3 deficiency, DNA-PK deficiency	Gemcitabine
	Krajewska et al., 2015 [59]		Cell lines	HRR deficient, Rad51 deficiency	—
	Sakurikar et al., 2016 [82]	MK-8776	Cell lines	CDK2 activation in S phase	—
	Dietlein et al., 2015 [96]	PF477736	Cell lines, cell-line xenografts, primary cells	KRAS mutation	MK2 inhibitor

Abbreviations: ALL, acute lymphoblastic leukemia; AML, acute myeloid leukemia; SCC, squamous cell carcinoma; PDX, patient-derived xenografts.

therapeutic combinations with ATR pathway inhibitors are not confined to agents targeting DDR. ATR inhibitors are capable of enhancing the effect of a BET bromodomain inhibitor in Myc-driven lymphoma and a B-cell receptor signaling inhibitor in CLL [31,73].

10.6 CLINICAL TRIALS OF SMALL-MOLECULE INHIBITORS TARGETING THE ATR PATHWAY

10.6.1 Clinical Studies Involving Chk1 Inhibitors

To explore the use of selective Chk1 inhibitors in oncology practice, several clinical studies have been carried out on AZD7762, LY2603618, LY2606368, and MK-8776 (Sch900776). These were phase I or II trials with the aim to assess safety and tolerability, to establish the maximum tolerable dose and, within certain studies, to determine response rate and durability. Unfortunately, the clinical development of AZD7762 was hampered by drug-related cardiac toxicity, which was dose limiting in some cases [97,98]. In contrast, LY2603618, LY2606368, and MK-8776 (Sch900776) had an acceptable toxicity profile with rare serious adverse events [99–102]. Moreover, early evidence of clinical activity was seen with these agents. In particular, eight (out of 21) refractory AML patients achieved complete remission and 15 (out of 30) individuals with advanced solid malignancy had partial response or stable disease following treatment with MK-8776 (Sch900776), either alone or in combination with chemotherapy [101,102]. Likewise, partial response was seen in two patients with advanced cancer following LY2606368 monotherapy [100].

Despite this, two phase II trials investigating LY2603618 in advanced lung and pancreatic cancer reported nonsuperiority of LY2603618, used in combination with premetrexed or gemcitabine, respectively, when compared to premetrexed or gemcitabine alone [103,104]. However, since the patients enrolled on these trials were unselected, a substantial proportion would unlikely have harbored high RS phenotypes or lesions that would yield synthetic lethality with ATR/Chk1 inhibition. Indeed, genomic analysis of an exceptional responder, who was cured of her invasive ureteric cancer with AZD7762 in combination with irinotecan, revealed mutation within *RAD50* which attenuated ATM signaling and enforced dependence on the ATR pathway [105]. This underscores the importance of the use of ATR pathway inhibitors for individuals who are likely to benefit.

Unfortunately, the clinical development of AZD7762, LY2603618, and MK-8776 has been suspended. Currently, the Chk1 inhibitors under active clinical investigation include LY2606368 and CCT245737, as detailed in Table 10.3.

10.6.2 Clinical Studies Involving ATR Inhibitors

With regard to ATR inhibitors, both AZD6738 and VX-970 have now entered phase I/II clinical testing. However, the majority of these studies have been initiated recently, and no results have yet been reported. As of March 2017, 11 clinical trials on ATR inhibitors are ongoing for a range of malignancies. These studies examine the use of the ATR inhibitor either alone or in combination with a range of conventional and novel therapeutic agents, including cisplatin, carboplatin, gemcitabine, etoposide, irinotecan, the PARP inhibitors olaparib and veliparib, and the anti-PD-L1 immune checkpoint inhibitor durvalumab.

A summary of the completed and ongoing clinical trials involving Chk1 and ATR inhibitors is presented in Table 10.3.

10.7 TARGETING THE ATM PATHWAY: IS THIS OF ANY VALUE IN OVERCOMING CHEMORESISTANCE IN CANCER?

There is considerable cross-talk between the ATR and ATM signaling pathways. Given the central importance of ATM in DDR, it is reasonable to also examine whether there is potential utility in inhibiting ATM to overcome chemoresistance, particularly since a number of selective ATM inhibitors are now available, including KU-55933, KU-60019, KU-59403, and AZD0156. In comparison with the plethora of literature on ATR pathway targeting, the number of studies on ATM targeting is relatively limited. It is generally recognized that ATM-deficient cells are hypersensitive to ionizing radiation (IR) on account of their inability to repair DSBs. Therefore, historical studies have focused on the use of ATM inhibitors to sensitize tumor cells to radiotherapy. Indeed, ATM inhibitors are potent radiosensitizers, both *in vitro* and *in vivo* [106–108].

More recently, Batey et al. [109] showed that ATM inhibitors can also be used to potentiate the cytotoxic effect of topoisomerase inhibitors such as camptothecin, etoposide, and doxorubicin, but lacked single-agent activity. However, the prolonged use of ATM inhibition remains questionable. In contrast to ATR inhibitors with target predilection for p53- and HRR-deficient cells, the lack of such predilection of ATM inhibitors raises the possibility of toxicity

TABLE 10.3 Clinical Studies Involving Chk1 and ATR Inhibitors

Inhibitor	Phase	Study	Treatment	Malignancy	Pt no	Endpoint	Conclusions
AZD7762	I	Seto et al., 2013 [98]	Monotherapy; combination with gemcitabine	Advanced solid tumor	20	Safety, pharmacokinetics, preliminary efficacy	Cardiac and hematologic toxicity was dose limiting at 30 mg. Maximum tolerable dose of AZD7762 in combination with gemcitabine was 21 mg
		Sausville et al., 2014 [97]	Monotherapy; combination with gemcitabine	Advanced solid tumor	42	Safety, tolerability	Cardiac toxicity was dose limiting
LY2603618	I	Doi et al., 2015 [99]	Monotherapy; combination with gemcitabine	Advanced solid tumor	17	Safety, tolerability	Acceptable safety and tolerability in combination with gemcitabine. Partial response in four patients. Most common high-grade toxicities were hematological
		Calvo et al., 2016 [136]	Monotherapy; combination with gemcitabine	Advanced/ metastatic solid tumor	50	Safety, tolerability	Adverse events include fatigue, thrombocytopenia, neutropenia, nausea, and anemia. Maximum tolerated dose was 200 mg/m^2 when combined with gemcitabine. Proceed to phase II trial with a fixed dose of 230 mg
	II	Scagliotti et al., 2016 [103]	Combination with pemetrexed	Metastatic non-small cell lung cancer	53	Response rate, pharmacokinetics, safety	Partial response in five patients. No association between p53 status and response. Efficacy and safety similar to pemetrexed monotherapy
		Laguente et al., 2017 [104]	Combination with gemcitabine	Metastatic pancreatic cancer	99	Response rate, PFS, pharmacokinetics, safety	Maximum pharmacodynamic response was achieved after 230 mg of LY2603618. Combination of gemcitabine with LY2603618 was not superior to gemcitabine alone

LY2606368	I	Hong et al., 2016 [100]	Monotherapy	Advanced cancer	45	Safety, pharmacokinetics, pharmacodynamics	Dose-limiting toxicities in seven patients were all hematological. Partial response was seen in two patients. Phase II study will utilize LY2606368 dose of 105 mg/m^2
MK-8776 (Sch900776)	I	Karp et al., 2012 [101]	Combination with cytarabine	Refractory acute leukemia	24	Safety, tolerability, preliminary efficacy	Complete response seen in eight patients (33%). Dose-limiting toxicities were cardiac. Recommended dose of Sch900776 for phase II study is 100 mg
		Daud et al., 2015 [102]	Monotherapy; combination with gemcitabine	Advanced solid tumor	43	Safety, pharmacokinetics, pharmacodynamics	Well tolerated as single agent and in combination with gemcitabine. Of 30 evaluable patients, 2 showed partial response and 13 had stable disease. Recommended phase II dose was 200 mg MK-8776 + 1000 mg/m^2 gemcitabine

(Continued)

TABLE 10.3 Clinical Studies Involving Chk1 and ATR Inhibitors (*cont.*)

Inhibitor		Phase	Start Date	Malignancy	Treatment	Pt no	ClinicalTrials.gov Identifier
ATR	AZD6738	I	2014	Advanced solid tumor	Monotherapy; combination with palliative radiotherapy	100	NCT02223923
				Advanced solid tumor with ATM deficiency	Monotherapy; combination with carboplatin, olaparib, or durvalumab	114	NCT02264678
			2015	Advanced solid tumor	Combination with paclitaxel	21	NCT02630199
	VX-970	I	2012	Advanced solid tumor	Combination with gemcitabine, cisplatin, etoposide, and/or carboplatin	205	NCT02157792
			2016	Advanced head and neck squamous cell carcinoma	Combination with cisplatin/X-ray therapy	45	NCT02567422
				Non-small cell lung cancer with brain metastases	Combination with whole brain radiotherapy	46	NCT02589522
				Advanced solid tumor	Combination with irinotecan	51	NCT02595931
				Advanced solid tumor	Combination with veliparib and cisplatin	60	NCT02723864
		I/II	2015	Small cell lung cancer	Combination with topotecan	55	NCT02487095
		II	2016	Metastatic urothelial cancer	Combination with cisplatin/gemcitabine (vs cisplatin/gemcitabine alone)	90	NCT02567409
				Platinum-resistant recurrent ovarian or fallopian tube cancer	Combination with gemcitabine (vs gemcitabine alone)	70	NCT02595892
Chk1	CCT 245737	I	2016	Advanced solid tumor	Monotherapy	40	NCT02797964
				Advanced solid tumor	Monotherapy	70	NCT02797977
	LY 2606368	I	2017	Advanced pediatric solid tumors	Monotherapy	65	NCT02808650
		II	2014	BRCA mutant or high-grade breast and ovarian cancer, metastatic prostate cancer	Monotherapy	144	NCT02203513
			2016	Solid tumors with replication stress or HRR deficiency	Monotherapy	38	NCT02873975

Published studies are listed in the upper panel, and ongoing studies in the lower panel. Abbreviations: Pt no, patient number; PFS, progression-free survival.

to healthy cells [109]. ATM inhibitors may be more effective in certain tumor types, such as glioblastoma, where tumor-associated overexpression of platelet-derived growth factor alpha (PDGFRA) can be normalized with ATM inhibition [110]. While there is no published clinical data on ATM inhibition, a phase I study of AZD0156, alone and in combination with chemotherapy or olaparib, is currently underway (NCT02588105). Therefore, the clinical utility of ATM inhibition remains to be seen.

10.8 CONSIDERATIONS, CAVEATS, AND UNRESOLVED QUESTIONS IN ATR TARGETING

10.8.1 ATR vs Chk1 targeting

While ATR and Chk1 act in the same pathway, ATR is upstream of Chk1 and therefore controls additional downstream processes. Targeting ATR may therefore produce more wide-ranging effects than targeting Chk1, and this is reflected in the ability of ATR inhibitors to potentiate a broader range of chemotherapeutic agents. It remains to be determined whether ATR inhibitors are clinically more effective than Chk1 inhibitors. This might vary from one malignancy to another, or indeed from one tumor phenotype to another. For example, tumors with ATM deficiency might benefit more from ATR inhibition than from Chk1 inhibition, given the functional cross-talk between the ATM and ATR.

10.8.2 Potential Mechanisms of Resistance to ATR Pathway Inhibition

A fundamental assumption underpinning models of synthetic lethality is that there are only two major pathways regulating a process. Therefore, if one pathway is defective, cellular demise is assured when the other pathway is blocked. However, this notion is inevitably an oversimplification of the myriad of collateral pathways regulating a cellular process, many of which are hitherto underappreciated or unknown. When one collateral pathway is therapeutically inhibited, tumor cells may upregulate an alternative collateral pathway to mitigate the effects of the initial block, thereby resulting in therapeutic resistance.

Alongside ATM, DNA-dependent protein kinase (DNA-PK) plays a role in DSB repair through control of NHEJ. Recently, there have been reports pointing to a possible functional redundancy between ATR and DNA-PK in regulating downstream Chk1 activity [111]. Buisson et al. [112], in particular, demonstrated the existence of a DNA-PK-Chk1 backup pathway that can mediate resistance to ATR inhibitors. In this model, ATR inhibition is cytotoxic to a proportion of tumor cells with the highest levels of replication stress, but those with moderate levels are protected through the backup pathway [112]. Simultaneous targeting of ATR and Chk1 or DNA-PK could potentially overcome this. Indeed, a potentiating interaction between ATR and Chk1 inhibitors has been reported, suggesting that the combined use of these inhibitors could be more efficacious than either agent alone [113]. However, the potential toxicities and adverse effects of any combined use of ATR and Chk1 inhibitors will need to be properly assessed.

Finally, both ATR and Chk1 inhibitors are kinase inhibitors. As observed from the small-molecule kinase inhibitors currently in clinical use, point mutations in the target kinase can develop. This constitutes another potential source of therapeutic resistance that physicians and investigators will need to anticipate and overcome.

10.8.3 Precision Medicine and the Selection of Patients for ATR/Chk1 Targeting

The long-term effects and toxicity of ATR and Chk1 inhibition are unknown and this is often a source of contention. Targeting a process of such importance to genome integrity can be potentially dangerous if it also affects healthy cells. Moreover, sublethal targeting of tumor cells with ATR pathway inhibitors can inadvertently promote tumorigenesis and clonal evolution, as these cells accumulate replication stress and DNA damage but escape death [114]. It is therefore of paramount importance that patients are selected for ATR pathway inhibitor treatment according to whether their tumor cells are differentially sensitive to these inhibitors. By selecting patients whose tumor is hypersensitive to these inhibitors, tumor killing can be maximized while minimizing treatment duration, thereby reducing both toxicity to healthy tissues and the risk of tumor evolution and escape. While a number of biomarkers of sensitivity to ATR pathway inhibition have been discussed earlier, it is likely that many others have yet to be discovered. The onus is on investigators to continue to identify predictive biomarkers of sensitivity to ATR and Chk1 inhibitors, specific for each tumor type, in order to enable accurate therapeutic stratification.

10.8.4 The Use of ATR/Chk1 Inhibitors with Other Therapeutic Agents

Finally, should ATR/Chk1 inhibitors be administered as a single agent or in combination with chemotherapy? If so, what is the ideal therapeutic regimen to be used? An argument in favor of combined use with chemotherapy is that the addition of the latter would increase the potency of ATR pathway inhibitors, but this needs to be balanced by a possible increase in toxicity. In addition, the use of therapeutic combinations may potentially reduce the likelihood of developing resistance to single agents. The ideal chemotherapeutic agent to be used in combination with different tumor types will need to be ascertained through mechanistic and clinical studies. In addition, ATR/Chk1 inhibitors can potentially be combined with other novel small-molecule inhibitors or monoclonal antibodies. This may be of particular relevance to a number of malignancies where a substantial proportion of tumor cells is quiescent. These cells are unlikely to be susceptible to ATR pathway inhibition and may give rise to a pool of residual cells from which therapeutic resistance can arise. Therefore, combination of ATR pathway inhibitors with agents targeting pathways unrelated to replication could allow the simultaneous eradication of both proliferating and quiescent tumor populations. The scheduling and sequencing of ATR pathway inhibitors within therapeutic regimens will require continuous optimization.

10.8.5 Concluding Remarks

The availability of ATR pathway inhibitors offer a beacon of hope that chemoresistance, which is a barrier to effective cancer treatment, could finally be overcome. Preclinical results of ATR and Chk1 inhibitors have shown remarkable promise. However, our understanding of the mechanistic basis of ATR pathway inhibition is by no means complete, and the clinical development of ATR pathway inhibitors is still in its infancy. Identification of predictive biomarkers of response underpinned by sound mechanistic understanding remains a research priority, and the rational selection of patients into clinical studies based on these biomarkers could translate into better therapeutic outcomes. In conclusion, the development of ATR pathway inhibitors epitomizes the emerging role of precision medicine in cancer treatment, through which vulnerabilities in cancer are systematically identified and targeted.

Acknowledgments

We would like to acknowledge Bloodwise UK for continuous support of our studies.

Conflict of Interest: No potential conflicts of interest were disclosed.

References

[1] Jackson SP, Bartek J. The DNA-damage response in human biology and disease. Nature 2009;461(7267):1071–8.
[2] Castedo M, Perfettini JL, Roumier T, Andreau K, Medema R, Kroemer G. Cell death by mitotic catastrophe: a molecular definition. Oncogene 2004;23(16):2825–37.
[3] Vitale I, Galluzzi L, Castedo M, Kroemer G. Mitotic catastrophe: a mechanism for avoiding genomic instability. Nat Rev Mol Cell Biol 2011;12(6):385–92.
[4] O'Connor MJ. Targeting the DNA damage response in cancer. Mol Cell 2015;60(4):547–60.
[5] Cimprich KA, Cortez D. A.T.R: an essential regulator of genome integrity. Nat Rev Mol Cell Biol 2008;9(8):616–27.
[6] Ge XQ, Blow JJ. Chk1 inhibits replication factory activation but allows dormant origin firing in existing factories. J Cell Biol 2010;191(7):1285–97.
[7] Chen YH, Jones MJ, Yin Y, et al. ATR-mediated phosphorylation of FANCI regulates dormant origin firing in response to replication stress. Mol Cell 2015;58(2):323–38.
[8] Marechal A, Zou L. RPA-coated single-stranded DNA as a platform for post-translational modifications in the DNA damage response. Cell Res 2015;25(1):9–23.
[9] Cotta-Ramusino C, McDonald ER III, Hurov K, Sowa ME, Harper JW, Elledge SJ. A DNA damage response screen identifies RHINO, a 9-1-1 and TopBP1 interacting protein required for ATR signaling. Science 2011;332(6035):1313–7.
[10] Haahr P, Hoffmann S, Tollenaere MA, et al. Activation of the ATR kinase by the RPA-binding protein ETAA1. Nat Cell Biol 2016;18(11):1196–207.
[11] Bass TE, Luzwick JW, Kavanaugh G, et al. ETAA1 acts at stalled replication forks to maintain genome integrity. Nat Cell Biol 2016;18(11):1185–95.
[12] Lee YC, Zhou Q, Chen J, Yuan J. RPA-binding protein ETAA1 is an ATR activator involved in DNA replication stress response. Curr Biol 2016;26(24):3257–68.
[13] Dungrawala H, Rose KL, Bhat KP, et al. The replication checkpoint prevents two types of fork collapse without regulating replisome stability. Mol Cell 2015;59(6):998–1010.
[14] Neelsen KJ, Lopes M. Replication fork reversal in eukaryotes: from dead end to dynamic response. Nat Rev Mol Cell Biol 2015;16(4):207–20.
[15] Berti M, Vindigni A. Replication stress: getting back on track. Nat Struct Mol Biol 2016;23(2):103–9.
[16] Ammazzalorso F, Pirzio LM, Bignami M, Franchitto A, Pichierri P. ATR and ATM differently regulate WRN to prevent DSBs at stalled replication forks and promote replication fork recovery. EMBO J 2010;29(18):3156–69.
[17] Thangavel S, Berti M, Levikova M, et al. DNA2 drives processing and restart of reversed replication forks in human cells. J Cell Biol 2015;208(5):545–62.
[18] Couch FB, Bansbach CE, Driscoll R, et al. ATR phosphorylates SMARCAL1 to prevent replication fork collapse. Genes Dev 2013;27(14):1610–23.
[19] Betous R, Mason AC, Rambo RP, et al. SMARCAL1 catalyzes fork regression and Holliday junction migration to maintain genome stability during DNA replication. Genes Dev 2012;26(2):151–62.

[20] Singh TR, Ali AM, Paramasivam M, et al. ATR-dependent phosphorylation of FANCM at serine 1045 is essential for FANCM functions. Cancer Res 2013;73(14):4300–10.
[21] Schwab RA, Blackford AN, Niedzwiedz W. ATR activation and replication fork restart are defective in FANCM-deficient cells. EMBO J 2010;29(4):806–18.
[22] Davies SL, North PS, Hickson ID. Role for BLM in replication-fork restart and suppression of origin firing after replicative stress. Nat Struct Mol Biol 2007;14(7):677–9.
[23] Iannascoli C, Palermo V, Murfuni I, Franchitto A, Pichierri P. The WRN exonuclease domain protects nascent strands from pathological MRE11/EXO1-dependent degradation. Nucleic Acids Res 2015;43(20):9788–803.
[24] Petermann E, Woodcock M, Helleday T. Chk1 promotes replication fork progression by controlling replication initiation. Proc Natl Acad Sci USA 2010;107(37):16090–5.
[25] Fragkos M, Ganier O, Coulombe P, Mechali M. DNA replication origin activation in space and time. Nat Rev Mol Cell Biol 2015;16(6):360–74.
[26] Peng CY, Graves PR, Thoma RS, Wu Z, Shaw AS, Piwnica-Worms H. Mitotic and G2 checkpoint control: regulation of 14-3-3 protein binding by phosphorylation of Cdc25C on serine-216. Science 1997;277(5331):1501–5.
[27] Sorensen CS, Syljuasen RG, Falck J, et al. Chk1 regulates the S phase checkpoint by coupling the physiological turnover and ionizing radiation-induced accelerated proteolysis of Cdc25A. Cancer Cell 2003;3(3):247–58.
[28] Zeman MK, Cimprich KA. Causes and consequences of replication stress. Nat Cell Biol 2014;16(1):2–9.
[29] Mankouri HW, Huttner D, Hickson ID. How unfinished business from S-phase affects mitosis and beyond. EMBO J 2013;32(20):2661–71.
[30] Toledo LI, Altmeyer M, Rask MB, et al. ATR prohibits replication catastrophe by preventing global exhaustion of RPA. Cell 2013;155(5):1088–103.
[31] Kwok M, Davies N, Agathanggelou A, et al. ATR inhibition induces synthetic lethality and overcomes chemoresistance in TP53- or ATM-defective chronic lymphocytic leukemia cells. Blood 2016;127(5):582–95.
[32] Macheret M, Halazonetis TD. DNA replication stress as a hallmark of cancer. Annu Rev Pathol 2015;10:425–48.
[33] Hills SA, Diffley JF. DNA replication and oncogene-induced replicative stress. Curr Biol 2014;24(10):R435–44.
[34] Brown EJ, Baltimore D. Essential and dispensable roles of ATR in cell cycle arrest and genome maintenance. Genes Dev 2003;17(5):615–28.
[35] Zighelboim I, Schmidt AP, Gao F, et al. ATR mutation in endometrioid endometrial cancer is associated with poor clinical outcomes. J Clin Oncol 2009;27(19):3091–6.
[36] Chen CF, Ruiz-Vega R, Vasudeva P, et al. ATR mutations promote the growth of melanoma tumors by modulating the immune microenvironment. Cell Rep 2017;18(10):2331–42.
[37] de Klein A, Muijtjens M, van Os R, et al. Targeted disruption of the cell-cycle checkpoint gene ATR leads to early embryonic lethality in mice. Curr Biol 2000;10(8):479–82.
[38] Ruzankina Y, Pinzon-Guzman C, Asare A, et al. Deletion of the developmentally essential gene ATR in adult mice leads to age-related phenotypes and stem cell loss. Cell Stem Cell 2007;1(1):113–26.
[39] O'Driscoll M, Ruiz-Perez VL, Woods CG, Jeggo PA, Goodship JA. A splicing mutation affecting expression of ataxia-telangiectasia and Rad3-related protein (ATR) results in Seckel syndrome. Nat Genet 2003;33(4):497–501.
[40] Murga M, Bunting S, Montana MF, et al. A mouse model of ATR-Seckel shows embryonic replicative stress and accelerated aging. Nat Genet 2009;41(8):891–8.
[41] Schoppy DW, Ragland RL, Gilad O, et al. Oncogenic stress sensitizes murine cancers to hypomorphic suppression of ATR. J Clin Invest 2012;122(1):241–52.
[42] Murga M, Campaner S, Lopez-Contreras AJ, et al. Exploiting oncogene-induced replicative stress for the selective killing of Myc-driven tumors. Nat Struct Mol Biol 2011;18(12):1331–5.
[43] McLornan DP, List A, Mufti GJ. Applying synthetic lethality for the selective targeting of cancer. N Engl J Med 2014;371(18):1725–35.
[44] Shaheen M, Allen C, Nickoloff JA, Hromas R. Synthetic lethality: exploiting the addiction of cancer to DNA repair. Blood 2011;117(23):6074–82.
[45] Menendez D, Inga A, Resnick MA. The expanding universe of p53 targets. Nat Rev Cancer 2009;9(10):724–37.
[46] Muller PA, Vousden KH. p53 mutations in cancer. Nat Cell Biol 2013;15(1):2–8.
[47] Nghiem P, Park PK, Kim Y, Vaziri C, Schreiber SL. ATR inhibition selectively sensitizes G1 checkpoint-deficient cells to lethal premature chromatin condensation. Proc Natl Acad Sci USA 2001;98(16):9092–7.
[48] Sangster-Guity N, Conrad BH, Papadopoulos N, Bunz F. ATR mediates cisplatin resistance in a p53 genotype-specific manner. Oncogene 2011;30(22):2526–33.

[49] Ruzankina Y, Schoppy DW, Asare A, Clark CE, Vonderheide RH, Brown EJ. Tissue regenerative delays and synthetic lethality in adult mice after combined deletion of Atr and Trp53. Nat Genet 2009;41(10):1144–9.
[50] Ma CX, Cai S, Li S, et al. Targeting Chk1 in p53-deficient triple-negative breast cancer is therapeutically beneficial in human-in-mouse tumor models. J Clin Invest 2012;122(4):1541–52.
[51] Toledo LI, Murga M, Zur R, et al. A cell-based screen identifies ATR inhibitors with synthetic lethal properties for cancer-associated mutations. Nat Struct Mol Biol 2011;18(6):721–7.
[52] Marechal A, Zou L. DNA damage sensing by the ATM and ATR kinases. Cold Spring Harb Perspect Biol 2013;5(9). a012716.
[53] Choi M, Kipps T, Kurzrock R. ATM mutations in cancer: therapeutic implications. Mol Cancer Ther 2016;15(8):1781–91.
[54] Austen B, Powell JE, Alvi A, et al. Mutations in the ATM gene lead to impaired overall and treatment-free survival that is independent of IGVH mutation status in patients with B-CLL. Blood 2005;106(9):3175–82.
[55] Austen B, Skowronska A, Baker C, et al. Mutation status of the residual ATM allele is an important determinant of the cellular response to chemotherapy and survival in patients with chronic lymphocytic leukemia containing an 11q deletion. J Clin Oncol 2007;25(34):5448–57.
[56] Skowronska A, Parker A, Ahmed G, et al. Biallelic ATM inactivation significantly reduces survival in patients treated on the United Kingdom Leukemia Research Fund Chronic Lymphocytic Leukemia 4 trial. J Clin Oncol 2012;30(36):4524–32.
[57] Shiloh Y, Ziv Y. The ATM protein kinase: regulating the cellular response to genotoxic stress, and more. Nat Rev Mol Cell Biol 2013;14(4):197–210.
[58] Reaper PM, Griffiths MR, Long JM, et al. Selective killing of ATM- or p53-deficient cancer cells through inhibition of ATR. Nat Chem Biol 2011;7(7):428–30.
[59] Krajewska M, Fehrmann RS, Schoonen PM, et al. ATR inhibition preferentially targets homologous recombination-deficient tumor cells. Oncogene 2015;34(26):3474–81.
[60] Middleton FK, Patterson MJ, Elstob CJ, et al. Common cancer-associated imbalances in the DNA damage response confer sensitivity to single agent ATR inhibition. Oncotarget 2015;6(32):32396–409.
[61] Min A, Im SA, Jang H, et al. AZD6738, a novel oral inhibitor of ATR, induces synthetic lethality with ATM-deficiency in gastric cancer cells. Mol Cancer Ther 2017;16(4):566–77.
[62] Yazinski SA, Comaills V, Buisson R, et al. ATR inhibition disrupts rewired homologous recombination and fork protection pathways in PARP inhibitor-resistant BRCA-deficient cancer cells. Genes Dev 2017;31(3):318–32.
[63] Mohni KN, Kavanaugh GM, Cortez D. ATR pathway inhibition is synthetically lethal in cancer cells with ERCC1 deficiency. Cancer Res 2014;74(10):2835–45.
[64] Sultana R, Abdel-Fatah T, Perry C, et al. Ataxia telangiectasia mutated and Rad3 related (ATR) protein kinase inhibition is synthetically lethal in XRCC1 deficient ovarian cancer cells. PLoS ONE 2013;8(2):e57098.
[65] Hocke S, Guo Y, Job A, et al. A synthetic lethal screen identifies ATR-inhibition as a novel therapeutic approach for POLD1-deficient cancers. Oncotarget 2016;7(6):7080–95.
[66] Williamson CT, Miller R, Pemberton HN, et al. ATR inhibitors as a synthetic lethal therapy for tumours deficient in ARID1A. Nat Commun 2016;7:13837.
[67] Flynn RL, Cox KE, Jeitany M, et al. Alternative lengthening of telomeres renders cancer cells hypersensitive to ATR inhibitors. Science 2015;347(6219):273–7.
[68] Charrier JD, Durrant SJ, Golec JM, et al. Discovery of potent and selective inhibitors of ataxia telangiectasia mutated and Rad3 related (ATR) protein kinase as potential anticancer agents. J Med Chem 2011;54(7):2320–30.
[69] Foote KM, Blades K, Cronin A, et al. Discovery of 4-{4-[(3R)-3-methylmorpholin-4-yl]-6-[1-(methylsulfonyl) cyclopropyl]pyrimidin-2-y l}-1*H*-indole (AZ20): a potent and selective inhibitor of ATR protein kinase with monotherapy *in vivo* antitumor activity. J Med Chem 2013;56(5):2125–38.
[70] McNeely S, Beckmann R, Bence Lin AK. CHEK again: revisiting the development of CHK1 inhibitors for cancer therapy. Pharmacol Ther 2014;142(1):1–10.
[71] Ma CX, Janetka JW, Piwnica-Worms H. Death by releasing the breaks: CHK1 inhibitors as cancer therapeutics. Trends Mol Med 2011;17(2):88–96.
[72] Cottini F, Hideshima T, Suzuki R, et al. Synthetic lethal approaches exploiting DNA damage in aggressive myeloma. Cancer Discov 2015;5(9):972–87.
[73] Muralidharan SV, Bhadury J, Nilsson LM, Green LC, McLure KG, Nilsson JA. BET bromodomain inhibitors synergize with ATR inhibitors to induce DNA damage, apoptosis, senescence-associated secretory pathway and ER stress in Myc-induced lymphoma cells. Oncogene 2016;35(36):4689–97.

[74] Pires IM, Olcina MM, Anbalagan S, et al. Targeting radiation-resistant hypoxic tumour cells through ATR inhibition. Br J Cancer 2012;107(2):291–9.
[75] Ma J, Li X, Su Y, et al. Mechanisms responsible for the synergistic antileukemic interactions between ATR inhibition and cytarabine in acute myeloid leukemia cells. Sci Rep 2017;7:41950.
[76] Kim HJ, Min A, Im SA, et al. Anti-tumor activity of the ATR inhibitor AZD6738 in HER2 positive breast cancer cells. Int J Cancer 2017;140(1):109–19.
[77] Morgado-Palacin I, Day A, Murga M, et al. Targeting the kinase activities of ATR and ATM exhibits antitumoral activity in mouse models of MLL-rearranged AML. Sci Signal 2016;9(445):ra91.
[78] Zhang Y, Lai J, Du Z, et al. Targeting radioresistant breast cancer cells by single agent CHK1 inhibitor via enhancing replication stress. Oncotarget 2016;7(23):34688–702.
[79] Magnussen GI, Emilsen E, Giller Fleten K, et al. Combined inhibition of the cell cycle related proteins Wee1 and Chk1/2 induces synergistic anti-cancer effect in melanoma. BMC Cancer 2015;15:462.
[80] Yuan LL, Green A, David L, et al. Targeting CHK1 inhibits cell proliferation in FLT3-ITD positive acute myeloid leukemia. Leuk Res 2014;38(11):1342–9.
[81] Kim H, George E, Ragland RL, et al. Targeting the ATR/CHK1 axis with PARP inhibition results in tumor regression in BRCA mutant ovarian cancer models. Clin Cancer Res 2016;23(12):3097–108.
[82] Sakurikar N, Thompson R, Montano R, Eastman A. A subset of cancer cell lines is acutely sensitive to the Chk1 inhibitor MK-8776 as monotherapy due to CDK2 activation in S phase. Oncotarget 2016;7(2):1380–94.
[83] Zhao J, Niu X, Li X, et al. Inhibition of CHK1 enhances cell death induced by the Bcl-2-selective inhibitor ABT-199 in acute myeloid leukemia cells. Oncotarget 2016;7(23):34785–99.
[84] Duan L, Perez RE, Hansen M, Gitelis S, Maki CG. Increasing cisplatin sensitivity by schedule-dependent inhibition of AKT and Chk1. Cancer Biol Ther 2014;15(12):1600–12.
[85] Sarmento LM, Povoa V, Nascimento R, et al. CHK1 overexpression in T-cell acute lymphoblastic leukemia is essential for proliferation and survival by preventing excessive replication stress. Oncogene 2015;34(23): 2978–90.
[86] Chila R, Basana A, Lupi M, et al. Combined inhibition of Chk1 and Wee1 as a new therapeutic strategy for mantle cell lymphoma. Oncotarget 2015;6(5):3394–408.
[87] Bryant C, Rawlinson R, Massey AJ. Chk1 inhibition as a novel therapeutic strategy for treating triple-negative breast and ovarian cancers. BMC Cancer 2014;14:570.
[88] Zhou ZR, Yang ZZ, Wang SJ, et al. The Chk1 inhibitor MK-8776 increases the radiosensitivity of human triple-negative breast cancer by inhibiting autophagy. Acta Pharmacol Sin 2017;.
[89] Hall AB, Newsome D, Wang Y, et al. Potentiation of tumor responses to DNA damaging therapy by the selective ATR inhibitor VX-970. Oncotarget 2014;5(14):5674–85.
[90] Vendetti FP, Lau A, Schamus S, Conrads TP, O'Connor MJ, Bakkenist CJ. The orally active and bioavailable ATR kinase inhibitor AZD6738 potentiates the anti-tumor effects of cisplatin to resolve ATM-deficient non-small cell lung cancer *in vivo*. Oncotarget 2015;6(42):44289–305.
[91] Mohni KN, Thompson PS, Luzwick JW, et al. A synthetic lethal screen identifies DNA repair pathways that sensitize cancer cells to combined ATR inhibition and cisplatin treatments. PLoS ONE 2015;10(5):e0125482.
[92] Josse R, Martin SE, Guha R, et al. ATR inhibitors VE-821 and VX-970 sensitize cancer cells to topoisomerase i inhibitors by disabling DNA replication initiation and fork elongation responses. Cancer Res 2014;74(23):6968–79.
[93] Chaudhuri L, Vincelette ND, Koh BD, et al. CHK1 and WEE1 inhibition combine synergistically to enhance therapeutic efficacy in acute myeloid leukemia *ex vivo*. Haematologica 2014;99(4):688–96.
[94] Russell MR, Levin K, Rader J, et al. Combination therapy targeting the Chk1 and Wee1 kinases shows therapeutic efficacy in neuroblastoma. Cancer Res 2013;73(2):776–84.
[95] Mak JP, Man WY, Chow JP, Ma HT, Poon RY. Pharmacological inactivation of CHK1 and WEE1 induces mitotic catastrophe in nasopharyngeal carcinoma cells. Oncotarget 2015;6(25):21074–84.
[96] Dietlein F, Kalb B, Jokic M, et al. A synergistic interaction between Chk1- and MK2 inhibitors in KRAS-mutant cancer. Cell 2015;162(1):146–59.
[97] Sausville E, Lorusso P, Carducci M, et al. Phase I dose-escalation study of AZD7762, a checkpoint kinase inhibitor, in combination with gemcitabine in US patients with advanced solid tumors. Cancer Chemother Pharmacol 2014;73(3):539–49.
[98] Seto T, Esaki T, Hirai F, et al. Phase I dose-escalation study of AZD7762 alone and in combination with gemcitabine in Japanese patients with advanced solid tumours. Cancer Chemother Pharmacol 2013;72(3):619–27.

[99] Doi T, Yoshino T, Shitara K, et al. Phase I study of LY2603618, a CHK1 inhibitor, in combination with gemcitabine in Japanese patients with solid tumors. Anticancer Drugs 2015;26(10):1043–53.

[100] Hong D, Infante J, Janku F, et al. Phase I study of LY2606368, a checkpoint kinase 1 inhibitor, in patients with advanced cancer. J Clin Oncol 2016;34(15):1764–71.

[101] Karp JE, Thomas BM, Greer JM, et al. Phase I and pharmacologic trial of cytosine arabinoside with the selective checkpoint 1 inhibitor Sch 900776 in refractory acute leukemias. Clin Cancer Res 2012;18(24):6723–31.

[102] Daud AI, Ashworth MT, Strosberg J, et al. Phase I dose-escalation trial of checkpoint kinase 1 inhibitor MK-8776 as monotherapy and in combination with gemcitabine in patients with advanced solid tumors. J Clin Oncol 2015;33(9):1060–6.

[103] Scagliotti G, Kang JH, Smith D, et al. Phase II evaluation of LY2603618, a first-generation CHK1 inhibitor, in combination with pemetrexed in patients with advanced or metastatic non-small cell lung cancer. Invest New Drugs 2016;34(5):625–35.

[104] Laquente B, Lopez-Martin J, Richards D, et al. A phase II study to evaluate LY2603618 in combination with gemcitabine in pancreatic cancer patients. BMC Cancer 2017;17(1):137.

[105] Al-Ahmadie H, Iyer G, Hohl M, et al. Synthetic lethality in ATM-deficient RAD50-mutant tumors underlies outlier response to cancer therapy. Cancer Discov 2014;4(9):1014–21.

[106] Hickson I, Zhao Y, Richardson CJ, et al. Identification and characterization of a novel and specific inhibitor of the ataxia-telangiectasia mutated kinase ATM. Cancer Res 2004;64(24):9152–9.

[107] Golding SE, Rosenberg E, Valerie N, et al. Improved ATM kinase inhibitor KU-60019 radiosensitizes glioma cells, compromises insulin, AKT and ERK prosurvival signaling, and inhibits migration and invasion. Mol Cancer Ther 2009;8(10):2894–902.

[108] Biddlestone-Thorpe L, Sajjad M, Rosenberg E, et al. ATM kinase inhibition preferentially sensitizes p53-mutant glioma to ionizing radiation. Clin Cancer Res 2013;19(12):3189–200.

[109] Batey MA, Zhao Y, Kyle S, et al. Preclinical evaluation of a novel ATM inhibitor, KU59403, in vitro and in vivo in p53 functional and dysfunctional models of human cancer. Mol Cancer Ther 2013;12(6):959–67.

[110] Blake SM, Stricker SH, Halavach H, et al. Inactivation of the ATMIN/ATM pathway protects against glioblastoma formation. Elife 2016;5. e08711.

[111] Lin YF, Shih HY, Shang Z, Matsunaga S, Chen BP. DNA-PKcs is required to maintain stability of Chk1 and Claspin for optimal replication stress response. Nucleic Acids Res 2014;42(7):4463–73.

[112] Buisson R, Boisvert JL, Benes CH, Zou L. Distinct but concerted roles of ATR, DNA-PK, and Chk1 in countering replication stress during S phase. Mol Cell 2015;59(6):1011–24.

[113] Sanjiv K, Hagenkort A, Calderon-Montano JM, et al. Cancer-specific synthetic lethality between ATR and CHK1 kinase activities. Cell Rep 2016;17(12):3407–16.

[114] Gilad O, Nabet BY, Ragland RL, et al. Combining ATR suppression with oncogenic Ras synergistically increases genomic instability, causing synthetic lethality or tumorigenesis in a dosage-dependent manner. Cancer Res 2010;70(23):9693–702.

[115] Teng PN, Bateman NW, Darcy KM, et al. Pharmacologic inhibition of ATR and ATM offers clinically important distinctions to enhancing platinum or radiation response in ovarian, endometrial, and cervical cancer cells. Gynecol Oncol 2015;136(3):554–61.

[116] Leszczynska KB, Dobrynin G, Leslie RE, et al. Preclinical testing of an Atr inhibitor demonstrates improved response to standard therapies for esophageal cancer. Radiother Oncol 2016;121(2):232–8.

[117] Menezes DL, Holt J, Tang Y, et al. A synthetic lethal screen reveals enhanced sensitivity to ATR inhibitor treatment in mantle cell lymphoma with ATM loss-of-function. Mol Cancer Res 2015;13(1):120–9.

[118] Huntoon CJ, Flatten KS, Wahner Hendrickson AE, et al. ATR inhibition broadly sensitizes ovarian cancer cells to chemotherapy independent of BRCA status. Cancer Res 2013;73(12):3683–91.

[119] Prevo R, Fokas E, Reaper PM, et al. The novel ATR inhibitor VE-821 increases sensitivity of pancreatic cancer cells to radiation and chemotherapy. Cancer Biol Ther 2012;13(11):1072–81.

[120] Fokas E, Prevo R, Pollard JR, et al. Targeting ATR *in vivo* using the novel inhibitor VE-822 results in selective sensitization of pancreatic tumors to radiation. Cell Death Dis 2012;3:e441.

[121] Wang WT, Catto JW, Meuth M. Differential response of normal and malignant urothelial cells to CHK1 and ATM inhibitors. Oncogene 2015;34(22):2887–96.

[122] Tang Y, Hamed HA, Poklepovic A, Dai Y, Grant S, Dent P. Poly(ADP-ribose) polymerase 1 modulates the lethality of CHK1 inhibitors in mammary tumors. Mol Pharmacol 2012;82(2):322–32.

[123] Martino-Echarri E, Henderson BR, Brocardo MG. Targeting the DNA replication checkpoint by pharmacologic inhibition of Chk1 kinase: a strategy to sensitize APC mutant colon cancer cells to 5-fluorouracil chemotherapy. Oncotarget 2014;5(20):9889–900.

[124] Origanti S, Cai SR, Munir AZ, White LS, Piwnica-Worms H. Synthetic lethality of Chk1 inhibition combined with p53 and/or p21 loss during a DNA damage response in normal and tumor cells. Oncogene 2013;32(5):577–88.

[125] Gadhikar MA, Sciuto MR, Alves MV, et al. Chk1/2 inhibition overcomes the cisplatin resistance of head and neck cancer cells secondary to the loss of functional p53. Mol Cancer Ther 2013;12(9):1860–73.

[126] Barker HE, Patel R, McLaughlin M, et al. CHK1 inhibition radiosensitizes head and neck cancers to paclitaxel-based chemoradiotherapy. Mol Cancer Ther 2016;15(9):2042–54.

[127] Bartucci M, Svensson S, Romania P, et al. Therapeutic targeting of Chk1 in NSCLC stem cells during chemotherapy. Cell Death Differ 2012;19(5):768–78.

[128] Zemanova J, Hylse O, Collakova J, et al. Chk1 inhibition significantly potentiates activity of nucleoside analogs in TP53-mutated B-lymphoid cells. Oncotarget 2016;7(38):62091–106.

[129] Pei XY, Dai Y, Felthousen J, et al. Circumvention of Mcl-1-dependent drug resistance by simultaneous Chk1 and MEK1/2 inhibition in human multiple myeloma cells. PLoS ONE 2014;9(3):e89064.

[130] Dai Y, Chen S, Shah R, et al. Disruption of Src function potentiates Chk1-inhibitor-induced apoptosis in human multiple myeloma cells *in vitro* and *in vivo*. Blood 2011;117(6):1947–57.

[131] Cole KA, Huggins J, Laquaglia M, et al. RNAi screen of the protein kinome identifies checkpoint kinase 1 (CHK1) as a therapeutic target in neuroblastoma. Proc Natl Acad Sci USA 2011;108(8):3336–41.

[132] Sankunny M, Parikh RA, Lewis DW, Gooding WE, Saunders WS, Gollin SM. Targeted inhibition of ATR or CHEK1 reverses radioresistance in oral squamous cell carcinoma cells with distal chromosome arm 11q loss. Genes Chromosomes Cancer 2014;53(2):129–43.

[133] Vance S, Liu E, Zhao L, et al. Selective radiosensitization of p53 mutant pancreatic cancer cells by combined inhibition of Chk1 and PARP1. Cell Cycle 2011;10(24):4321–9.

[134] Engelke CG, Parsels LA, Qian Y, et al. Sensitization of pancreatic cancer to chemoradiation by the Chk1 inhibitor MK8776. Clin Cancer Res 2013;19(16):4412–21.

[135] McNeely S, Conti C, Sheikh T, et al. Chk1 inhibition after replicative stress activates a double strand break response mediated by ATM and DNA-dependent protein kinase. Cell Cycle 2010;9(5):995–1004.

[136] Calvo E, Braiteh F, Von Hoff D, et al. Phase I study of CHK1 inhibitor LY2603618 in combination with gemcitabine in patients with solid tumors. Oncology 2016;91(5):251–60.

CHAPTER

11

PARP Inhibition to Enhance Response to Chemotherapy

Naoko Takebe, Mary Quinn*, Garima Gupta**, Alice P. Chen**

***National Cancer Institute, National Institutes of Health, Bethesda, MD, United States**
****Jewish Hospital of Cincinnati, Cincinnati, OH, United States**

Abstract

The inhibition of poly(ADP-ribose) polymerase (PARP) enzymes is a relatively new anticancer therapeutic strategy designed to impair the ability of tumor cells to repair DNA damage. PARP inhibitors induce synthetic lethality in cells lacking the ability to repair DNA breaks through homologous recombination, such as those with mutations in *BRCA1/2* genes. Furthermore, the combination of DNA-damaging chemotherapies with PARP inhibitors may potentiate the DNA damage caused by these agents and lead to enhanced antitumor activity. An additional mechanism of action of PARP inhibitors, known as PARP trapping, has been shown to result in double-strand breaks in DNA, further expanding the potential of these drugs in causing DNA damage accumulation and ultimately apoptosis. While clinical trials testing the combination of PARP inhibitors with chemotherapeutic agents have been challenging due to enhanced toxicity, a number of studies have demonstrated clinical benefits using this strategy. This chapter will review the current status of PARP inhibitor development and discuss the lessons that have been learned from novel chemotherapy combination studies.

ABBREVIATIONS

AML	Acute myeloid leukemia
A-T	Ataxia-telangiectasia
ATM	Ataxia-telangiectasia mutated
ATR	Ataxia-telangiectasia and Rad3 related
AUC	Area under the curve; BER, Base excision repair
CHK	Checkpoint kinase
CR	Complete response
CTEP	Cancer Therapy Evaluation Program
DLT	Dose-limiting toxicity
DSB	Double-strand break
FA	Fanconi anemia
FANCD2	Fanconi anemia complementation group D2

Targeting Cell Survival Pathways to Enhance Response to Chemotherapy. http://dx.doi.org/10.1016/B978-0-12-813753-6.00011-1
Copyright © 2018 Elsevier Inc. All rights reserved.

FDA	Food and Drug Administration
FLT3-ITD	fms-like tyrosine kinase 3-internal tandem duplication
GBM	Glioblastoma multiforme
HGSOC	High-grade serous ovarian cancer
HI	Hematologic improvement
HR	Homologous recombination
HRD	Homologous recombination deficiency
H2AX	Histone 2A family member X
IV	Intravenous
LOH	Loss of heterozygosity
MGMT	O6-methylguanine DNA-methyltransferase
MMR	Mismatch repair
MTD	Maximum tolerated dose
NER	Nucleotide excision repair
NHEJ	Nonhomologous end-joining
NSCLC	Non-small cell lung cancer
OS	Overall survival
OvCa	Ovarian cancer
PARG	Poly(ADP-ribose) glycohydrolase
PARP	Poly(ADP-ribose) polymerase
PBMC	Peripheral blood mononuclear cells
pCR	Pathological complete response
PD	Pharmacodynamic
PFS	Progression-free survival
PK	Pharmacokinetic
PR	Partial response
PSA	Prostate-specific antigen
RR	Response rate
SCLC	Small cell lung cancer
SD	Stable disease
SSB	Single-strand break
TDP1	Tyrosyl-DNA-phosphodiesterase
TMZ	Temozolomide
TNBC	Triple-negative breast cancer

11.1 INTRODUCTION

Poly(ADP-ribose) polymerase (PARP) inhibitors have been under development for the last 10 years, with the first FDA approval in 2014 for olaparib (Astra Zeneca-Lynparza®, AZD 2281, KU 0059436), quickly followed by approval for rucaparib (Clovis-Rubraca®, PF-01367338, AGO14699) and niraparib (Tesaro-Zejula®, MK 4827). The success of these inhibitors is based on the synthetic lethality they induce in *BRCA1/2*-mutant cells. This effect was first demonstrated in preclinical models by two groups in 2005 [1,2]. Subsequently, this finding has been further advanced in human clinical trials, including in a phase I trial that analyzed the pharmacokinetic and pharmacodynamic characteristics of olaparib in a study population enriched for *BRCA1/2* mutation carriers [3]. The success in this population led to olaparib becoming the first PARP inhibitor to obtain regulatory approval as the fourth or later treatment line for germline *BRCA*-mutated high-grade serous ovarian cancer (HGSOC) [4].

In parallel to targeting existing homologous recombination deficiency (HRD), such as *BRCA1/2* mutation, to induce synthetic lethality, efforts have been made to combine PARP inhibitors with other drugs to potentiate the DNA damage caused by cytotoxic chemotherapies or to induce HRD by targeted agents. Despite promising preclinical data, the combination of PARP inhibitors with cytotoxic chemotherapy has been quite challenging due to poor tolerability in patients, mainly because of myelosuppression. This unexpected result reflects the difficulty in predicting clinical tolerability using preclinical models of combination therapy. The Cancer Therapy Evaluation Program of the National Cancer Institute has sponsored a substantial number of PARP inhibitor combination trials, mostly with veliparib (ABT-888), to investigate novel chemotherapy combination strategies with PARP inhibitors based on sound preclinical data. This chapter focuses on reviewing the current status of PARP inhibitors and the progress that has been made in exploring potential combinations with chemotherapy, radiotherapy, and targeted agents as an anticancer therapeutic strategy. It will also provide perspectives regarding this strategy in the area of PARP inhibitor drug development.

11.2 TARGETING PARP TO INHIBIT DNA REPAIR

PARP-1 is a nuclear enzyme that catalyzes the formation of PAR polymers at DNA damage sites, promoting recruitment of repair proteins to single-strand DNA breaks and facilitating base excision repair (BER) [5–8]. More precisely, upon DNA strand breaks, the DNA-binding domain of PARP-1, which contains two zinc finger motifs, recognizes and binds to damaged sites. This binding results in the activation of the enzyme to generate long poly(ADP)-ribose from NAD^+ [9,10]. Polymer elongation leads to the catalysis and degradation activity of nuclear enzyme poly(ADP-ribose) glycohydrolase (PARG), which cleaves glycosidic bonds between ADP-ribose units, resulting in the removal and inactivation of PARP in preparation for further DNA damage [9].

When DNA damage occurs, DNA glycosylases remove damaged bases and generate apurinic/apyrimidinic sites, which are cleaved by endonucleases, leaving single-strand breaks (SSBs). As damaged base replacement involves recruitment of a DNA repair complex through PARP enzymatic activity, SSBs persist if PARP-1 is inhibited [10]. In the state of cell replication, an SSB will be converted into a double-strand break (DSB) at the replication fork, eventually inducing apoptosis if multiple DSBs persist in a cell. Homologous recombination (HR) is a major pathway for DSB repair and involves recruitment of ataxia-telangiectasia (A-T) mutated (ATM) or A-T and Rad3 related (ATR) proteins to the DSB. HR also involves activation and phosphorylation of a cascade of proteins, including checkpoint kinase (CHK)-1 and 2, histone H2A family member X (H2AX), the Fanconi anemia (FA) complementation group D2 (FANCD2) protein, and BRCA1 and BRCA2 [9]. If replicating cells have DSB repair deficiencies due to HR pathway mutations, these cells undergo apoptosis.

As a result of this role of PARP in DNA damage repair, PARP became an important target for pharmacological inhibition. When poly-ADP-ribosylation (PARylation) is inhibited, impairment of BER occurs and unrepaired SSBs accumulate, leading to widespread DSB formation and replication fork collapse. As mentioned above, DSBs are normally repaired by the HR repair pathway; this notion led to the idea of clinical development of PARP inhibitors in patients with germline *BRCA1/2* mutations. Of 17 known members of the PARP nuclear superfamily, only PARP-1 and PARP-2 are predominantly involved in DNA repair and can impair the process if disrupted [8,11]. In addition, PARP enzymes have been implicated in several nuclear processes besides DNA repair, such as DNA replication, transcription, and modulation of chromatin structure [12].

Moreover, increasing efforts have been made in broadening the synthetic lethality concept beyond germline *BRCA* mutations. For example, in the setting of clinical trials, exploratory efforts have been made in identifying patients with defects in other proteins involved the HR pathway or in proteins that sense DNA damage and initiate or coordinate DNA repair [13–15]. This chapter will not discuss such PARP inhibitor development as monotherapy in these patients, but rather discuss the approach of combining PARP inhibitors with chemotherapy, which is intended to induce synergistic DNA damage or sensitize cells to chemotherapy through PARP inhibition.

Prior to the discussion of preclinical evidence for this combination approach, it is important to recognize an additional consequence of PARP inhibition. It has been shown that the binding of PARP to DNA strand breaks prevents the recombination machinery from working until PARP disassociates [16]. This finding has developed into the currently well-recognized concept of PARP trapping, the phenomenon in which, as a consequence of PARP inhibition, PARP does not disengage from the DNA break site, leading not only to unrepaired SSBs, but also to replication-dependent DSBs. It was elegantly demonstrated that PARP inhibitors have varying degrees of PARP trapping capability, with talazoparib being the strongest trapper [17]. The overall rationale of combining PARP inhibitors with chemotherapy stems from the idea that inhibition of PARP sensitizes tumor cells to chemotherapies that induce DNA damage. Damage that would normally be repaired through the BER system cannot be repaired in the presence of PARP inhibition. This rationale, which has driven multiple clinical trials, will be further discussed below.

11.3 RATIONALE FOR COMBINING WITH CHEMOTHERAPY

11.3.1 Chemosensitization Through PARP Inhibition by Impairing SSB or BER

Chemotherapy sensitization by causing enhanced DNA damage through PARP inhibition was considered as early as 1980 [18]. Notably, the rationale for the combination of a topoisomerase I (Topo1) inhibitor and a PARP inhibitor was extensively studied. Topo1 stabilizes the topoisomerase–DNA cleavable complexes in the nicked formation at the stage in which DNA breaks occur. Repair of DNA damage involves BER, in which the key BER protein XRCC1 is recruited to Topo1-dependent DNA breaks in association with PARP enzymes [19]. Consequently, tyrosyl-DNA-phosphodiesterase (TDP1), which removes Topo1 from DNA, is recruited to the break site [20]. Since PARP-1 is capable of interacting with Topo1 and repairing Topo1-dependent SSBs, inhibition of PARP enzymes sensitizes cells to Topo1 inhibition [21]. Topo1 inhibition leads to a need for DNA repair to prevent cell death, but this repair is impeded in cells in which PARP activity is inhibited [22,23].

Several studies have demonstrated the potentiation of Topo1 inhibitors in the presence of PARP inhibitors [24,25]. For example, Thomas et al. evaluated a panel of 42 potent PARP inhibitors for potential chemosensitization of temozolomide and topotecan using human colorectal cells *in vitro* and mouse xenograft models. The investigation to further delineate the mechanism of action for PARP inhibitor sensitization on topotecan cytotoxicity was expanded to multiple cell-line models. It was revealed that this sensitization occurs at veliparib concentrations far below those required to substantially inhibit PARP synthesis and at least an order of magnitude lower than those involved in selective killing of HRD cells [26]. This preclinical work led to a phase I combination study of veliparib and topotecan in solid tumors and a phase I study of veliparib and irinotecan in solid tumors [27,28].

Temozolomide is an alkylating agent that has been recognized as a candidate for a PARP inhibitor combination. Temozolomide is a DNA alkylator that adds methyl adducts to N7-guanine, N3-adenine, and O6-guanine positions [29,30]. O6-methylguanine, which comprises 5% of the alkylated DNA products after temozolomide treatment, triggers a continuous cycle of futile mismatch repair (MMR), subsequently leading to apoptosis [31,32]. As O6-methylguanine is removed by O6-methylguanine DNA-methyltransferase (MGMT), MGMT overexpression is considered a potential mechanism of temozolomide resistance. However, N7-methylguanine and N3-methyladenine represent 80% of the alkylated DNA after temozolomide treatment, and these adducts are not susceptible to MGMT removal; they are instead excised by the BER pathway. Therefore, increased activity of the BER pathway may also contribute to temozolomide drug resistance. As PARP recruits and activates DNA repair proteins for the BER and nonhomologous end-joining (NHEJ) pathways [33], PARP inhibition is expected to enhance tumor sensitivity to temozolomide through inhibition of BER. For example, potentiation of temozolomide and topotecan in an *in vitro* cytotoxicity assay and in a clonogenic assay with PARP inhibitors was demonstrated in human tumor cell lines including lung, colon, ovary, and breast cancer [34]. In the B16F10 subcutaneous murine melanoma model, veliparib strongly potentiated the temozolomide antitumor effect [35]. In the 9L orthotopic rat glioma model, veliparib in combination with temozolomide significantly slowed tumor progression [35].

Veliparib and temozolomide combination efficacy was also tested in hematological malignancy preclinical models such as an AML model [36]. Both MMR-proficient and -deficient leukemia cells with varying MGMT activity were used to evaluate the veliparib sensitization effect on temozolomide growth inhibition. Interestingly, veliparib potentiation was most effective in MMR-deficient cells with low MGMT activity, whereas veliparib also potentiated temozolomide activity in MMR-deficient cells with elevated MGMT activity. A phase I clinical study to test this combination in AML patients was completed; the details of this study are discussed in Section 11.5.2 [37]. In solid tumors and lymphoma, veliparib and temozolomide combination efficacy was tested in multiple xenograft models representing various human tumors with different responses to temozolomide [38]. These xenograft tumors, including subcutaneous and orthotopic models and metastatic sites, were assessed by tumor burden, expression of poly(ADP-ribose) polymer, and MGMT. The combination had activity to various degrees across a broad histologic spectrum, including B-cell lymphoma, small and non-small-cell lung, pancreatic, ovarian, breast, and prostate cancer xenografts. More interestingly, efficacy in otherwise temozolomide-nonresponsive tumors suggests that temozolomide resistance may be overcome by PARP inhibition. However, the degree of sensitivity to this combination did not correlate with tumor MGMT, MMR status, or poly(ADP-ribose) (PAR) polymer levels.

Another class of agent that may be successfully combined with PARP inhibitors is platinum-based therapies. Preclinical evidence has demonstrated the potentiation of effects of platinum by PARP inhibitors, as well as synergism between PARP inhibitors and platinum compounds, in *BRCA1*- or *BRCA2*-associated breast cancer models and in non-small cell lung cancer (NSCLC) models [39–41]. Particularly, the combination of carboplatin and a PARP inhibitor has been shown to result in enhanced DNA damage and antitumor activity, leading to a more significant reduction in tumor growth than cisplatin alone. These and other findings led to a clinical combination study of carboplatin, taxol, and veliparib in advanced solid tumors, since carboplatin and taxol are used in combination as standard of care [42]. Another triple combination of topotecan, carboplatin, and veliparib was tested in a phase I clinical trial in AML patients, based on the hypothesis that these agents act synergistically [43].

The concept of veliparib being combined with cyclophosphamide was also derived from preclinical data using the MX-1 breast xenograft model (*BRCA1* deletion and *BRCA2* mutation), in which veliparib in combination with cyclophosphamide showed tumor regression of established tumors, whereas cyclophosphamide alone resulted in only modest tumor inhibition [35]. Alkylating agents such as cyclophosphamide are known to form inter-strand cross-links, which are thought to result in cell death due to the formation of SSBs or DSBs during the DNA damage repair process. PARP inhibitors interfere with DNA repair and may potentiate the antitumor effects of cyclophosphamide.

By inhibiting BER, PARP inhibitors also enhance the cytotoxicity of floxuridine [44], an agent previously shown to have clinical activity in ovarian cancer.

11.3.2 Chemosensitization Through PARP Inhibition by Enhancing DNA Trapping

In addition to NAD(+)-competitive catalytic inhibition, which was the mechanism of action first defined for these agents, PARP inhibitors were recently found to also exhibit cytotoxic trapping of PARP–DNA complexes [17]. Although PARP inhibitors are potent catalytic inhibitors, PARP trapping is drug-specific. Murai et al. evaluated the cytotoxicity and molecular mechanisms of the combination of olaparib or veliparib with multiple agents: the Topo1 inhibitor camptothecin, the alkylating agent temozolomide, the cross-linking agent cisplatin, and the Topo2 inhibitor etoposide. PARP–DNA trapping and catalytic PARP inhibition were assessed in the genetically modified chicken B-cell lymphoma line DT40, human prostate DU145, and glioblastoma SF295 cancer cells. Topo1 inhibitor camptothecin combined with either veliparib or olaparib demonstrated highly synergistic effects due to catalytic PARP inhibition. However, PARP trapping was critical for the temozolomide combination, in which olaparib was more effective than veliparib. For other chemotherapeutic agent combinations, including cisplatin and etoposide, the olaparib combination was not effective due to the lack of PARP involvement in DNA repair following cisplatin- or etoposide-induced DNA damage. This paper elegantly illustrates the highly effective combination of catalytic PARP inhibitors, such as veliparib, with camptothecins, vs the effective combination of PARP-trapping inhibitors with temozolomide [45].

PARP inhibitors in combination with temozolomide have been shown to enhance antitumor activity in Ewing sarcoma xenograft and orthotopic models [46]. In these models, PARP inhibitor sensitivity is not due to an apparent HR DNA repair defect, but instead results from hypersensitivity to trapped PARP-1:DNA complexes. During cell replication, PARP-1:DNA complexes cause accumulation of DNA damage, subsequently leading to apoptosis; the potentiation of PARP inhibitors by temozolomide is associated with this process. Other tumor cell lines, such as a subset of glioma, neuroblastoma, and melanoma cells, were also identified as particularly sensitive to a combination of temozolomide and PARP inhibitors by this mechanism [47]. Temozolomide potentiation of PARP inhibitor trapping was further investigated in a phase II three-arm randomized trial in breast cancer comparing veliparib plus temozolomide vs veliparib plus paclitaxel/carboplatin vs placebo plus paclitaxel/carboplatin [48].

11.4 PARP INHIBITORS

11.4.1 Iniparib

Iniparib (BSI 201) initially was thought to be a PARP inhibitor, whose inhibition of single-strand DNA break repair was dependent on PARP-1 expression. The initial excitement over this agent was driven by the activity of iniparib combined with carboplatin and gemcitabine in triple negative breast cancer, reported by O'Shaughnessy et al. [49]. While this phase II trial provided early evidence of the benefit of adding a PARP inhibitor to chemotherapy, the phase III trial did not meet the predefined criteria for the coprimary endpoints of progression-free survival (PFS) and overall survival (OS) [50]. As the phase III trial was commencing, data were released suggesting that iniparib may not be a true PARP inhibitor [51,52].

A phase II study was conducted to assess efficacy and tolerability of iniparib in patients with recurrent, germline *BRCA1/2*-mutated, advanced epithelial ovarian, fallopian tube, or primary peritoneal cancer [53]. Among the 12 patients who received the study drug, no objective responses were seen [53]. The investigators admitted to learning two important lessons from this trial, which guided future early-phase studies. First, attention should be paid to seek out and understand the preclinical data prior to clinical investigations. Although there were basic science data reported from environmental and chemistry laboratories suggesting PARP-1 inhibition by iniparib prior to the clinical development of the drug, data from preclinical cancer models were not available. Studies showing the lack of efficacy of PARP inhibition with iniparib were reported in 2012, well after the completion of the phase II chemotherapy combination study. Second, although there had been favorable activity with the PARP inhibitor olaparib in heavily pretreated (3–4 regimens) germline *BRCA* mutation-associated ovarian cancer patients [54], similar results were not observed in the iniparib study, possibly due to the patients being further along in their disease course or the iniparib mechanism of action differing from direct PARP inhibition [53].

11.4.2 Olaparib

Olaparib is an oral PARP inhibitor approved in the United States as a monotherapy for patients with germline *BRCA*-mutated advanced ovarian cancer who have been treated with three or more prior lines of chemotherapy. It is known that women with germline mutations in *BRCA1*, *BRCA2*, or both have an increased risk of ovarian cancer, with the most common type being HGSOC. Approximately 15% of epithelial ovarian cancers are deficient in HR repair due to mutations in *BRCA1/2*, and olaparib induces synthetic lethality in these HR-deficient tumor cells [4,55]. Clinical evidence of the efficacy of olaparib in patients with germline *BRCA* mutation-associated ovarian cancer was provided by several phase II trials [56–58]. Although the clinical benefits of olaparib are more dramatic in patients with *BRCA1/2* mutations, the drug was shown to have clinical activity in HGSOC patients regardless of *BRCA* status [58]. In 2014, olaparib was approved in the United States for the treatment of adult patients with deleterious or suspected deleterious germline *BRCA*-mutated advanced ovarian cancer who have been treated with three or more prior lines of chemotherapy.

A randomized, double-blind phase II trial (Study 19) was also conducted to evaluate the efficacy of olaparib monotherapy as maintenance treatment in patients with platinum-sensitive relapsed HGSOC who had achieved a response to their most recent platinum-based chemotherapy. Patients received olaparib 400 mg BID or placebo within 8 weeks after completion of chemotherapy. The primary endpoint was PFS, defined as the time from randomization to objective disease progression or death. PFS was significantly longer in the olaparib group, with a PFS of 8.4 months, compared to 4.8 months in the placebo group. The majority of adverse events were grade 1 or 2 and included nausea, vomiting, fatigue, and anemia; the most common adverse events leading to drug interruption or dose reduction were nausea, vomiting, and fatigue [4]. An updated analysis of this study was reported in 2016 and demonstrated that, although there was not a statistically significant improvement in OS, patients who received olaparib maintenance therapy seemed to have longer survival than patients who received placebo, without experiencing any additional safety issues [59]. The group deriving the most significant survival benefit was the *BRCA*-mutated group.

The SOLO-2 trial was designed to prospectively confirm the findings seen in Study 19 and was performed using a tablet form of olaparib, offering patients a reduced daily pill burden. The dose administered was 300 mg BID in tablets (four tablets per day) vs olaparib 400 mg BID in capsules (16 capsules per day). Overall, the safety profile of the tablet was shown to be similar to the capsule formulation. The only major difference in toxicity was an increased rate of anemia with the tablets, thought to be due to longer olaparib exposure in SOLO-2 compared to Study 19 (median duration of exposure of 588 days vs 206 days). Results from SOLO-2 confirmed an improvement in PFS with olaparib maintenance therapy in patients with *BRCA1/2*-mutated platinum-sensitive relapsed ovarian cancer [60]. In August 2017, the FDA approved olaparib tablets as a second indication for maintenance treatment of adult patients with recurrent epithelial ovarian, fallopian tube, or primary peritoneal cancer who are in a complete or partial response to platinum-based chemotherapy [61–63].

11.4.3 Veliparib

The Division of Cancer Treatment and Diagnosis at the National Cancer Institute conducted the first phase 0 clinical trial in oncology of a therapeutic agent under the Exploratory Investigational New Drug Guidance of the FDA. It was a first-in-human study of the oral PARP inhibitor veliparib in patients with advanced malignancies. The trial looked not to determine a maximum tolerated dose (MTD) but to define a dose range and time course over which veliparib inhibited PARP activity. This analysis was done using a validated pharmacodynamic (PD) assay for PAR, a product of PARP. PAR measurements were performed on tumor cells and peripheral blood mononuclear cells (PBMCs), and pharmacokinetic (PK) data were collected as well. A novel statistical approach was also developed, specifically for phase 0 trials, in which the endpoints are based on PD measurements rather than toxicity [64].

In the phase 0 trial, patients received a single oral dose of veliparib. There were five dose levels, with three patients on each dose level: 10, 25, 50, 100, and 150 mg. Planned serial blood sampling, pre- and post-drug administration, was performed for PD and PK analyses. Tumor biopsies were performed once significant inhibition of PARP activity was observed in PBMCs; this process reduced the likelihood of obtaining tumor biopsies from patients receiving doses unlikely to show drug effect. Statistically significant reductions in PAR levels were observed in both tumor and PBMC samples at the 25 and 50 mg dose levels. Veliparib was well tolerated, with no significant adverse effects reported. This study was pivotal in that, in about 5 months, data were available that showed molecular proof-of-mechanism: target inhibition by veliparib in patient tumor cells. Along with PK and PD data, these results were the foundation for subsequent combination studies of veliparib with DNA-damaging agents [64].

Veliparib is currently under extensive clinical development, both as a single agent and in combination trials. It is well suited for use in combination trials, as it has modest hematopoietic toxicity [15]. The first single-agent phase I/II trial of veliparib was performed in patients with germline *BRCA*-mutated or platinum-refractory ovarian cancer and *BRCA*-wild-type basal-like breast cancer. Greater clinical activity was demonstrated in patients with *BRCA*-mutated disease than in those with wild-type disease [65]. The toxicity profile of veliparib was found to be favorable and the phase II dose was established at 400 mg BID. A subsequent phase II trial using single-agent oral veliparib was reported by Coleman et al. [66], examining clinical activity in ovarian cancer patients with a germline *BRCA1* or *BRCA2* mutation. The overall response rate was 26%, and responses were seen in both platinum-sensitive and -resistant disease.

Veliparib is also being evaluated in other solid tumors, including rectal, pancreatic, head and neck, and testicular cancer. In November 2016, the FDA granted Orphan Drug Designation to veliparib for advanced squamous NSCLC [61]. Novel combination regimens of veliparib with different radiotherapy or chemotherapy treatments are being explored in various tumor types, such as with whole-brain radiation in brain metastasis, temozolomide in metastatic melanoma, cisplatin and etoposide in extensive stage SCLC, whole-abdominal radiation in peritoneal carcinomatosis, and carboplatin and paclitaxel in NSCLC, all with promising results [67–71]. However, reports of results from phase III veliparib–chemotherapy combination studies in NSCLC and triple-negative breast cancer (TNBC) have not indicated improvement in clinical activity for the veliparib combination arms [61,72].

11.4.4 Niraparib

Niraparib is a highly selective inhibitor of PARP-1/2, and its antitumor activity was initially demonstrated in a phase I dose escalation study, with the MTD being 300 mg daily. Objective clinical responses were observed in patients with ovarian cancer, with good tolerability [73].

These results led to the multinational phase III NOVA trial evaluating the efficacy and safety of niraparib vs placebo as maintenance treatment in platinum-sensitive recurrent ovarian cancer. There were two cohorts, defined by the presence or absence of a germline *BRCA* mutation [74]. Archival tissue from the non-*BRCA* group was further analyzed using the myChoice HRD test (Myriad Genetics) to define the population of patients with HRD, as decreased rates of HR are associated with inefficient DNA repair. The primary endpoint of the study was the duration of PFS. The three predefined primary efficacy populations were: (1) the germline *BRCA* cohort, (2) a subgroup of the nongermline *BRCA* cohort positive for HRD, and (3) the overall nongermline *BRCA* cohort. The most common adverse events included thrombocytopenia, anemia, and neutropenia, all managed with dose reductions. However, the incidence of myelodysplastic syndrome was 1.4% in patients receiving niraparib, and there were three deaths occurring during the follow-up period (one MDS and one AML), which were considered to be treatment-related. The results demonstrated that niraparib provided significant clinical benefit, regardless of *BRCA* status or HRD status, in this group of platinum-sensitive patients. Niraparib is the first PARP inhibitor approved by the FDA for use regardless of *BRCA* mutation status; it is approved for the maintenance treatment of patients with recurrent epithelial ovarian, fallopian tube, or primary peritoneal cancer who have a partial or complete response to platinum-based therapy without regard to *BRCA* status [74,75].

11.4.5 Rucaparib

Rucaparib is an oral PARP inhibitor that, in a phase I/II trial, demonstrated an objective response in 67% of women with platinum-sensitive relapsed high-grade ovarian carcinoma, who harbored a germline *BRCA* mutation [14].

ARIEL2 was a clinical trial that sought to help predict which *BRCA* wild-type cancers would respond to PARP inhibitor therapy. Several clinical trials reported antitumor activity and improvement in PFS with PARP inhibitors in patients without a *BRCA* mutation, implying that other HR deficiencies may be present. In ARIEL2, patients were classified into one of the three predefined HRD subgroups on the basis of tumor genetic analysis: (1) *BRCA* mutant (deleterious germline or somatic), (2) *BRCA* wild-type and loss of heterozygosity-high group (LOH-high), or (3) *BRCA* wild-type and LOH-low (LOH-low group) [14].

Patients in the ARIEL2 trial received rucaparib 600 mg BID for 28 days, with the primary endpoint being PFS. The most common grade 3 adverse events were decreased hemoglobin and elevations in alanine aminotransferase or aspartate aminotransferase. In patients with platinum-sensitive ovarian carcinoma tumors classified as *BRCA* mutant or *BRCA* wild-type and LOH-high, PFS was longer than in those patients with *BRCA*-wild type and LOH-low tumors. This suggests that assessment of tumor LOH could identify *BRCA* wild-type patients who might benefit from rucaparib, extending the benefit of this agent beyond those with *BRCA* mutation-associated tumors [14]. In 2016, rucaparib was granted FDA approval for treatment of patients with deleterious *BRCA* (germline and/or somatic) mutation-associated advanced ovarian cancer who have been treated with two or more chemotherapies.

In September 2017, data were reported from ARIEL3, a study in which patients with high-grade ovarian cancer who had responded to platinum-based therapy in the second or third line of treatment were randomized 2:1 to rucaparib maintenance therapy or placebo. The primary endpoint was PFS, which was measured sequentially in the following three groups if benefit was found in the previous group: *BRCA* mutant, HR-deficient (*BRCA* mutant or *BRCA* wild-type with high LOH), or the entire study population. Rucaparib produced statistically significant improvement in PFS in all three groups, with the greatest improvement in the *BRCA*-mutated group. However, patients without BRCA mutations were divided based on LOH, and while patients with high LOH demonstrated more improvement in PFS than those with low LOH, rucaparib performed statistically better than placebo in both groups. Therefore, the LOH test was not able to successfully differentiate responders from nonresponders as the investigators had hoped [76].

11.4.6 Talazoparib

The PARP-1/2 inhibitor talazoparib was brought to a phase I first-in-human trial because it was demonstrated to be the most potent agent in terms of both *in vitro* PARP inhibition and *in vitro* trapping of PARP–DNA complexes at sites of single-strand DNA breaks. The phase I dose-escalation trial was done in patients with advanced malignancies, whereas the expansion cohort included patients with tumors harboring germline *BRCA1/2* mutations or predicted to be potentially sensitive to PARP inhibition based on preclinical data. Tumors in the expansion cohort included TNBC, HGSOC, fallopian tube or peritoneal cancers, and castration-resistant prostate and pancreatic cancers [77].

At the MTD of 1 mg daily, confirmed responses were observed in 7 of 14 patients with *BRCA*-mutated breast cancer, with one complete response (CR). In the ovarian cancer cohort, 5 of 12 responses occurred in patients with deleterious germline *BRCA1/2* mutations, including one CR. It should be noted that for all *BRCA*-mutated ovarian cancer patients treated at any dose level, the overall response rate was 55% in platinum-sensitive patients, when compared with 20% in platinum-resistant patients. Of the 13 pancreatic patients treated, two had a partial response (PR), one with a *BRCA2* mutation and one with a *PALB2* mutation.

Patients with Ewing sarcoma and SCLC were also included in the expansion cohort based on the strong preclinical rationale for use of a PARP inhibitor in these tumor types. The 23 SCLC patients were all treated with 1 mg/day, with two PRs. These two patients also had had an objective response to their last prior platinum therapy. No objective responses were seen in Ewing sarcoma patients [77].

Talazoparib was well tolerated overall, with the dose-limiting toxicity being thrombocytopenia. Myelosuppression was transient and reversible, managed with dose interruptions and/or dose reductions [77]. The findings from this study demonstrated the efficacy of single-agent talazoparib in patients with *BRCA1/2* or *PALB2* mutations in ovarian, breast, small-cell lung, and pancreatic cancers.

11.5 PARP INHIBITOR AND CHEMOTHERAPY COMBINATIONS

Based on the strong rationale and preclinical evidence supporting the combination of PARP inhibitors with chemotherapy, various clinical trials to evaluate patient response have been performed or are ongoing (Tables 11.1 and 11.2). However, these studies have been challenging due to severe adverse events, particularly myelosuppression. This increased frequency and severity of myelosuppression observed with combination treatment, especially with the

TABLE 11.1 Combination Studies of PARP Inhibitors With Other Agents: Phase I Trials

Trial Identifier	Disease	PARP Inhibitor	Other Agent(s)	Cycle Length	Combination MTD	Results[a]
NCT00553189	Refractory solid tumors and lymphomas	Veliparib	Topotecan	21 days	Veliparib 10 mg BID on days 1–5 + topotecan 0.6 mg/m^2/day on days 1–5	>75% reduction in PAR levels in all three paired tumor biopsies; >50% reduction in PAR levels in PBMCs from 19 of 23 patients with measurable levels (no data on clinical benefit reported) [27]
NCT01012817 (ongoing)	Metastatic or unresectable solid tumors	Veliparib	Topotecan	28 days	Veliparib 300 mg BID on days 1–3, 8–10, and 15–17 + topotecan 3 mg/m^2 on days 2, 9, and 16	Of 52 evaluable patients, 1 CR (2%), 3 PR (6%), and 6 SD for ≥8 cycles (12%); no response correlation to *BRCA* status [78]
NCT01445522	Refractory solid tumors and lymphomas	Veliparib	Cyclo-phosphamide	21 days	Veliparib 60 mg QD + cyclophosphamide 50 mg QD	Of 35 pts, 7 PR (20%) and 6 SD for ≥6 cycles (17%) [79]
NCT01085422	Metastatic castration-resistant prostate cancer	Veliparib	Temozolomide	28 days	Veliparib 40 mg BID on days 1–7 + TMZ 150–200 mg/m^2 QD on days 1–5	Of 25 pts, two had PSA response (8%; decline ≥30%), 13 had stable PSA (52%) [80]
NCT01139970	Acute myeloid leukemia	Veliparib	Temozolomide	28 days	Veliparib 150 mg BID × 9 days + TMZ 200 mg/m^2 QD × 7 days (veliparib beginning the day after temozolomide initiation)	Of 48 pts, 8 CR (17%) and 8 SD/HI (17%) [37]
NCT01642251	Extensive stage SCLC	Veliparib	Cisplatin and etoposide	21 days	Veliparib 100 mg BID on days 1–7 + cisplatin (75 mg/m^2 on day 1) and etoposide (100 mg/m^2 on days 1–3)	Of 7 pts, 1 CR (14%), 4 PR (57%), and 2 SD (29%) [69]
NCT01749397 (ongoing)	Metastatic epithelial ovarian, primary peritoneal cavity, or fallopian tube cancer	Veliparib	Floxuridine	21 days	MTD not yet established (testing treatment schedule: veliparib BID on days 1–10 + floxuridine on days 3–5)	Not yet available

NCT00678132	Advanced solid tumors	Olaparib	Cisplatin and gemcitabine	21 days	Olaparib 100 mg QD on day 1 + cisplatin 60 mg/m^2 on day 1 + gemcitabine 500 mg/m^2 on days 1 and 8	Of 21 pts, 2 PR (10%) and 13 SD (62%; median duration of 20 weeks) [81]
NCT00782574	Advanced solid tumors	Olaparib	Cisplatin	21 days	MTD not established; olaparib 50 mg BID on days 1–5 + cisplatin 60 mg/m^2 on day 1 deemed tolerable	Of 46 total pts, 19 PR (41%) and 19 SD (41%); of 25 pts with *BRCA* mutation, 15 PR (60%) and 10 SD (40%) [82]
NCT01445418	*BRCA*1/2-mutated ovarian and breast cancer	Olaparib	Carboplatin	21 days	MTD not reached; highest tested dose was olaparib 400 mg BID on days 1–7 + carboplatin AUC5 once per cycle	Of 42 pts, 1 CR (2%) and 21 PR (50%) [83]
NCT00707707	Metastatic TNBC	Olaparib	Paclitaxel	28 days	MTD not established; initial tested dose was olaparib 200 mg BID + paclitaxel 90 mg/m^2 on days 1, 8, and 15	Of 19 pts, 7 PR (37%) and 6 SD for ≥7 weeks (32%) [84]
NCT00516802	Advanced solid tumors	Olaparib	Dacarbazine	21 days	Olaparib 100 mg BID on days 1–7 + dacarbazine 600 mg/m^2 on day 1	Of 40 pts, 2 PR (5%; both in melanoma) and 8 SD (20%) [85]
NCT01116648	Recurrent ovarian or metastatic TNBC	Olaparib	Cediranib	28 days	Olaparib 200 mg BID + cediranib 30 mg QD	Of 18 OvCa pts, 1 CR (6%), 7 PR (39%), and 3 SD for ≥24 weeks (17%); of 11 *BRCA*-mutated OvCa pts, 1 CR (9%) and 4 PR (36%); of 7 TNBC pts, no responses and 2 SD for ≥24 weeks (29%) [86]
NCT01009190	Advanced solid tumors	Rucaparib	Carboplatin	21 days	Rucaparib (oral) 240 mg QD + carboplatin AUC 5	Of 33 pts, 3 PR (9%) and 18 SD for ≥12 weeks (55%) [87]

Abbreviations: AUC, area under the curve; BID, twice daily; CR, complete response; HI, hematologic improvement; MTD, maximum tolerated dose; NSCLC, non-small cell lung cancer; OvCa, ovarian cancer; PAR, poly(ADP-ribose); PMBC, peripheral blood mononuclear cell; PR, partial response; PSA, prostate-specific antigen; pts, patients; QD, once daily; SCLC, small cell lung cancer; SD, stable disease; TMZ, temozolomide; TNBC, triple negative breast cancer.

[a] *The patient population used to determine the denominator (number of total patients) is defined differently depending on the study. In some studies, this number refers to evaluable patients in the expansion cohort only, whereas in other studies the number refers to the total number of patients in the trial, including all dose levels. Please see the cited reference for this information.*

TABLE 11.2 Phase II of combinations of PARP with other agents

Trial	Disease	PARP Inhibitor	Other Agent(s)	Cycle Length	Study Design/ Treatment Dose	Results
NCT00804908	Stage III unresectable or stage IV metastatic melanoma	Veliparib	Temozolomide	28 days	Randomized, double-blind TMZ (150–200 mg/m^2 QD on days 1–5) + placebo OR TMZ (150–200 mg/m^2 QD on days 1–5) + veliparib (20 mg or 40 mg BID on days 1–7)	Veliparib 20 and 40 mg increased median PFS vs placebo, but the difference did not reach statistical significance—3.7 months PFS in the 20 mg veliparib arm—3.6 months PFS in the 40 mg veliparib arm—2 months PFS in the placebo arm [68]
NCT01306032	*BRCA*-mutated ovarian (or high-grade serous ovarian, primary peritoneal, or fallopian tube cancer) triple-negative breast cancer	Veliparib	Cyclo-phosphamide	21 days	Randomized, open-label cyclo-phosphamide (50 mg QD) alone or combined with veliparib (60 mg QD)	Veliparib added to cyclophosphamide did not improve response rate or PFS In *BRCA*-mutated ovarian: cyclophosphamide alone (36 pts): 1 CR (3%), 6 PR (17%), 6 SD for ≥6 cycles (17%)—combination (34 pts): 1 CR (3%); 3 PR (9%), 5 SD for ≥6 cycles (15%) [88] In TNBC: cyclophosphamide alone (18 pts): 1 PR (6%), 1 SD for ≥6 cycles (6%); combination (21 pts): 2 PR (10%), 2 SD for ≥6 cycles (10%) [89]
NCT01042379 (I-SPY 2)	Stage II or III breast cancer (with a tumor ≥2.5 cm in diameter)	Veliparib	Carboplatin and paclitaxel	7 days (for 12 weeks)	Randomized, open-label paclitaxel (80 mg/m^2, weekly for 12 weeks) alone (control group) or combined with veliparib (50 mg BID for 12 weeks) + carboplatin (AUC 6 every 3 weeks)	Veliparib + carboplatin added to standard therapy resulted in higher rates of pCR in triple-negative patients: 51% pCR in the veliparib + carboplatin group; 26% pCR in the control group [90]

NCT01506609 (BROCADE)	Locally recurrent or metastatic *BRCA*-mutated breast cancer	Veliparib	Temozolomide or carboplatin and paclitaxel	28 days for TMZ arm; 21 days for other arms	Randomized, double-blind (48) veliparib (40 mg BID on days 1–7) + TMZ (150–200 mg/m^2 QD on days 1–5) OR veliparib (120 mg BID days 1–7) or placebo + carboplatin (AUC 6 on day 3) + paclitaxel (175 mg/m^2 on day 3)	Interim results demonstrated that differences in PFS and OS between the arms are not statistically significant [91]
NCT01116648	Platinum-sensitive, relapsed HGSOC or endometrioid ovarian, fallopian tube, or primary peritoneal cancer, or *BRCA*-mutated cancer	Olaparib	Cediranib	28 days	Randomized, open-label Olaparib (400 mg BID) alone or olaparib (200 mg BID) + cediranib (30 mg QD)	Not yet available; interim results demonstrated that cediranib plus olaparib improves PFS compared to olaparib monotherapy—9 months PFS in the olaparib arm—17.7 months PFS in the combination arm [92]
NCT number not available	Chemotherapy-naïve, metastatic melanoma	Rucaparib	Temozolomide	28 days	Single-arm rucaparib 12 mg/m^2 (IV) on days 1–5 + TMZ 150–200 mg/m^2 QD on days 1–5	Of 46 pts, 8 PR (17.4%) and 8 SD for ≥24 weeks (17.4%); median PFS of 3.5 months [93]

Abbreviations: BID, twice daily; CR, complete response; HGSOC, high-grade serous ovarian cancer; IV, intravenous; OS, overall survival; pCT, pathological complete response; PFS, progression-free survival; PR, partial response; Pts, patients; QD, once daily; RR, response rate; SD, stable disease; TMZ, temozolomide.

chemotherapies topotecan, doxorubicin, and gemcitabine, is the finding most suggestive of synergism between PARP inhibitors and chemotherapeutic agents. It is important to note, however, that not all chemotherapies carry this same risk; most PARP inhibitors, veliparib, in particular, do not yield increased incidence of myelosuppression when combined with platinum. The duration of therapy is an important factor when determining the PARP inhibitor dose that could be given without inducing severe toxicity—the longer the duration of the PARP inhibitor therapy, the greater is the potential for significant myelosuppression. One key unanswered question is the duration of DNA repair and when it is initiated. A better understanding of DNA repair temporal dynamics would be helpful in knowing when to start and how long to continue the administration of PARP inhibitor for optimal benefit and minimal toxicity.

11.5.1 Enhanced Toxicity

The first PARP inhibitor trial to indicate some synergism of this drug class with chemotherapy was a study of rucaparib given intravenously with temozolomide in melanoma patients [93]. Pharmacodynamic analysis showed PARP inhibition in PBMCs, and a clinical response rate of 17.4% was observed. The median OS was 9.9 months, and 36% of patients were progression-free at 6 months. This study predates the activity that has since been seen in melanoma with BRAF inhibitors and immunotherapy. Fifty-four percent of the patients in this trial required a dose reduction of the temozolomide due to myelosuppression. This was a single-arm study, and no subsequent trials were conducted to show the benefit of rucaparib in this combination.

More significant evidence of enhanced activity comes from the combination of PARP inhibition with topotecan. Based on preclinical data, veliparib's enhanced cytotoxicity with topotecan is thought to result from PARP trapping. In the phase I trial of topotecan given on a 5-day schedule, veliparib was only able to be administered at 10 mg BID on days 1–5 [27]. Also, the topotecan dose, 0.6 mg/m^2 on days 1–5, was 50% of the standard dose. This is the most significant compromise of the backbone chemotherapy in combination with PARP inhibitor. Interestingly, when veliparib was tested with topotecan given on a weekly schedule [78], veliparib was tolerable at close to the single-agent MTD. The veliparib MTD in this latter combination study is 300 mg BID given orally from 1 day before through 1 day after the topotecan dose; topotecan is given weekly at 3 mg/m^2 on days 2, 8, and 16, every 28 days. Similar to the previous topotecan–veliparib combination study, the DLT for this study is myelosuppression. Though veliparib is thought to be the weakest PARP trapper, results from the combination with topotecan given daily for 5 days indicate significant synergy. The increased cytotoxicity for this combination is not due to any pharmacokinetic interaction of the two agents. In the daily schedule, a >75% reduction in tumor PAR levels was observed in all three sets of paired biopsy samples, and a >50% reduction of PAR in PBMCs was also observed in 19 of 23 patients [27]. The circulating tumor cells and PBMCs collected also exhibited increased levels of γH2AX, a marker of apoptosis.

Olaparib also raised substantial toxicity issues in combination therapy. It was combined with gemcitabine and cisplatin in a phase I trial [81]. The initial starting dose was olaparib administered at 100 mg PO BID days 1–4, with cisplatin given at 60 mg/m^2 on day 3 and gemcitabine 500 mg/m^2 given on days 3 and 10. Significant neutropenia and thrombocytopenia were seen at this dose level. The trial was modified several times to try to improve the marrow tolerance, including changing the days of cisplatin and gemcitabine administration to days 1 and 8, dosing only on day 1, and reducing cisplatin and gemcitabine dose. The final recommended phase II dose was olaparib 100 mg BID day 1, cisplatin 60 mg/m^2 day 1, and gemcitabine 500 mg/m^2 days 1 and 8. The increased myelosuppression observed was thought to be secondary to the pharmacokinetics of gemcitabine and olaparib. When administered together, gemcitabine seemed to increase the elimination half-life of olaparib. This, in the presence of cisplatin, may have increased the frequency of double-strand DNA breaks and therefore of myelosuppression. The timing of administration may also be relevant; in the single dosing studies, PAR values return to baseline 36 h after administration of the last olaparib dose, implying that a longer dosing schedule would be preferable. Two PRs were seen in this study despite the less-than-optimal dosing of cisplatin and gemcitabine [81].

Oral rucaparib has been tested in combination with carboplatin. With a dose of carboplatin AUC 5, 240 mg of oral rucaparib could be given daily for 14 days [87]. The DLTs were grade 4 neutropenia and grade 4 thrombocytopenia. No pharmacokinetic interaction was detected. The doses of carboplatin and rucaparib in this combination study are reduced relative to single-agent dosing of both agents. Despite the reduced dose, activity was seen in 3 out of 33 patients. The toxicity and activity imply potential synergy between the two agents.

The Pediatric Preclinical Testing Program compared the combination of talazoparib with temozolomide or topotecan *in vitro* and *in vivo* [94], and temozolomide showed great synergy with talazoparib, especially in Ewing sarcoma cell lines. This synergy was not lost despite the low dose of temozolomide used. This study was the basis for the development of a clinical trial (NCT02116777). Given concerns of potential myelosuppression, as the single-agent DLT is thrombocytopenia, the phase I trial was designed to start with low-dose temozolomide, representing a change in paradigm.

11.5.2 Clinical Benefits

Despite the extensive clinical research that has been performed with PARP inhibitors, there are few trials comparing the effects of chemotherapy with or without combination PARP inhibitor treatment. These types of studies are needed to more clearly define the beneficial role PARP inhibition plays, but a major obstacle has been the fact that, due to myelosuppression, most combination regimens require a reduction in the chemotherapy dose to make PARP inhibition tolerable [95]. While most of the available data on combination treatment are from studies with veliparib, the first agent to have undergone testing of chemotherapy with or without a PARP inhibitor is iniparib. The initial phase II results suggested a role for iniparib, but the phase III data did not support this finding [50,96], and this agent was subsequently reported not to be a PARP inhibitor [97,98]. The most common chemotherapeutic agents that have shown promising preclinical activity in combination with PARP inhibition are temozolomide, platinum, and topoisomerase inhibitors, likely due to the impact of these agents on DNA repair mechanisms.

Temozolomide has been tested in combination with veliparib, rucaparib, and talazoparib in various cancers, including metastatic melanoma and SCLC. In melanoma, although the results were encouraging, the timing of introduction into the clinic was poor in that it coincided with the release of BRAF inhibitors. A phase II double-blind, placebo-controlled, randomized trial evaluated veliparib 20 or 40 mg BID for 7 days given with temozolomide 150–200 mg/m^2 QD for days 1–5 in patients with metastatic melanoma. The median PFS was 3.7 and 3.6 months in the veliparib 20 and 40 mg combination arms, respectively, compared to PFS of 2 months in the temozolomide plus placebo group. While the PFS nearly doubled with the PARP inhibitor, the duration of PFS was very short, and these differences did not reach statistical significance [68]. However, in the exploratory subgroup analysis, patients with low expression of ERCC1 treated with veliparib had PFS of 5.6 months, compared to 1.9 months in the group treated with placebo [68]. Another interesting finding was that patients with detectable p16 had doubled PFS with veliparib treatment (3.8 months vs 1.8 months with placebo). This finding is suggestive of synthetic lethality, i.e., impairment of BER enhancing the effects of the interference of other pathways [99].

Temozolomide (150–200 mg/m^2/day) was combined with 40 mg veliparib vs placebo in relapsed SCLC as well; a randomized, double-blind phase II trial demonstrated a significantly higher response rate with veliparib–temozolomide compared to temozolomide alone, although an improvement in PFS was not observed [100]. Similarly, a single-arm phase II study evaluating rucaparib and temozolomide in patients with metastatic melanoma showed an improvement in the response rate and PFS compared with historical controls [93].

The BROCADE trial evaluates the concept of PARP inhibition vs PARP trapping [48]. Though PARP trapping has not been demonstrated in humans due to the lack of a clinical assay, preclinical studies suggest that platinum is not affected by PARP trapping, while temozolomide is an ideal candidate for synergy with a PARP trapper. The trial population is patients with *BRCA*-associated breast cancer, so a positive response demonstrating the benefit of adding veliparib is anticipated. Interestingly, in preclinical studies, veliparib is the weakest PARP trapper [17]. This trial is ongoing and will help elucidate possible mechanism of action for PARP inhibitors.

The dose of veliparib is heavily dependent on whether inhibition or trapping is the predominant mechanism; as noted, the tolerable dose of veliparib with platinum is greater than two times the dose used when veliparib is combined with alkylating agents or topoisomerase inhibitors. The major side effect of the combination for PARP trapping-based synergy is increased frequency and severity of myelosuppression. When myelosuppression has not been used to define DLT, the dose of veliparib used in combination has been significantly higher. In a phase I trial of veliparib with temozolomide in acute myeloid leukemia, for example, 150 mg of veliparib was given in combination with 200 mg/m^2 of temozolomide for 7 days [37]. The DLT was gastrointestinal, and greater doses of temozolomide and veliparib were administered.

The I-SPY 2 (Investigation of Serial Studies to Predict Your Therapeutic Response through Imaging and Molecular Analysis 2; NCT01042379) trial also suggested significant benefits for a PARP inhibitor being administered in combination with chemotherapy treatments. I-SPY 2 is the second phase of a neoadjuvant breast cancer trial testing various investigative combinations, with a primary endpoint of pathological complete response (pCR). Ten relevant biomarker classifications were used to define groups based on HER-2 status, hormone receptor status, and risk within each arm: all, hormone-receptor-positive, hormone-receptor-negative, HER2-positive, HER2-negative, high-risk category 2 on the

70-gene MammaPrint assay, HER2-positive and hormone-receptor-positive, HER2-positive and hormone-receptor-negative, HER2-negative and hormone-receptor-positive, and triple-negative (HER2-negative, estrogen-receptor-negative, and progesterone-receptor-negative). The patient population that has shown benefit from the addition of veliparib–carboplatin to standard chemotherapy is the TNBC group, with a pCR rate of almost double that of the group that received the control treatment [90]. However, this is the same group that had shown benefit from carboplatin in the neoadjuvant setting in the GeparSixto and CALGB 40603 trials [101,102]. The other tumor types in which the veliparib–carboplatin regimen was evaluated, HER2-negative and hormone-receptor-positive/HER2-negative, did not demonstrate the same improvement. The I-SPY 2 trial was not designed to show the isolated benefit of veliparib alone, but despite this lack of evidence regarding single-agent veliparib, the advantage of the combination in the triple-negative population is an important finding.

11.6 NONCHEMOTHERAPY COMBINATIONS

In addition to the evidence supporting the combination of PARP inhibitors with chemotherapy, there are preclinical data to indicate a possible synergy with nonchemotherapy agents as well. Antiangiogenic drugs, for example, may enhance the effects of PARP inhibition. Hypoxia and VEGFR3 inhibition downregulate homologous recombination repair genes, such as *BRCA* and *RAD51*, creating a favorable environment for PARP inhibitor activity [103–105]. Based on these data, a phase I trial of olaparib and cediranib was completed, with activity observed in HGSOC patients. Interestingly, responses were seen in *BRCA* wild-type patients and in cisplatin-resistant patients [86]. No response was seen in the TNBC group. A phase II randomized trial was initiated in cisplatin-sensitive recurrent high-grade serous or endometrioid epithelial ovarian, fallopian tube, or primary peritoneal cancer, comparing olaparib vs olaparib and cediranib [92]. The trial was not restricted to *BRCA* mutation patients. In a post hoc analysis, the *BRCA* wild-type patient group benefited most from the combination vs olaparib alone, with PFS 16.5 months vs 5.7 months, respectively. Though this trial was designed to evaluate the addition of cediranib rather than to demonstrate the benefit of adding olaparib, it did broaden the activity of PARP inhibitor into *BRCA* wild-type patients.

Another interesting combination is the administration of a PARP inhibitor with trabectedin. Both agents are involved in the BER and nucleotide excision repair DNA damage repair pathways. In studies of rucaparib, synergy was observed in soft-tissue sarcoma cell lines, and *in vivo* tumor growth inhibition in a mouse model of liposarcoma was greater with the combination than with either of the single agents [106].

In addition to the prototypical synthetic lethality induction approach that has been demonstrated (cell death from loss of *PARP-1* in cells with *BRCA1* and *BRCA2* mutations), PARP inhibitors may also be combined with compounds that inactivate HR repair through other mechanisms. Inhibition of the proteasome by bortezomib or genetic inactivation of the proteasome inhibited DNA repair complex recruitment and led to suppression of HR in mammalian cells [107]. Furthermore, Neri et al. [108] used bortezomib to induce "BRCAness" to sensitize multiple myeloma cells to PARP inhibitors. The veliparib and bortezomib combination resulted in sustained levels of H2AX and enhanced cell kill, further suggesting that the proteasome is involved in HR repair.

Because radiation induces DNA strand breaks and replication stress, PARP inhibitors, especially veliparib, have been studied in combination with radiation. However, determination of the appropriate dosing and schedule for such combinations is complex; because PARP inhibitors alone may not have single-agent activity in non-HR-defective cells, any decrease in radiation dose or duration may have detrimental effects, especially in very radiosensitive tumors. As previously discussed, the combination of veliparib and radiation has been evaluated in several trials. In one such study, low-dose, fractionated whole-abdominal radiation was paired with veliparib. Fatigue and myelosuppression were the major toxicities, and veliparib was tolerable at a reasonable dose of 250 mg BID [109]. In a study of locally advanced rectal cancer, veliparib was given with capecitabine and radiation, and the MTD was 400 mg BID, identical to the single-agent dose. After treatment, patients underwent surgery, and 29% of patients exhibited a pathological CR [110]. The treatment of primary or secondary brain tumors with veliparib and radiation has also been studied, with mixed results. For example, in NSCLC patients with brain metastases, veliparib (50 and 200 mg) was tested in combination with whole-brain radiation; no significant difference in OS, relative to the whole-brain radiation plus placebo arm, was observed [111]. In glioblastoma multiforme (GBM), veliparib was given at 10 mg BID with radiation and temozolomide. This was hindered by myelosuppression, suggestive of an additive effect, according to the authors [112].

Lastly, no area of investigation is complete today without considering the addition of immunotherapy. Immune checkpoint inhibitors represent a new class of anticancer therapy that has had promising results in various solid tumor types. These agents block signaling through immune checkpoints, regulatory pathways that typically act to downregulate cytotoxic T-cell function and can be co-opted by malignant cells to evade immune destruction. Immune checkpoint inhibitor therapies, such as those targeting PD-1 or CTLA-4, allow T-cells to become activated and mount an immune response against the tumor [113,114]. Preclinical evidence in a *BRCA1*-deficient ovarian cancer model suggests that combining CTLA-4 blockade therapy with a PARP inhibitor may have clinical benefit [115], as PARP may, in addition to mediating DNA repair, play a role in immunomodulation [116,117]. There are also preclinical data that suggest an association between *BRCA1/2* mutational status and neoantigen load, tumor-infiltrating lymphocytes, and the expression of immune checkpoint inhibitors. Specifically, *BRCA1/2*-mutated cancers may express higher levels of neoantigens and therefore be more immunogenic [118]. Several clinical trials evaluating the combination of PARP inhibitors with immune checkpoint blockade agents in various BRCA-deficient reproductive system cancers are ongoing [15].

11.7 CONCLUSION

Several PARP inhibitors are currently available in the clinic, and some of them have received regulatory approval in specific diseases and indications. Recently, PARP trapping was identified as an additional mechanism of action of PARP inhibitors. The effect of PARP binding directly to DNA causes DNA DSBs, and this activity seems to correlate with myeloid and platelet suppression effects *in vivo*. This strong myelosuppression effect makes it difficult to combine PARP inhibitors with chemotherapies without compromising the dose of either component. Veliparib is a potent PARP enzyme catalytic inhibitor but a weak PARP trapper, which seems to allow combination with chemotherapy agents to achieve enhanced clinical activity. However, the veliparib dose in these combinations is less than 50% of the single-agent MTD in several combinations, especially with topoisomerase inhibitors. On the other hand, in cases where PARP trapping is not an important component of synergism, as with platinum, PARP inhibitor combinations may provide an improved therapeutic window. This combination window is associated with a broad range of tumor types and combinations and may yield activity outside of HR-deficient patients. Another promising area outside of HR deficiencies is hematologic malignancies; for example, in AML, both veliparib with temozolomide and veliparib with carboplatin/topotecan seem to be well tolerated and have shown promise in phase I trials. However, the most significant advances with the PARP inhibitors have focused on the HR-deficient setting, with continued widening of the beneficial population beyond *BRCA*-mutated patients.

PARP inhibitors are only just starting to become extremely useful tools in the treatment of cancer. In the area of HR deficiency, continued work in defining the optimal population is ongoing, both in terms of genomic testing as well as using other agents to induce tumor "BRCAness". In the realm of chemotherapy combinations, the chemotherapy backbone of interest and its area of activity should define the population that will potentially benefit most from each combination, as in the case of carboplatin and ovarian cancer. Meanwhile, PARP inhibitor combinations with other targeted therapies and immunotherapy are still in the early stages of exploration. PARP inhibitors are infants in their journey to becoming powerful oncologic therapeutic tools.

Acknowledgments

The authors thank Dr. Laura K. Fogli, Kelly Services, for medical writing support.

Conflict of Interest: No potential conflicts of interest were disclosed.

References

[1] Bryant HE, Schultz N, Thomas HD, Parker KM, Flower D, Lopez E, et al. Specific killing of BRCA2-deficient tumours with inhibitors of poly(ADP-ribose) polymerase. Nature 2005;434(7035):913–7.
[2] Farmer H, McCabe N, Lord CJ, Tutt AN, Johnson DA, Richardson TB, et al. Targeting the DNA repair defect in BRCA mutant cells as a therapeutic strategy. Nature 2005;434(7035):917–21.
[3] Fong PC, Boss DS, Yap TA, Tutt A, Wu PJ, Mergui-Roelvink M, et al. Inhibition of poly(ADP-ribose) polymerase in tumors from BRCA mutation carriers. New Engl J Med 2009;361(2):123–34.
[4] Ledermann J, Harter P, Gourley C, Friedlander M, Vergote I, Rustin G, et al. Olaparib maintenance therapy in platinum-sensitive relapsed ovarian cancer. New Engl J Med 2012;366(15):1382–92.
[5] D'Amours D, Desnoyers S, D'Silva I, Poirier GG. Poly(ADP-ribosyl)ation reactions in the regulation of nuclear functions. Biochem J 1999;342(Pt 2):249–68.
[6] Herceg Z, Wang ZQ. Functions of poly(ADP-ribose) polymerase (PARP) in DNA repair, genomic integrity and cell death. Mutat Res 2001;477(1–2):97–110.
[7] Memisoglu A, Samson L. Base excision repair in yeast and mammals. Mutat Res 2000;451(1–2):39–51.
[8] Schreiber V, Amé J-C, Dollé P, Schultz I, Rinaldi B, Fraulob V, et al. Poly(ADP-ribose) polymerase-2 (PARP-2) is required for efficient base excision DNA repair in association with PARP-1 and XRCC1. J Biol Chem 2002;277(25):23028–36.

[9] Plummer ER, Calvert H. Targeting poly(ADP-ribose) polymerase: a two-armed strategy for cancer therapy. Clin Cancer Res 2007;13(21):6252–6.

[10] Schreiber V, Dantzer F, Ame JC, de Murcia G. Poly(ADP-ribose): novel functions for an old molecule. Nat Rev Mol Cell Biol 2006;7(7):517–28.

[11] de Murcia JM, Ricoul M, Tartier L, Niedergang C, Huber A, Dantzer F, et al. Functional interaction between PARP-1 and PARP-2 in chromosome stability and embryonic development in mouse. EMBO J 2003;22(9):2255–63.

[12] Shall S, de Murcia G. Poly(ADP-ribose) polymerase-1: what have we learned from the deficient mouse model? Mutat Res 2000;460(1):1–15.

[13] Mateo J, Carreira S, Sandhu S, Miranda S, Mossop H, Perez-Lopez R, et al. DNA-repair defects and olaparib in metastatic prostate cancer. New Engl J Med 2015;373(18):1697–708.

[14] Swisher EM, Lin KK, Oza AM, Scott CL, Giordano H, Sun J, et al. Rucaparib in relapsed, platinum-sensitive high-grade ovarian carcinoma (ARIEL2 Part 1): an international, multicentre, open-label, phase 2 trial. Lancet Oncol 2017;18(1):75–87.

[15] O'Sullivan Coyne G, Chen AP, Meehan R, Doroshow JH. PARP inhibitors in reproductive system cancers: current use and developments. Drugs 2017;77(2):113–30.

[16] Lindahl T, Satoh MS, Poirier GG, Klungland A. Post-translational modification of poly(ADP-ribose) polymerase induced by DNA strand breaks. Trends Biochem Sci 1995;20(10):405–11.

[17] Murai J, Huang SY, Das BB, Renaud A, Zhang Y, Doroshow JH, et al. Trapping of PARP1 and PARP2 by clinical PARP inhibitors. Cancer Res 2012;72(21):5588–99.

[18] Kummar S, Chen A, Parchment RE, Kinders RJ, Ji J, Tomaszewski JE, et al. Advances in using PARP inhibitors to treat cancer. BMC Med 2012;10:25.

[19] El-Khamisy SF, Masutani M, Suzuki H, Caldecott KW. A requirement for PARP-1 for the assembly or stability of XRCC1 nuclear foci at sites of oxidative DNA damage. Nucleic Acids Res 2003;31(19):5526–33.

[20] Plo I, Liao ZY, Barcelo JM, Kohlhagen G, Caldecott KW, Weinfeld M, et al. Association of XRCC1 and tyrosyl DNA phosphodiesterase (Tdp1) for the repair of topoisomerase I-mediated DNA lesions. DNA Repair (Amst) 2003;2(10):1087–100.

[21] Malanga M, Althaus FR. The role of poly(ADP-ribose) in the DNA damage signaling network. Biochem Cell Biol 2005;83(3):354–64.

[22] Pommier Y, Tanizawa A, Kohn KW. Mechanisms of topoisomerase I inhibition by anticancer drugs. Adv Pharmacol 1994;29B:73–92.

[23] Xu Y, Her C. Inhibition of topoisomerase (DNA) I (TOP1): DNA damage repair and anticancer therapy. Biomolecules 2015;5(3):1652–70.

[24] Calabrese CR, Almassy R, Barton S, Batey MA, Calvert AH, Canan-Koch S, et al. Anticancer chemosensitization and radiosensitization by the novel poly(ADP-ribose) polymerase-1 inhibitor AG14361. J Natl Cancer Inst 2004;96(1):56–67.

[25] Thomas HD, Calabrese CR, Batey MA, Canan S, Hostomsky Z, Kyle S, et al. Preclinical selection of a novel poly(ADP-ribose) polymerase inhibitor for clinical trial. Mol Cancer Ther 2007;6(3):945–56.

[26] Patel AG, Flatten KS, Schneider PA, Dai NT, McDonald JS, Poirier GG, et al. Enhanced killing of cancer cells by poly(ADP-ribose) polymerase inhibitors and topoisomerase I inhibitors reflects poisoning of both enzymes. J Biol Chem 2012;287(6):4198–210.

[27] Kummar S, Chen A, Ji J, Zhang Y, Reid JM, Ames M, et al. Phase I study of PARP inhibitor ABT-888 in combination with topotecan in adults with refractory solid tumors and lymphomas. Cancer Res 2011;71(17):5626–34.

[28] LoRusso PM, Li J, Burger A, Heilbrun LK, Sausville EA, Boerner SA, et al. Phase I safety, pharmacokinetic, and pharmacodynamic study of the poly(ADP-ribose) polymerase (PARP) inhibitor veliparib (ABT-888) in combination with irinotecan in patients with advanced solid tumors. Clin Cancer Res 2016;22(13):3227–37.

[29] Newlands ES, Stevens MF, Wedge SR, Wheelhouse RT, Brock C. Temozolomide: a review of its discovery, chemical properties, pre-clinical development and clinical trials. Cancer Treat Rev 1997;23(1):35–61.

[30] Wyatt MD, Pittman DL. Methylating agents and DNA repair responses: methylated bases and sources of strand breaks. Chem Res Toxicol 2006;19(12):1580–94.

[31] Beranek DT. Distribution of methyl and ethyl adducts following alkylation with monofunctional alkylating agents. Mutat Res 1990;231(1):11–30.

[32] Zhang H, Marra G, Jiricny J, Maher VM, McCormick JJ. Mismatch repair is required for O(6)-methylguanine-induced homologous recombination in human fibroblasts. Carcinogenesis 2000;21(9):1639–46.

[33] Ruscetti T, Lehnert BE, Halbrook J, Le Trong H, Hoekstra MF, Chen DJ, et al. Stimulation of the DNA-dependent protein kinase by poly(ADP-ribose) polymerase. J Biol Chem 1998;273(23):14461–7.

[34] Delaney CA, Wang LZ, Kyle S, White AW, Calvert AH, Curtin NJ, et al. Potentiation of temozolomide and topotecan growth inhibition and cytotoxicity by novel poly(adenosine diphosphoribose) polymerase inhibitors in a panel of human tumor cell lines. Clin Cancer Res 2000;6(7):2860–7.

[35] Donawho CK, Luo Y, Luo Y, Penning TD, Bauch JL, Bouska JJ, et al. ABT-888, an orally active poly(ADP-ribose) polymerase inhibitor that potentiates DNA-damaging agents in preclinical tumor models. Clin Cancer Res 2007;13(9):2728–37.

[36] Horton TM, Jenkins G, Pati D, Zhang L, Dolan ME, Ribes-Zamora A, et al. Poly(ADP-ribose) polymerase inhibitor ABT-888 potentiates the cytotoxic activity of temozolomide in leukemia cells: influence of mismatch repair status and O6-methylguanine-DNA methyltransferase activity. Mol Cancer Ther 2009;8(8):2232–42.

[37] Gojo I, Beumer JH, Pratz KW, McDevitt MA, Baer MR, Blackford AL, et al. A Phase 1 study of the PARP inhibitor veliparib in combination with temozolomide in acute myeloid leukemia. Clin Cancer Res 2017;23(3):697–706.

[38] Palma JP, Wang YC, Rodriguez LE, Montgomery D, Ellis PA, Bukofzer G, et al. ABT-888 confers broad *in vivo* activity in combination with temozolomide in diverse tumors. Clin Cancer Res 2009;15(23):7277–90.

[39] Clark CC, Weitzel JN, O'Connor TR. Enhancement of synthetic lethality via combinations of ABT-888, a PARP inhibitor, and carboplatin *in vitro* and *in vivo* using BRCA1 and BRCA2 isogenic models. Mol Cancer Ther 2012;11(9):1948–58.

[40] Maag D, Solomon L, Hopkins T, Bukhofzer G, Shi Y, Donawho C, et al. Rationale for the combination of veliparib with platinum-based chemotherapy. J Clin Oncol 2015;33(15):2556.

[41] Miknyoczki SJ, Jones-Bolin S, Pritchard S, Hunter K, Zhao H, Wan W, et al. Chemopotentiation of temozolomide, irinotecan, and cisplatin activity by CEP-6800, a poly(ADP-ribose) polymerase inhibitor. Mol Cancer Ther 2003;2(4):371–82.

[42] Appleman LJ, Beumer JH, Jiang Y, Puhalla S, Lin Y, Owonikoko TK, et al. A phase I study of veliparib (ABT-888) in combination with carboplatin and paclitaxel in advanced solid malignancies. J Clin Oncol 2012;35(15):1015.

[43] Pratz KW, Kaufmann SH, Litzow MR, Ji J, Chen A, Rudek MA, et al. Phase I trial of the oral poly(ADP-ribose) polymerase (PARP) inhibitor veliparib (ABT-888, V) combined with topoecan (T) and carboplatin (C) for adults with relapsed and refractory acute leukemias. Blood 2011;118(21):3634.

[44] Huehls AM, Wagner JM, Huntoon CJ, Geng L, Erlichman C, Patel AG, et al. Poly(ADP-ribose) polymerase inhibition synergizes with 5-fluorodeoxyuridine but not 5-fluorouracil in ovarian cancer cells. Cancer Res 2011;71(14):4944–54.

[45] Murai J, Zhang Y, Morris J, Ji J, Takeda S, Doroshow JH, et al. Rationale for poly(ADP-ribose) polymerase (PARP) inhibitors in combination therapy with camptothecins or temozolomide based on PARP trapping versus catalytic inhibition. J Pharmacol Exp Ther 2014;349(3):408–16.

[46] Brenner JC, Feng FY, Han S, Patel S, Goyal SV, Bou-Maroun LM, et al. PARP-1 inhibition as a targeted strategy to treat Ewing's sarcoma. Cancer Res 2012;72(7):1608–13.

[47] Gill SJ, Travers J, Pshenichnaya I, Kogera FA, Barthorpe S, Mironenko T, et al. Combinations of PARP inhibitors with temozolomide drive PARP1 trapping and apoptosis in Ewing's sarcoma. PLoS ONE 2015;10(10):e0140988.

[48] Isakoff SJ, Puhalla S, Domchek SM, Friedlander M, Kaufman B, Robson M, et al. A randomized phase II study of veliparib with temozolomide or carboplatin/paclitaxel versus placebo with carboplatin/paclitaxel in BRCA1/2 metastatic breast cancer: design and rationale. Future Oncol 2017;13(4):307–20.

[49] O'Shaughnessy J, Osborne C, Pippen JE, Yoffe M, Patt D, Rocha C, et al. Iniparib plus chemotherapy in metastatic triple-negative breast cancer. New Engl J Med 2011;364(3):205–14.

[50] O'Shaughnessy J, Schwartzberg L, Danso MA, Miller KD, Rugo HS, Neubauer M, et al. Phase III study of iniparib plus gemcitabine and carboplatin versus gemcitabine and carboplatin in patients with metastatic triple-negative breast cancer. J Clin Oncol 2014;32(34):3840–7.

[51] Ji J, Lee MP, Kadota M, Zhang Y, Parchment RE, Tomaszewski JE, et al. Abstract 4527 pharmacodynamic and pathway analysis of three presumed inhibitors of poly(ADP-ribose) polymerase: ABT-888, AZD2281, and BSI201. Cancer Res 2011;71(8 Suppl.):4527.

[52] Patel AG, De Lorenzo SB, Flatten KS, Poirier GG, Kaufmann SH. Failure of iniparib to inhibit poly(ADP-Ribose) polymerase *in vitro*. Clin Cancer Res 2012;18(6):1655–62.

[53] Bell-McGuinn KM, Konner JA, Tew WP, Hensley ML, Iasonos A, Charpentier E, et al. A phase 2, single arm study of iniparib in patients with BRCA1 or BRCA2 associated advanced epithelial ovarian, fallopian tube, or primary peritoneal cancer. Int J Gynecol Cancer 2016;26(2):255–60.

[54] Yap T, Boss D, Fong P, Roelvink M, Tutt A, Carmichael J, et al. First in human phase I pharmacokinetic (PK) and pharmacodynamic (PD) study of KU-0059436 (Ku), a small molecule inhibitor of poly ADP-ribose polymerase (PARP) in cancer patients (p), including BRCA1/2 mutation carriers. J Clin Oncol 2007;25(18 Suppl.):3529.
[55] Helleday T. The underlying mechanism for the PARP and BRCA synthetic lethality: clearing up the misunderstandings. Mol Oncol 2011;5(4):387–93.
[56] Audeh MW, Carmichael J, Penson RT, Friedlander M, Powell B, Bell-McGuinn KM, et al. Oral poly(ADP-ribose) polymerase inhibitor olaparib in patients with BRCA1 or BRCA2 mutations and recurrent ovarian cancer: a proof-of-concept trial. Lancet 2010;376(9737):245–51.
[57] Tutt A, Robson M, Garber JE, Domchek SM, Audeh MW, Weitzel JN, et al. Oral poly(ADP-ribose) polymerase inhibitor olaparib in patients with BRCA1 or BRCA2 mutations and advanced breast cancer: a proof-of-concept trial. Lancet 2010;376(9737):235–44.
[58] Gelmon KA, Tischkowitz M, Mackay H, Swenerton K, Robidoux A, Tonkin K, et al. Olaparib in patients with recurrent high-grade serous or poorly differentiated ovarian carcinoma or triple-negative breast cancer: a phase 2, multicentre, open-label, non-randomised study. Lancet Oncol 2011;12(9):852–61.
[59] Ledermann JA, Harter P, Gourley C, Friedlander M, Vergote I, Rustin G, et al. Overall survival in patients with platinum-sensitive recurrent serous ovarian cancer receiving olaparib maintenance monotherapy: an updated analysis from a randomised, placebo-controlled, double-blind, phase 2 trial. Lancet Oncol 2016;17(11):1579–89.
[60] Pujade-Lauraine E, Ledermann JA, Selle F, Gebski V, Penson RT, Oza AM, et al. Olaparib tablets as maintenance therapy in patients with platinum-sensitive, relapsed ovarian cancer and a BRCA1/2 mutation (SOLO2/ENGOT-Ov21): a double-blind, randomised, placebo-controlled, phase 3 trial. Lancet Oncol 2017;18(9):1274–84.
[61] United States Food and Drug Administration. FDA approves olaparib tablets for maintenance treatment in ovarian cancer. Available at: https://www.fda.gov/drugs/informationondrugs/approveddrugs/ucm572143.htm [Accessed September 19, 2017].
[62] Pujade-Lauraine E, Ledermann JA, Penson RT, Oza AM, Korach J, Huzarski T, et al. Treatment with olaparib monotherapy in the maintenance setting significantly improves progression-free survival in patients with platinum-sensitive relapsed ovarian cancer: results from the phase III SOLO2 study. Gynecol Oncol 2017;145:219–20.
[63] Ledermann J, Harter P, Gourley C, Friedlander M, Vergote I, Rustin G, et al. Olaparib maintenance therapy in platinum-sensitive relapsed ovarian cancer. New Engl J Med 2012;366(15):1382–92.
[64] Kummar S, Kinders R, Gutierrez ME, Rubinstein L, Parchment RE, Phillips LR, et al. Phase 0 clinical trial of the poly(ADP-ribose) polymerase inhibitor ABT-888 in patients with advanced malignancies. J Clin Oncol 2009;27(16):2705–11.
[65] Final results of a phase 1 study of single-agent veliparib (V) in patients (pts) with either BRCA1/2-mutated cancer (BRCA+), platinum-refractory ovarian, or basal-like breast cancer (BRCA-wt). In: Puhalla S, Beumer JH, Pahuja S, Appleman LJ, Tawbi HA-H, Stoller RG, editors. ASCO Annual Meeting Proceedings; 2014.
[66] Coleman RL, Sill MW, Bell-McGuinn K, Aghajanian C, Gray HJ, Tewari KS, et al. A phase II evaluation of the potent, highly selective PARP inhibitor veliparib in the treatment of persistent or recurrent epithelial ovarian, fallopian tube, or primary peritoneal cancer in patients who carry a germline BRCA1 or BRCA2 mutation—an NRG Oncology/Gynecologic Oncology Group study. Gynecol Oncol 2015;137(3):386–91.
[67] Mehta MP, Wang D, Wang F, Kleinberg L, Brade A, Robins HI, et al. Veliparib in combination with whole brain radiation therapy in patients with brain metastases: results of a phase 1 study. J Neurooncol 2015;122(2):409–17.
[68] Middleton MR, Friedlander P, Hamid O, Daud A, Plummer R, Falotico N, et al. Randomized phase II study evaluating veliparib (ABT-888) with temozolomide in patients with metastatic melanoma. Ann Oncol 2015;26(10):2173–9.
[69] Owonikoko TK, Dahlberg SE, Khan SA, Gerber DE, Dowell J, Moss RA, et al. A phase 1 safety study of veliparib combined with cisplatin and etoposide in extensive stage small cell lung cancer: a trial of the ECOG-ACRIN Cancer Research Group (E2511). Lung Cancer 2015;89(1):66–70.
[70] Reiss KA, Herman JM, Zahurak M, Brade A, Dawson LA, Scardina A, et al. A phase I study of veliparib (ABT-888) in combination with low-dose fractionated whole abdominal radiation therapy in patients with advanced solid malignancies and peritoneal carcinomatosis. Clin Cancer Res 2015;21(1):68–76.

[71] Ramalingam SS, Blais N, Mazieres J, Reck M, Jones CM, Juhasz E, et al. Randomized, placebo-controlled, phase II study of veliparib in combination with carboplatin and paclitaxel for advanced/metastatic non-small cell lung cancer. Clin Cancer Res 2016;23(8):1937–44.

[72] Geyer CE, O'Shaughnessy J, Untch M, Sikov W, Rugo HS, McKee MD, et al. Phase 3 study evaluating efficacy and safety of veliparib (V) plus carboplatin (Cb) or Cb in combination with standard neoadjuvant chemotherapy (NAC) in patients (pts) with early stage triple-negative breast cancer (TNBC). J Clin Oncol 2017;35(15 Suppl.):520.

[73] Sandhu SK, Schelman WR, Wilding G, Moreno V, Baird RD, Miranda S, et al. The poly(ADP-ribose) polymerase inhibitor niraparib (MK4827) in BRCA mutation carriers and patients with sporadic cancer: a phase 1 dose-escalation trial. Lancet Oncol 2013;14(9):882–92.

[74] Mirza MR, Monk BJ, Herrstedt J, Oza AM, Mahner S, Redondo A, et al. Niraparib maintenance therapy in platinum-sensitive recurrent ovarian cancer. New Engl J Med 2016;375(22):2154–64.

[75] Scott LJ. Niraparib: first global approval. Drugs 2017;77(9):1029–34.

[76] The ASCO Post. ESMO 2017: Rucaparib boosts progression-free survival in BRCA-mutant recurrent ovarian cancer. Available at: http://www.ascopost.com/News/58021 [Accessed September 19, 2017].

[77] de Bono J, Ramanathan RK, Mina L, Chugh R, Glaspy J, Rafii S, et al. Phase I, dose-escalation, two-part trial of the PARP inhibitor talazoparib in patients with advanced germline BRCA1/2 mutations and selected sporadic cancers. Cancer Discov 2017;7(6):620–9.

[78] Kaufmann SH, Hendrickson AEW, Harrell MI, Menefee ME, Tanner EJ, Mutch DG, et al. Abstract IS04: PARP inhibitor combinations for the treatment of ovarian cancer. AACR 2017;.

[79] Kummar S, Ji J, Morgan R, Lenz HJ, Puhalla SL, Belani CP, et al. A phase I study of veliparib in combination with metronomic cyclophosphamide in adults with refractory solid tumors and lymphomas. Clin Cancer Res 2012;18(6):1726–34.

[80] Hussain M, Carducci MA, Slovin S, Cetnar J, Qian J, McKeegan EM, et al. Targeting DNA repair with combination veliparib (ABT-888) and temozolomide in patients with metastatic castration-resistant prostate cancer. Invest New Drugs 2014;32(5):904–12.

[81] Rajan A, Carter CA, Kelly RJ, Gutierrez M, Kummar S, Szabo E, et al. A phase I combination study of olaparib with cisplatin and gemcitabine in adults with solid tumors. Clin Cancer Res 2012;18(8):2344–51.

[82] Balmana J, Tung NM, Isakoff SJ, Grana B, Ryan PD, Saura C, et al. Phase I trial of olaparib in combination with cisplatin for the treatment of patients with advanced breast, ovarian and other solid tumors. Ann Oncol 2014;25(8):1656–63.

[83] Lee JM, Hays JL, Annunziata CM, Noonan AM, Minasian L, Zujewski JA, et al. Phase I/Ib study of olaparib and carboplatin in BRCA1 or BRCA2 mutation-associated breast or ovarian cancer with biomarker analyses. J Natl Cancer Inst 2014;106(6):dju089.

[84] Dent RA, Lindeman GJ, Clemons M, Wildiers H, Chan A, McCarthy NJ, et al. Phase I trial of the oral PARP inhibitor olaparib in combination with paclitaxel for first- or second-line treatment of patients with metastatic triple-negative breast cancer. Breast Cancer Res 2013;15(5):R88.

[85] Khan OA, Gore M, Lorigan P, Stone J, Greystoke A, Burke W, et al. A phase I study of the safety and tolerability of olaparib (AZD2281, KU0059436) and dacarbazine in patients with advanced solid tumours. Br J Cancer 2011;104(5):750–5.

[86] Liu JF, Tolaney SM, Birrer M, Fleming GF, Buss MK, Dahlberg SE, et al. A phase 1 trial of the poly(ADP-ribose) polymerase inhibitor olaparib (AZD2281) in combination with the anti-angiogenic cediranib (AZD2171) in recurrent epithelial ovarian or triple-negative breast cancer. Eur J Cancer 2013;49(14):2972–8.

[87] Wilson RH, Evans TJ, Middleton MR, Molife LR, Spicer J, Dieras V, et al. A phase I study of intravenous and oral rucaparib in combination with chemotherapy in patients with advanced solid tumours. Br J Cancer 2017;116(7):884–92.

[88] Kummar S, Oza AM, Fleming GF, Sullivan DM, Gandara DR, Naughton MJ, et al. Randomized trial of oral cyclophosphamide and veliparib in high-grade serous ovarian, primary peritoneal, or fallopian tube cancers, or BRCA-mutant ovarian cancer. Clin Cancer Res 2015;21(7):1574–82.

[89] Kummar S, Wade JL, Oza AM, Sullivan D, Chen AP, Gandara DR, et al. Randomized phase II trial of cyclophosphamide and the oral poly(ADP-ribose) polymerase inhibitor veliparib in patients with recurrent, advanced triple-negative breast cancer. Invest New Drugs 2016;34(3):355–63.

[90] Rugo HS, Olopade OI, DeMichele A, Yau C, van 't Veer LJ, Buxton MB, et al. Adaptive randomization of veliparib-carboplatin treatment in breast cancer. New Engl J Med 2016;375(1):23–34.

[91] Efficacy and tolerability of veliparib (V; ABT-888) in combination with carboplatin (C) and paclitaxel (P) vs placebo (Plc)+ C/P in patients (pts) with BRCA1 or BRCA2 mutations and metastatic breast cancer: a randomized, phase 2 study. In: Han H, Diéras V, Robson M, Palácová M, Marcom P, Jager A, editors. 39th San Antonio Breast Cancer Symposium (SABCS); 2016San Antonio.
[92] Liu JF, Barry WT, Birrer M, Lee JM, Buckanovich RJ, Fleming GF, et al. Combination cediranib and olaparib versus olaparib alone for women with recurrent platinum-sensitive ovarian cancer: a randomised phase 2 study. Lancet Oncol 2014;15(11):1207–14.
[93] Plummer R, Lorigan P, Steven N, Scott L, Middleton MR, Wilson RH, et al. A phase II study of the potent PARP inhibitor, Rucaparib (PF-01367338, AG014699), with temozolomide in patients with metastatic melanoma demonstrating evidence of chemopotentiation. Cancer Chemother Pharmacol 2013;71(5):1191–9.
[94] Smith MA, Reynolds CP, Kang MH, Kolb EA, Gorlick R, Carol H, et al. Synergistic activity of PARP inhibition by talazoparib (BMN 673) with temozolomide in pediatric cancer models in the pediatric preclinical testing program. Clin Cancer Res 2015;21(4):819–32.
[95] Chen A. PARP inhibitors: its role in treatment of cancer. Chin J Cancer 2011;30(7):463–71.
[96] Telli ML, Jensen KC, Vinayak S, Kurian AW, Lipson JA, Flaherty PJ, et al. Phase II study of gemcitabine, carboplatin, and iniparib as neoadjuvant therapy for triple-negative and BRCA1/2 mutation-associated breast cancer with assessment of a tumor-based measure of genomic instability: PrECOG 0105. J Clin Oncol 2015;33(17):1895–901.
[97] Sinha G. Downfall of iniparib: a PARP inhibitor that doesn't inhibit PARP after all. J Natl Cancer Inst 2014;106(1):djt447.
[98] Liu X, Shi Y, Maag DX, Palma JP, Patterson MJ, Ellis PA, et al. Iniparib nonselectively modifies cysteine-containing proteins in tumor cells and is not a bona fide PARP inhibitor. Clin Cancer Res 2012;18(2):510–23.
[99] Lord CJ, Ashworth A. PARP inhibitors: synthetic lethality in the clinic. Science 2017;355(6330):1152–8.
[100] Pietanza MC, Krug LM, Waqar SN, Dowlati A, Hann CL, Chiappori A, et al. A multi-center, randomized, double-blind phase II study comparing temozolomide (TMZ) plus either veliparib (ABT-888), a PARP inhibitor, or placebo as 2nd or 3rd-line therapy for patients (Pts) with relapsed small cell lung cancers (SCLCs). American Society of Clinical Oncology 2016;.
[101] Sikov WM, Berry DA, Perou CM, Singh B, Cirrincione CT, Tolaney SM, et al. Impact of the addition of carboplatin and/or bevacizumab to neoadjuvant once-per-week paclitaxel followed by dose-dense doxorubicin and cyclophosphamide on pathologic complete response rates in stage II to III triple-negative breast cancer: CALGB 40603 (Alliance). J Clin Oncol 2015;33(1):13–21.
[102] von Minckwitz G, Schneeweiss A, Loibl S, Salat C, Denkert C, Rezai M, et al. Neoadjuvant carboplatin in patients with triple-negative and HER2-positive early breast cancer (GeparSixto; GBG 66): a randomised phase 2 trial. Lancet Oncol 2014;15(7):747–56.
[103] Bindra RS, Gibson SL, Meng A, Westermark U, Jasin M, Pierce AJ, et al. Hypoxia-induced down-regulation of BRCA1 expression by E2Fs. Cancer Res 2005;65(24):11597–604.
[104] Bindra RS, Schaffer PJ, Meng A, Woo J, Maseide K, Roth ME, et al. Down-regulation of Rad51 and decreased homologous recombination in hypoxic cancer cells. Mol Cell Biol 2004;24(19):8504–18.
[105] Lim JJ, Yang K, Taylor-Harding B, Wiedemeyer WR, Buckanovich RJ. VEGFR3 inhibition chemosensitizes ovarian cancer stemlike cells through down-regulation of BRCA1 and BRCA2. Neoplasia 2014;16(4):343–53. e1–e2.
[106] Laroche A, Chaire V, Le Loarer F, Algeo MP, Rey C, Tran K, et al. Activity of trabectedin and the PARP inhibitor rucaparib in soft-tissue sarcomas. J Hematol Oncol 2017;10(1):84.
[107] Murakawa Y, Sonoda E, Barber LJ, Zeng W, Yokomori K, Kimura H, et al. Inhibitors of the proteasome suppress homologous DNA recombination in mammalian cells. Cancer Res 2007;67(18):8536–43.
[108] Neri P, Ren L, Gratton K, Stebner E, Johnson J, Klimowicz A, et al. Bortezomib-induced "BRCAness" sensitizes multiple myeloma cells to PARP inhibitors. Blood 2011;118(24):6368–79.
[109] Reiss KA, Herman JM, Armstrong D, Zahurak M, Fyles A, Brade A, et al. A final report of a phase I study of veliparib (ABT 888) in combination with low-dose fractionated whole abdominal radiation therapy (LDFWAR) in patients with advanced solid malignancies and peritoneal carcinomatosis with a dose escalation in ovarian and fallopian tube cancers. Gynecol Oncol 2017;144(3):486–90.
[110] Czito BG, Deming DA, Jameson GS, Mulcahy MF, Vaghefi H, Dudley MW, et al. Safety and tolerability of veliparib combined with capecitabine plus radiotherapy in patients with locally advanced rectal cancer: a phase 1b study. Lancet Gastroenterol Hepatol 2017;2(6):418–26.

[111] Chabot P, Hsia T-C, Ryu J-S, Gorbunova V, Belda-Iniesta C, Ball D, et al. Veliparib in combination with whole-brain radiation therapy for patients with brain metastases from non-small cell lung cancer: results of a randomized, global, placebo-controlled study. J Neuro-Oncol 2017;131(1):105–15.

[112] Kleinberg LR, Blakeley J, Mikkelsen T, Stevens G, Ye X, Ryu S, et al. An American Brain Tumor Consortium (ABTC) phase 1 trial of veliparib (ABT-888), temozolomide (TMZ), and concurrent radiation therapy (RT) for newly diagnosed glioblastoma multiforme (GBM): need for modification to approach for combining these therapies. Int J Radiat Oncol Biol Phys 2012;84(3):S7–8.

[113] Korman AJ, Peggs KS, Allison JP. Checkpoint blockade in cancer immunotherapy. Adv Immunol 2006;90:297–339.

[114] Pardoll DM. The blockade of immune checkpoints in cancer immunotherapy. Nat Rev Cancer 2012;12(4):252–64.

[115] Higuchi T, Flies DB, Marjon NA, Mantia-Smaldone G, Ronner L, Gimotty PA, et al. CTLA-4 blockade synergizes therapeutically with PARP inhibition in BRCA1-deficient ovarian cancer. Cancer Immunol Res 2015;3(11):1257–68.

[116] Laudisi F, Sambucci M, Pioli C. Poly(ADP-ribose) polymerase-1 (PARP-1) as immune regulator. Endocr Metab Immune Disord Drug Targets 2011;11(4):326–33.

[117] Rosado MM, Bennici E, Novelli F, Pioli C. Beyond DNA repair, the immunological role of PARP-1 and its siblings. Immunology 2013;139(4):428–37.

[118] Strickland KC, Howitt BE, Shukla SA, Rodig S, Ritterhouse LL, Liu JF, et al. Association and prognostic significance of BRCA1/2-mutation status with neoantigen load, number of tumor-infiltrating lymphocytes and expression of PD-1/PD-L1 in high grade serous ovarian cancer. Oncotarget 2016;7(12):13587–98.

CHAPTER

12

Autophagy Inhibition and Chemosensitization in Cancer Therapy

*Emmanuel K. Cudjoe, Jr.**,**, *S. Lauren Kyte**,**, *Tareq Saleh**,**, *Joseph W. Landry**,**, *David A. Gewirtz**,**

*Virginia Commonwealth University, Richmond, VA, United States

**Massey Cancer Center, Virginia Commonwealth University, Richmond, VA, United States

Abstract

Autophagy is a cellular degradation mechanism involving protein turnover as well as the recycling of excessive and dysfunctional organelles. Autophagy is generally considered to occur at basal levels in cells, but may be induced to higher levels under conditions of stress, such as starvation, hypoxia, ionizing radiation treatment, or chemotherapy. Aberrations in the autophagic machinery may represent either cause or effect in various diseases. In cancer, autophagy has been shown to be one of the mechanisms affecting the response to chemotherapy and/or radiation treatment. In this chapter, the different types and functional forms of autophagy, the autophagy signaling pathway, as well as the role of autophagy in transformation and tumor suppression are discussed. Modulation of autophagy for improving tumor cell chemosensitivity and radiosensitivity is currently being explored in clinical trials; the potential impact and limitations of these studies are discussed. We also briefly review the intersection between autophagy and the immune system with regard to the effectiveness of chemotherapy. The chapter concludes with future directions in cancer treatment in relation to novel therapies and potential drug targets for autophagy modulation.

ABBREVIATIONS

AMBRA1	Autophagy/Beclin-1 regulator 1
AMPK	AMP kinase
Atg	Autophagy-related
AV	autophagic vacuole
BMPCs	Bone marrow plasma cells
CTCs	Circulating tumor cells
CMA	Chaperone-mediated autophagy

Targeting Cell Survival Pathways to Enhance Response to Chemotherapy. http://dx.doi.org/10.1016/B978-0-12-813753-6.00012-3

Copyright © 2018 Elsevier Inc. All rights reserved.

CQ	Chloroquine
DAMP	Danger associated molecular pattern
ER	Endoplasmic reticulum
FDA	Food and Drug Administration
HCQ	Hydroxychloroquine
HIF-1a	Hypoxia-induced factor 1a
HMGB1	High-mobility group box 1
ICD	Immunogenic cell death
LC3	Microtubule-associated protein light chain 3
mTOR	Mammalian target of rapamycin
NK	Natural killer
PBMC	Peripheral blood mononuclear cell
PtdIns3K	Phosphatidylinositol 3-kinase
ROS	Reactive oxygen species
SMERs	Small-molecule enhancers of rapamycin
SR-2	Sigma-2 receptor
STAT3	Signal transducer and activator of transcription
Ubl	Ubiquitin-like
ULK	Unc-51-like kinase
UVRAG	Ultraviolet radiation resistance-associated gene
Vps34	Vacuolar sorting protein 34

12.1 INTRODUCTION

Cancer currently constitutes one of the highest disease burdens in the United States and the world at large. In addition to surgery and radiation therapy, chemotherapy remains one of the primary therapeutic options for this disease. Unfortunately, there are various strategies whereby cancer cells are able to evade chemotherapy-induced cell death and continue to proliferate. These include, but are not limited to, the upregulation of multidrug efflux pumps, enhanced DNA repair capability, evasion of apoptosis, and autophagy, which is the focus of this chapter [1]. This chapter will also address how autophagy may, in theory and in practice, be manipulated to improve the response to conventional or targeted cancer therapy.

Over the last few decades, autophagy has become a widely studied cellular phenomenon, the discovery of which earned Dr. Yoshinori Ohsumi the 2016 Nobel Prize in Physiology or Medicine. Autophagy is a conserved homeostatic cellular process that is present primarily in eukaryotic cells [2]. Autophagy has been defined as a cytoplasmic protein and organelle degradation mechanism involving the lysosomal machinery in which dysfunctional, damaged and/or excess cellular components are recycled for homeostasis, survival, and the production of energy and metabolic precursors [1–5]. Autophagy is also induced above basal levels in various cells in response to stress such as starvation, radiation, and chemotherapy.

12.2 AUTOPHAGY FUNCTION AND ROLE IN CANCER

12.2.1 The Autophagic Machinery

Autophagy has been classified into three main classes, specifically microautophagy, macro-autophagy, and chaperone-mediated autophagy (CMA) [4,6]. This classification is based largely on the method of delivery of the substrates to the lysosomes, where the substrates are degraded. For example, in CMA, the substrate protein is first selectively identified by a chaperone molecule in the cytosol, then delivered to the surface of the lysosome where it unfolds and crosses the lysosomal membrane [7]. Unlike macro- and microautophagy, CMA has only been identified in mammalian cells, is always targeted to selective cargo, and is used to degrade soluble proteins [4,7,8]. It is known that the proteins targeted for CMA have a motif of amino acids in their sequence that makes them recognizable to their chaperones [9]. In contrast, macro- and microautophagy may be selective or nonselective. The terminology "macroautophagy" is used interchangeably with autophagy in many publications and represents the large-scale autophagy that involves the formation of a double-membrane phagophore which matures into an autophagosome. The autophagosome then fuses with a lysosome to form an autolysosome in which the contents are degraded by hydrolases. Finally, in microautophagy, there is no direct involvement of a delivery machinery and the substrates are engulfed directly by the lysosome where they are degraded by hydrolases [4,10]. Microautophagy, among other processes, is stimulated primarily through cellular nitrogen depletion [11].

The molecular mechanisms of autophagy have been elucidated primarily utilizing yeast models. These studies have led to the discovery of about 36 Atg (autophagy-related) genes, 34 of which are listed by Klionsky et al. [12,13]. The majority of the Atg family proteins discovered in yeast have been found to have mammalian homologs which have essentially similar roles in yeast and in human cells. The initial phase of the (macro)autophagy process has four stages, specifically the induction, nucleation, elongation, and completion steps [8]. When cells are deprived of nutrients via starvation, a classical approach for inducing autophagy, the induction step involves the mammalian target of rapamycin (mTOR—a protein that inhibits autophagy via the inhibition of the class III PI3 kinase/Beclin-1 complex formation), which is inhibited by AMP-kinase, a protein that responds to energy changes within the cell [14–16]. Inactivation of mTOR in turn leads to its dissociation from and hence activation of the Unc-51-like kinases ULK1 and ULK2. The activated kinases then phosphorylate Atg13 and FIP200 following which Atg101 binds and stabilizes Atg13 in the ULK-Atg13-FIP200 complex [17] in the nucleation step. The phosphatidylinositol 3-kinase (PtdIns3K) complex which comprises vacuolar sorting protein 34 (Vps34—a class III PI3 kinase), p150, Atg14, and Beclin-1 is required for the initiation of vesicle nucleation [15,17]. The class III PI3 kinase Vps34, the only one in mammalian cells, is known to associate with Beclin-1, which facilitates its phosphorylation of phosphatidylinositol to phosphatidylinositol-3-kinase (PtdIns3K) [15,18]. Beclin-1 has several binding partner proteins which may serve to promote the process of autophagy, as is the case with Atg14L and the UV radiation resistance-associated gene (UVRAG); conversely, binding of Beclin-1 to Bcl-2 and Bcl-XL results in inhibition of autophagy [15]. After the nucleation phase, elongation is mediated by conjugation of two ubiquitin-like (Ubl) protein complexes—Atg5–Atg12–Atg16 and Atg8–phosphatidylethanolamine—which then leads to the completion step that involves the autophagy-related (Atg) proteins 3, 5, 7, 10, 12, and 16L and microtubule-associated protein light chain 3 (LC3) [8,17,19]. The formation of the Atg5–Atg12–Atg16 complex plays a central role in this process and leads to the conversion of cytosolic LC3-I to the lipidated membrane-bound isoform, LC3-II [19]. Completion of autophagy involves the fusion of the autophagosome with the lysosome to form the autolysosome where the cargo (contents of the autophagosome) are broken down by hydrolases to generate energy, amino acids for protein synthesis, or other precursors of metabolism.

It should be noted that the mTOR signaling pathway is not the exclusive pathway for the induction of autophagy. As demonstrated by Sarkar et al. [15,20], there are a number of small molecules that may induce autophagy independent of rapamycin and mTOR. These molecules were shown to enhance the effects of rapamycin-induced growth arrest in yeast and were thus termed small-molecule enhancers of rapamycin (SMERs) by Sarkar et al. [20]. It was also previously believed that Atg5 and Atg7 are necessary for autophagy in mammalian cells but an alternative pathway independent of Atg5/Atg7 has been identified [21].

The importance of the ULK1 kinase complex—Vsp34 axis—has been recognized over the last few years. The Beclin1–Atg14L–Vps34–Vps15 complex that has been shown to be important in autophagosome formation has been found to be regulated by ULK1. In recent studies of the relationship between ULK1 and Vps34, it was found that amino acid or glucose starvation resulted in increased activity of Vps34 lipid kinases associated with Atg14L, suggestive of a direct role of ULK1 in the regulation of autophagy through the phosphorylation of Beclin-1 [18,22].

12.2.2 The Functions of Autophagy

Autophagy is primarily considered to be a protective mechanism owing to its recycling and homeostatic functions at basal levels. It is largely believed that cells undergo autophagy as a defense mechanism against death by starvation and other sources of stress. However, there is substantial evidence to indicate that autophagy is not always protective. Other forms of autophagy that have been reported include one that is toxic (also referred to as autophagic cell death), one that results in cell stasis, and one that is nonprotective [23–29]. Briefly, autophagy is cytoprotective when its induction directly results in survival of the cell (i.e., resistance to treatment), in which case its inhibition, either by genetic or pharmacologic means, results in increased sensitivity to treatment and hence cell death. Conversely, cytotoxic autophagy is determined to occur when the promotion of autophagy leads to cell death. This form of autophagy is not obligatorily associated with apoptosis. Confirmation of cytotoxic autophagy is based on the observation that inhibition or suppression of the autophagy promotes cell survival [23]. Autophagy is termed cytostatic when its induction leads to cell growth arrest without pronounced cell death. The fourth and most controversial or least understood functional form of autophagy is the nonprotective form. It is unclear what benefit this form of autophagy might be to the cell in terms of the response to chemotherapy or radiation treatment as sensitivity of cells to treatment is not enhanced with suppression of nonprotective autophagy either through pharmacological or genetic means (i.e., the autophagy by definition does not provide a protective advantage) [26,30,31].

Autophagy has traditionally been considered as only a catabolic process but recent studies have provided evidence to indicate that it has additional functions including cellular secretion (via both the conventional pathway involving the ER/Golgi apparatus and nonconventional pathways involving alternate pathways independent of the ER/Golgi machinery). In cancer, autophagy has been found to have both anti- and protumorigenic roles. It is widely believed that moderate induction of autophagy is protective in most cancer cells, whereas the excessive induction of autophagy in some cancer cells following chemotherapy and/or radiation treatment promotes autophagy-mediated cell death [19].

12.2.3 Autophagy in Transformation and Tumor Suppression

As mentioned previously, autophagy occurs in cells at normal or basal levels and is believed to be part of the cell homeostatic machinery that ensures that excess, damaged, or dysfunctional macromolecules that may induce cellular transformation are removed. Cellular transformation involves the transition of normal cells into the tumorigenic state and is accompanied by alterations in cell morphology as well as cell function, particularly the acquisition of the capacity for uninhibited growth. Autophagy therefore represents one of the checks and balances in place against processes that may lead to uncontrolled proliferation and tumorigenesis. In mice with heterozygous knockdown of the essential autophagy gene Beclin-1, tumorigenesis was significantly higher than in mice with wild-type Beclin-1; similar results have been seen with Atg5, Atg7, and autophagy/Beclin-1 regulator 1 [32]. Also, in many tumor cells, the autophagy pathway is downregulated due to the upregulation of autophagy inhibitors such as the class I PI3 kinase pathway proteins [33]. This lends credence to the assertion that autophagy is a tumor suppression mechanism that limits cellular transformation.

Various tumor cells differ in the extent of their dependence on autophagy. In some, such as pancreatic tumor cells, there is a high level of basal autophagy necessary to maintain the high metabolic demand of proliferating cells [34,35]. Autophagy protects some tumors from cell death by eliminating damaged proteins and preventing the generation of reactive oxygen species (ROS) that lead to DNA damage in developing tumors [31]. In such cases, defects in autophagy have been found to be associated with increased sensitivity to metabolic stress and DNA damage [19,31]. This is evident in studies of pancreatic and breast cancer tumors where inhibition of autophagy or defective autophagy (in association with defective apoptosis) was observed to promote ROS production, DNA damage, and tumorigenesis [35–37]. The antitumorigenic effects, however, stem from its ability to protect cells from the DNA damage (that leads to transformation into cancer cells) via damaged protein clearance and the blockage of ROS generation [31].

The fundamental aim of cancer treatment strategies is to eliminate the tumor cells by the promotion of apoptosis or alternative cell death pathways. Where this is not feasible, interfering with tumor cell proliferation is the less desirable alternative. In efforts to achieve these ends, the logical approach has been to identify and/or develop agents that will cause irrevocable injury to the tumor cells and to abrogate cellular integrity. Protective autophagy, one of the processes that serves to maintain tumor cell viability and tumor progression, may be inhibited to facilitate tumor cell death. In tumor cells that are unable to undergo autophagy, promoting autophagy would be counterintuitive since it would prevent metabolic stress, ultimately resulting in survival [38]. It is for this reason that clinical trials have been initiated to evaluate the influence of pharmacological autophagy manipulation on sensitivity to chemotherapy.

12.3 CURRENT APPROACHES IN AUTOPHAGY MODULATION

During the course of cancer therapy, it is highly uncertain as to the potential consequences of autophagy inhibition. Dysfunctional autophagy can lead to defective organelle degradation and turnover, ROS production, mutated or damaged DNA, and inflammation [39].

In addition, impairing autophagy has been reported to prevent the release of ATP and the subsequent recruitment of dendritic cells and T lymphocytes to the tumor microenvironment, potentially resulting in a diminished immune response and attenuation of tumor cell killing [40]. Conversely, inhibiting autophagy *that is cytoprotective* should sensitize tumor cells to cancer chemotherapy, such as DNA-damaging compounds (doxorubicin, temozolomide, etoposide), as well as to radiation therapy [41,42]. If tumor cells survive treatment, their ability to undergo autophagy could allow for tumor dormancy and recurrence [43]. Therefore, it may be beneficial to inhibit autophagy both during and after cancer therapy.

The outcome of autophagy inhibition during cancer treatment is currently under investigation in a multitude of clinical trials. All of these studies utilize either chloroquine (CQ), a clinically used antimalarial drug, or hydroxychloroquine (HCQ), an FDA-approved analog of CQ, for autophagy inhibition. The mechanism by which HCQ and CQ inhibit autophagy is not fully understood; however, it is known that these compounds are weak bases that undergo ion trapping when located in acidic environments, such as lysosomes [44]. Because HCQ and CQ are unable to cross plasma membranes once ionized, they will accumulate and subsequently increase the lysosomal pH. A basic pH can inhibit the fusion of autophagosomes with lysosomes, thereby preventing the late-stage autophagy wherein the cellular cargo is degraded.

Clinical trials have been completed with HCQ alone or in combination with chemotherapy and/or radiotherapy to determine the clinical outcome of autophagy inhibition in cancer patients. Phase I trials, those primarily concerned with the safety of investigative new drugs, have revealed that HCQ can be safely administered chronically at doses up to 1200 mg per day [45–48]. Determining the highest tolerated dose regimen is critical since steady-state concentrations of HCQ can only be achieved after weeks of administration, rather than days [49]. The Phase I trials were conducted in patients with advanced solid tumors, melanoma, relapsed/refractory myeloma, or metastatic colorectal cancer [45–48]. The secondary outcome, overall survival, was also assessed in some of the Phase I studies. For example, it was found that HCQ in combination with temsirolimus allows for a median progression-free survival of 3.5 months in patients with advanced solid tumors [45]. Phase I/II and II trials have also been performed to determine the efficacy of HCQ in prolonging the survival of cancer patients. In patients with glioblastoma, HCQ in combination with both radiation and temozolomide resulted in a median survival of approximately 16 months [50]. However, when given alone, HCQ presented negligible efficacy in metastatic pancreatic adenocarcinoma patients [51]. CQ has performed similarly in clinical trials with high tolerance in a wide range of cancer patients, including those with colorectal, breast, non-small cell lung, ovarian, and renal cancers, and has been shown to increase median survival in glioblastoma patients by 13 months [52,53].

Overall, HCQ is capable of promoting partial responses to therapy and prolonging a stable disease state. When comparing the outcomes of the various cancer patient populations, it appears that a subset of cancer types may be more sensitive to HCQ or CQ. It has also been noted that cancer cells are more susceptible to HCQ-induced autophagy inhibition than normal, nontransformed cells and peripheral blood mononuclear cells (PBMCs); yet, the mechanism behind this selectivity remains to be determined [54].

Despite these promising findings, the toxicity of HCQ and CQ must be considered. Fortunately, HCQ is less toxic than CQ, with most clinical trials observing a lack of dose-limiting side effects with the administration of 1200 mg/day for up to 1 year [55]. However, HCQ has been found to cause retinopathy, an irreversible disease of the retina that leads to impairment or loss of vision, in as much as 7.5% of patients treated with the drug [56]. Those with a high risk for developing retinopathy, such as patients with renal disease or who are receiving tamoxifen, can be given HCQ in low doses (200–400 mg/day) for up to 10 years. However, the efficacy of HCQ may be compromised by reducing the dose. In addition, fatigue, gastrointestinal side effects, and myelosuppression have also been observed [50]. Yet, the main concern for HCQ use is the variability in toxicity when combined with different cancer therapy regimens. Therefore, it is doubtful that a safe, standard dose of HCQ can be determined; individual clinical trials may be necessary for each chemotherapy drug and/or radiotherapy regimen administered in combination with HCQ. In the meantime, increasing the frequency of screening for changes in vision and more accurate calculations of HCQ for real patient weight, rather than ideal weight, can lower the risk for retinopathy.

Despite the progress being made in active clinical trials for autophagy inhibitors, the interpretation of outcomes will remain a challenge for a number of reasons. First, the promotion of autophagy is being assessed by the extent of autophagic vacuole (AV) formation; yet, the presence of AVs in patient tissues does not explicitly identify the functional state of autophagy; that is whether autophagy progression has been compromised in the presence of HCQ or if autophagy is simply being induced by chemotherapy and/or radiation. Greater clarity can be obtained by determining the ratio of LC3-I to LC3-II expression. Secondly, there are currently no unequivocally accurate clinical biomarkers of autophagy. The most common indication of autophagy inhibition in the tumor is actually determined by evaluating the level of autophagy in PBMCs, which frequently exhibit less autophagy inhibition than tumor cells [52]. Other cell types, such as bone marrow plasma cells (BMPCs), have been found to possess AVs following HCQ administration when none is detected in PBMCs [47]. Another option for detecting autophagy is by comparing the formation of AVs in circulating tumor cells (CTCs) vs in the solid tumor to determine if CTCs are the best surrogate tissue, especially for metastatic cancers [49]. Lastly, autophagy-associated secreted proteins in the blood plasma can be measured as an indicator of intratumoral autophagy [57]. The establishment of an accurate clinical biomarker is crucial in order to determine if the use of a specific cancer therapy to treat a particular cancer is inducing cytoprotective autophagy, which can then be modulated by autophagy inhibition.

In some ways, testing autophagy inhibitors in a clinical setting may actually be premature in the absence of accurate information relating to the roles of autophagy in patient tumors. It would be beneficial to identify which cancers are sensitive to autophagy inhibition to better determine the appropriate patient population and therapeutic approach. On the other hand, moving forward clinically may be the best option in that by progressing to phase III trials the effect of the autophagy inhibitor on tumor sensitivity to chemotherapy and/or radiation can be evaluated. These clinical trials would also provide an opportunity to explore whether intermittent high levels or continuous low levels of autophagy inhibition are more appropriate in modulating the efficacy of chemotherapy and/or radiation. There is also a need to assess not only the toxicity of the autophagy-inhibiting drugs, but also the consequences of chronic autophagy inhibition. It has been proposed that modulating autophagy can initiate the development of tumors [54]. However, the current use of autophagy inhibitors is being investigated in patients with advanced cancers; therefore, hindering the progression of the life-threatening cancer is a greater priority than the possible development of secondary, and potentially, benign tumors.

12.4 AUTOPHAGY AND THERAPY-INDUCED IMMUNE RESPONSE

While the importance of the antitumor immune response to the effectiveness of cancer therapeutics has been recognized for over 100 years [58], it has only recently been appreciated as a reliable means to suppress tumor growth [59]. Therefore, any discussion of the consequences of cancer therapeutics must include consideration of the contribution of the antitumor immune response. In the context of antitumor immunity, autophagy has been reported to have contradictory functions, both stimulatory and inhibitory. Hypoxia-induced autophagy, a common occurrence in large tumors, has immune protective functions [60]. As a consequence of hypoxia-induced autophagy, the signal transducer and activator of transcription 3 (STAT3) is activated through a hypoxia-induced factor 1a (HIF-1a)-regulated pathway [61,62]. Once activated, STAT3 can stimulate tumor cell secretion of the immune suppressive cytokines IL-10, IL-23, and TGF-beta [63]. Inhibition of antitumor immunity through hypoxia-induced autophagy has been confirmed using the B16-F10 TRP2 vaccine tumor model where tumors exposed to the pharmacological autophagy inhibitor, HCQ, exhibited a significant decrease in tumor growth in nonvaccinated and vaccinated mice [62]. This model shows that autophagy suppresses CD8 T-cell antitumor immunity. These effects could be relevant to the observation that CQ or HCQ increased long-term survival and enhanced immune cell proliferation and tumor infiltration using MC38 and B16-F10 tumor models [62,64].

In addition to activating STAT3, hypoxia-induced autophagy inhibits NK cell-mediated antitumor immunity. This occurs in two ways. First, via autophagosome fusion with endocytosed cytolytic perforin and granzymes released from NK cells [65], limiting their activities. Second, hypoxia-induced autophagy can disrupt the NK cell–tumor cell immune synapse through hypoxic stress inducing the endocytosis and degradation of connexin 43, a major component of gap junctions [66]. Because perforin and granzymes and the immune synapse are also relevant to CD8 cytotoxic lymphocytes, these inhibitory activities would presumably be relevant to suppression of T-cell function, as well.

In contrast to these immune inhibitory activities, there is evidence that autophagy can promote antitumor immunity [40]. A limited subset of chemotherapeutic agents has been shown to stimulate the antitumor immune response through a process called immunogenic cell death (ICD). Although it is not understood why some therapies induce ICD, this appears to be mediated through an autophagy and apoptosis-dependent process [67]. Several reports have identified danger-associated molecular pattern (DAMP) molecules as having critical roles in stimulating therapy-induced ICD [68]. These DAMPs are not normally accessible to the immune system but become accessible during ICD. The most well-studied DAMPs include the nuclear high-mobility group box 1 (HMGB1), cellular ATP, and endoplasmic reticulum localized chaperone calreticulin [69]. While calreticulin is exposed on the plasma membrane in the early stages of therapy-induced ICD, later stages of ICD promote the secretion of HMGB1 and ATP through compromised plasma membranes [70]. DAMPs stimulate immune cell migration to the tumor [71], tumor cell phagocytosis by antigen presenting cells, and the antitumor activity of T-cells and NK cells [72] largely through binding receptors on immune cells [73]. The consequence of DAMP secretion and ICD are an enhanced adaptive and innate antitumor immune response, with adaptive immune cell memory.

The positive correlations between antitumor immunity and autophagy are not reserved to animal models only and are also observed in the clinic. As a group, patients with solid or hematologic malignancies with elevated autophagy have lived longer [74–76] and their tumors have been shown to retain elevated levels of CD8 T-cells [77]. These observations provide some evidence that induction of autophagy could prolong survival and improve antitumor immunity. In addition, fasting and other less aggressive caloric restriction regimens induce autophagy organism-wide in animal tumor models with therapeutic benefits to ICD-inducing therapies [78]. While effective in animal models, a caloric restriction strategy is not likely to be plausible in humans because it could exacerbate life-threatening cancer-induced cachexia [79].

Like many of its functions in cancer biology, autophagy's roles in regulating the antitumor immune response are complex, with both pro- and antitumor activities. Roles for hypoxia-induced autophagy generally appear to promote tumor survival and therefore might benefit from a strategy to inhibit autophagy. Conversely, therapy-induced autophagy, which would likely be experienced by many cancer patients, has antitumor action and therefore autophagy inhibition might be detrimental to patients. This raises serious reservations relating to current clinical trials combining CQ or hydroxylchloroquine with chemotherapy (or radiation). These trials are largely based on data from cell culture and tumor xenograft experiments where inhibition of the cytoprotective form of autophagy sensitizes the tumor cells to therapy. However, in the clinical situation where the immune system is presumably functional, interference with autophagy could be counterproductive if, in fact, this results in immunosuppression.

12.5 FUTURE IMPLICATIONS, NOVEL THERAPIES, AND DRUG TARGETS

As discussed above, the finding that conventional cytotoxic cancer therapy can promote autophagy has provided the basis for autophagy modulation in therapy. Many cancer therapeutics as well as radiation can induce autophagy by the activation of multiple pathways including the PI3K/AKT/mTOR pathway as well as the DNA damage response [80]. Most efforts to modulate autophagy in cancer cells are focused on inhibiting autophagy as a protective mechanism that facilitates survival. For this purpose, the use of classical autophagy inhibitors, such as HCQ, has been widely investigated. However, a number of limitations to the current approach have been described. These include multiple off-target effects of the lysosomotropic agents and the uncertainty of sufficient autophagy inhibition in humans, which is complicated by the absence of a widely accepted approach for evaluating autophagy inhibition in patients' tumors. Furthermore, the antitumor effects of CQ or HCQ can be attributed to mechanisms other than autophagy (reflecting the nonspecificity of their antitumor effects) in addition to unfavorably interfering with autophagy in nontransformed cells [81].

Accordingly, there is a compelling need for developing novel autophagy modulators that can affect autophagy more specifically in patients' tumors. Current efforts are directed toward developing more specific and potent lysosomotropic agents. Despite the fact that HCQ, which has a safer therapeutic profile than CQ, has FDA approval and is used in many clinical trials, high-dose regimens must be administered in order to achieve sufficient autophagy inhibition in patients, presumably due to the drug's lack of lysosomal specificity. Other antimalarial drugs, such as mefloquine and quinacrine, have been shown to be more potent as lysosomal inhibitors, possibly resulting in more effective or sustained sensitization to cancer therapy [82]. Several other compounds have been shown to affect lysosomal function. For example, siramesine, an S2R (sigma-2 receptor) agonist and an anxiolytic compound, was found to induce lysosomal dysfunction in a manner similar to CQ where it passively accumulates in the lysosome, resulting in an increased pH [83]. In addition, the active metabolite of clomipramine, a tricyclic antidepressant, has been shown to block the autophagosome–lysosome fusion, while lucanthone, an antischistosomal drug with potential antineoplastic activity, can inhibit autophagy by increasing lysosomal membrane permeability, leading to a significant increase in cathepsin D cytosolic levels, and ultimately, apoptosis [84].

Derivatives of CQ, such as the analog Lys05, have also been developed, which can accumulate to extremely high levels within the lysosome, causing a large increase in lysosomal pH. The resulting inhibition of the autophagic flux was found to sensitize several cancer cell lines to therapy more effectively than HCQ. In addition, Lys05 showed a significant capability to interfere with autophagy *in vivo*, allowing for the development of a potent lysosomotropic agent that can be used at lower concentrations than HCQ in patients [85]. ARN5187 is a dual inhibitor of REV-ERBβ and autophagolysosome formation that exerts improved cytotoxic activity, which is yet to be evaluated *in vivo* [86]. VATG-027 was identified via high-throughput screening of antimalarial compounds as another autophagy inhibitor which had antitumor effects in melanoma cells [87].

A better understanding of the different pathways that regulate autophagy has allowed for the identification of novel, promising molecular targets. The ability to modify these signaling pathways should provide more specific means to modulate autophagy, as opposed to lysosomotropic agents which can sensitize tumor cells to therapy independent of autophagy. Potential autophagy targets include Atg7, Atg4, BECLIN1, and ULK-1. ULK-1 is the initial upstream kinase in the autophagic pathway [54,88]. Genetic knockdown of ULK-1 results in dramatic dysfunction in the autophagic response [18]. MRT68921 is a kinase inhibitor that was found to significantly interfere with autophagy primarily due to its effect on ULK-1 [89]. SBI-0206965 was also presented as an additional catalytic inhibitor of ULK-1 that might possibly serve as a new small molecule to modulate autophagy [90]. These agents are still in the initial phases of development and their effect on therapy-induced autophagy in tumor cells has not been elucidated. Furthermore, there have not been any clinical trials for any of the ULK-1 inhibitors.

Vsp34 is emerging as an important target for novel anticancer therapy. Vsp34 is downstream to ULK-1 and is involved in many vesicular trafficking pathways including autophagy [18]. Genetic deletion of Vsp34 results in the failure of autophagosome formation. Spautin-1 inhibits two ubiquitin-specific proteases USP10 and USP13, resulting in the degradation of Vps34 and consequently, the inhibition of autophagy [91]. Initial studies showed a synergistic effect of Spautin-1 with imatinib, suggesting a potential role of autophagy inhibition in the treatment of chronic myeloid leukemia [92]. A novel small molecule, SAR405, was found to inhibit autophagy by preventing the catalytic activity of the two Vps34 complexes, Atg14L and UVRAG-containing Vps34 complexes, providing an additional means of selective autophagy inhibition [93]. Lastly, PIK-III, a kinase inhibitor of the bisaminopyrimidine family, inhibits the degradation of autophagy adapters, such as p62, following catalytic inhibition of mTOR. PIK-III can also interfere with mitophagy, the autophagic degradation of mitochondria [94].

Both ULK-1 and Vsp34 are not specific to the autophagic pathway; however, they provide more selective drug targets that should replace the highly nonspecific pan-PI3K inhibitors, such as 3-methyladenine. Continuous efforts to improve these novel small molecules or to identify other targets in the autophagic machinery is under way. However, another challenge for inhibiting autophagy in tumor cells arises from the identification of multiple roles or function of autophagy induced by cancer therapy [95]. Consequently, even with a more specific target, inhibition of cytotoxic autophagy might have deleterious therapeutic outcomes. The strategy of autophagy inhibition would greatly benefit from the identification of autophagy-related biomarkers suggestive of a cytoprotective role of autophagy or by restricting these approaches to tumors with an established dependence on autophagy sometimes referred to as "addiction to autophagy".

12.6 CONCLUSION

The roles of autophagy in cancer and cancer treatment are diverse and remain to be completely elucidated. Thus far, it has been found that autophagy can play a tumor-suppressive role by preventing oncogenic mutations, as well as dysfunctional mitochondria and the subsequent release of genotoxic ROS. On the other hand, autophagy can adopt a tumor-promoting role following tumor formation by allowing tumor cells to adapt to the hypoxic and nutrient-deprived tumor microenvironment. It has been proposed that these various functions of autophagy are dependent upon multiple factors including the specific oncogenic mutation(s) (e.g., p53), the functional and/or impaired signaling pathways, such as the mTOR and PI3K/Akt pathways, and the level of cellular stress. With regard to the response to chemotherapy and/or radiation, autophagy can also manifest and play various roles, each of which contribute in a unique way to either cell survival or death. The current FDA-approved modulators of autophagy, HCQ and CQ, could prove to be effective in prolonging the survival of cancer patients; however, their limited efficacy and potential toxicities have hindered their progression toward becoming a common component of cancer therapy. The search continues for an accurate method for detecting autophagy in the tumors of patients, for identifying the nature of that autophagy as a basis for therapeutic intervention, and for a specific and potent autophagy inhibitor that can successfully sensitize tumor cells to the established cancer chemotherapy and radiation regimens.

Acknowledgments

Work in Dr. Gewirtz's laboratory has been supported by the Office of the Assistant Secretary of Defense for Health Affairs through the Breast Cancer Research Program [grant no. W81XWH-14-1-0088] and Massey Center Support Grant P30 CA016059.

Conflict of Interest: No potential conflicts of interest were disclosed.

References

[1] Hanahan D, Weinberg RA. Hallmarks of cancer: the next generation. Cell 2011;144(5):646–74. doi: 10.1016/j.cell.2011.02.013.

[2] Yang Z, Klionsky DJ. Eaten alive: a history of macroautophagy. Nat Cell Biol 2010;12(9):814–22. doi: 10.1038/ncb0910-814.

[3] Mizushima N. Autophagy: process and function. Genes Dev 2007;21:2861–73. doi: 10.1101/gad.1599207.eralize.

[4] Mizushima N, Levine B, Cuervo AM, Klionsky DJ. Autophagy fights disease through cellular self-digestion. Nature 2008;451(7182):1069–75. doi: 10.1038/nature06639.

[5] Klionsky DJ, Ohsumi Y. Vacuolar import of proteins and organelles from the cytoplasm. Annu Rev Cell Dev Biol 1999;15:1–32.

[6] Su Z, Yang Z, Xu Y, Chen Y, Yu Q. Apoptosis, autophagy, necroptosis, and cancer metastasis. Mol Cancer 2015;14(1):48. doi: 10.1186/s12943-015-0321-5.

[7] Kaushik S, Cuervo AM. Chaperone-mediated autophagy: a unique way to enter the lysosome world. Trends Cell Biol 2013;22(8):407–17. doi: 10.1016/j.tcb.2012.05.006.

[8] Janku F, McConkey DJ, Hong DS, Kurzrock R. Autophagy as a target for anticancer therapy. Nat Rev Clin Oncol 2011;8:528–39. doi: 10.1038/nrclinonc.2011.71.

[9] Cuervo AM, Wong E. Chaperone-mediated autophagy: roles in disease and aging. Cell Res 2014;24(1):92–104. doi: 10.1038/cr.2013.153.

[10] Mony VK, Benjamin S, O'Rourke EJ. A lysosome-centered view of nutrient homeostasis. Autophagy 2016;12(4):619–31.

[11] Li WW, Li J, Bao JK. Microautophagy: lesser-known self-eating. Cell Mol Life Sci 2012;69(7):1125–36. doi: 10.1007/s00018-011-0865-5.

[12] Klionsky DJ, Baehrecke EH, Brumell JH, et al. A comprehensive glossary of autophagy-related molecules and processes (2nd edition). Autophagy 2011;7(11):1273–94. doi: 10.4161/auto.7.11.17661.

[13] Wong PM, Puente C, Ganley IG, Jiang X. The ULK1 complex sensing nutrient signals for autophagy activation. Autophagy 2013;9(2):124–37. doi: 10.4161/auto.23323.

[14] Gibson SB. Investigating the role of reactive oxygen species in regulating autophagy. 1st ed. Methods in enzymology, 528. Elsevier Inc.; 2013. 217-235.

[15] Pyo J-O, Nah J, Jung Y-K. Molecules and their functions in autophagy. Exp Mol Med 2012;44(2):73. doi: 10.3858/emm.2012.44.2.029.

[16] Sui X, Chen R, Wang Z, et al. Autophagy and chemotherapy resistance: a promising therapeutic target for cancer treatment. Cell Death Dis 2013;4:e838. doi: 10.1038/cddis.2013.350.

[17] He C, Klionsky DJ. Regulation mechanisms and signaling pathways of autophagy. Annu Rev Genet 2010;43(68):67–93. doi: 10.1146/annurev-genet-102808-114910.

[18] Russell RC, Tian Y, Yuan H, et al. ULK1 induces autophagy by phosphorylating Beclin-1 and activating VPS34 lipid kinase. Nat Cell Biol 2013;15(7):741–50. doi: 10.1038/ncb2757.

[19] Yang ZJ, Chee CE, Huang S, Sinicrope FA. The role of autophagy in cancer: therapeutic implications. Mol Cancer Ther 2011;10(9):1533–41. doi: 10.1158/1535-7163.MCT-11-0047.

[20] Sarkar S, Perlstein EO, Imarisio S, et al. Small molecules enhance autophagy and reduce toxicity in Huntington's disease models. Nat Chem Biol 2009;3(6):331–8. doi: 10.1038/nchembio883.

[21] Nishida Y, Arakawa S, Fujitani K, et al. Discovery of Atg5/Atg7-independent alternative macroautophagy. Nature 2009;461(7264):654–8. doi: 10.1038/nature08455.

[22] Nazarko VY, Zhong Q. ULK1 targets Beclin-1 in autophagy Volodymyr. Nat Cell Biol 2013;15(7):727–8. doi: 10.1002/aur.1474.

[23] Sharma K, Le N, Alotaibi M, Gewirtz DA. Cytotoxic autophagy in cancer therapy. Int J Mol Sci 2014;15(6):10034–51. doi: 10.3390/ijms150610034.

[24] Dou Q, Chen HN, Wang K, et al. Ivermectin induces cytostatic autophagy by blocking the PAK1/Akt Axis in breast cancer. Cancer Res 2016;76(15):4457–69. doi: 10.1158/0008-5472.CAN-15-2887.

[25] Sharma K, Goehe RW, Di X, et al. A novel cytostatic form of autophagy in sensitization of non-small cell lung cancer cells to radiation by vitamin D and the vitamin D analog, EB 1089. Autophagy 2014;10(12):2346–61. doi: 10.4161/15548627.2014.993283.

[26] Gewirtz DA. The four faces of autophagy: implications for cancer therapy. Cancer Res 2014;74(3):647–51. doi: 10.1158/0008-5472.CAN-13-2966.

[27] Shen HM, Codogno P. Autophagic cell death: Loch Ness monster or endangered species? Autophagy 2011;7(5):457–65. doi: 10.4161/auto.7.5.14226.
[28] Chen Y, Klionsky DJ. The regulation of autophagy—unanswered questions. J Cell Sci 2011;124(Pt 2):161–70. doi: 10.1242/jcs.064576.
[29] Maiuri MC, Zalckvar E, Kimchi A, Kroemer G. Self-eating and self-killing: crosstalk between autophagy and apoptosis. Nat Rev Mol Cell Biol 2007;8:741–52. doi: 10.1038/nrm2239.
[30] Bristol ML, Emery SM, Maycotte P, Thorburn A, Chakradeo S, Gewirtz D. Autophagy inhibition for chemosensitization and radiosensitization in cancer: do the preclinical data support this therapeutic strategy? J Pharmacol Exp Ther 2013;344(3):544–52. doi: 10.1124/jpet.112.199802.
[31] Gewirtz DA. Autophagy and senescence in cancer therapy. J Cell Physiol 2014;229:6–9. doi: 10.1002/jcp.24420.
[32] Galluzzi L, Pietrocola F, Bravo-San Pedro JM, et al. Autophagy in malignant transformation and cancer progression. EMBO J 2015;34(7):856–80. doi: 10.15252/embj.201490784.
[33] Brech A, Ahlquist T, Lothe RA, Stenmark H. Autophagy in tumour suppression and promotion. Mol Oncol 2009;3(4):366–75. doi: 10.1016/j.molonc.2009.05.007.
[34] Mancias JD, Wang X, Gygi SP, Harper JW, Kimmelman AC. Quantitative proteomics identifies NCOA4 as the cargo receptor mediating ferritinophagy. Nature 2014;508(7498):105–9. doi: 10.1038/nature13148.
[35] Yang S, Wang X, Contino G, et al. Pancreatic cancers require autophagy for tumor growth. Genes Dev 2011;25(7):717–29. doi: 10.1101/gad.2016111.
[36] Karantza-Wadsworth V, Patel S, Kravchuk O, et al. Autophagy mitigates metabolic stress and genome damage in mammary tumorigenesis. Genes Dev 2007;21(13):1621–35. doi: 10.1101/gad.1565707.
[37] Wang Y, Wang XD, Lapi E, et al. Autophagic activity dictates the cellular response to oncogenic RAS. Proc Natl Acad Sci 2012;109(33):13325–30. doi: 10.1073/pnas.1120193109.
[38] Mathew R, Karantza-Wadsworth V, White E. Role of autophagy in cancer. Nat Rev Cancer 2010;7(12):961–7. doi: 10.1038/nrc2254.
[39] Ozpolat B, Benbrook DM. Targeting autophagy in cancer management—strategies and developments. Cancer Manag Res 2015;7:291–9.
[40] Michaud M, Martins I, Sukkurwala AQ, et al. Autophagy-dependent anticancer immune responses induced by chemotherapeutic agents in mice. Science 2011;334:1573–8.
[41] Dalby KN, Tekedereli I, Lopez-Berestein G, Ozpolat B. Targeting the prodeath and prosurvival functions of autophagy as novel therapeutic strategies in cancer. Autophagy 2010;6(3):322–9.
[42] Wu WKK, Coffelt SB, Cho CH, et al. The autophagic paradox in cancer therapy. Oncogene 2012;31(8):939–53. doi: 10.1038/onc.2011.295.
[43] Lu Z, Luo RZ, Lu Y, et al. The tumor suppressor gene ARHI regulates autophagy and tumor dormancy in human ovarian cancer cells. J Clin Invest 2008;118(12):3917–29. doi: 10.1172/JCI35512.
[44] Poole B, Ohkuma S. Effect of weak bases on the intralysosomal pH in mouse peritoneal macrophages. J Cell Biol 1981;90:665–9.
[45] Rangwala R, Chang YC, Hu J, et al. Combined MTOR and autophagy inhibition Phase I trial of hydroxychloroquine and temsirolimus in patients with advanced solid tumors and melanoma. Autophagy 2014;10(8):1391–402. doi: 10.4161/auto.29119.
[46] Rangwala R, Leone R, Chang YC, et al. Phase I trial of hydroxychloroquine with dose-intense temozolomide in patients with advanced solid tumors and melanoma Phase I trial of hydroxychloroquine with dose-intense temozolomide in patients with advanced solid tumors and melanoma. Autophagy 2014;10(8):1369–79. doi: 10.4161/auto.29118.
[47] Vogl DT, Stadtmauer EA, Tan K, et al. Combined autophagy and proteasome inhibition: a phase 1 trial of hydroxychloroquine and bortezomib in patients with relapsed / refractory myeloma. Autophagy 2014;10(8):1380–90. doi: 10.4161/auto.29264.
[48] Loaiza-Bonilla A, O'Hara MH, Redlinger M, et al. Phase II trial of autophagy inhibition using hydroxychloroquine (HCQ) with FOLFOX/bevacizumab in the first-line treatment of advanced colorectal cancer. J Clin Oncol 2015;33(15 Suppl.):3614. doi: 10.1200/jco.2015.33.15_suppl.3614.
[49] Poklepovic A, Gewirtz DA. Outcome of early clinical trials of the combination of hydroxychloroquine with chemotherapy in cancer. Autophagy 2014;10(8):1478–80. doi: 10.4161/auto.29428.
[50] Rosenfeld MR, Ye X, Supko JG, et al. A phase I/II trial of hydroxychloroquine in conjunction with radiation therapy and concurrent and adjuvant temozolomide in patients with newly diagnosed glioblastoma multiforme. Autophagy 2014;10(8):1359–68. doi: 10.4161/auto.28984.

[51] Wolpin BM, Rubinson DA, Wang X, et al. Phase II and pharmacodynamic study of autophagy inhibition using hydroxychloroquine in patients with metastatic pancreatic adenocarcinoma. Oncologist 2014;19:637–8.
[52] Mahalingam D, Mita M, Sarantopoulos J, et al. Combined autophagy and HDAC inhibition: a phase I safety, tolerability, pharmacokinetic, and pharmacodynamic analysis of hydroxychloroquine in combination with the HDAC inhibitor vorinostat in patients with advanced solid tumors. Autophagy 2014;10(8):1403–14.
[53] Sotelo J, Briceno E, Lopez-Gonzalez MA. Adding chloroquine to conventional treatment for glioblastoma multiforme: a randomized, double-blind, placebo-controlled trial. Ann Intern Med 2006;144(5):337–44.
[54] Rebecca VW, Amaravadi RK. Emerging strategies to effectively target autophagy in cancer. Oncogene 2015;35(1):1–11. doi: 10.1038/onc.2015.99.
[55] Shi T, Xiao S, Yu X, Yan L, Xiao H. Research progress of hydroxychloroquine and autophagy inhibitors on cancer. Cancer Chemother Pharmacol 2016;79(2):287–94. doi: 10.1007/s00280-016-3197-1.
[56] Melles RB, Marmor MF. The risk of toxic retinopathy in patients on long-term hydroxychloroquine therapy. JAMA Ophthalmol 2014;132(12):1453–60. doi: 10.1001/jamaophthalmol.2014.3459.
[57] Kraya AA, Piao S, Xu X, et al. Identification of secreted proteins that reflect autophagy dynamics within tumor cells. Autophagy 2015;11(1):60–74. doi: 10.4161/15548627.2014.984273.
[58] Coley WB. The treatment of malignant tumors by repeated inoculations of erysipelas. With a report of ten original cases. 1893. Clin Orthop Relat Res 1991;(262):3–11.
[59] Rosenberg SA. Decade in review-cancer immunotherapy: entering the mainstream of cancer treatment. Nat Publ Gr 2014;11(11):630–2. doi: 10.1038/nrclinonc.2014.174.
[60] Janji B. Hijacker of the antitumor immune response: autophagy is showing its worst facet. Front Oncol 2016;6:1–10. doi: 10.3389/fonc.2016.00246.
[61] Noman MZ, Buart S, Pelt J, Van, et al. The cooperative induction of hypoxia-inducible factor-1α and STAT3 during hypoxia induced an impairment of tumor susceptibility to CTL-mediated cell lysis. J Immunol 2017;182(6):3510–21. doi: 10.4049/jimmunol.0800854.
[62] Noman MZ, Janji B, Kaminska B, et al.. Blocking hypoxia-induced autophagy in tumors restores cytotoxic T-cell activity and promotes regression 2011;71(7):5976–87. doi: 10.1158/0008-5472.CAN-11-1094.
[63] Wang T, Niu G, Kortylewski M, et al. Regulation of the innate and adaptive immune responses by Stat-3 signaling in tumor cells. Nat Med 2004;10(1):48–54. doi: 10.1038/nm976.
[64] Liang X, Vera De ME, Buchser WJ, et al. Inhibiting systemic autophagy during interleukin 2 immunotherapy promotes long-term tumor regression. Cancer Res 2012;72(11):2791–801. doi: 10.1158/0008-5472. CAN-12-0320.
[65] Baginska J, Viry E, Berchem G, Poli A, Zaeem M, Moer K, et al. Granzyme B degradation by autophagy decreases tumor cell susceptibility to natural killer-mediated lysis under hypoxia. Proc Natl Acad Sci USA 2013;110(43):17450–5. doi: 10.1073/pnas.1304790110.
[66] Janji B, Moer K Van, Noman MZ, Chouaib S. The selective degradation of synaptic connexin 43 protein by hypoxia-induced autophagy impairs natural killer cell-mediated tumor cell killing. J Biol Chem 2015;290(39):23670–9. doi: 10.1074/jbc.M115.651547.
[67] Galluzzi L, Buque A, Kepp O, Zitvogel L, Kroemer G. Immunological effects of conventional chemotherapy and targeted anticancer agents. Cancer Cell Rev 2015;28(6):690–714. doi: 10.1016/j.ccell.2015.10.012.
[68] Matzinger P. The danger model: a renewed sense of self. Science 1975;296(5566):301–5.
[69] Nie Y, Yang D, Oppenheim JJ. Alarmins and antitumor immunity. Clin Ther 2016;38(5):1042–53. doi: 10.1016/j.clinthera.2016.03.021.
[70] Galluzzi L, Buqué A, Kepp O, Zitvogel L, Kroemer G. Immunogenic cell death in cancer and infectious disease. Nat Publ Gr 2016;17(2):97–111. doi: 10.1038/nri.2016.107.
[71] Yang D, Wei F, Tewary P, Howard OMZ, Oppenheim J. Alarmin-induced cell migration. Eur J Immunol 2013;43(6):1412–8. doi: 10.1002/eji.201343670.
[72] Zelenay S, Reis C. Adaptive immunity after cell death. Trends Immunol 2013;34(7):329–35. doi: 10.1016/j.it.2013.03.005.
[73] Deretic V. Autophagy as an innate immunity paradigm: expanding the scope and repertoire of pattern recognition receptors. Curr Opin Immunol 2012;24(1):21–31. doi: 10.1016/j.coi.2011.10.006.
[74] Yang M, Zhao H, Guo L, Zhang Q, Zhao L, Bai S. Autophagy-based survival prognosis in human colorectal carcinoma. Oncotarget 2015;6(9):7084–103.
[75] Jung G, Lee H. Autophagic markers BECLIN 1 and LC3 are associated with prognosis of multiple myeloma. Acta Haematol 2015;134(1):17–24. doi: 10.1159/000368848.

[76] Ladoire S, Penault-llorca F, Senovilla L, et al. Combined evaluation of LC3B puncta and HMGB1 expression predicts residual risk of relapse after adjuvant chemotherapy in breast cancer Combined evaluation of LC3B puncta and HMGB1 expression predicts residual risk of relapse after adjuvant chemotherapy in breast cancer. Autophagy 2017;11(10):1878–90. doi: 10.1080/15548627.2015.1082022.

[77] Ladoire S, Enot D, Senovilla L, et al. The presence of LC3B puncta and HMGB1 expression in malignant cells correlate with the immune infiltrate in breast cancer. Autophagy 2016;12(5):864–75. doi: 10.1080/15548627.2016.1154244.

[78] Pietrocola F, Pol J, Vacchelli E, et al. Caloric restriction mimetics enhance anticancer immunosurveillance. Cancer Cell 2016;30(1):147–60. doi: 10.1016/j.ccell.2016.05.016.

[79] Argilés JM, Busquets S, Stemmler B, López-Soriano FJ. Cancer cachexia: understanding the molecular basis. Nat Publ Gr 2014;14(11):754–62. doi: 10.1038/nrc3829.

[80] Thorburn A, Thamm DH, Gustafson DL. Autophagy and cancer therapy. Mol Pharmacol 2014;85(6):830–8.

[81] Solitro AR, Mackeigan JP. Leaving the lysosome behind: novel developments in autophagy inhibition. Future Med Chem 2016;8:73–86.

[82] Sharma N, Thomas S, Golden EB, et al. Inhibition of autophagy and induction of breast cancer cell death by mefloquine, an antimalarial agent. Cancer Lett 2012;326(2):143–54. doi: 10.1016/j.canlet.2012.07.029.

[83] Ostenfeld MS, Høyer-Hansen M, Bastholm L, et al. Anti-cancer agent siramesine is a lysosomotropic detergent that induces cytoprotective autophagosome accumulation. Autophagy 2008;4(4):487–99. doi: 10.4161/auto.5774.

[84] Carew JS, Espitia CM, Ii JAE, et al. Lucanthone is a novel inhibitor of autophagy that induces cathepsin D-mediated apoptosis. J Biol Chem 2011;286(8):6602–13. doi: 10.1074/jbc.M110.151324.

[85] Ravi AK, Jeffrey WD. Lys05: a new lysosomal autophagy inhibitor. Autophagy 2012;8(9):1383–4.

[86] Mei C De, Ercolani L, Parodi C, et al. Dual inhibition of REV-ERBβ and autophagy as a novel pharmacological approach to induce cytotoxicity in cancer cells. Oncogene 2015;34(20):2597–608. doi: 10.1038/onc.2014.203.

[87] Goodall ML, Wang T, Martin KR, et al. Development of potent autophagy inhibitors that sensitize oncogenic BRAF V600E mutant melanoma tumor cells to vemurafenib. Autophagy 2014;10(6):1120–36.

[88] Wang C, Hu Q, Shen H. Pharmacological inhibitors of autophagy as novel cancer therapeutic agents. Pharmacol Res 2016;105:164–75. doi: 10.1016/j.phrs.2016.01.028.

[89] Petherick KJ, Conway OJL, Mpamhanga C, et al. Pharmacological inhibition of ULK1 kinase blocks mammalian target of rapamycin (mTOR)-dependent autophagy. J Biol Chem 2015;290(18):11376–83. doi: 10.1074/jbc.C114.627778.

[90] Substrates ULK, Egan DF, Chun MGH, et al. Small molecule inhibition of the autophagy kinase article small molecule inhibition of the autophagy kinase ULK1 and identification of ULK1 substrates. Mol Cell 2015;59(2):285–97. doi: 10.1016/j.molcel.2015.05.031.

[91] Zhang L, Yu J, Pan H, et al. Small molecule regulators of autophagy identified by an image-based high-throughput screen. Proc Natl Acad Sci USA 2007;104(48):19023–8. doi: 10.1073/pnas.0709695104.

[92] Shao S, Li SU, Qin Y, Wang X, Yang Y. Spautin-1, a novel autophagy inhibitor, enhances imatinib-induced apoptosis in chronic myeloid leukemia. Int J Oncol 2014;1661–8. doi: 10.3892/ijo.2014.2313.

[93] Ronan B, Flamand O, Vescovi L, et al. A highly potent and selective Vps34 inhibitor alters vesicle trafficking and autophagyalters vesicle trafficking and autophagy. Nat Chem Biol 2014;10(12):1013–9. doi: 10.1038/nchembio.1681.

[94] Honda A, Harrington E, Cornella-taracido I, et al. Potent, selective, and orally bioavailable inhibitors of VPS34 provide chemical tools to modulate autophagy *in vivo*. ACS Med Chem Lett 2015;7(1):72–6. doi: 10.1021/acsmedchemlett.5b00335. (Figure 1).

[95] Gewirtz DA. The challenge of developing autophagy inhibition as a therapeutic strategy. Cancer Res 2016;76(19):5610–4. doi: 10.1158/0008-5472.CAN-16-0722.

CHAPTER

13

Targeting Necroptosis in Antitumor Therapy

Francis Ka-Ming Chan, Himani Nailwal*, Kenta Moriwaki***

*University of Massachusetts Medical School, Worcester, MA, USA
**Osaka University, Japan

Abstract

Necroptosis is an emerging mode of programmed cell death that is defined at the molecular level through involvement of receptor interacting protein kinase 3 (RIPK3) and mixed lineage kinase domain-like (MLKL). In contrast to apoptosis, phosphorylation is a key event in necroptosis. Necroptosis can be triggered in response to various cell-surface receptors, such as death receptors in the tumor necrosis factor receptor family, Toll-like receptors, and interferon receptors. Because apoptotic machinery is often impaired in cancer cells, necroptosis has garnered attention as a promising strategy to treat apoptosis-resistant cancer. Moreover, since necroptosis leads to release of intracellular immunogenic contents and is therefore highly immunogenic, it has the potential to promote robust antitumor adaptive immune response. In this chapter, we describe current understanding of the role of necroptosis in cancer progression and discuss strategies through which we can harness the power of necroptosis in cancer therapy.

ABBREVIATIONS

CYLD	Cylindromatosis
DAMPs	Damage-associated molecular patterns
DC	Dendritic cell
LUBAC	Linear ubiquitin chain assembly complex
MHC	Major histocompatibility complex
MLKL	Mixed lineage kinase domain-like
MPT	Mitochondrial permeability transition
RHIM	RIP homotypic interaction motif
RIPK3	Receptor interacting protein kinase 3
RNAi	RNA interference
SNP	Single-nucleotide polymorphism
TLR	Toll-like receptor
TNFR	Tumor necrosis factor receptor

Targeting Cell Survival Pathways to Enhance Response to Chemotherapy. http://dx.doi.org/10.1016/B978-0-12-813753-6.00013-5
Copyright © 2018 Elsevier Inc. All rights reserved.

13.1 INTRODUCTION

Until recently, cell death research has primarily focused on studies of apoptosis. Apoptosis is characterized by rounding up of the cell and retraction of pseudopodia, followed by reduction in cell volume (pyknosis), chromatin condensation, nuclear fragmentation (karyorrhexis), plasma membrane blebbing, and *in vivo* engulfment by resident phagocytes. Apoptosis can be triggered via death receptors in the tumor necrosis factor receptor (TNFR) superfamily (extrinsic apoptosis), or through direct activation of mitochondrial apoptosis effectors (intrinsic apoptosis). The programmed and regulated nature of apoptosis stands in sharp contrast to necrosis induced by stress or trauma. Rapid clearance of apoptotic cells by professional phagocytes essentially limits inflammation. Hence, apoptosis is an immunologically silent form of cell death. On the other hand, necrosis is often marked by an increase in cell volume (oncosis) and swelling and rupture of the cell membrane. Rupture of the plasma membrane causes release of damage-associated molecular patterns (DAMPs), which elicit immune responses [1]. As such, lytic cell death is primarily linked to inflammation, infection, and trauma-induced cellular insults.

Besides accidental necrosis, recent evidence indicates that necrosis can also occur in a regulated manner. Regulated necrosis includes necroptosis, parthanatos, ferroptosis or oxytosis, mitochondrial permeability transition (MPT)-dependent necrosis, pyroptosis and pyronecrosis, and NETosis or ETosis [2]. Among these, necroptosis is the best characterized form of regulated necrosis. Similar to apoptosis, necroptosis is a tightly regulated process that can be induced in response to stimulation with death receptors in the TNF receptor family, as well as certain Toll-like receptors (TLRs). The serine/threonine kinase receptor interacting protein kinase 3 (RIPK3) is a key adaptor in necroptosis. One of the most well-known functions of necroptosis is to promote antimicrobial inflammation. In addition, deregulation of the necroptotic signaling pathway components is observed in a number of disease models such as TNF-mediated hypothermia and systemic inflammation [3], ischemic reperfusion injury [4], and Gaucher's disease [5]. In particular, there is growing evidence of dysfunction of necroptosis in different human cancers [6]. These findings highlight the tantalizing possibility of targeting necroptosis in therapies.

13.2 NECROPTOSIS SIGNALING PATHWAY

Death domain-containing receptors of the TNF receptor superfamily (i.e., TNFR1, Fas/CD95/APO-1, and TRAIL-R), Toll-like receptor 3 (TLR3), and TLR4 are the main receptors that trigger necroptosis. TNF/TNFR1 stimulation causes formation of a plasma membrane-associated complex termed Complex I [7], which contains the adaptor proteins TRADD, TRAF2, the E3 ubiquitin ligases cIAP1 and cIAP2, the serine/threonine kinase RIPK1, and the linear ubiquitin chain assembly complex (LUBAC). LUBAC promotes linear ubiquitination of NEMO, RIPK1 and other adaptors, whereas cIAP1 and cIAP2 mainly mediate K63-linked ubiquitination. The ubiquitin scaffold generated in Complex I is a pivotal checkpoint for induction of NF-κB, apoptosis, and necroptosis. This is mostly achieved through recruitment and activation of TAK1 and the IKK complex. IKKα/β within the IKK complex phosphorylates IκBα, leading to its K48-linked ubiquitination and proteasomal degradation. In addition, IKKα/β can directly phosphorylate RIPK1 to inhibit its death-inducing function [8].

Phosphorylation of IκBα exposes the nuclear localization signal on NF-κB, resulting in its translocation to the nucleus. NF-κB is the most widely known to drive expression of inflammatory mediators. In addition, NF-κB also stimulates expression of pro-survival genes such as c-FLIP, an inactive homolog of Caspase-8 that lacks protease activity. Although the long isoform of c-FLIP (c-FLIP$_L$) forms a heterodimer with Caspase-8 to promote its activity [9], it can also inhibit Caspase-8 and apoptosis when overexpressed. It is therefore noteworthy that overexpression of c-FLIP$_L$ has been reported in many tumors [10–12]. In addition to c-FLIP$_L$, the E3 ligases cIAP1 and cIAP2, which ubiquitinate RIPK1 to sterically interfere with binding to other death-inducing signal adaptors, are also well-known targets of NF-κB. The induction of pro-survival factors explains why inhibition of NF-κB is required to achieve optimal cell death in response to TNF (Fig. 13.1).

The membrane-associated Complex I is short-lived and is quickly internalized into the cytosol within the first hour. Internalization of Complex I leads to dissociation of TNFR1 and recruitment of FADD and Caspase-8 to the complex. This cytosolic complex, termed Complex II (Fig. 13.1) [7], is the key signaling node for cell death. The mechanism that regulates the transition of Complex I to Complex II is still under investigation, but is believed to involve de-ubiquitination of RIPK1 in Complex I [13]. This process critically requires the de-ubiquitinating enzyme cylindromatosis (CYLD), which also regulates RIPK1 ubiquitination in Complex II [14]. In addition, IKKα/β directly phosphorylates RIPK1 to impede downstream Complex II assembly [15]. Interestingly, a similar complex called the ripoptosome is assembled in response to chemotherapeutic agents independent of receptor stimulation [16,17]. Hence, Complex II or the ripoptosome can be assembled in response to multiple stimuli.

Complex II is normally an apoptosis-inducing complex (Complex IIa). When Caspase-8 is inhibited, such as that during certain viral infections [18], RIPK1 recruits RIPK3 via the RIP homotypic interaction motif (RHIM) to form the necrosome or Complex IIb (Fig. 13.1). The RHIM is a protein interaction motif marked by a highly conserved tetrapeptide core and flanking β-strand dominant residues. The RHIMs of RIPK1 and RIPK3 form an amyloid-like complex that facilitates downstream necroptosis signaling [19]. The requirement for caspase inhibition for necroptosis is explained by the fact that Caspase-8 cleaves RIPK1 and RIPK3 at the boundary of the kinase domains [20,21]. Thus, cleavage of RIPK1 and RIPK3 separates the kinase domain from the RHIM and prevents kinase activation within Complex II. Although RIPK1 kinase activity is essential for RIPK3 phosphorylation and activation in response to TNF, it is noteworthy that TRIF and DAI/ZBP-1, two other RHIM-containing adaptors that promote necroptosis, are not kinases. Hence, it is likely that RIPK1 does not activate RIPK3 through direct phosphorylation. Rather, RIPK1 kinase activity may play a more important role in transitioning of Complex I to Complex II.

Once RIPK3 is phosphorylated, it recruits and phosphorylates its downstream effector mixed lineage kinase domain-like (MLKL). MLKL is a pseudokinase that has no enzymatic activity [22]. Upon phosphorylation by RIPK3, MLKL assembles into an oligomer and translocates to the plasma membrane and other cellular membranes [23–25] (Fig. 13.1). Recent studies reveal that MLKL oligomers form channels on membranes, suggesting that it may directly trigger the plasma membrane rupture that is characteristic of necroptosis [25–27]. Besides TNFR-1 and related death receptors, RIPK3 can also be activated by TRIF or DAI/ZBP-1 in response to TLR3/TLR4 stimulation or herpesvirus infection, respectively [28,29]. Although the upstream activating signals are different, the downstream mechanisms stimulating MLKL appear to be conserved for all the necroptosis stimuli.

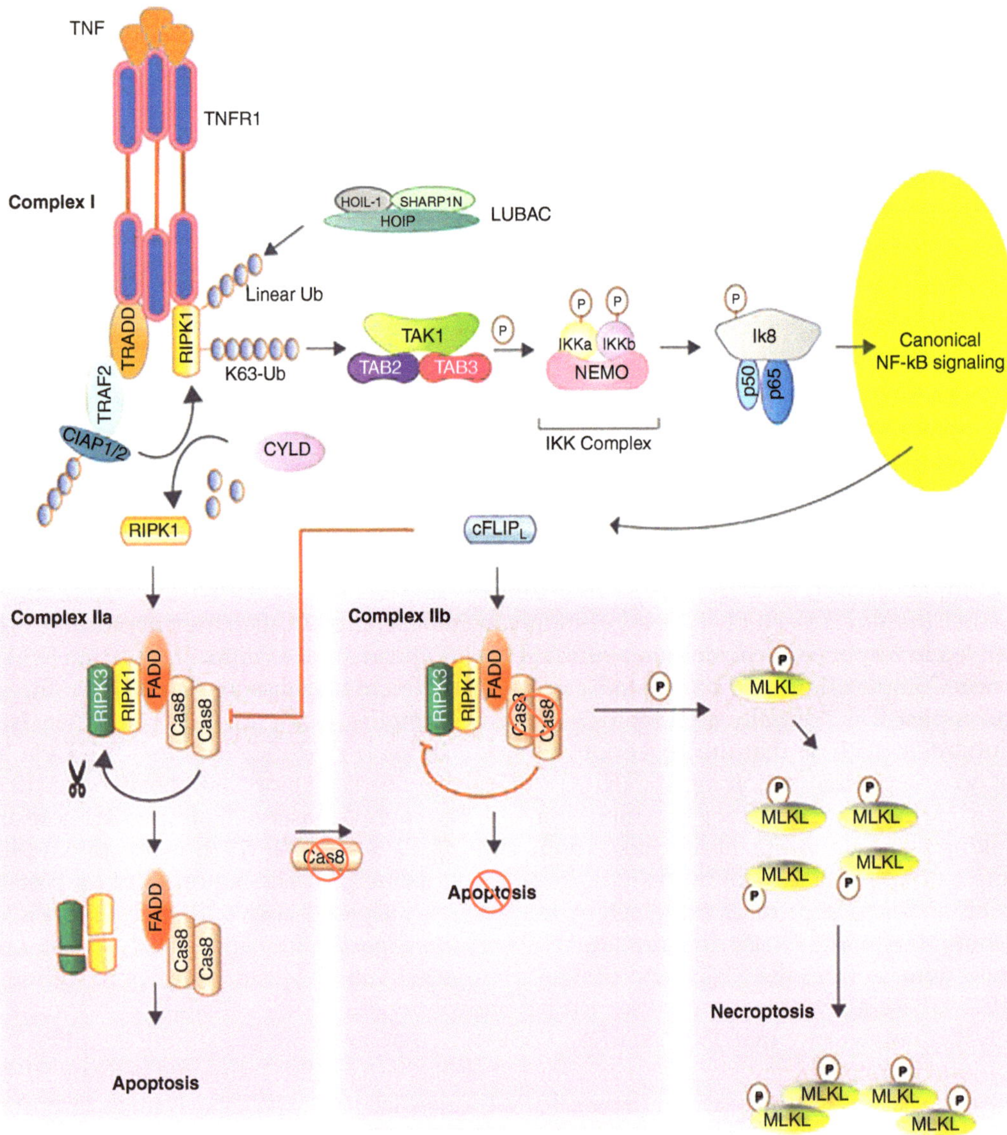

FIG. 13.1 **Molecular mechanism of TNF-induced necroptosis.** TNF stimulation leads to formation of a TNFR1-associated complex whose main function is to activate the latent transcription factor NF-κB. Poly-ubiquitination of adaptors such as RIPK1 in this complex promotes recruitment of the kinase TAK1 through the adaptors TAB2 and TAB3, which in turn activates the IKK complex, IκBα phosphorylation and degradation, and NF-κB dependent gene transcription. De-ubiquitination of RIPK1 by CYLD marks the transition of Complex I to cytosolic Complex II. Active Caspase-8 in Complex IIa is responsible for apoptosis induction and inhibition of necroptosis through cleavage of RIPK1 and RIPK3. NF-κB-dependent expression of survival factors such as cFLIP$_L$ inhibits apoptosis by suppressing Caspase-8 activity. When Caspase-8 is inactive, RIPK1 and RIPK3 interact through their RHIM domains, leading to Complex IIb assembly and MLKL phosphorylation. Phospho-MLKL undergoes oligomerization and translocates to the plasma membrane to trigger membrane rupture and necroptosis. Abbreviations: Casp8, Caspase-8; cFLIP$_L$,

13.3 NECROPTOSIS AND CANCER

13.3.1 Differential Expression of Necroptosis Regulators in Normal and Cancer Cells

Resistance to apoptosis is a major hallmark of cancers. This resistance is often acquired by reduced expression/function of pro-apoptotic molecules. Likewise, several reports showed that pro-necroptotic molecules are similarly downregulated in cancer cells. For instance, expression of RIPK3 and CYLD was found to be significantly reduced in various cancer tissues compared to adjacent normal tissues [30–33]. In addition to RIPK3 and CYLD, MLKL expression was also reduced in primary leukemia [34–39]. Similar loss of RIPK3 expression is found in many cancer cell lines commonly used in the laboratory [31,32]. Various mechanisms such as DNA hypermethylation or hypoxia have been attributed to cause downregulation of RIPK3 expression in cancer cells (Fig. 13.2) [31,32]. In addition, mutations in critical amino acids of pro-necroptotic molecules, such as D156N in the DLG motif of the kinase domain of RIPK1, V458M in the RHIM of RIPK3, and L291P in the pseudokinase domain of MLKL, have been found in cancer tissues [40]. In addition to these mutations, single-nucleotide polymorphism (SNP) analysis revealed strong correlation between SNPs in the *Ripk3* gene and non-Hodgkin lymphoma [6]. The reduced expression of MLKL and CYLD is associated with poor prognosis in pancreatic adenocarcinoma, cervical squamous cell carcinoma, melanoma, and leukemia patients [36,38,41,42]. Germline loss-of-function mutations of CYLD were linked to familial cylindromatosis, Brooke–Spiegler syndrome, and multiple familial trichoepithelioma, all of which are autosomal dominant genetic disorders marked by multiple skin tumors [43–48]. Consistent with the data from human patients, $Ripk3^{-/-}$ mice were more prone to develop inflammation-driven colorectal cancer [49,50]. These results strongly suggest that necroptosis limits tumor generation and progression.

Despite the large number of studies suggesting a tumor-suppressive role for necroptosis, several recent reports show that RIPK3 can also promote tumor growth. For example, necroptosis-dependent release of the chemokine CXCL1 appears to promote pancreatic ductal adenocarcinoma [51]. Two recent studies reported that RIPK3 expression in endothelial cells promotes extravasation of cancer cells and metastasis [52,53], although these reports differ on whether RIPK3 exerts this function through necroptosis. How can we reconcile these disparate reports on the role of RIPK3 and necroptosis in cancer? One possible explanation lies in the recent discovery that RIPK3 can promote inflammatory cytokine expression independent of cell death [54]. Thus, it is important to remember that these cell death-independent functions of RIPK3, as well as necroptosis, can both contribute to tumor progression and metastasis [52].

cellular FLICE-like inhibitory protein; cIAP, cellular inhibitor of apoptosis 1; CYLD, cylindromatosis; FADD, Fas-associated via death domain; IKK, inhibitor of κB kinase; LUBAC, linear ubiquitin chain assembly complex; MLKL, mixed lineage kinase domain-like; NEMO, NF-κB essential modulator; P, phosphorylation; TAB, TAK1-binding protein; TAK, TGFβ-activated kinase; TNF, tumor necrosis factor; TNFR1, TNF receptor 1; TRADD, TNF receptor-associated death domain; TRAF, TNF receptor-associated factor 2; Ub, ubiquitin.

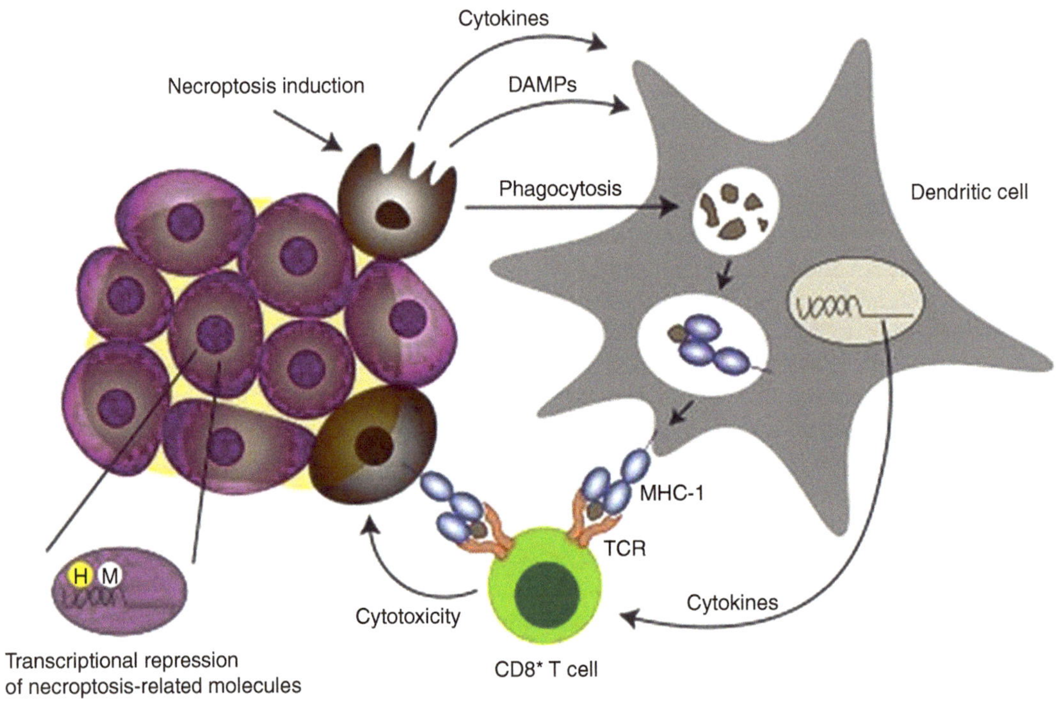

FIG. 13.2 **Schematic overview of necroptosis in cancer.** Necroptotic cells release immuno-stimulating molecules called DAMPs through the ruptured plasma membrane. In addition, necroptotic cells can actively secrete cytokines through NF-κB-dependent transcription. Both these factors promote functional maturation of DCs. The remnants of necroptotic cells are engulfed by DCs and processed in the endosomes. Tumor-associated peptides are loaded on MHC class I (MHC-I) complex and recognized by T-cell receptors (TCRs) expressed on naïve $CD8^+$ T-cells. This cross-presentation generates effector $CD8^+$ T-cells with cytotoxic function against cancer cells. On the other hand, cancer cells at the center of the tumor tissues that is under hypoxic stress (H, yellow circle) can lead to transcriptional silencing of RIPK3. In addition, RIPK3 expression can also be suppressed by DNA methylation (M, white circle). These competing events ultimately determine whether the developing tumor survives or is eliminated by immune surveillance mechanisms. Abbreviation: DAMPs, damage-associated molecular patterns.

13.3.2 Necroptosis and Anticancer Drug

Necroptosis is mainly induced by cell-surface receptors such as TNFR-1. Certain conventional anticancer agents have been reported to stimulate this canonical necroptotic pathway by stimulating autocrine TNF production. For example, cisplatin induces cancer cell necroptosis through TNF-dependent and -independent mechanisms [55,56]. Autocrine TNF also plays a key role in necroptosis of colon carcinoma in response to 5-FU and pan-caspase inhibitor [57]. Neoalbaconol, a compound extracted from fungus, also induced necroptosis through autocrine TNF production [58].

Although many anticancer agents cause necroptosis through autocrine TNF production, there are also examples in which tumoricidal compounds can trigger necroptosis independent of TNF. For instance, the Bcl-2 inhibitor obatoclax and the novel chalcone derivative chalcone-24 potently induced necroptotic cancer cell death [59–63]. The kinase inhibitors sorafenib and staurosporine induced necroptosis in lymphoma and prostate cancer cells [64–66]. Interestingly, the DNA-damaging agents 5-FU, etoposide, and camptothecin induced MLKL-dependent, but RIPK3-independent necroptosis in colon cancer cells that lack caspase 3 [67]. The combination of simvastatin and metformin induced necroptosis in metastatic castration-resistant prostate cancer cells [68]. Although TNF is not involved in these cases, other mechanisms such as autophagy, reactive oxygen species production, and degradation of inhibitors of apoptosis protein have also been implicated. How the necroptosis machinery interacts with these processes at the molecular level, however, is not clear. In this light, it is noteworthy that many of these studies relied on the RIPK1 kinase inhibitor necrostatin-1 to determine whether cell death was caused by necroptosis [62–64,66,69,70]. Although RIPK1 kinase activity is crucial for TNF-induced necroptosis, it is also required for certain types of apoptosis [16,17]. Moreover, off-target effects have been reported on necrostatin-1 [71,72]. Hence, necrostatin-1 and other RIPK1 kinase inhibitors are not the best criteria to determine if necroptosis is the causative cell death modality. In fact, RNAi-mediated knockdown of RIPK3 expression may also be insufficient to determine whether necroptosis is involved, since RIPK3 can promote apoptosis in a kinase-independent manner in certain situations [15,73–77]. These caveats highlight the importance of using multiple criteria and approach to interrogate the role of necroptosis in cancer cell death.

13.3.3 Necroptosis in Antitumor Immunity

The immune system not only protects the body from infectious, nonself agents, but also limits abnormal proliferation of cancer cells that can be viewed as noninfectious "self". Immunotherapy against cancer has been shown to be effective and emerged as a promising alternative to current standard chemotherapy [78]. $CD8^+$ cytotoxic T-cells play a major role in the immune surveillance and elimination of cancer cells. Activation of tumor-specific $CD8^+$ T-cells requires tumor antigen cross-presentation by professional antigen-presenting cells [79]. This can be achieved through phagocytosis of necrotic tumor cells by dendritic cells (DCs) or macrophages [80]. Indeed, this notion is corroborated by recent reports that necroptotic tumor cells taken up by DCs can efficiently cross-prime cytotoxic $CD8^+$ T-cell response against tumor grafts [81,82]. Interestingly, RIPK1-dependent NF-κB activation and gene induction appears to further enhance efficient cross-priming of tumor-specific $CD8^+$ T-cells (Fig. 13.2) [81]. In addition to immunization with necroptotic tumor cells, TLR3 stimulation could directly induce IL-1α secretion and IL-1α-induced necroptosis of cervical cancer cells. In this case, the dying cancer cells stimulated DC-mediated IL-12 production, suggesting an indirect adjuvant effect of the necroptotic cells [83]. These results are in agreement with the widely accepted notion that DAMPs released from necroptotic cells stimulate innate inflammation, which in turn promotes a robust adaptive immune response against cancer cells. Based on these results, immunization with necroptotic cells appears to be a promising strategy for therapeutic cancer vaccine.

13.4 CONCLUDING REMARKS

Intensive studies in the last decade have established a critical role for the phosphorylation-driven RIPK3-MLKL pathway in cell death and inflammation. In contrast, our knowledge on the role of necroptosis in cancer development, progression, and metastasis is still evolving.

Since inflammation is a double-edged sword that can promote or inhibit tumor growth [84], the effect of necroptosis on carcinogenesis is likely going to be context dependent. Having said that, given the reduction of essential necroptosis regulators RIPK3 and MLKL in many cancers, we can postulate that necroptosis has a general tumor-suppressive function. It is noteworthy that while many cancer cell lines are normally refractory to necroptosis, the combination of agents that primes cells to necroptosis and other standard chemotherapeutic agents can often break this resistance to necroptosis [85]. Hence, both experimental results and observations in human patients give credence to the idea that one can harness the power of necroptosis in anticancer therapy. Besides elimination of the tumor itself, necroptosis offers the additional advantage of stimulating tumor-specific cytotoxic T-cell responses. Hence, it will be interesting to determine whether necroptosis-sensitizing agents can also enhance the efficacy of immune checkpoint inhibitors in cancer therapy.

Acknowledgment

We thank members of the Chan lab for discussion. This work is supported by NIH grant AI 119030 (to F.K. Chan).

Conflict of Interest: No potential conflicts of interest were disclosed.

References

[1] Kroemer G, Galluzzi L, Kepp O, Zitvogel L. Immunogenic cell death in cancer therapy. Annu Rev Immunol 2013;31:51–72.
[2] Vanden Berghe T, Linkermann A, Jouan-Lanhouet S, Walczak H, Vandenabeele P. Regulated necrosis: the expanding network of non-apoptotic cell death pathways. Nat Rev Mol Cell Biol 2014;15(2):135–47.
[3] Duprez L, Takahashi N, Van Hauwermeiren F, Vandendriessche B, Goossens V, Vanden Berghe T, et al. RIP kinase-dependent necrosis drives lethal systemic inflammatory response syndrome. Immunity 2011;35(6):908–18.
[4] Degterev A, Huang Z, Boyce M, Li Y, Jagtap P, Mizushima N, et al. Chemical inhibitor of nonapoptotic cell death with therapeutic potential for ischemic brain injury. Nat Chem Biol 2005;1(2):112–9.
[5] Vitner EB, Salomon R, Farfel-Becker T, Meshcheriakova A, Ali M, Klein AD, et al. RIPK3 as a potential therapeutic target for Gaucher's disease. Nat Med 2014;20(2):204–8.
[6] Cerhan JR, Ansell SM, Fredericksen ZS, Kay NE, Liebow M, Call TG, et al. Genetic variation in 1253 immune and inflammation genes and risk of non-Hodgkin lymphoma. Blood 2007;110(13):4455–63.
[7] Micheau O, Tschopp J. Induction of TNF receptor I-mediated apoptosis via two sequential signaling complexes. Cell 2003;114(2):181–90.
[8] Dondelinger Y, Jouan-Lanhouet S, Divert T, Theatre E, Bertin J, Gough PJ, et al. NF-kappaB-independent role of IKKalpha/IKKbeta in preventing RIPK1 kinase-dependent apoptotic and necroptotic cell death during TNF signaling. Mol Cell 2015;60(1):63–76.
[9] LaCasse EC, Mahoney DJ, Cheung HH, Plenchette S, Baird S, Korneluk RG. IAP-targeted therapies for cancer. Oncogene 2008;27(48):6252–75.
[10] Safa AR, Pollok KE. Targeting the anti-apoptotic protein c-FLIP for cancer therapy. Cancers (Basel) 2011;3(2):1639–71.
[11] Ullenhag GJ, Mukherjee A, Watson NF, Al-Attar AH, Scholefield JH, Durrant LG. Overexpression of FLIPL is an independent marker of poor prognosis in colorectal cancer patients. Clin Cancer Res 2007;13(17):5070–5.
[12] Wang W, Wang S, Song X, Sima N, Xu X, Luo A, et al. The relationship between c-FLIP expression and human papillomavirus E2 gene disruption in cervical carcinogenesis. Gynecol Oncol 2007;105(3):571–7.
[13] O'Donnell MA, Legarda-Addison D, Skountzos P, Yeh WC, Ting AT. Ubiquitination of RIP1 regulates an NF-kappaB-independent cell-death switch in TNF signaling. Curr Biol 2007;17(5):418–24.
[14] Moquin DM, McQuade T, Chan FK. CYLD deubiquitinates RIP1 in the TNFalpha-induced necrosome to facilitate kinase activation and programmed necrosis. PLoS ONE 2013;8(10):e76841.

[15] Dondelinger Y, Aguileta MA, Goossens V, Dubuisson C, Grootjans S, Dejardin E, et al. RIPK3 contributes to TNFR1-mediated RIPK1 kinase-dependent apoptosis in conditions of cIAP1/2 depletion or TAK1 kinase inhibition. Cell Death Differ 2013;20(10):1381–92.
[16] Tenev T, Bianchi K, Darding M, Broemer M, Langlais C, Wallberg F, et al. The Ripoptosome, a signaling platform that assembles in response to genotoxic stress and loss of IAPs. Mol Cell 2011;43(3):432–48.
[17] Feoktistova M, Geserick P, Kellert B, Dimitrova DP, Langlais C, Hupe M, et al. cIAPs block Ripoptosome formation, a RIP1/caspase-8 containing intracellular cell death complex differentially regulated by cFLIP isoforms. Mol Cell 2011;43(3):449–63.
[18] Upton JW, Chan FK. Staying alive: cell death in antiviral immunity. Mol Cell 2014;54(2):273–80.
[19] Li J, McQuade T, Siemer AB, Napetschnig J, Moriwaki K, Hsiao YS, et al. The RIP1/RIP3 necrosome forms a functional amyloid signaling complex required for programmed necrosis. Cell 2012;150(2):339–50.
[20] Lin Y, Devin A, Rodriguez Y, Liu ZG. Cleavage of the death domain kinase RIP by caspase-8 prompts TNF-induced apoptosis. Genes Dev 1999;13(19):2514–26.
[21] Feng S, Yang Y, Mei Y, Ma L, Zhu DE, Hoti N, et al. Cleavage of RIP3 inactivates its caspase-independent apoptosis pathway by removal of kinase domain. Cell Signal 2007;19(10):2056–67.
[22] Murphy JM, Czabotar PE, Hildebrand JM, Lucet IS, Zhang JG, Alvarez-Diaz S, et al. The pseudokinase MLKL mediates necroptosis via a molecular switch mechanism. Immunity 2013;39(3):443–53.
[23] Zhao J, Jitkaew S, Cai Z, Choksi S, Li Q, Luo J, et al. Mixed lineage kinase domain-like is a key receptor interacting protein 3 downstream component of TNF-induced necrosis. Proc Natl Acad Sci USA 2012;109(14):5322–7.
[24] Cai Z, Jitkaew S, Zhao J, Chiang HC, Choksi S, Liu J, et al. Plasma membrane translocation of trimerized MLKL protein is required for TNF-induced necroptosis. Nat Cell Biol 2014;16(1):55–65.
[25] Chen X, Li W, Ren J, Huang D, He WT, Song Y, et al. Translocation of mixed lineage kinase domain-like protein to plasma membrane leads to necrotic cell death. Cell Res 2014;24(1):105–21.
[26] Wang H, Sun L, Su L, Rizo J, Liu L, Wang LF, et al. Mixed lineage kinase domain-like protein MLKL causes necrotic membrane disruption upon phosphorylation by RIP3. Mol Cell 2014;54(1):133–46.
[27] Dondelinger Y, Declercq W, Montessuit S, Roelandt R, Goncalves A, Bruggeman I, et al. MLKL compromises plasma membrane integrity by binding to phosphatidylinositol phosphates. Cell Rep 2014;7(4):971–81.
[28] Upton JW, Kaiser WJ, Mocarski ES. DAI/ZBP1/DLM-1 complexes with RIP3 to mediate virus-induced programmed necrosis that is targeted by murine cytomegalovirus vIRA. Cell Host Microbe 2012;11(3):290–7.
[29] Kaiser WJ, Sridharan H, Huang C, Mandal P, Upton JW, Gough PJ, et al. Toll-like receptor 3-mediated necrosis via TRIF, RIP3, and MLKL. J Biol Chem 2013;288(43):31268–79.
[30] Hellerbrand C, Bumes E, Bataille F, Dietmaier W, Massoumi R, Bosserhoff AK. Reduced expression of CYLD in human colon and hepatocellular carcinomas. Carcinogenesis 2007;28(1):21–7.
[31] Moriwaki K, Bertin J, Gough PJ, Orlowski GM, Chan FK. Differential roles of RIPK1 and RIPK3 in TNF-induced necroptosis and chemotherapeutic agent-induced cell death. Cell Death Dis 2015;6:e1636.
[32] Koo GB, Morgan MJ, Lee DG, Kim WJ, Yoon JH, Koo JS, et al. Methylation-dependent loss of RIP3 expression in cancer represses programmed necrosis in response to chemotherapeutics. Cell Res 2015;25(6):707–25.
[33] Karami-Tehrani F, Malek AR, Shahsavari Z, Atri M. Evaluation of RIP1K and RIP3K expressions in the malignant and benign breast tumors. Tumour Biol 2016;37(7):8849–56.
[34] Liu P, Xu B, Shen W, Zhu H, Wu W, Fu Y, et al. Dysregulation of TNFalpha-induced necroptotic signaling in chronic lymphocytic leukemia: suppression of CYLD gene by LEF1. Leukemia 2012;26(6):1293–300.
[35] Hockendorf U, Yabal M, Herold T, Munkhbaatar E, Rott S, Jilg S, et al. RIPK3 restricts myeloid leukemogenesis by promoting cell death and differentiation of leukemia initiating cells. Cancer Cell 2016;30(1):75–91.
[36] Wu W, Zhu H, Fu Y, Shen W, Xu J, Miao K, et al. Clinical significance of down-regulated cylindromatosis gene in chronic lymphocytic leukemia. Leuk Lymphoma 2014;55(3):588–94.
[37] Kuphal S, Shaw-Hallgren G, Eberl M, Karrer S, Aberger F, Bosserhoff AK, et al. GLI1-dependent transcriptional repression of CYLD in basal cell carcinoma. Oncogene 2011;30(44):4523–30.
[38] Massoumi R, Kuphal S, Hellerbrand C, Haas B, Wild P, Spruss T, et al. Down-regulation of CYLD expression by Snail promotes tumor progression in malignant melanoma. J Exp Med 2009;206(1):221–32.
[39] Nugues AL, El Bouazzati H, Hetuin D, Berthon C, Loyens A, Bertrand E, et al. RIP3 is downregulated in human myeloid leukemia cells and modulates apoptosis and caspase-mediated p65/RelA cleavage. Cell Death Dis 2014;5:e1384.
[40] Forbes SA, Bhamra G, Bamford S, Dawson E, Kok C, Clements J, et al. The catalogue of somatic mutations in cancer (COSMIC). Curr Protoc Hum Genet 2008;doi: 10.1002/0471142905.hg1011s57. Chapter 10:Unit 10 11.

[41] Colbert LE, Fisher SB, Hardy CW, Hall WA, Saka B, Shelton JW, et al. Pronecrotic mixed lineage kinase domain-like protein expression is a prognostic biomarker in patients with early-stage resected pancreatic adenocarcinoma. Cancer 2013;119(17):3148–55.
[42] Ruan J, Mei L, Zhu Q, Shi G, Wang H. Mixed lineage kinase domain-like protein is a prognostic biomarker for cervical squamous cell cancer. Int J Clin Exp Pathol 2015;8(11):15035–8.
[43] Biggs PJ, Wooster R, Ford D, Chapman P, Mangion J, Quirk Y, et al. Familial cylindromatosis (turban tumour syndrome) gene localised to chromosome 16q12-q13: evidence for its role as a tumour suppressor gene. Nat Genet 1995;11(4):441–3.
[44] Bignell GR, Warren W, Seal S, Takahashi M, Rapley E, Barfoot R, et al. Identification of the familial cylindromatosis tumour-suppressor gene. Nat Genet 2000;25(2):160–5.
[45] Zheng G, Hu L, Huang W, Chen K, Zhang X, Yang S, et al. CYLD mutation causes multiple familial trichoepithelioma in three Chinese families. Hum Mutat 2004;23(4):400.
[46] Salhi A, Bornholdt D, Oeffner F, Malik S, Heid E, Happle R, et al. Multiple familial trichoepithelioma caused by mutations in the cylindromatosis tumor suppressor gene. Cancer Res 2004;64(15):5113–7.
[47] Hu G, Onder M, Gill M, Aksakal B, Oztas M, Gurer MA, et al. A novel missense mutation in CYLD in a family with Brooke–Spiegler syndrome. J Invest Dermatol 2003;121(4):732–4.
[48] Zhang XJ, Liang YH, He PP, Yang S, Wang HY, Chen JJ, et al. Identification of the cylindromatosis tumor-suppressor gene responsible for multiple familial trichoepithelioma. J Invest Dermatol 2004;122(3):658–64.
[49] Moriwaki K, Balaji S, Chan FK. Border security: the role of RIPK3 in epithelium homeostasis. Front Cell Dev Biol 2016;4:70.
[50] Bozec D, Iuga AC, Roda G, Dahan S, Yeretssian G. Critical function of the necroptosis adaptor RIPK3 in protecting from intestinal tumorigenesis. Oncotarget 2016;7(29):46384–400.
[51] Seifert L, Werba G, Tiwari S, Giao Ly NN, Alothman S, Alqunaibit D, et al. The necrosome promotes pancreatic oncogenesis via CXCL1 and Mincle-induced immune suppression. Nature 2016;532(7598):245–9.
[52] Hanggi K, Vasilikos L, Valls AF, Yerbes R, Knop J, Spilgies LM, et al. RIPK1/RIPK3 promotes vascular permeability to allow tumor cell extravasation independent of its necroptotic function. Cell Death Dis 2017;8(2):e2588.
[53] Strilic B, Yang L, Albarran-Juarez J, Wachsmuth L, Han K, Muller UC, et al. Tumour-cell-induced endothelial cell necroptosis via death receptor 6 promotes metastasis. Nature 2016;536(7615):215–8.
[54] Moriwaki K, Chan FK. The inflammatory signal adaptor RIPK3: functions beyond necroptosis. Int Rev Cell Mol Biol 2017;328:253–75.
[55] Xu Y, Ma HB, Fang YL, Zhang ZR, Shao J, Hong M, et al. Cisplatin-induced necroptosis in TNFalpha dependent and independent pathways. Cell Signal 2017;31:112–23.
[56] Xu Y, Lin Z, Zhao N, Zhou L, Liu F, Cichacz Z, et al. Receptor interactive protein kinase 3 promotes Cisplatin-triggered necrosis in apoptosis-resistant esophageal squamous cell carcinoma cells. PLoS ONE 2014;9(6):e100127.
[57] Oliver Metzig M, Fuchs D, Tagscherer KE, Grone HJ, Schirmacher P, Roth W. Inhibition of caspases primes colon cancer cells for 5-fluorouracil-induced TNF-alpha-dependent necroptosis driven by RIP1 kinase and NF-kappaB. Oncogene 2016;35(26):3399–409.
[58] Yu X, Deng Q, Li W, Xiao L, Luo X, Liu X, et al. Neoalbaconol induces cell death through necroptosis by regulating RIPK-dependent autocrine TNFalpha and ROS production. Oncotarget 2015;6(4):1995–2008.
[59] Bonapace L, Bornhauser BC, Schmitz M, Cario G, Ziegler U, Niggli FK, et al. Induction of autophagy-dependent necroptosis is required for childhood acute lymphoblastic leukemia cells to overcome glucocorticoid resistance. J Clin Invest 2010;120(4):1310–23.
[60] Basit F, Cristofanon S, Fulda S. Obatoclax (GX15-070) triggers necroptosis by promoting the assembly of the necrosome on autophagosomal membranes. Cell Death Differ 2013;20(9):1161–73.
[61] He W, Wang Q, Srinivasan B, Xu J, Padilla MT, Li Z, et al. A JNK-mediated autophagy pathway that triggers c-IAP degradation and necroptosis for anticancer chemotherapy. Oncogene 2014;33(23):3004–13.
[62] Urtishak KA, Edwards AY, Wang LS, Hudome A, Robinson BW, Barrett JS, et al. Potent obatoclax cytotoxicity and activation of triple death mode killing across infant acute lymphoblastic leukemia. Blood 2013;121(14):2689–703.
[63] Sulkshane P, Teni T. BH3 mimetic Obatoclax (GX15-070) mediates mitochondrial stress predominantly via MCL-1 inhibition and induces autophagy-dependent necroptosis in human oral cancer cells. Oncotarget 2016;8(36):60060–79.
[64] Locatelli SL, Cleris L, Stirparo GG, Tartari S, Saba E, Pierdominici M, et al. BIM upregulation and ROS-dependent necroptosis mediate the antitumor effects of the HDACi Givinostat and Sorafenib in Hodgkin lymphoma cell line xenografts. Leukemia 2014;28(9):1861–71.

[65] Dunai ZA, Imre G, Barna G, Korcsmaros T, Petak I, Bauer PI, et al. Staurosporine induces necroptotic cell death under caspase-compromised conditions in U937 cells. PLoS ONE 2012;7(7):e41945.
[66] Kharaziha P, Chioureas D, Baltatzis G, Fonseca P, Rodriguez P, Gogvadze V, et al. Sorafenib-induced defective autophagy promotes cell death by necroptosis. Oncotarget 2015;6(35):37066–82.
[67] Brown MF, Leibowitz BJ, Chen D, He K, Zou F, Sobol RW, et al. Loss of caspase-3 sensitizes colon cancer cells to genotoxic stress via RIP1-dependent necrosis. Cell Death Dis 2015;6:e1729.
[68] Babcook MA, Sramkoski RM, Fujioka H, Daneshgari F, Almasan A, Shukla S, et al. Combination simvastatin and metformin induces G1-phase cell cycle arrest and Ripk1- and Ripk3-dependent necrosis in C4-2B osseous metastatic castration-resistant prostate cancer cells. Cell Death Dis 2014;5:e1536.
[69] Han W, Li L, Qiu S, Lu Q, Pan Q, Gu Y, et al. Shikonin circumvents cancer drug resistance by induction of a necroptotic death. Mol Cancer Ther 2007;6(5):1641–9.
[70] Wada N, Kawano Y, Fujiwara S, Kikukawa Y, Okuno Y, Tasaki M, et al. Shikonin, dually functions as a proteasome inhibitor and a necroptosis inducer in multiple myeloma cells. Int J Oncol 2015;46(3):963–72.
[71] Takahashi N, Duprez L, Grootjans S, Cauwels A, Nerinckx W, DuHadaway JB, et al. Necrostatin-1 analogues: critical issues on the specificity, activity and *in vivo* use in experimental disease models. Cell Death Dis 2012;3:e437.
[72] Cho Y, McQuade T, Zhang H, Zhang J, Chan FK. RIP1-dependent and independent effects of necrostatin-1 in necrosis and T cell activation. PLoS ONE 2011;6(8):e23209.
[73] Moriwaki K, Bertin J, Gough PJ, Chan FKM. A RIPK3-Caspase 8 complex mediates atypical pro-IL-1b processing. J Immunol 2015;194(4):1938–44.
[74] Moriwaki K, Chan FK. Regulation of RIPK3- and RHIM-dependent necroptosis by the proteasome. J Biol Chem 2016;291(11):5948–59.
[75] Nogusa S, Thapa RJ, Dillon CP, Liedmann S, Oguin TH III, Ingram JP. RIPK3 activates parallel pathways of MLKL-driven necroptosis and FADD-mediated apoptosis to protect against influenza A virus. Cell Host Microbe 2016;20(1):13–24.
[76] Mandal P, Berger SB, Pillay S, Moriwaki K, Huang C, Guo H, et al. RIP3 induces apoptosis independent of pronecrotic kinase activity. Mol Cell 2014;56(4):481–95.
[77] Newton K, Dugger DL, Wickliffe KE, Kapoor N, de Almagro MC, Vucic D, et al. Activity of protein kinase RIPK3 determines whether cells die by necroptosis or apoptosis. Science 2014;343(6177):1357–60.
[78] Khalil DN, Smith EL, Brentjens RJ, Wolchok JD. The future of cancer treatment: immunomodulation, CARs and combination immunotherapy. Nat Rev Clin Oncol 2016;13(5):273–90.
[79] Fehres CM, Unger WW, Garcia-Vallejo JJ, van Kooyk Y. Understanding the biology of antigen cross-presentation for the design of vaccines against cancer. Front Immunol 2014;5:149.
[80] Sancho D, Joffre OP, Keller AM, Rogers NC, Martinez D, Hernanz-Falcon P, et al. Identification of a dendritic cell receptor that couples sensing of necrosis to immunity. Nature 2009;458(7240):899–903.
[81] Yatim N, Jusforgues-Saklani H, Orozco S, Schulz O, Barreira da Silva R, Reis e Sousa C, et al. RIPK1 and NF-kappaB signaling in dying cells determines cross-priming of CD8(+) T cells. Science 2015;350(6258):328–34.
[82] Aaes TL, Kaczmarek A, Delvaeye T, De Craene B, De Koker S, Heyndrickx L, et al. Vaccination with necroptotic cancer cells induces efficient anti-tumor immunity. Cell Rep 2016;15(2):274–87.
[83] Schmidt SV, Seibert S, Walch-Ruckheim B, Vicinus B, Kamionka EM, Pahne-Zeppenfeld J, et al. RIPK3 expression in cervical cancer cells is required for PolyIC-induced necroptosis, IL-1alpha release, and efficient paracrine dendritic cell activation. Oncotarget 2015;6(11):8635–47.
[84] Shalapour S, Karin M. Immunity, inflammation, and cancer: an eternal fight between good and evil. J Clin Invest 2015;125(9):3347–55.
[85] Chromik J, Safferthal C, Serve H, Fulda S. Smac mimetic primes apoptosis-resistant acute myeloid leukaemia cells for cytarabine-induced cell death by triggering necroptosis. Cancer Lett 2014;344(1):101–9.

Index

FSC
www.fsc.org
MIX
Papier | Fördert gute Waldnutzung
FSC® C083411

Kolibri 360 GmbH
Ferdinand-Jühlke-Straße 7
99095 Erfurt, Deutschland
produktsicherheit@kolibri360.de